THE OCEAN BASINS AND MARGINS

Volume 2
The North Atlantic

THE OCEAN BASINS AND MARGINS

THE OCEAN BASINS AND MARGINS

Edited by

Alan E. M. Nairn

Department of Geology
University of South Carolina
Columbia, South Carolina

and

Francis G. Stehli

Department of Geology
Case Western Reserve University
Cleveland, Ohio

Volume 2
The North Atlantic

PLENUM PRESS • NEW YORK-LONDON

Library of Congress Cataloging in Publication Data

Nairn, A   E   M
   The ocean basins and margins.

   Includes bibliographies.
   CONTENTS: v. 1. The South Atlantic. v. 2. The North Atlantic.
   1. Submarine geology. 2. Continental margins.
   I. Stehli, Francis Greenough, joint author. II. Title.
   QE39.N27   551.4'608   72-83046
   ISBN 0-306-37772-1 (v. 2)

© 1974 Plenum Press, New York
A Division of Plenum Publishing Corporation
227 West 17th Street, New York, N.Y. 10011

United Kingdom edition published by Plenum Press, London
A Division of Plenum Publishing Company, Ltd.
4a Lower John Street, London W1R 3PD, England

Printed in the United States of America

CONTRIBUTORS TO THIS VOLUME

T. Birkelund
Institute of Historical Geology and Paleontology
Østervoldgade 10, Copenhagen, Denmark

D. Bridgwater
The Geological Survey of Greenland
Østervoldgade 10, Copenhagen, Denmark

John F. Dewey
Department of Geological Sciences
State University of New York at Albany
Albany, New York

William P. Dillon
U.S. Geological Survey
Woods Hole Oceanographic Institute
Woods Hole, Massachusetts

Frank John Fitch
Birkbeck College
University of London, U. K.

A. K. Higgins
The Geological Survey of Greenland
Østervoldgade 10, Copenhagen, Denmark

M. J. Keen
Dalhousie University
Halifax, Nova Scotia

M. J. Kennedy
Department of Geology
Memorial University of Newfoundland
St. Johns, Newfoundland, Canada

D. J. MacFarlane
Shell-BP
Lagos, Nigeria

John Arthur Miller
Department of Geodesy and Geophysics
University of Cambridge, U. K.

A. E. M. Nairn
Department of Geology
University of South Carolina
Columbia, South Carolina

E. R. W. Neale
Department of Geology
Memorial University of Newfoundland
St. Johns, Newfoundland, Canada

Robin Nicholson
Department of Geology
The University
Manchester, England

Arne Noe-Nygaard
Mineralogisk Museum
Copenhagen, Denmark

H. C. Noltimier
Department of Geology
Ohio State University
Columbus, Ohio

Thomas Richard Owen
Department of Geology
University College
Swansea, Great Britain

K. Perch-Nielsen
Institute of Historical Geology and Paleontology
Østervoldgade 10, Copenhagen, Denmark

W. I. Ridley
Lunar Science Institute
Houston, Texas

Jean M. A. Sougy
Laboratoire de Géologie Structurale
Université de Provence
Centre de St. Jérôme
Marseille, France

Francis G. Stehli
Department of Geology
Case Western Reserve University
Cleveland, Ohio

M. Vigneaux
Institut de Géologie du Bassin d'Aquitaine
Université de Bordeaux
Talence, France

Diana Mildred Warrell
Department of Geodesy and Geophysics
University of Cambridge, U. K.

N. D. Watkins
Graduate School of Oceanography
University of Rhode Island
Kingston, Rhode Island

Harold Williams
Department of Geology
Memorial University of Newfoundland
St. Johns, Newfoundland, Canada

Susan Carole Williams
Birkbeck College
University of London, U. K.

CONTENTS

Chapter 3. **The Continental Margin of Eastern North America, Florida to Newfoundland**

M. J. Keen

Chapter 4. **The Northeastward Termination of the Appalachian Orogen**

Harold Williams, M. J. Kennedy, and E. R. W. Neale

Chapter 5.   **An Outline of the Geology of the Atlantic Coast of Greenland**

T. Birkelund, D. Bridgwater, A. K. Higgins, and K. Perch-Nielsen

Chapter 8. The Geology of the Western Approaches

Thomas Richard Owen

Chapter 9. The Geology and Sedimentation History of the Bay of Biscay

M. Vigneaux

Chapter 10. **Geology of West Africa and Canary and Cape Verde Islands**

William P. Dillon and Jean M. A. Sougy

Chapter 11. **Cenozoic to Recent Volcanism in and around the North
             Atlantic Basin**

Arne Noe-Nygaard (with contributions from Kent Brooks, Sveinn
Jakobsson, and Asger Ken Pedersen)

Chapter 12.   The Oceanic Islands: Azores

W. I. Ridley, N. D. Watkins, and D. J. MacFarlane

Chapter 13.   Tectonic and Radiometric Age Comparisons

Frank John Fitch, John Arthur Miller, Diana Mildred Warrel, and
Susan Carole Williams

Chapter 14. The Geophysics of the North Atlantic Basin

H. C. Noltimier

Chapter 1

A MODEL FOR THE NORTH ATLANTIC*

A. E. M. Nairn
Department of Geology
University of South Carolina
Columbia, South Carolina

and

F. G. Stehli
Department of Geology
Case Western Reserve University
Cleveland, Ohio

I. INTRODUCTION

Arbitrarily, but rather conventionally we have chosen to consider the North Atlantic as the region bounded by North America on the west and Europe and North Africa on the east, with the Arctic Circle and Tropic of Cancer as the northern and southern limits. These limits are only approximate, and where it has seemed appropriate regions extending beyond them have been considered, for instance, Greenland (Birkelund *et al.*, this volume) and Scandinavia (Nicholson, this volume) both reach well beyond the Arctic Circle and coverage of West Africa (Dillon and Sougy, this volume) includes regions south of the Tropic of Cancer.

* Contribution No. 92, Department of Geology, Case Western Reserve University, Cleveland, Ohio.

Around the North Atlantic, coastline similarities are not so obvious as in the South Atlantic, and perhaps for this reason the first suggestions of relative motion in and around the basin were based on tectonic arguments (Taylor, 1910) rather than on a Wegenerian approach. Tectonic arguments continued dominant (Carey, 1955), if not fully convincing, until geophysics revitalized the concept of continental drift through the interpretation of paleomagnetic results from the continents bordering the North Atlantic (Collinson and Runcorn, 1960; Nairn, 1956; Creer, Irving, and Runcorn 1957). Later attention was accorded to fitting continental margins, and Bullard, Everett, and Smith (1965) essayed a reconstruction of the North and South Atlantic margins which placed Northwestern Africa against the eastern coast of the United States. It was pointed out at the time that this fit involved large areas of overlap in Central America and in the Blake–Bahama–Florida platform region and that it ignored data for the Caribbean. The regions of overlap and omission certainly represent constraints which cannot be ignored and lead either to a rejection of the model or to modifications which increase its complexity. Most subsequent investigators have preferred to modify rather than reject the model, for instance, Freeland and Dietz (1971) provide a modification in which Yucatan, Honduras–Nicaragua, and Oaxaca were originally located in the Gulf of Mexico and the Blake–Bahama–Florida platform was displaced eastward to its present position along a shear whose position is marked by the Straits of Florida and northeast Providence Channel. The modifications proposed must each be tested against the backlog of geological information available on the continents and within the ocean basin and against the wealth of accumulating geophysical data before it can be determined whether the general model of coastline fitting can be retained for the North Atlantic basin.

We believe for the North Atlantic, as for the South Atlantic, that the best available model is one based on geophysical (mostly magnetic) data. The model is more complex than that for the South Atlantic—in part because the available data are more complete, but in part as well because three crustal plates are involved: the North American, the Eurasian, and the African. Interaction of the North American plate with the Caribbean region is also pertinent but so complex that it will be considered in a subsequent volume of this series. The model that we adopt is essentially that proposed by Pitman and Talwani (1972), although as Noltimier (this volume) points out in a detailed review of the geophysical data, a number of important questions relating to this model remain unresolved. Unfortunately, only future research can resolve them, and this volume can hope only to point to some of the problems and perhaps to directions that may lead toward their solution.

The geophysical model tentatively accepted is temporally limited. The data on which it is largely predicated deal only with the last one-third of Phanerozoic time. Geological data on the earlier history of the margins of the North Atlantic are available, however, and seem to suggest major departures from the simple model. The development of a geophysical and tectonic model capable of satisfying these older data is still in its early stages with as yet little sign of consensus among various workers. Nonetheless, the work of Bird and Dewey (1970), though highly speculative, suggests intriguing avenues of approach.

In general, it appears that the South Atlantic provides us with a classic example of continental displacements explicable in terms of plate tectonics, while the North Atlantic, vastly more complex in its history, may become the classic area for the establishment of the concept of recurrent plate displacements. If a cogent long-term history of the region can be pieced together, the North Atlantic may well show us that plate tectonics are as important an element in overall geological history as the long-studied orogenic movements of which indeed they may be the cause.

II. THE NORTH ATLANTIC MODEL

A. General Considerations

Serious consideration of displacement models for the northern hemisphere was slow in coming, and the reasons are not without importance in the context of model evaluation. The margins of the North Atlantic Ocean basin subject to nearly two centuries of geological observation are the best known regions of the world. Nevertheless, all this study failed to force a displacement interpretation for the North Atlantic, and the geological evidence is in fact at best merely permissive of some displacement models while clearly not requiring them. The margins of the basin at one point or another show development of marine strata virtually throughout the Phanerozoic, in strong contrast to the margins of the South Atlantic. Close similarities between coastal basins on the two shores of the North Atlantic are not as obvious as they are between Africa and Brazil; for instance, the Argana basin of Morocco, while containing Triassic volcanics and sediments and trending parallel to the coast, cannot be matched with the Triassic basin of eastern North America to the extent that basins of Gabon may be matched with those in Brazil. Even within the North Atlantic basin itself there is greater heterogeneity than in the South Atlantic, for some islands appear to be of continental type (e.g., Rockall) while others are of normal volcanic aspect (e.g., Jan Mayen, Iceland, and the Faeroes), and yet others which span the continent–ocean transition appear of mixed aspect (e.g., Canary Islands).

Because the peripheral geology and even the outlines of the bounding continental masses seem no more than permissive of continental displacements, interest in models which suggested them was low until new evidence from the oceans and from geophysical measurements forced a reconsideration. This new data has proven sufficiently compelling to result in the promulgation of a variety of models that do require displacement. No model yet seems wholly adequate, and the one considered by us to be most probable on the basis of present evidence is incomplete in that it deals with but the last short interval of geologic time, i.e., the last phase in the evolution of the North Atlantic. The observation (Wilson, 1966) that recurrent intervals of displacement may be required for a complete model indicates the amount of work remaining to be done if understanding of the North Atlantic and its margins is to be achieved. It is clear that each modification and extension of the model will require intensive testing against geological and geophysical evidence. It is also clear that the very aspects of the North Atlantic Ocean which led to formulation of the current model may be nearly mute in its extension backwards in time, for if the model is correct, an unequivocal record of the events of the early phases of displacement no longer exists within the ocean basin.

B. The Model

Although historically the first unambiguous argument in favor of continental drift in the North Atlantic province was paleomagnetic, and reconstructions based upon paleomagnetic data were prepared—the latest use of data being by Roy (1972) and Phillips and Forsyth (1972)—the method is not used here save as a control. A more precise model based upon the use of the linear magnetic anomalies recorded in the sea floor is preferred. Not only is the matching of anomalies (isochrons) better than the use of virtual pole positions, but it also provides a means of displaying relative continental positions for a variety of time intervals. In this context it is important to note that a complex sea-floor spreading model is consistent with the two limiting conditions: full reassembly and present position. It is thus consistent with the paleomagnetic data and the physiographic coastal matching data (Bullard, Everett, and Smith, 1965).

In its essentials as devised by Morgan (1968) and LePichon (1968) the sea-floor spreading or, more appropriately, the plate model assumes no internal plate deformation, that is, each plate may be regarded as moving in a rigid manner. In the North Atlantic, as we shall see, this simple concept requires modification to take into account the proposed rotation of the Iberian peninsula, usually regarded as part of the Eurasian plate. In the absence of a trench or subduction zone and of major orogenic activity since early Mesozoic times around the North Atlantic seaboard (excepting the Caribbean region), our concern is with the generation of new crust only.

The motion of the North Atlantic plates is most commonly described in terms of movements about a pole of rotation, such that the transform faults parallel small circles. In the early models of the evolution of the North Atlantic Ocean (LePichon, 1968; Dietz and Holden, 1970) a single pole of rotation was assumed. However, it was soon recognized that there were at least two poles of rotation (Fox *et al.*, 1969; Phillips and Luyendyk, 1970; Pitman and Talwani, 1971) and McGregor and Krause (1972) dated one change in the position of the pole of rotation as Late Cretaceous based upon a study of the Corners seamount area. LePichon and Fox (1971) suggest that such a re-organization with the migration of the pole of rotation becomes possible once the plates have separated sufficiently not to be constrained by marginal effects. Pitman and Talwani's (1972) model allows for the migration of the poles of

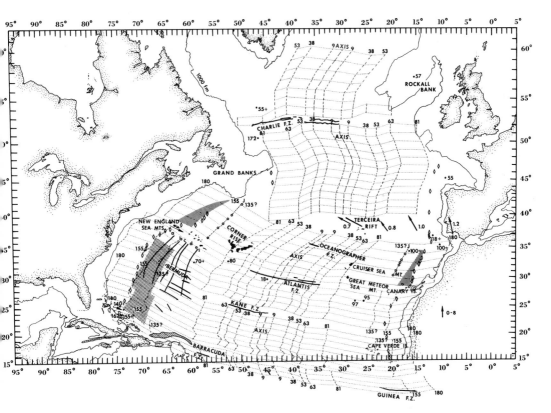

Fig. 1. Flow lines describing the motions of the continents on either side of the ridge with respect to the ridge axis. The flow lines form small circles about the pole of relative motion and are thus parallel to the fracture zones. A change in direction of the flow lines (or fracture zones) reflects a shift in the pole of relative motion. Stars define the quiet zone boundary with the Keathley anomaly sequence, lying on the seaward side, indicated by a dotted zone. Key magnetic anomalies are numbered. The ages determined paleontologically for JOIDES drill sites (shown as solid dots) are given in millions of years. [Simplified from Pitman and Talwani (1972); reproduced with permission of the authors and the Geological Society of America.]

rotation. This migration is well displayed by the construction of flow lines (Fig. 1) which describe the motion of the plates on either side with respect to the ridge axis. The flow lines are parallel to fracture zones and are constructed by generating new plate segments about successively older rotation poles (see Pitman and Talwani, 1972, p. 631) using Eurasian and African data for segments east of the ridge and American data for regions to the west. It is also possible to use the magnetic anomalies as isochrons, and matching isochrons in a manner analogous to Bullard *et al.* (1965) coastline fitting shows the relative position of the continents around the North Atlantic at various stages of separation. This is illustrated in Fig. 2, which is drawn with a coordinate system fixed with respect to North America.

A full discussion of the geophysical data upon which the model is based is given by Noltimier (this volume), and it is our purpose merely to point to a number of questions and problems noted by Pitman and Talwani and brought into focus by the use of their model. A measure of the success of the model is provided by its ability to reveal critical questions that can be brought to test.

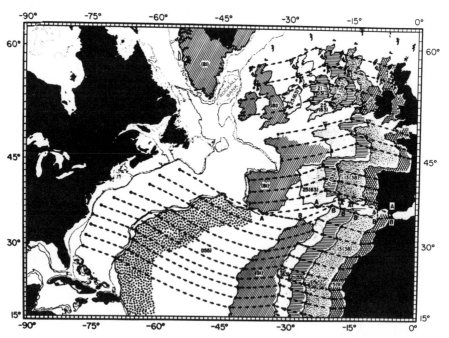

Fig. 2. The relative position of Europe and Africa with respect to North America. Blacked in continents represent present positions; the dates for earlier positions are given on the diagram. Greenland, at 81 m.y., is assumed to have been attached to Europe. The 200 and 1800 m isobaths are shown for the relative positions at 81 m.y. The relative motions of Africa and Europe are indicated by a line joining a point in Spain to a point in North Africa. [Reproduced from Pitman and Talwani (1972) by permission of the authors and the Geological Society of America.]

The major problem common to all models based upon the interpretation of the linear magnetic anomaly patterns, accepting as we have done the basis of the models, is the correct assessment of the age of the different anomalies (Heirtzler *et al.*, 1968). The established paleomagnetic reversal chronology extends back to only 4.5 m.y., and the problem is to establish the age of the identified (and numbered) anomalies by a variety of techniques, all of which until JOIDES drilling were indirect. As a result, there are some differences in the ages assigned to different anomalies, although the differences are generally small. It is necessary to assume, with little direct evidence, that spreading between dated anomalies is linear and this leads to problems in reconciling the resulting uniform movement with geological evidence of intermittent activity. The problems are compounded when anomalies are poorly defined and hard to correlate, as occurs in some parts of the North Atlantic.

Pitman and Talwani (1972) concluded that the first rifting of Africa from North America dated from some 200 m.y. ago, with the formation of volcanic rocks of the Newark Series in grabens marginal to eastern North America, but that active separation began some 180 m.y. A similar delay between graben formation and the period of physical separation is found in the South Atlantic (Nairn and Stehli, 1973). This may reflect the preparatory stage during which fracturing uplift and erosion, together perhaps with some subcrustal erosion (Sleep, 1971), occurred. However, a repetition of this phase prior to the separation of Eurasia and North America is not known.

The initial rifting of Eurasia from America may have occurred during the Jurassic, with the major phase of drift dating from the Late Cretaceous. Thus at all times since the Late Triassic separation between North America and Africa has differed from that between North America and Eurasia, although even if the rates of separation are different there is a general correspondence between the times of rapid separation. The line of separation between the Eurasian and African plates is a narrow, wedge-shaped region stretching from its tip southeast of Grand Banks to Gibraltar. While there is little evidence of major disturbances along the Azores–Gibraltar fracture zone, the effects in the Mediterranean are profound according to Hsü (1971). It is possible that detailed study along the Azores–Gibraltar fracture zone will show evidence of adjustment similar to that found by Christoffel (1971) off New Zealand.

Although one of the most striking features of the oceanic magnetic anomalies is symmetry across the ridge axis, evidence is accumulating that models must take into account two possible forms of asymmetry. These are asymmetrical spreading, first clearly recorded in a segment of ridge south of Australia (Weissel and Hayes, 1971), and the possibility of ridge migration. In the North Atlantic, Kasameyer *et al.* (1972) point out that between the lat. 43° N fracture zone and lat. 49° N there are areas of the ridge which are asymmetric and where neither the ridge shape nor heat flow is consistent with the adopted model.

The principal asymmetry in the North Atlantic, however, lies in the different widths of the magnetic quiet zone (see Fig. 1) on opposite sides of the Atlantic. There has been considerable discussion concerning the nature of the quiet zone (Vogt *et al.*, 1970), and Pitman and Talwani (1972) give their reasons for considering its margin to be an isochron. Seaward of the quiet zone lies an anomaly sequence referred to as the Keathley sequence (Vogt *et al.*, 1970). Along the boundary between the two, and actually within the quiet zone, is a series of diapirs suggesting perhaps not only halokinetic activity but also the presence of troughs (? rifts) in which the evaporites were initially deposited. The isochrons of Pitman and Talwani (1972) would assign an age of 110–140 m.y. to the Keathley sequence, whereas Vogt *et al.* (1969) propose an age of 160–190 m.y. Thus the quiet zone may be of Lower Jurassic age and be explained as crust formed during a period of predominantly normal magnetization. The asymmetry in the widths of the eastern and western quiet zones may be due to chance or, as suggested, by Noltimier (this volume) to an eastward ridge migration occurring prior to the formation of the Upper Jurassic Keathley sequence and essentially coeval with the separation of Eurasia from North America. However, the problem of the nature of the magnetic quiet zone is still far from clear.

Some anomalous ages have been reported from materials dredged from fracture zones. As the fracture zones form small circles normal to the movement of the plates, they do not spread with the adjoining plates. If the zone is sufficiently wide it may trap sediment and thus give the anomalous date (Bonatti and Honnorez, 1971).

The failure as yet to clearly recognize and identify the anomaly pattern in the Labrador Sea and Norwegian and Greenland seas precludes an absolute determination of the position of Greenland with respect to the other two continental blocks. If the axial anomaly of the inactive Labrador Sea ridge is anomaly 19 (dated as 47 m.y.) (Vogt *et al.*, 1969) and anomaly 24 (60 m.y.) is the oldest found in the Norwegian and Greenland seas then we can assume Greenland–Eurasia spreading began about 60 m.y. ago but that separation of Greenland from North America ceased at 47 m.y. Thus for a short period an independent Greenland plate may have existed before Greenland became part of the North American plate. The presence of Jurassic volcanic rocks suggest that the initial American separation from Eurasia took place during the Jurassic (Birkelund *et al.*, this volume).

The rotation of the Iberian peninsula (Carey, 1955) and the consequent opening of the Bay of Biscay, which seems well documented paleomagnetically (Van der Voo, 1969), still poses a problem. According to Watkins and Richardson (1967) as much as 22° of counterclockwise rotation could be post-Eocene;

however, according to Van der Voo (1969), since there is no significant difference between the Late Cretaceous and Early Tertiary virtual pole positions for the Iberian peninsula and stable Europe, Cenozoic rotation seems unlikely. Pitman and Talwani (1972) conclude that the opening occurred prior to the Late Cretaceous, consistent with the occurrence of Upper Cretaceous rocks on the Cantabria seamount. The occurrence of strongly folded Cretaceous flysch along the northern margin of Spain (Vigneaux, this volume) places limits on the possible timing. Nevertheless, the occurrence of attapulgite in the Paleocene recovered from the Cantabrian seamount (Latouche, 1971) indicates that normal marine conditions were absent at that time so that there is still need for caution in the interpretation of the results.

Intuitively the cause for the rotation of Spain might be sought in the relative motions of the Eurasian and African plates. Matthews and Williams (1968) reported a pattern of magnetic lineations with an east–west and east–southeast trend in the Bay of Biscay, interpreted as being generated by the rotation of Spain but which obviously do not match up with the Atlantic anomaly pattern. Conceivably they may represent local restricted spreading associated with what is primarily a fault plate boundary much as Christoffel (1971) describes off New Zealand. The alternative may be to regard Spain as a now detached piece of Africa.

Apart from west of the Iberian peninsula and along the north coast of Spain, the only other young deformation is found off the African coast south of Agadir. Intensely deformed Jurassic to Eocene sediments occur in the Cape Verde Islands, and Jurassic to Eocene rocks in the easternmost of the Canary Islands, Fuerteventura, are also deformed. Pitman and Talwani (1972) record this data and mention minor deformation of rocks of Cenozoic age at Cape Blanca, suggesting that it might be due to independent motion of Africa south of the South Atlas fault. The suggestion is not satisfactory, but we have as yet no other alternative.

III. THE EARLY HISTORY OF THE NORTH ATLANTIC

The North Atlantic model of Pitman and Talwani (1972), based as it is on linear magnetic anomalies, reaches its limit as a guide to interpretation some 200 m.y. before the present. Interpretation of the prior history of the region requires other techniques and the development of new models. The standard technique of paleomagnetism, as measured on the persistent continental areas rather than in the ephemeral oceans, provides the most valuable source of geophysical data telling us of early relative movements.

By excluding the Caribbean region it was possible in both the North and South Atlantic to deal with the simplest case of plate tectonics, plate growth persisting over the last 200 m.y. Models designed to deal with the earlier history of the Atlantic must involve the more complex case in which plates are destroyed at subduction zones, for if mountain ranges are indeed coupled to these zones, we have ample evidence around the North Atlantic that they existed prior to 200 m.y. ago. The idea of a necessary couple between mountains and subduction zones prompted Wilson's (1966) provocative article on multiple opening and closing of the Atlantic, though Nairn (1960, 1963) had earlier proposed a long history of displacements in this region. Dewey and Bird (1970) expanded this concept with particular emphasis on the Appalachians, and suggested several mountain building mechanisms. They regard collisions between continents, and those between continents and island arcs, as mechanically driven, while island arc–cordilleran systems are thought to be thermally driven. If we accept the Dewey and Bird hypothesis rather than the views of Cady (1972), then the Appalachian–Caledonian chain would seem likely to be the reflection of an Ordovician island arc–cordilleran mechanism followed by a Devonian continental collision. There is an obvious need to test these conclusions by a paleomagnetic study of lower Paleozoic rocks on both sides of the Atlantic. Under this concept an earlier Atlantic Ocean closed to produce the so-called Old Red Sandstone Continent, as Wilson (1966) suggested.

However, seldom in the North Atlantic reassembly is the significance of the Hercynian chain considered. Johnson (1973) reviewed some geological arguments for supposing the existence of a post-Devonian ocean stretching from Ireland towards central Poland which was closed by the northward movement of the region now forming Central Europe, along a line now represented by the Hercynian chain. The movement, if it occurred, was presumably not great in view of similarities of the Devonian (Erben, 1964) and older faunas on both sides of the seaway. Can this represent the collision which added African to the Laurasian unit, and thus be the birth of Pangea? Such a hypothesis, while speculative, might explain the departures in paleomagnetic results from lower Paleozoic rocks in Bohemia (Bucha, 1961) and Great Britain (Creer, Irving, and Runcorn 1957) and certainly suggests the degree of complexity that may be required of a satisfactory model for the earlier history of the North Atlantic. If true it suggests Pangea was a short-lived union from Carboniferous to Late Triassic times only.

The Pitman and Talwani model prepares us to expect strike–slip plate junctions as well as the types noted above, but in the older record of the Atlantic they may be very difficult to recognize and to interpret. Irving et al. (1972) have tentatively suggested, on the basis of paleomagnetic work, that the Grenville front of Canada may represent such a junction. While all older models are as yet very tentative and poorly constrained, they are already sufficient to allow experiments designed to provide understanding of possible mechanisms for the movements that appear required (Knopoff *et al.*, 1972; Lliboutry, 1972). It is clear that highly imaginative integration of geological and geophysical data will be required to produce a satisfactory model for the early North Atlantic history, but it is also clear that attempts to produce the model will greatly enhance our understanding of the possibilities for and constraints upon plate motions.

IV. CONCLUSIONS

The model of Pitman and Talwani (1972) as adopted here, or one like it, provides a generally satisfactory basis upon which to explain the major geological features on both sides of the North Atlantic Ocean. It provides for the juxtaposition of provinces of the same age and alignment of mountain belts with a similar history. At the same time, there are relative weaknesses in the model, areas where further research is required, and it is by pointing to such areas that the model demonstrates its real value rather than in the reconciliation of data, essential though the latter is.

The model, however, only provides an explanation and interpretation of events of the last one-third of Phanerozoic time, and by the very constraints used in its adoption is incapable of further extension. For earlier periods in time it is therefore necessary to derive models from alternate sources with the use of paleomagnetism replacing the use of anomaly patterns as the principal source of data. As a constraint on the paleomagnetic model we may use the pattern and ages of orogenic belts which may be related to the existence of former subduction zones whose identity otherwise may no longer be easily discerned.

In this way the study of the North Atlantic indicates that the adoption of an initially unnecessary hypothesis in that area leads necessarily to the acceptance of probable plate motions throughout geological time. If only the continental part of the former plates are preserved, then wholly oceanic plates may apparently disappear without a trace. There is therefore no means of indicating the former extent of the plates. Nor, it must be added, is there any reason to suppose that the continental fragments of former plates need necessarily bear any relation to the size and distribution of present plates. It is also worth a word of caution not to necessarily assume every orogenic belt represents a plate margin; however, the presence of the "Steinman Trinity," ophiolites, serpentinites, and radiolarian cherts, or the recognition of the presence of an old subduction zone beneath an orogen, should suffice to establish its nature. It is a dynamic and exciting way to investigate the history of the earth with continental areas far from being the all important elements of continental drift, appearing instead as the forlorn relics of much more extensive displacements, displacements occurring at least at all times of the geologic past where our records are sufficiently precise to detail them.

We are still far from an understanding of the mechanism responsible for the movements recorded, although some possibilities have been suggested. It is at least clear that the model of ponderous deep-seated mantle convection cells is an unsatisfactory basis for an explanation of the rotation of Iberia when related to the mid-Atlantic ridge movement.

REFERENCES

Bird, J. M. and Dewey, J. F., 1970, Lithosphere plate–continental margin tectonics and the evolution of Appalachian orogen: *Bull. Geol. Soc. Am.*, v. 81, p. 1031–1060.

Birkelund, T., Bridgwater, D., Higgins, A. K., and Perch-Nielsen, K., 1974, An outline of the Geology of the Atlantic coast of Greenland, in: *The Ocean Basins and Margins. 2. The North Atlantic*, Nairn, A. E. M. and Stehli, F. G., eds.: Plenum Press, New York, Chap. 5.

Bonatti, E. and Honnorez, J., 1971, Nonspreading crustal blocks at the mid-Atlantic ridge: *Science*, v. 174, p. 1329–1331.

Bucha, V., 1961, Palaeomagnetic pole positions in the Precambrian and Palaeozoic periods investigated from Czechoslovak rocks: *Studia Geophysica et Geodaetica Cesk Akad Ved*, v. 5, p. 269–273.

Bullard, E. C., Everett, J. E., and Smith, A. G., 1965, The fit of the continents around the Atlantic: *Phil. Trans. Roy. Soc. (London)*, v. 258A, p. 41–51.

Cady, W. M., 1972, Are the Ordovician northern Appalachians and the Mesozoic cordilleran system homologous?: *J. Geophys. Res.*, v. 77, p. 3806–3815.

Carey, S. W., 1955, The orocline concept in geotectonics: *Papers & Proc. Roy. Soc. Tas.*, v. 89, p. 255–288.

Christoffel, D. A., 1971, Motion of the New Zealand Alpine fault deduced from the pattern of sea-floor spreading: *Bull. Roy. Soc. (N.Z.)*, v. 9, p. 25–30.

Collinson, D. W. and Runcorn, S. K., 1960, Polar wandering and continental drift: evidence from paleomagnetic observations in the United States: *Bull. Geol. Soc. Am.*, v. 71, p. 915–958.

Creer, K. M., Irving, E., and Runcorn, S. K., 1957, Geophysical interpretation of palaeomagnetic results from Great Britain: *Phil. Trans. Roy. Soc. A.*, v. 250, p. 144–156.

Dewey, J. F. and Bird, J. M., 1970, Mountain belts and the new global tectonics: *J. Geophys. Res.*, v. 75, p. 2625–2647.

Dietz, R. S. and Holden, J. C., 1970, Reconstruction of Pangea: breakup and dispersion of continents, Permian to present: *J. Geophys. Res.*, v. 75, p. 4939–4956.

Dillon, W. P. and Sougy, J. M. A., 1974, The northwestern margin of Africa and its relationship with the North American continent, in: *The Ocean Basins and Margins. 2. The North Atlantic*, Nairn, A. E. M. and Stehli, F. G., eds.: Plenum Press, New York.

Erben, H. K., 1964, Facies developments in the marine Devonian of the Old World: *Proc. Ussher Soc.*, v. 1, p. 92–118.

Fox, P. J., Pitman, W. C., and Shephard, F., 1969, Crustal plates in the Central Atlantic: evidence for at least two poles of rotation: *Science*, v. 165, p. 487–489.

Freeland, G. L., Dietz, R. S., 1971, Plate tectonic evolution of Caribbean–Gulf of Mexico region: *Nature*, v. 232, p. 20–23.

Heirtzler, J. R., Dickson, G. O., Herron, E. M., Pitman, W. C., and LePichon, X., 1968, Marine magnetic anomalies, geomagnetic field reversals, and motions of the ocean floor and continents: *J. Geophys. Res.*, v. 73, p. 2119–2136.

Hsü, K. J., 1971, Origin of the Alps and Western Mediterranean: *Nature*, v. 233, p. 44–48.

Irving, E., Park, J. K., and Roy, J. L., 1972, Palaeomagnetism and the origin of the Grenville Front: *Nature*, v. 236, p. 344–346.

Johnson, G. A. L., 1973, Crustal margins and plate tectonics during the Carboniferous. VII International Carboniferous Conference, Krefeld, 1971.

Kasameyer, P. W., Von Herzen, R. P., Simmons, G., 1972, Heat flow, bathymetry and the mid-Atlantic ridge at 43°N: *J. Geophys. Res.*, v. 77, p. 2535–2542.

Knopoff, L., Poehls, K. A., and Smith, R. C., 1972, Drift of continental rafts with asymmetric heating: *Science*, v. 176, p. 1023–1024.

Latouche, C., 1971, Découverte d'attapulgite dans des sediments carottés sur le dôme Cantabria (Golfe de Gascogne) consequences paléogéographiques: *c.r. Acad. Sci.*, v. 272, p. 2064–2066.

Le Pichon, X., 1968, Sea-floor spreading and continental drift: *J. Geophys. Res.*, v. 73, p. 3661–3697.

LePichon, X. and Fox, P. J., 1971, Marginal offsets, fracture zones, and the early opening of the North Atlantic: *J. Geophys. Res.*, v. 76, p. 6294–6308.

Lliboutry, L., 1972, The driving mechanism, its source of energy, and its evolution studied with a three-layer model: *J. Geophys. Res.*, v. 77, p. 3759–3770.

Matthews, D. H. and Williams, C. A., 1968, Linear magnetic anomalies in the Bay of Biscay; a qualitative interpretation: *Earth Planet. Sci. Letters*, v. 4, p. 315–320.

McGregor, B. A. and Krause, D. C., 1972, Evolution of the sea floor in the Corner seamounts area: *J. Geophys. Res.*, v. 77, p. 2526–2534.

Morgan, W. J., 1968, Rises, trenches, great faults, and crustal blocks: *J. Geophys. Res.*, v. 73, p. 1959–1982.

Nairn, A. E. M., 1956, Relevance of palaeomagnetic studies of Jurassic rocks to continental drift: *Nature*, v. 178, p. 935–936.

Nairn, A. E. M., 1960, Paleomagnetic results from Europe: *J. Geol.*, v. 68, p. 385–406.

Nairn, A. E. M., 1963, A review of the variation in position of land masses during geological times: *Uzitá Geofyz.*, v. 1, p. 97–108.

Nairn, A. E. M. and Stehli, F. G., 1973, A model for the South Atlantic, in: *The Ocean Basin and Margins. 1. The South Atlantic*, Nairn, A. E. M. and Stehli, F. G., eds.: Plenum Press, New York.

Nicholson, R., 1973, The Scandinavian Caledonides, in: *The Ocean Basins and Margins. 2. The North Atlantic*, Nairn, A. E. M. and Stehli, F. G., eds.: Plenum Press, New York, Chap. 6.

Noltimier, H. C., 1973, The geophysics of the North Atlantic basin, in: *The Ocean Basins and Margins. 2. The North Atlantic*, Nairn, A. E. M. and Stehli, F. G., eds.: Plenum Press, New York, Chap. 14.

Phillips, J. D. and Forsyth, D., 1972, Plate tectonics, palaeomagnetism, and the opening of the Atlantic: *Bull. Geol. Soc. Am.*, v. 83, p. 1579–1600.

Phillips, J. D. and Luyendyk, B. P., 1970, Central North Atlantic plate motions over the last 40 million years: *Science*, v. 170, p. 727–729.

Pitman, W. C., and Talwani, M., 1971, Central North Atlantic plate motions: *Science*, v. 174, p. 845–846.

Pitman, W. C. and Talwani, M., 1972, Sea-floor spreading in the North Atlantic: *Bull. Geol. Soc. Am.*, v. 83, p. 619–646.

Roy, J. L., 1972, A pattern of rupture of the eastern North American–Western European paleoblock: *Earth Planet. Sci. Letters*, v. 14, p. 103–114.

Sleep, N. H., 1971, Thermal effects of the formation of Atlantic continental margins by continental break-up: *Geophys. Jour.*, v. 24, p. 325–350.

Taylor, F. B., 1910, Bearing of the Tertiary mountain belt on the origin of the earth's plan: *Bull. Geol. Soc. Am.*, v. 21, p. 179–226.

Van der Voo, R., 1969, Paleomagnetic evidence for the rotation of the Iberian peninsula: *Tectonophysics*, v. 7, p. 5–66.

Vigneaux, M., 1973, The geology and sedimentation history of the Bay of Biscay, in: *The Ocean Basins and Margins. 2. The North Atlantic*, Nairn, A. E. M. and Stehli, F. G., eds.: Plenum Press, New York, Chap. 9.

Vogt, P. R., Anderson, C. N., Bracey, D. R., and Schneider, E. D., 1970, North Atlantic magnetic smooth zones: *J. Geophys. Res.*, v. 75, p. 3955–3967.

Vogt, P. R., Avery, O. E., Schneider, E. D., Anderson, C. N., and Bracey, D. R., 1969, Discontinuities in sea-floor spreading: *Tectonophysics*, v. 8, p. 285–317.

Watkins, N. D. and Richardson, A., 1967, Paleomagnetism of the Lisbon volcanics: *Geophys. Jour.*, v. 15, p. 287–304.

Weissel, J. K. and Hayes, D. E., 1971, Asymmetric sea-floor spreading south of Australia: *Nature*, v. 231, p. 518–522.

Wilson, J. T., 1966, Did the Atlantic Ocean close and then reopen?: *Nature*, v. 211, p. 676–681.

Chapter 2

THE GEOLOGY OF THE BAHAMA–BLAKE PLATEAU REGION*

Francis G. Stehli

Department of Geology
Case Western Reserve University
Cleveland, Ohio

I. INTRODUCTION

The fit of the continental blocks around the Atlantic achieved by Bullard *et al.* (1965) while generally striking contains features which militate against its too facile acceptance. The most obvious shortcoming is failure to deal with the whereabouts of much of Mexico and all of Central America. The second major difficulty is posed by the considerable overlap which occurs between the Florida–Blake plateau–Bahama region on the southeastern extremity of North America and the Senegal–Gambia–Portuguese Guinea–Cape Verde Islands region of Africa. This overlap signals a forbidden configuration—if the crust in these regions is as old or older than the time of the suggested fit. This chapter attempts to summarize what is known of the Bahama–Blake plateau area that may prove pertinent to an understanding of the history and origin of the North Atlantic Ocean.

* Contribution No. 84, Department of Geology, Case Western Reserve University, Cleveland, Ohio.

15

II. REGIONAL SETTING

The largely submerged Bahama–Blake plateau province lies in the focus of a great crescent formed by the main continental mass of North America, the Florida peninsula, and the Antillean mobile belt (Fig. 1). It consists of two principal elements: the wholly submerged Blake plateau lying between Cape Hatteras, North Carolina, and about lat. 27°30′ N, and the partially emergent Bahamas which consist of a topographically complex assemblage of shallow- and deep-water elements. Eastward, the Bahama–Blake plateau province is bounded by the Blake–Bahama escarpment and the deep Atlantic, where 17,000 ft or more of water overlies the Hatteras Abyssal Plain. The outlines of the geology of each of the peripheral regions is significant in understanding the Bahama–Blake plateau province and will therefore be considered briefly below.

A. The Geology of the North American Coastal Region

Between Cape Hatteras on the north and northern Florida on the south, the coastal area of the North American mainland consists of a prism of coastal plain sediments of Mesozoic and Cenozoic age overlapping much older

Fig. 1. Index map of the region under consideration.

sedimentary, igneous, and metamorphic rocks. To the north, the coastal plain sediments consist largely of terrigenous clastics which thicken, become finer grained, and begin to include some carbonates seaward. It appears evident that the terrigenous detritus was derived from the adjacent and long consolidated Appalachian mobile belt. Farther south along the coastal plain the rocks show a decreasing terrigenous detrital content, and a concomitantly increasing carbonate content, until in peninsular Florida they are almost wholly carbonates.

A slightly submerged seaward continuation of the coastal plain constitutes the continental shelf, which is narrow off Cape Hatteras, reaches a width of 150 km off the Georgia coast, and then narrows to virtual disappearance just north of Miami, Florida. A combination of sparker profiles and wells drilled at sea indicate that at least back into the Cretaceous the continental shelf has been a constructional feature (Emery & Zarudzki, 1967). Below the Cretaceous there probably occur Jurassic sediments underlain by grabens filled with Triassic sediment similar to those exposed onshore beyond the coastal plain overlap and then a sedimentary, metamorphic, and igneous basement surface which has been warped to some extent. However, we have no direct knowledge on the submerged continental shelf of the basement upon which post-Triassic sediments were deposited. The probable age of these basement rocks is considered separately below.

B. The Geology of Florida

Emergent peninsular Florida exposes what is essentially a suite of coastal plain sediments in almost exclusively carbonate facies. The rocks exposed at the surface are of Cenozoic age, but extensive drilling has revealed the presence of a thick sheet of Mesozoic carbonates beneath them, and in some regions both Paleozoic sediments and igneous and metamorphic rocks presumably representing basement have been encountered by the drill (Applin, 1951). Structural contours on top of the Cretaceous (Fig. 2) serve to show the general features of the peninsula. A sag, the Suwannee Channel, diagonally crossing the peninsula near the Florida–Georgia border, sets peninsular Florida off from the main body of the North American continent. The backbone of the peninsula is formed by the Peninsular Arch running southeastward toward the Bahamas. Southern Florida is underlain by the deeply subsident east–west-oriented South Florida basin which shoals eastward. The southern flank of this basin is poorly known, but apparently structure contours rise again there.

Along the Peninsular Arch pre-Mesozoic rocks have been encountered as far south as Lake Okeechobee. Drilling has encountered rocks of latest Jurassic or earliest Cretaceous age overstepping this pre-Mesozoic basement

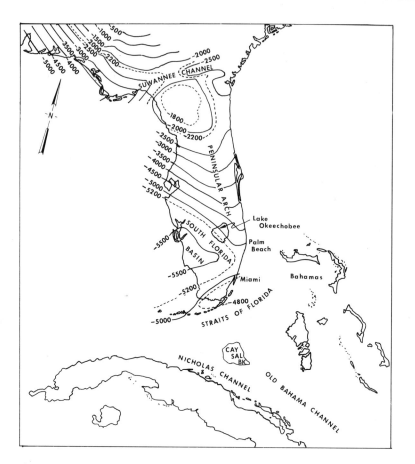

Fig. 2. Structure contours on top of the Cretaceous showing the principal positive and negative features of the Florida peninsula. (Modified from Chen, 1965.)

surface (Applin and Applin, 1965). To judge from the presence of thick marine Jurassic rocks in the Gulf Coast, Cuba, and offshore maritime Canada, older sediments in the coastal plain sequence may exist in the South Florida basin where the drill has not reached basement. While the age of the basement is considered separately, it may be noted here that a basement complex encountered in St. Lucie County, Florida, has been dated as Cambrian (Bass, 1969).

The seaward edge of the continental shelf impinges upon the coast near Palm Beach, Florida, and both here and to the south the peninsula is bordered by the deep waters (ca. 1000 m) of the Florida straits. Eastward the straits separate Florida and the Bahamas, while to the south they separate Florida from Cay Sal Bank and Cuba. Faulting may have played some part in the formation of the straits (Malloy and Hurley, 1970), but a major portion of the topographic relief is due to carbonate buildup on a subsiding platform by organisms during the Cenozoic (Sheridan *et al.*, 1966). Pre-Eocene rocks appear to extend across the straits into the Bahamas without significant disruption.

C. The Antillean Region

Bordering the Bahama–Blake plateau province on the south are the major islands of the Antillean mobile belt: Cuba and Hispaniola (Fig. 3). Cuba, geologically the best known of the two islands, is composed primarily of deformed and in part metamorphosed and intruded rocks generally equivalent in age to the coastal plain rocks to the north. In facies, however, they are quite different, at least from those of Florida and the Bahamas, in containing abundant terrigenous clastics, many of basic or intermediate volcanic origin but some that are quartzose (Khudoley, 1964). As yet, there has been little reliable radiometric dating done on the older rocks of Cuba, and rocks as old as Paleozoic may exist. Rocks bearing Jurassic fossils are found in several places, and it is clear that much of the island is underlain by Jurassic or older sediments, metasediments, and igneous rocks. According to Khudoley (1964) the Jurassic rocks include quartzose silts and some quartzites, so it appears that

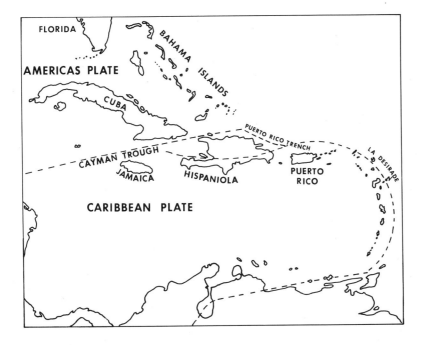

Fig. 3. Map of Antillean region showing principal islands and the probable boundaries of the Caribbean and Americas plates in this region. (Plate boundaries after Molnar and Sykes 1969.)

in the Jurassic Cuba either existed as continental crust or was in close proximity to some region from which quartzose sediments could be derived. The presence of plant-bearing beds also shows us that Cuba was at least in part above sea level in the Jurassic. Radiometric dating of zircons from the Jurassic quartzites would be of considerable significance in determining where the island might have been located in the Jurassic, but has not yet been undertaken. The geologic evidence, incomplete as it still is, together with the geophysical evidence of thick crust beneath the Antillean islands and, for an oceanic area, unusually thick crust in the Caribbean (Edgar *et al.*, 1971) suggests that these islands may have had a long continental history.

A topographic depression separates the Antillean mobile belt from the Florida–Bahama–Blake plateau region. The Straits of Florida, with water depths in this region of more than 1000 m and a width of about 120 km, separates Florida from Cuba. The 35-km-wide Nicholas Channel, with depths of about 1200 m on the west, shoals to about 600 m on the east and separates Cuba from Cay Sal Bank where the Jurassic and younger section is undeformed (Khudoley, 1964). The Old Bahama Channel between Cuba and the Bahamas, is narrow and shallow to the west, but deepens eastward off Cuba and Hispaniola and continues into the very deep Puerto Rico trench. Off the coast of Cuba seismicity is slight along the bounding topographic trench, but a region of moderate seismicity following the Cayman trough within the Caribbean, passes between Cuba and Hispaniola and then appears to be associated with the depression marginal to Hispaniola and Puerto Rico. Molnar and Sykes (1969) consider that this zone of seismicity separates a Caribbean plate from an American plate, though there seems little compelling evidence from the general geologic history of Cuba and Hispaniola, insofar as it is known, to place them on different plates. However, the fault motions determined by Molnar and Sykes (1969) are generally consistent with the hypothesis that Atlantic crust is being thrust below the Caribbean plate. Chase and Bunce (1969) have shown seismic reflection profiles farther southeast which can be interpreted to suggest underthrusting from the Atlantic, though the data they present may also be explained as due to depositional phenomena, and if that is the case only vertical movements are required.

The topographic depression between the Antillean mobile belt with strongly deformed sediment and the Cay Sal Bank–Florida–Bahama–Blake plateau region, with more or less flat-lying sediments of the same age, hides a tectonic boundary of great interest. Particularly between Cuba and Cay Sal Bank where the intervening distance is only 35 km, it would appear desirable to drill a series of deep holes in the course of the DSDP work. At present it is not clear that strike–slip faulting or any persistent and systematic pattern of underthrusting of the Bahamas beneath the Caribbean plate has occurred, nor can it be shown that it has not occurred.

D. The Deep Sea East of the Bahama–Blake Plateau Province

The eastern boundary of the Bahama–Blake plateau province is formed by the continental slope, sometimes called the Blake–Bahama escarpment. Beyond this escarpment lies the deep-sea floor characterized by topographic highs, such as the Blake outer ridge (Fig. 4), and topographic lows in part formed by abyssal plains. Seismic profiling in the region east of the Bahamas has revealed the presence of several reflectors of considerable persistence within the sedimentary cover on the sea floor. Examination of the distribution of these reflectors suggested that relatively old sediments might crop out on the sea floor north and east of San Salvador (Ewing *et al.*, 1966; Windisch *et al.*,

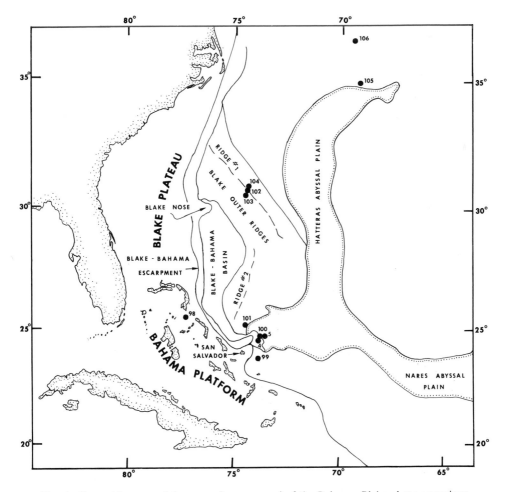

Fig. 4. General features of the ocean floor seaward of the Bahama–Blake plateau province, and of the continental margin. DSDP holes in this region are also shown.

1968). Piston core sampling in this region recovered rocks of Cretaceous age close to the surface and sites 4 and 5 of leg 1 of DSDP were subsequently drilled in this area (Fig. 4). At both of these sites rocks of Late Jurassic–Early Cretaceous age were encountered (Ewing *et al.*, 1969). Leg 11 of DSDP returned to this area and drilled at sites 99, 100, and 101. Each of these holes reached the Jurassic, and at site 100 rocks near the Middle Jurassic–Upper Jurassic boundary were encountered before drilling was terminated in what is presumably a basalt flow, since it has a glassy upper surface but is not reported to have metamorphosed overlying sediments (Ewing *et al.*, 1970). It is most unfortunate that drilling was abandoned here in favor of what were certainly less valuable sites elsewhere, and it is to be hoped that high priority will be accorded return of the *Glomar Challenger* to this region with orders to drill a number of holes to the limit of capability of the equipment.

A second group of three sites was drilled on the Blake–Bahama outer ridge, a thick sedimentary accumulation. These holes penetrated as much as 660 m, but did not apparently reach below sediments of middle Miocene age (Ewing *et al.*, 1970). However, the demonstration that this topographic high is composed of soft sediments suggests the presence of currents on the deep-sea floor adequate for both large-scale constructional and erosional activity.

Still farther north, sites 105 and 106 were drilled on the continental rise. Drilling at site 105 yielded Jurassic (Oxfordian) limestones resting on a basalt sill which metamorphosed the overlying sediments. It thus appears probable that despite failure of the drill to penetrate deeply on the Blake–Bahama outer ridge, Oxfordian sediments are likely to be continuous at this distance from the continent, at least from lat. 35° N to lat. 24° N. Rocks of this age are presumed to extend beneath the Bahama–Blake plateau province, though they have not been unequivocally identified in published reports of drilling. It is also clear that none of the holes can be considered to have reached basement and that older sediments may be expected in deeper drilling, though probably increasing numbers of flows and sills will impede drilling at greater depths.[*]

E. Geology of Cay Sal Bank

A deep well has been drilled on Cay Sal Bank, a detached element of the carbonate platform characteristic of the Bahamas and Florida and lying west of the Bahamas between Florida and Cuba (Fig. 1). The only published account of this well seems to be that given by Khudoley (1964) which indicates penetra-

---

[*] There has been a most unfortunate tendency in DSDP reports to equate "acoustic basement" with true basement. Now that acoustic basement has been drilled through several times to reveal more sediments below, it is hoped that this imprecise and misleading usage can be terminated.

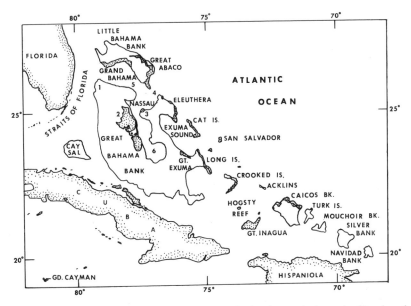

Fig. 5. Bahama region showing principal islands, banks, and channels. Numbered localities are as follows: (1) Bimini Island; (2) Andros Island; (3) New Providence Island; (4) Northeast Providence Channel; (5) Northwest Providence Channel; (6) Tongue of the Ocean.

tion to 5761 m through limestones, dolomites, and evaporites. The oldest rocks encountered in this flat-lying sequence appear to be either very early Cretaceous or Jurassic.

F. Geology of the Bahamas

Our direct knowledge of the pre-Pleistocene geology of the Bahamas rests on slender evidence indeed. Significant sources of data are: (1) the Superior Oil Company well at Stafford Creek on Andros Island,* and (2) DSDP drilling in Northeast Providence Channel and north and east of San Salvador on the deep ocean floor (Fig. 5).

The Superior Oil Company well reached 14,585 ft, and though there is some doubt about the age of the oldest rocks penetrated, it appears clear that rocks at least as old as early Albian were encountered (Spencer, 1967). Samples from the well were studied by Goodell and Garman (1969) who concluded that the entire sequence represents shallow-water deposition. This accords

---

* A well has been drilled to a depth of 17,557 ft on Long Island by Gulf, Chevron, and Mobil, but no information has been released, and as of September, 1971, Chevron was drilling below 11,580 ft on Bimini, but again no information has been released.

well with environmental interpretations made of the carbonate sequence encountered in drilling in Florida, and with the seismic refraction work of Sheridan *et al.* (1966). It corroborates the suggestion of Hess (1933, 1960) and Newell (1959) that the present banks and islands in the Bahamas are due to upbuilding largely controlled by reef-building organisms operating on a subsiding platform.

Site 98 of the DSDP was drilled in 2769 m of water in Northeast Providence Channel between Great Abaco Island on the north and New Providence Island on the south. This hole provided a direct test of the already substantially established upbuilding hypothesis. Drilling penetrated only 357 meters but reached lower Campanian rocks. The drill penetrated to total depth in pelagic carbonate oozes interpreted as representing open-ocean deposition (Ewing *et al.*, 1970).

Farther to the southeast on Hogsty Reef (near lat. 21°31′ N, long. 74° W) Milliman (1967) has reported the presence of what appeared to be fossils of shoal-water corals on carbonate banks now submerged to depths of 500–700 m. It thus appears probable that the pattern of reef-controlled upward growth of carbonate banks is general in the Bahamas and that the decreasing relative area of shoal water as one follows the Bahamas southeastward may mean either more rapid subsidence or fewer favorable sites for reef growth on the initial platform in this direction.

Seismic reflection profiles in the southeastern Bahamas have been reported on by Uchupi *et al.* (1971). Here one encounters a deep layer showing few internal reflecting horizons, and believed to be reef-complex carbonates, and a shallower sequence showing numerous internal reflections and filling topographic lows in the highly irregular deeper surface. The upper sequence is thought to consist of carbonate sediments transported into deep water or formed there as planktonic oozes. Based on what is known from drilling, one can construct a very general cross section running from the Cowles Magazines well in St. Lucie County, Florida, via the Superior well on Andros, and site 98 in Northeast Providence Channel to site 100 northeast of San Salvador for a dog-legged distance of 700 km. Along this traverse, basement has been encountered only in the Cowles Magazines well in Florida, but not anywhere in the Bahamas nor in the deep-ocean sites of the DSDP. DSDP site 100 encountered Oxfordian rocks at 5650 m, and it therefore appears probable, but can not be proven, that the Upper Jurassic section, like the Cretaceous, thickens and changes facies westward beneath the Bahamas into a reef complex. In the Superior well on Andros the Campanian would appear to occur in shallow-water facies somewhere near 3000 m, though the only information available places the top of the Lower Cretaceous at about 3080 m and the top of the Upper Cretaceous at about 2600 m (Spencer, 1967). DSDP site 98 also encountered the Campanian near 3100 m below sea level, but according to Ewing *et al.* (1970) the rocks here exist in a deep-water facies. One would surmise from this information that the Northeast Providence Channel together with its tributaries, the Northwest Providence Channel, and the Tongue of the Ocean existed as significant topographic depressions before the Campanian. The straits of Florida, on the other hand, apparently began to develop as a deep-water area only after Paleocene time (Sheridan *et al.*, 1966). The process of gradual drowning of some of the shoal areas evidently continued beyond the Paleocene since Milliman (1967) found recrystallized Pleistocene *Globigerina*-pteropod ooze atop the so-called Gerda Guyot now submerged to a depth of 670–690 m, though he was unable to determine the age of the presumably shallow-water substrate. At first thought one might ascribe the formation of deep-water areas in the Bahamas solely to an inherited erosional drainage pattern or to differential subsidence, but it is also possible that changes in circulation patterns across the region, affecting the availability of nutrients for reef builders, water temperature, and water chemistry, may have played a role and that the actual explanation may be complex.

G. The Geology of the Blake Plateau

The Blake plateau is somewhat better known than the Bahamas, thanks to seismic reflection profiling and JOIDES drilling. The two types of information have been integrated by Emery and Zarudzki (1967), and much additional information has been given by Ewing *et al.* (1966). JOIDES sites 1 and 2 were drilled on the continental shelf (Fig. 6). Site 5 was drilled on the Florida–Hatteras slope which separates the continental shelf from the more deeply submerged Blake plateau. Sites 4 and 6 were located on the Blake plateau itself, one (6) near the inner and the other (4) near the outer margin. Also of some interest are DSDP sites 102, 103, and 104 drilled on the Blake–Bahama outer ridge (Fig. 4). Information from wells drilled in the emergent coastal plain provide some additional information (Herrick and Wait, 1965; Leve, 1961). Figure 7, taken from Emery and Zarudzki (1967), shows a cross section of the shelf east of Jacksonville, Florida. The shelf is seen to be underlain by about 1 km of Cenozoic sediments, mostly of Eocene age, and to have prograded seaward throughout the Cenozoic. A thick sequence of Cretaceous rocks penetrated by drilling on land, but not in JOIDES drilling, can be followed seaward by means of reflectors beneath the Cenozoic deposits.

On the Blake plateau itself, the prism of Cenozoic sediments thins to about 0.5 km, with the principal thinning, relative to the shelf section, occurring in the Eocene rocks. Several kilometers of Cretaceous and possibly older rocks are seen from reflection profiles to exist beneath the Cenozoic section. The older rocks have not been penetrated by the drill, but early Cretaceous rocks have been dredged from the Blake escarpment (Heezen and

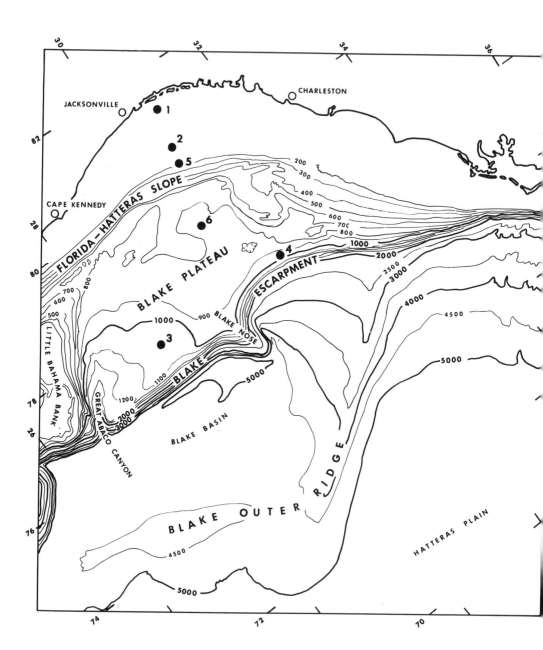

Fig. 6. Blake plateau region showing principal features of the submarine topography and the location of JOIDES holes 1–6. (Modified from Uchupi, 1968.)

Sheridan, 1966). Reflecting horizons occurring in what is probably a Cretaceous section show peculiarities at the eastern edge of the plateau thought by Ewing *et al.* (1966) to suggest the existence of a reef structure. These deep horizons then dip slightly to the west beneath the plateau. At its seaward edge the plateau terminates in the Blake escarpment. Some faulting is suggested by the Heezen and Sheridan (1966) report of early Cretaceous materials dredged from the escarpment. However, it appears likely that much of the relief is due to the growth of a reef barrier along the seaward edge of the plateau during parts of Cretaceous time. At a depth of 2500 m, Heezen and Sheridan recovered Aptian–Albian calcilutites dominated by planktonic foraminifera. At a depth of 3100 m and below, Aptian and older Cretaceous sediments were present but consisted of oolites and calcarenites which included algal fragments and are certainly of shallow-water origin. As Heezen and Sheridan note the evidence shows post-early Cretaceous subsidence of about 5000 m for the Blake plateau and indicates that, probably since the Aptian, upward growth has no longer kept pace with subsidence. In this respect the Blake plateau differs from both the Bahamas and Florida, in each of which carbonate deposition has kept up with subsidence over broad regions. Conceivably the difference is due to the lower temperature in the northern region, but this is at present speculative.

Ewing *et al.* (1966) have adduced evidence to show that a great deal of erosion has occurred on the Blake plateau, particularly along the axis of the Gulf Stream. They suggest that the Blake Nose may have been built partly by redeposition of the eroded materials. This supposition appears to be supported by drilling at DSDP sites 102, 103, and 104 which show Miocene and younger sediments exhibiting very rapid sedimentation rates (19 cm/1000 years, Fidé J. Ewing *et al.*, 1970).

III. DISCUSSION

A. Geophysical Inferences

Neither in the Bahamas nor in the Blake plateau do we have direct evidence of the existence of rocks older than Lower Cretaceous, though DSDP results suggest the presence of a Jurassic section beneath this province, and significant marine Jurassic sections occur on Cuba, La Désirade, the Gulf Coast, and offshore in the Maritime Province of Canada. Fortunately, a considerable amount of information on deeper horizons and crustal characteristics has been made available through the remarkable recent advances in marine geophysics. Valuable as these observations are, however they are not so satisfactory as direct evidence from drilling because they do not always lend themselves to

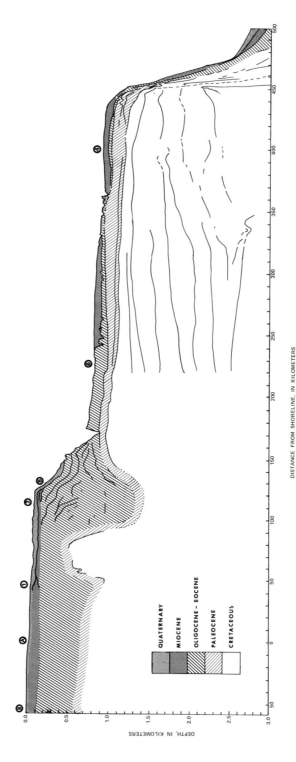

Fig. 7. Generalized cross section of coastal plain (wells A and B), the continental shelf (JOIDES holes 1, 2, and 5), and the Blake plateau (JOIDES holes 6 and 4 and reflectors from sparker profiling). (After Emery and Zarudski, 1967.)

unique interpretation in terms of lithology, depth or age. Much of the available geophysical information has been recently summarized and can be graphically presented (Figs. 7–11).

B. Thickness of Cenozoic Sediments

Most of the information on this subject comes from seismic reflection profiling, and it has therefore accumulated dramatically in the last few years. Data are usually presented as depth to a particular reflecting horizon, less the depth to the sea floor. Several assumptions are required for interpretation which can introduce both systematic and nonsystematic error. One must assume for thickness calculations that the velocity of acoustic waves through the section in question is known. One must assume that one is dealing with a continuous synchronously deposited acoustic reflector of uniform age, and one must assume the age of the reflector itself, unless nearby drilling has penetrated it. Despite these drawbacks, the data have proven to be of great value in understanding the general nature of the younger prisms of submarine sediment. Emery et al. (1970) have collated data for the Blake plateau, the northern Bahamas, and the adjacent sea floor. Ewing et al. (1966) have provided much data as well as interpretive suggestions for the Blake plateau. Uchupi et al. (1971) have provided data on the southeastern Bahamas and some refraction data has been provided by Ball et al. (1971), but in this region enormous local relief and rapid variations in thickness make meaningful mapping on a regional scale impossible. Figure 8 has been compiled from these and other sources. The surface shown is most reliable north of the Little Bahama bank and on the adjacent sea floor. Elsewhere a few numbers have been provided when data are available, but no attempt at contouring has been made and the reflector forming the lower boundary for thickness estimates probably varies in age from place to place so that only rather gross generalizations are implied.

On the Blake plateau several interesting features have been noted by Ewing et al. (1966). Northeast of the Little Bahama bank, Great Abaco Canyon is revealed as a feature which received little Cenozoic sedimentation and must predate the Cenozoic as a topographic element. Along the eastern edge of the Blake escarpment on the southern half of the plateau a marked thinning of the Cenozoic section occurs above a probable Cretaceous reef barrier, and a possible arch in the underlying basement (Ewing et al., 1966). The same authors have suggested that the zone of somewhat roughened surface with more or less east–west anomalies occurring north of the Blake Nose, is due largely to erosion and sweeping by the Gulf Stream and also suggest that the sediment removed may have been redeposited to form the Blake–Bahama outer ridge which has been shown by Emery et al. (1970) and by Ewing et al. (1970) to

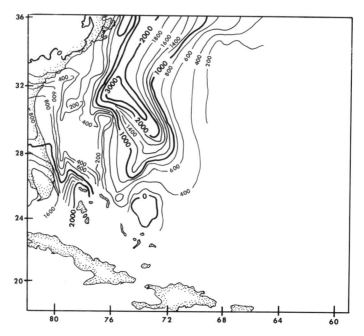

Fig. 8. Thickness of Cenozoic sediments shown in meters. (Modified
from Emory *et al.*, 1970.)

consist of a thick pile of Cenozoic sediments. Along the east side of the Peninsular Arch of Florida, the Cenozoic sediments also thicken toward the Bahamas, but the regional picture becomes less clear.

C. Depth to Basement

Emery *et al.* (1970) have provided a generalized map which indicates the
approximate depth of the basement. They define basement as consisting of pre-
Triassic rocks seaward of the continental slope and of ocean crust beneath the
continental rise and the abyssal plains. Salient features of their map (Fig. 9)
are the South Florida basin, where basement is shown as occurring at 11 km,
and the marked rise in the basement north and south of the basin to a depth
of 5 km or less. The effect of the Peninsular Arch is shown as extending in a
more easterly direction than other investigators have believed (e.g., Pressler,
1947; Sheridan *et al.*, 1966) and reaching to the Little Bahama bank. The
basement structure beneath the eastern edge of the southern Blake plateau
suggested by thinning of the Cenozoic section is clearly reflected in the basement surface, which slopes westward from this feature reaching a depth of 9 km
beneath the central portion of the plateau before rising toward its outcrop west
of the coastal plain. Basement appears also to be quite deeply depressed beneath

the Blake–Bahama outer ridge, where it is shown at a depth of 10 km. In the Bahamas there is little evidence of the depth to basement except that it lies below about 5 km, as is shown by drilling. Ball *et al.* (1971) suggests on the basis of seismic refraction work that beneath the Cat Island platform there is 5.5 km of material with a velocity of 4.6 km/sec, which they consider to be Cenozoic and Mesozoic carbonates, 17 km of material with a velocity of 5.7 km/sec, which they consider to include Lower Cretaceous carbonates and igneous basement rocks, but depth to the carbonate–igneous basement interface was not defined.

D. Depth to Mantle

Depth to the mantle in the Bahama–Blake plateau region has been investigated by a number of workers and summarized by Emery *et al.* (1970), primarily from data of Drake *et al.* (1968). Figure 10 provides a map of the available information based on seismic refraction observations taken at sea. For the most part the map shows a relatively simple transition from depths

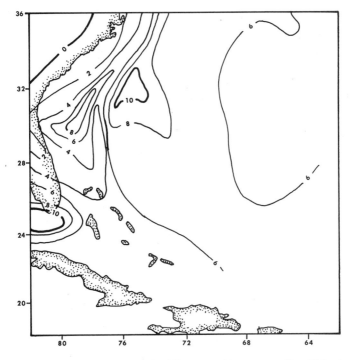

Fig. 9. Depth to basement in kilometers. Basement = "pre-Triassic rocks" under the coastal plain, continental shelf and slope, and oceanic basement beneath the continental rise and abyssal plains. (After Emory *et al.*, 1970.)

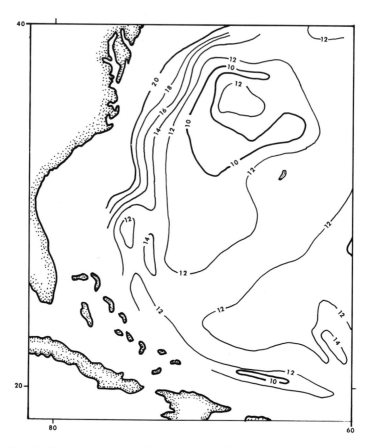

Fig. 10. Depth to mantle in kilometers. (Modified from Emory *et al.*, 1970.)

of 10–12 km in the clearly oceanic regions to 20 or more kilometers beneath
the continental rise. Additional data from Ball *et al.* (1971) allow extension
of the contours in the eastern Bahamas and indicate that here too the Blake–
Bahama escarpment follows a region in which depth to the mantle increases
rapidly to 20 km or more. However, Ball *et al.* (1971) have found the mantle
at more typically oceanic depths in the vicinity of Exuma Sound within the
Bahamas so that considerable local variation must exist.

Largely on the basis of gravity observations, Uchupi *et al.* (1971) have
concluded that the mantle beneath the southeastern Bahamas occurs at a depth
of about 20 km and that it is transitional to normal oceanic depth for about
100 km beyond the Bahama escarpment. The uncertainties in interpretation
of gravity data are such that a unique model is difficult to achieve and it is
not yet known with certainty whether the southeastern Bahamas are underlain
by oceanic or thin continental crust. It does appear probable, according to
Uchupi *et al.* (1971), that volcanic rocks played a part in the basement and may

have influenced the regions upon which carbonate platform development was initiated. They also appear to have definite evidence of the existence of a basement ridge beneath the Bahama escarpment which may be analogous to the ridge found by Ewing *et al.* (1966) beneath the eastern margin of the southern Blake plateau.

E. Magnetic Observations

Summaries of the distribution of magnetic anomalies off the east coast of North America have been provided by Drake *et al.* (1963), Bracey (1968), Emory *et al.* (1970), and Vogt *et al.* (1971). From their work it is clear that magnetic anomalies in the western North Atlantic are roughly parallel to the east coast of North America as far south as Florida (Fig. 11). In southwestern Florida it was shown by King (1959) and by Drake *et al.* (1963) that anomalies in this trend intersected essentially at right angles a southeast–northwest trend which is shown by Drake *et al.* (1963) as extending into the north-

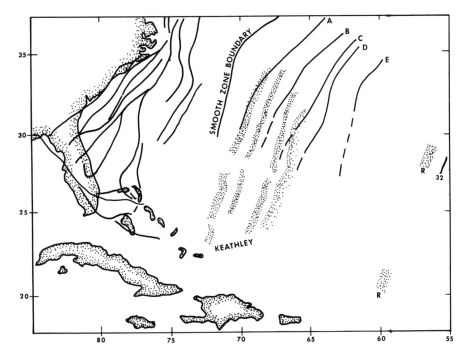

Fig. 11. Principal magnetic lineations and reversals showing the general parallelism to the coast except in southern and western Florida and the adjacent Bahamas. Anomalies A, B, C, D, and E together with quiet-zone boundary modified from Emory *et al.* (1970); Keathley Group of anomalies together with anomalies R and 32 modified from Vogt *et al.* (1971); coastal anomalies modified from Drake *et al.* (1963) and King (1959).

western Bahamas. Further east in the Bahamas, little has been published on regional magnetics, but Bracey (1968) and Vogt *et al.* (1971) show well-defined Atlantic anomalies approaching normal to the Bahama escarpment. If one follows Vogt *et al.* (1971) in assigning a Jurassic age to the Keathley group of anomalies, then the DSDP results at sites 99, 100, 101, and 105 are consistent with the sea-floor-spreading model. However, the only available age for the anomalies [as Vogt *et al.* (1971) note] is in fact provided by fossiliferous sediments in the DSDP holes, which cannot be shown to have reached basement. If the magnetic anomalies are correctly interpreted as due to sea-floor spreading, then it appears that the crust beneath the Bahamas must be older (possibly Permian if one follows the reasoning of Emery *et al.*, 1970) under the northwestern Bahamas than under the southeastern Bahamas, where it may possibly be Lower Cretaceous if one extrapolates from the conclusions of Vogt *et al.* (1971). Thus the critical piece of information in dealing with the significance of the Bahama–Blake platform region in the history and origin of the North Atlantic Ocean is one not yet available—the age of the basement.

F. Age of the Basement

In terms of an understanding of the history of the Atlantic Ocean, it is of great importance to establish the age of the basement beneath the Bahama–Blake plateau region. Should the basement here prove to be older than the Permian–Triassic interval during which North America and Africa are generally supposed to have exhibited the relationship shown in the Atlantic reconstruction of Bullard *et al.* (1965), it would constitute a considerable embarrassment to this model. If, on the other hand, it could be shown that basement in the Bahama–Blake plateau region were younger than Triassic, the reassembly is permitted, though it is not required.

Unfortunately, evidence of the age of the basement in the Bahama–Blake plateau region is indirect and inadequate to allow any save general speculations. It is possible only to establish a broad time interval within which the true age of the basement should fall. The evidence is derived from the age of the oldest rocks yet encountered to seaward of the region of interest, and from the youngest basement ages known in contiguous onshore regions.

Seaward of the Blake–Bahama escarpment, information is available from the preliminary study of materials obtained in the course of leg 11 of the DSDP. The significant holes for the problem at hand were drilled at sites 99, 100, and 101 north and east of the Bahamian island of San Salvador in a region shown by leg 1 to contain old sediments. At each site Jurassic sediments were encountered and the oldest materials are reported as being of Callovian(?)–Oxfordian age or close to the Middle Jurassic–Upper Jurassic boundary. At site 101 the hole bottomed in basalt after 4 m of penetration. The basalt may be reflecting horizon B (Ewing *et al.*, 1970) and is apparently a flow with a glassy upper surface. In any case, the available data indicate that the basement in the region just east of the Blake–Bahama escarpment is older than the Middle–Upper Jurassic boundary. Unfortunately, there is no other indication of how old it may be.

The other source of information potentially useful in constraining the age of the basement of the Bahama–Blake plateau region is geochronologic data for the southeastern margin of the North American continent. Two areas are significant: (1) the Cape Hatteras region of North Carolina which has yielded basement rocks within 40 miles of the seaward edge of the continental shelf, and (2) the basement rocks known to exist beneath the overstepping prism of Mesozoic sediments in peninsular Florida. Data from both areas will be considered critically, because it is an unfortunate fact that the ages provided through geochronologic work, while carefully qualified by their original authors, tend to become entrenched in the literature unburdened by any qualifications and soon come to carry a spurious aura of exactness which is not intended and often is not justified. In examining the dates available in the areas of interest, an attempt has been made to summarize those determinations made by the most reliable methods for the particular samples studied and to favor those studies which provide petrographic information on the samples dated.

West of Cape Hatteras along the North Carolina–Tennessee boundary and some 250 miles from the coast, Davis *et al.* (1962), on the basis of U–Pb dating of zircons, have shown the presence of crystalline rocks having an age of about 1000 m.y. Gneisses in the same region provide evidence for the presence of materials 1300 m.y. old, but Davis *et al.* (1962) suggest that the zircons in these rocks are detrital and reflect the age of an unknown source terrain. Further to the north, near Baltimore, Maryland, Tilton *et al.* (1958) reported U–Pb ages on zircons from gneiss domes suggesting an 1100 m.y. event, while K–Ar dates on biotite from the same rocks showed minimum ages of around 300 m.y. for a later metamorphic event. It appears from the available data that a belt of rocks near the western edge of the Piedmont between New Jersey and North Carolina, and perhaps beyond, contains rocks which became closed systems with respect to U–Pb about 1000 m.y. ago.

Working along the seaward margin of the exposed Piedmont some 120 miles from the coast, Fullagar (1971) has identified three episodes of plutonism by means of Rb–Sr whole-rock geochronology. Associated with the Carolina Slate Belt is a linear trend of plutons which became closed systems with respect to Rb–Sr between 595 and 520 m.y. ago, or in the Early–Middle Cambrian (Fig. 12). Slightly farther inland another linear trend of plutons having ages in the range 415–385 m.y. (Late Silurian–Middle Devonian) is associated with

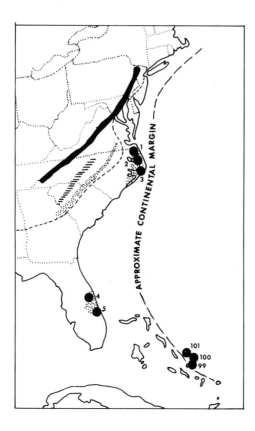

Fig. 12. Basement Age in eastern margin of North America. Solid black, Grenville ± 1000 m.y.; horizontal lines, 415–384 m.y.; stipple pattern, 590–520 m.y.; short dashed line, edge of coastal plain overlap on basement rocks. (1) Blair *et al.* #1 Wegerhaeuser; (2) Socony–Mobil #1 State; (3) Socony–Mobil #3 State; (4) Humble #1 Carrol; (5) Amarada #2 Cowles Magazine. 99–101 are DSDP holes bottoming in Jurassic.

the Charlotte belt of the Piedmont. This episode is shown by Fullager to be associated with the last and very pervasive metamorphic event in this area. Less clearly identified with any geologic belt and thus less linear in trend is a group of post-metamorphic plutons which yield ages around 300 m.y. (Middle Pennsylvanian). From the very comprehensive study of Fullagar we find that a major tectonic event associated with the Carolina Slate Belt occurred in Early-Middle Cambrian time and that the last episode involving both intrusion and regional metamorphism occurred around the Silurian–Devonian boundary. No data suggestive of plutons older than early Paleozoic was encountered.

A study by Denison *et al.* (1967) presents ages determined by several methods for basement rocks encountered in drilling for hydrocarbons in coastal North Carolina only 40 miles west of the seaward margin of the continental shelf. Two of the dates reported are whole-rock Rb–Sr determinations on seemingly fresh rocks, and probably yield the date of closure of the Rb–Sr system. Samples from the bottom of the Socony Mobil #1 State in Dare County yielded an age of 585 ± 40 m.y., while those from the Socony Mobil #3 State yielded an age of 610 ± 60 m.y. These determinations fall in the Early–Middle Cambrian event identified by Fullagar (1971) and the rocks dated are along a trend indicated by Sundelius (1970) to contain an eastern branch of the Carolina Slate Belt.

The data of Fullagar and Denison *et al.* taken together suggest that a broad region of the Piedmont both where it is exposed and where it is buried beneath a prism of Mesozoic and younger sediments, and extending to within 40 miles of the seaward edge of the continental shelf, was present very early in the Paleozoic. Fullagar (1971) has also considered $^{87}Sr/^{86}Sr$ ratios and concludes that the plutons he studied represent magmas derived from the lower part of the crust or upper mantle without significant anatexis. He believes that 300 through 600 m.y. ago in this area the lower crust had Rb–Sr ratios lower than did other continental regions. These data in turn suggest that the region had not had a long history as fully developed continental crust prior to 600 m.y. ago.

Additional data regarding basement ages are available from the work of Bass (1969) and of Milton and Grasty (1969) on rocks encountered in drilling for hydrocarbons in Florida. Unfortunately, the work of Milton and Grasty presents only whole-rock K–Ar dates and these are considered unreliable even for fresh rocks save in the case of very young basalts. Bass, however, reports the results of whole-rock and mineral Rb–Sr ages which appear likely to give the time at which the system became closed for Rb–Sr. In the Humble #1 Carroll well in Osceola County, Florida, a quartz monzonite yielded an age for intrusion or metamorphism of about 530 m.y. Considerably farther to the south in St. Lucie County, Florida, the Amarada #2 Coles Magazines well gave an age for amphibolite facies and retrograde zeolite facies metamorphism of what were probably basic or intermediate igneous rocks of about 530 m.y. From this work it appears that the basement beneath peninsular Florida shows evidence for the existence of intrusion into or metamorphism of an older basement in Middle Cambrian time.

Since the data both from Florida and North Carolina appear in agreement as to the age of the oldest intrusives and/or metamorphism (Lower–Middle Cambrian) one can tentatively conclude that the continental crust between this region and the continental slope is very likely to be of this age. Should this be true, it would follow that the basement beneath the Blake plateau is of Cambrian age, as is that beneath the northwestern Bahamas. These are, however, frankly speculative conclusions, and it appears to be extremely desirable that further drilling be undertaken in the Bahama region. Thus while it appears almost certain that Jurassic rocks underlie the Bahamas, the seaward edge of the Blake plateau, and the South Florida basin, we still lack any convincing evidence of the age of the basement. Therefore we can not obtain as yet from this region compelling evidence about the late Paleozoic–early Mesozoic interval so critical in understanding the history of the North Atlantic.

REFERENCES

Applin, P. L., 1951, Preliminary report on buried pre-Mesozoic rocks in Florida and adjacent states: U. S. Geol. Survey Circ. 91.

Applin, P. L. and Applin, E. R., 1965, The Comanche Series and associated rocks in the subsurface in Central and South Florida: U.S. Geol. Survey Prof. Pap. 447.

Ball, M. M., Dash, B. P., Harrison, C. G. A., and Ahmed, K. O., 1971, Refraction seismic measurements in the northeastern Bahamas: *Trans. Am. Geophys. Union*, v. 52, no. 4, p. 252 (abst).

Bass, M. N., 1969, Petrography and ages of crystalline basement rocks of Florida—some extrapolations, in: *Amer. Assoc. Petrol. Geol. Mem. 11*, Logan and McBirney, eds.

Bracey, D. R., 1968, Structural implications of magnetic anomalies north of the Bahama–Antilles islands: *Geophysics*, v. 33, p. 1950.

Bullard, E. D., Everett, J. E., and Smith, A. G., 1965, The fit of the continents around the Atlantic, in: A Symposium on Continental Drift, Blackett, Bullard, and Runcorn, eds.: *Phil. Trans. Roy. Soc. (London)*, v. A 258, p. 41–51.

Chase, R. L. and Bunce, E. T., 1969, Underthrusting of the eastern margin of the Antilles by the floor of the western North Atlantic Ocean and the origin of the Barbados ridge: *J. Geophys. Res.*, v. 74, p. 1413–1420.

Chen, C. S., 1965, The Regional Lithostratigraphic Analysis of Paleogene and Eocen Rocks of Florida: *Fla. Geol. Survey Geol. Bull.* 45, p. 1–105.

Davis, G. L., Tilton, G. R., and Wetherill, G. W., 1962, Mineral ages from the Appalachian Province in North Carolina and Tennessee: *J. Geophys. Res.*, v. 67, p. 1987–1996.

Denison, R. E., Raveling, H. P., and Rouse, J. T., 1967, Age and descriptions of subsurface basement rocks, Pamlico and Albemarle Sound areas, North Carolina: *Amer. Assoc. Petrol. Geol. Bull.*, v. 51, p. 268–272.

Drake, C. L., Ewing, J. I., and Stockard, H. P., 1968, The continental margin of the eastern United States: *Can. J. Earth Sci.*, v. 5, p. 933.

Drake, C. L., Heirtzler, J., and Hirshman, J., 1963, Magnetic anomalies off eastern North America: *J. Geophys. Res.*, v. 68, p. 5259–5276.

Edgar, N. T., Ewing, J. I., and Hennion, J., 1971, Seismic refraction and reflection in the Caribbean Sea: *Amer. Assoc. Petrol. Geol. Bull.*, v. 55, p. 833–870.

Emery, K. O., Uchupi, E., Phillips, J. D., Bowin, C. O., Bunce, E. T., and Knott, S. T., 1970, Continental rise off eastern North America: *Amer. Assoc. Petrol. Geol. Bull.*, v. 54, p. 44–108.

Emery, K. O. and Zarudski, E. F. K., 1967, Seismic reflection profiles along the drill holes on the Continental Margin off Florida: U.S. Geol. Survey Prof. Pap. 581-A.

Ewing, J., Ewing, M., and Leyden, R., 1966, Seismic-profiler survey of Blake Plateau: *Amer. Assoc. Petrol. Geol. Bull.*, v. 50, p. 1948–1971.

Ewing, M., *et al.*, 1969, Initial reports of the Deep Sea Drilling Project, v. 1: National Science Foundation Spec. Publ. 1.

Ewing, J., *et al.*, 1970, Deep Sea Drilling Project, Leg 11: *Geotimes*, v. 15, p. 14–16.

Fullagar, P. D., 1971, Age and origin of plutonic intrusions in the Piedmont of the Southeastern Appalachians: *Geol. Soc. Amer. Bull.*, v. 82, p. 2845–2862.

Goodell, H. G. and Garman, R. K., 1969, Carbonate geochemistry of Superior deep test well, Andros Island, Bahamas: *Amer. Assoc. Petrol. Geol. Bull.*, v. 53, p. 513–536.

Heezen, B. C. and Sheridan, R. E., 1966, Lower Cretaceous rocks (Neocomian–Albian) dredged from Blake Escarpment: *Science*, v. 154, p. 1644–1647.

Herrick, S. M. and Wait, R. L., 1965, Subsurface stratigraphy of coastal Georgia and South Carolina (abst.): *Amer. Assoc. Petrol. Geol. Bull.*, v. 49, p. 344.

Hess, H. H., 1933, Interpretation of geological and geophysical observations: U.S. Hydrographic Office, Navy–Princeton Gravity Expedition to the West Indies in 1932, p. 27–54.

Hess, H. H., 1960, The origin of the Tongue of the Ocean and other great valleys of the Bahama Bank: 2nd. Caribbean Geol. Conf. Trans., U. of Puerto Rico, Jan. 4–9, 1959, p. 160–161.

Khudoley, K. M., 1964, Principal features of Cuban geology: *Amer. Assoc. Petrol. Geol. Bull.*, v. 51, p. 668–677.

King, E. R., 1959, Regional magnetic map of Florida: *Amer. Assoc. Petrol. Geol. Bull.*, v. 43, p. 2844–2854.

Leve, G. W., 1961, Reconnaissance of the ground-water resources of the Fernandina Area, Nassau Co., Florida: Florida Geol. Survey Inf. Cirq. 28, p. 1–24.

Malloy, R. J. and Hurley, R. J., 1970, Geomorphology and geologic structure; Straits of Florida: *Geol. Soc. Amer. Bull.*, v. 81, p. 1947–1972.

Milliman, J., 1967, Guyot-like features in the southeastern Bahamas—a preliminary report: *Proc. Int. Conf. on Tropical Oceanography*: U. of Miami, Inst. Marine Sci. Press, p. 45–55.

Milton, C., and Grasty, R., 1969, "Basement" rocks of Florida and Georgia: *Amer. Assoc. Petrol. Geol. Bull.*, v. 53, p. 2483–2493.

Molnar, P. and Sykes, L. R., 1969, Tectonics of the Caribbean and Middle American regions from focal mechanisms and seismicity: *Geol. Soc. Amer. Bull.*, v. 80, p. 1639–1684.

Newell, N. D., 1959, The coral reefs, Pt. 1—Questions of coral reefs, Pt. 2—Biology of corals: *Nat. Hist.*, v. 68, p. 119–131, p. 226–235.

Pressler, E. D., 1947, Geology and occurrence of oil in Florida: *Amer. Assoc. Petrol. Geol. Bull.*, v. 31, p. 1851–1862.

Sheridan, R. E., Drake, C. L., Nafe, J. E. and Hennion, J., 1966, Seismic-refraction study of continental margin east of Florida: *Amer. Assoc. Petrol. Geol.*, v. 50, p. 1972–1991.

Spencer, M., 1967, Bahamas deep test: *Amer. Assoc. Petrol. Geol. Bull.*, v. 51, p. 263–268.

Sundelius, H. W., 1970, The Carolina Slate Belt, in: *Studies of Appalachian Geology, Central and Southern*, Fisher, *et al.*, eds.: Interscience, New York.

Tilton, G. R., Wetherill, G. W., Davis, G. L., and Hopson, C. A., 1958, Ages of minerals from the Baltimore gneiss near Baltimore, Maryland: *Geol. Soc. Amer. Bull.*, v. 69, p. 1469–1474.

Uchupi, E., 1968, Atlantic Continental Shelf and Slope of the United States – Physiography: U.S. Geol. Survey Prof. Pap. 529-C.

Uchupi, E., Milliman, J. D., Luyendyk, B. P., Bowin, C. O., and Emery, K. O., 1971, Structure and origin of the southeastern Bahamas: *Amer. Assoc. Petrol. Geol. Bull.*, v. 55, p. 687–704.

Vogt, P. R., Anderson, C. N., and Bracey, D. R., 1971, Mesozoic magnetic anomalies, sea-floor spreading, and geomagnetic reversals in the southwestern North Atlantic: *J. Geophys. Res.*, v. 76, p. 4796–4823.

Windisch, C. C., Leyden, R. J., Worzel, J. L., Saito, T., and Ewing, J., 1968, Investigation of horizon Beta: *Science*, v. 162, p. 1473.

Chapter 3

THE CONTINENTAL MARGIN OF EASTERN NORTH AMERICA, FLORIDA TO NEWFOUNDLAND

M. J. Keen

Department of Geology
Dalhousie University
Halifax, Nova Scotia

I. INTRODUCTION

It should be easy to write an account of the continental margin of eastern North America from Florida to Newfoundland because, by comparison with so many other margins, it has been studied in such detail. The task is made difficult, however, by the very wealth of information, and by the fact that, in spite of this wealth, many important geological questions still lie unanswered. For example, the origins of the magnetic "slope anomaly" and the magnetic "quiet zone" are not yet known, nor can we even describe the transition from the oceanic crust (Layers 2 and 3 of seismologists) to the continental crust. There are, perhaps, three objectives to meet in this chapter. First, to describe concisely what is known about this margin. Second, to relate these observations to the geological principles which may apply. Third, to show where there is still doubt and confusion, and suggest experiments which might be done to overcome this situation.

We can identify two phenomena which have led to the major features of the continental margin as we see it now. These are the rifting of the Atlantic Ocean in the early Mesozoic, and the changes in sea level in the Pleistocene. The model proposed by Schneider (1969) and Vogt (1970), considered in

greater depth by Sleep (1971), provides a convenient framework within which
to consider the effects of continental break up (Fig. 1). Uplift of the continental massif took place before rifting, associated perhaps with thermal
expansion. As rifting and ocean-floor spreading took place the continental
margins separated and gradually subsided and cooled as they moved farther
from the hot mid-ocean ridge. The history of the sedimentation reflects in part
the uplift and subsidence. The major morphology of the margin has its origin
in rifting and subsidence. The details of the morphology are, however, dominated by the effects of the Pleistocene and Holocene changes in sea level.
The latest, the Holocene transgression, has, for example, had the effect that
the margin as we see it now is not an "equilibrium" margin; the shelf is mantled
with relict sediments; much of our coastline has been drowned; and indeed
the very existence of the shelf, as we see it now, is anomalous. Good accounts
of these last points have been written by Curray (in Stanley *et al.*, 1969) and
by Swift (1970).

   The history of the margin from the point of view of rifting and spreading,

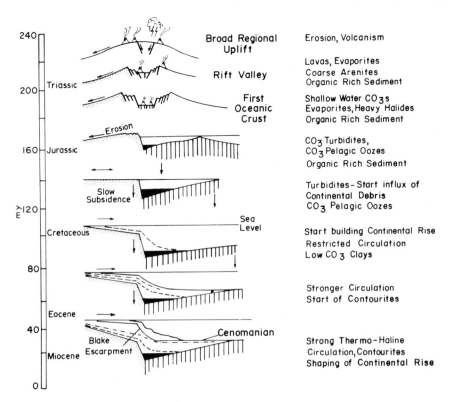

Fig. 1. Uplift and subsidence at the rifting of a continent followed by ocean-floor spreading.
Redrawn and reprinted by permission of copyright owners, Compass Publications, Inc., from
Undersea Technology (Schneider 1969).

uplift and subsidence, leads to one way of classifying some of the margin's morphological features. For example, Inman and Nordstrom (1971) proposed three major groupings of coastlines: collision coasts, those of the edges of colliding continents and island arcs; marginal sea coasts, those of continents protected by island arcs; and trailing edge coasts, those of continents nearer to the trailing edge of a plate spreading away from an active mid-ocean ridge. Clearly the eastern seaboard of North America and the western seaboard of Africa are examples of the last category. This classification is in a sense a genetic classification. A morphological classification can be established which considers the types of structures responsible for trapping sediment at the margin (Emery, 1969; Emery and Uchupi, 1972); tectonic dams, reef dams, and diapiric structures are possible barriers. Although Emery (1969) suggested that the margin of much of eastern North America was dam-free (north of reef dams off Florida), in fact, in places there must have been tectonic dams associated with rifting, and subsequently diapiric dams associated with uplift of evaporite bodies.

II. PHYSIOGRAPHY

The geography of the coastal regions reflects the underlying geology and the effects of differential changes of sea level relative to the land along the length of the margin (Figs. 2 and 3).

Rocks of the Coastal Plain Province, relatively soft and easily eroded, form the bedrock as far north as Cape Cod. They are separated from the more resistant Paleozoic rocks of the Appalachian Highlands by the Fall Line—the line of waterfalls resulting from retarded headward erosion associated with the change from one rock type to another. The Coastal Plain is wholly submerged north of Cape Cod, where it underlies a large part of the continental shelf. The differences in resistance to erosion of the bedrock leads to differences in the amount of solid material carried in suspension or in solution in the rivers. The northern rivers (such as the Susquehanna and Hudson) have higher rates of water discharge than the southern rivers, but the southern rivers (such as the Savannah) contain more suspended sediment (Fig. 4; Meade, 1969). This phenomenon, together with differential submergence and differential erosion, has led to some of the characteristics of the coastal morphology.

In the north we see drowned river valleys; in the south we do not. This is for a variety of reasons. River valleys in the north were more severely excavated in the Pleistocene, because of glacial erosion, or because the rivers' volumes were increased by glacial meltwater. The bedrock in the north is less easily eroded and less sediment is available to fill the drowned valleys. Some regions such as Nova Scotia have a relatively small drainage area. Glacial scouring removed debris during the Pleistocene from the northern regions. However, possibly another substantial contributor to the difference in mor-

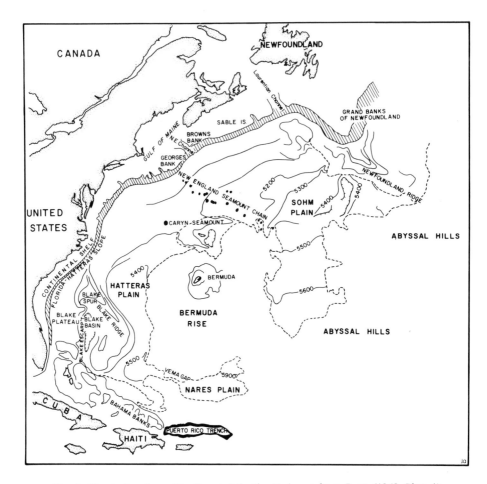

Fig. 2. Key to locations, Northwest Atlantic. Redrawn from Pratt (1968, Plate 1).

phology is the higher rate of rise in sea level in the north by comparison with the south (Grant, 1970). Grant showed that the rise in sea level in the past 5000 years has been approximately 13 m off eastern Nova Scotia; the rise in sea level in this time in southwestern Florida has been only 4 m (Fig. 5). Grant ascribes differences such as these to differential subsidence caused by greater and lesser loading on the shelves by the oceans, following A. L. Bloom's ideas. Where the shelf is wide and deep the loading is high, and where narrow and shallow the loading is low. Although the idea is neat, the analytical development of the idea has so far perhaps been crude.

Many characteristics of the morphology of the shelf, slope, and rise are due to the presence or absence of the Gulf Stream; the effects of glaciation and of the Holocene transgression; and, of course, the effects of local oceanographic conditions upon the shelf. The Gulf Stream is close to the margin in the southern part of the area, but relatively far from it in the north. The direct effects of glaciation are seen on the northern shelves, but not on the southern. The Holocene transgression has left much of the shelf mantled by relict material, only partly covered by a modern mud and sand blanket. As a result, the area can be divided for convenience into three morphological regions: a southerly region from the Florida Panhandle to just south of Cape Hatteras; a central region from Cape Hatteras to north of Long Island; and a northern region extending to the Grand Banks (Fig. 2; Emery, 1966; King, 1970; Uchupi, 1968).

The southern region does not possess a typical shelf, slope, and rise system,

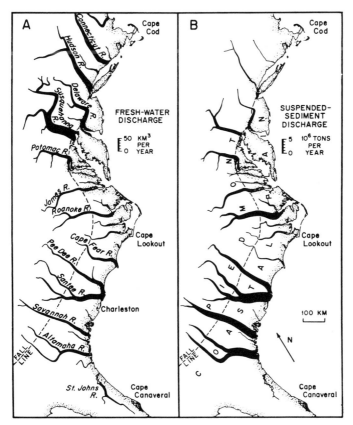

Fig. 4. Water and sediment discharged by rivers of the Atlantic drainage between Cape Cod and Cape Canaveral. (A) Fresh water discharged by rivers that deliver an average of more than 2 km³ per year. (B) Suspended sediment discharged by streams that deliver more than 100,000 tons per year. Reproduced from Meade (1969, Fig. 2); note the limitations of the data which he describes. They portray the situation in the period 1906-1907, and human activities have changed the picture since then.

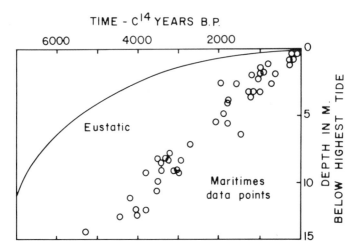

Fig. 5. Rise in sea level off the Maritime Provinces of Canada compared with the eustatic change in sea level. The open circles are data
points; the curve shows the world-wide or eustatic rise in sea level taken
from various sources. Redrawn from Grant (1970) where details will
be found. Reproduced from the *Canadian Journal of Earth Sciences* by
permission of the National Research Council of Canada.

the slope being partly occupied by a marginal plateau, the Blake plateau. This
is separated from the shelf by the Florida–Hatteras slope, and from the Hatteras Abyssal Plain by the Blake escarpment—whose gradient is steep, about
5°. The plateau's maximum width is approximately 275 km, and it becomes
narrower and shallower to the north, being no longer a conspicuous feature
off Cape Hatteras, where the central region begins. The Gulf Stream sweeps
over the plateau, and (a point made more fully in a later section) must be
responsible for many of its characteristics (Bunce *et al.*, 1965). The continental
shelf off Florida, like the shelf of the central region north of Cape Hatteras,
has many topographic features with a relief of a few tens of meters. The
narrow width and shallow overall level sets it apart from the northern region,
off Nova Scotia, for example.

The central region north of Cape Hatteras possesses the more "typical"
features of shelf, slope, and rise. The shelf is wider than off Florida, and
becomes deeper to the north, the shelf break being at 80 m off Cape Hatteras,
130 m off Long Island. The slope, with steeper gradient, merges into the rise,
with lesser gradient at greater depth than the same junction further north off
the Scotian shelf. No doubt this is due to the effects of glaciation with greater
Pleistocene deposition in the north. The slope is cut by numbers of submarine
canyons such as the Hatteras and Hudson canyons. The head of the Hudson
Canyon can be traced across the shelf toward the Hudson River.

The most characteristic feature of the northern region is the complexity of the shelf. Shallow banks are separated from one another by deep longitudinal troughs, as Georges bank and Browns bank are separated by the Northwest Channel, and Banquereau is separated from the Grand Banks by the Laurentian Channel. Longitudinal troughs, of which an example is the LaHave basin, are parallel to the coastline of Nova Scotia and separate an inner shelf from an outer shallow shelf. These features reflect the presence of the submerged coastal plain offshore, and the effects of glaciation.

The coastal plain extends on land as far north as the region of Cape Cod. From there it is not seen on land again along the whole of the Canadian eastern seaboard, but is wholly submerged as far north as northern Baffin Bay, 5000 km away. The boundary between the coastal plain and the underlying crystalline rocks, Paleozoic off Nova Scotia, Precambrian off Labrador, is marked by the longitudinal troughs, caused by glacial excavation. Those off Nova Scotia are the most southerly, and coincide approximately with the southern limit of glaciation. The coastal plain rocks of the outer shelf form mesa- and cuestalike features on the banks, mantled by Pleistocene and Holocene relict and modern sediments. The morphological forms suggest that the whole shelf was subjected to subaerial erosion at various times.

There are two sorts of contrasting features to be seen in the morphology of the shelf and of the mainland. The first is the one already described, between

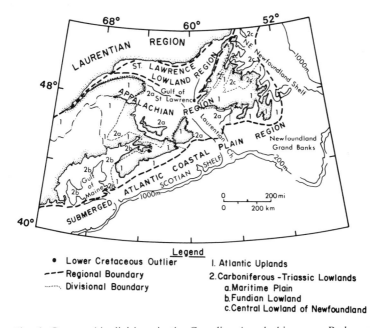

Legend

- Lower Cretaceous Outlier
- - - Regional Boundary
- ⋯⋯ Divisional Boundary

1. Atlantic Uplands
2. Carboniferous -Triassic Lowlands
   a. Maritime Plain
   b. Fundian Lowland
   c. Central Lowland of Newfoundland

Fig. 6. Geomorphic divisions in the Canadian Appalachian area. Redrawn after King (1972), from the *Bulletin of the Geological Society of America* and reproduced by permission.

the northern shelf and the shelves to the south; the second is between the Scotian shelf and its adjacent mainland. The mainland is occupied by Paleozoic and Triassic rocks principally, and its morphology is very different from that of the shelf (Fig. 6). It can be divided into regions of uplands formed from resistant rock, and lowlands carved into the softer rocks such as the Carboniferous and Triassic sediments. Evidence of the remnants of a peneplain can be seen in the accordant hilltops from the coast of Cape Breton to the Gaspé peninsula, which slopes to the east, toward the shelf. Nonmarine Lower Cretaceous deposits are found in the lowlands cut into the peneplain. Consequently, King (1972) points out that the development of the peneplain and therefore the elements of the mainland morphology is older, perhaps dating from the Jurassic. He associates this with uplift and erosion in the early phases of opening of the North Atlantic Ocean.

III. STRATIGRAPHY

Our knowledge of the stratigraphy of the margin is restricted largely to that of the shelf, with the exception of the Blake plateau. The summary which follows is based upon descriptions by Maher (1965) (the emerged Atlantic coastal plain), Charm et al. (1969) (the Florida shelf and Blake plateau), McIver (1972) (the Scotian shelf), and Bartlett and Smith (1971) (the Grand Banks). Summaries of pertinent aspects of the stratigraphy can be found in Table I, and in Figs. 7 and 8.

The stratigraphy can be described briefly as a wedge of Mesozoic sediments which thickens seaward, and which overlies a "basement" of Paleozoic and Triassic rocks. (Whether or not the Triassic sediments and volcanic rocks in which some wells bottom should be called "basement" is presumably an inconsequential matter of taste.) The sediments range in age from Jurassic to Holocene. They have a variable lithology; evaporites, shales, mudstones, sandstones, carbonates, and chalk are all found. They show that the Mesozoic sea entered the region of our present margin in the Jurassic, in restricted basins where evaporites formed; shallow-water conditions were replaced in time by deep-water conditions, to be succeeded in their turn by a return to shallow-water environments.

The basement upon which the sediment wedge formed is variable in its nature. Beneath the Scotian shelf granites and metamorphosed sediments akin to those of mainland Nova Scotia are found, whereas the Triassic Newark Group forms the basement beneath parts of the New Jersey coastal plain (McIver, 1972; Spangler and Peterson, 1950). Cape Hatteras wells bottomed in granitic rocks, and Paleozoic sediments and volcanic rocks form the basement in Georgia and Florida (Applin, 1951). At the present time no systematic pattern has been discerned within the basement of the margin.

TABLE I

Stratigraphic Terminology for the Jurassic and Cretaceous Sediments of the Scotian Shelf, Proposed by McIver (1972)[a]

Group	Formation	Member	Dominant lith.	Maximum thickness
Quaternary	* Sable Island		Sand & gravel	
	* La Have		Clay	
	* Sambro		Sand	
	* Emerald		Silt	
	* Scotian Shelf		Glacial drift	
The Gully Group	Banquereau		Mudstone	4000
	Wyandot		Chalk	750
	Dawson Canyon	Limestone Marker	Shale	3000
Nova Scotia Group	Logan Canyon		Sandstone & shale	800
		Sable Shale	Shale	500
			Sandstone & shale	2000
	Naskapi		Shale	750
	Missisauga	Limestone Marker	Sandstone	3700
Western Bank Group	Mic Mac		Calcareous shale	4000
	Verrill Canyon		Shale	>2000
	Abenaki	Baccaro	Limestone	2500
		Misaine	Calcareous shale	300
		Scatarie	Limestone	400
	Mohawk		Sandstone & shale	3500
	Iroquois		Dolomite	650
	Argo		Salt	>3000

[a] Formations Marked by an Asterisk are Informal Map Units Proposed by King (1970).

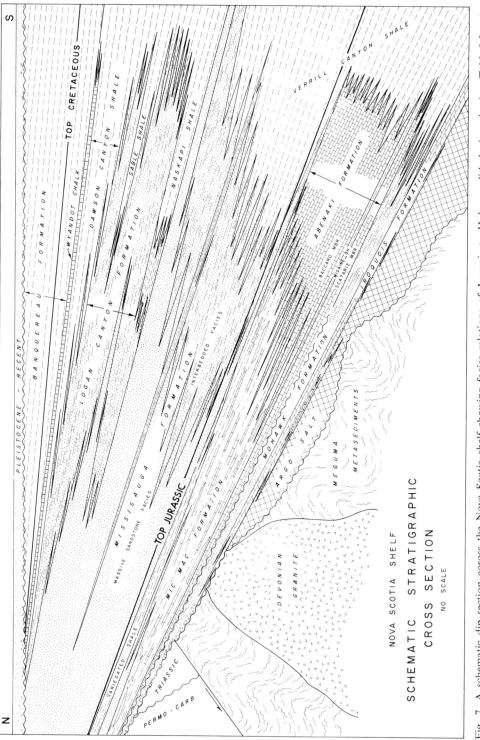

Fig. 7. A schematic dip section across the Nova Scotia shelf showing facies relations of Jurassic to Holocene lithologic units (see Table I for the stratigraphic terminology). Reproduced from McIver (1972, Fig. 5), from the *Canadian Journal of Earth Sciences*, by permission of the National Research Council of Canada.

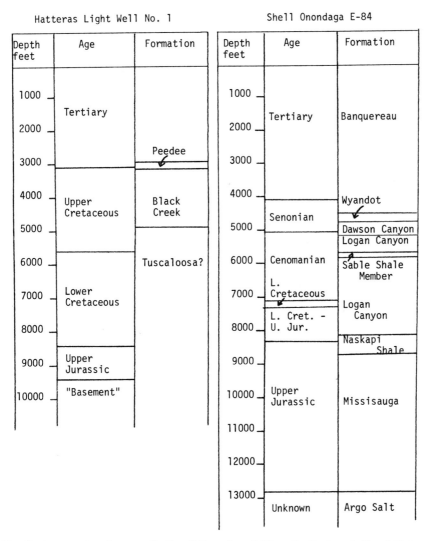

Fig. 8. A comparison between Shell well Onondaga E-84 on the Scotian shelf and Hatteras Light well No. 1.

The Triassic sediments and volcanic rocks of the eastern seaboard are found in fault-bounded basins, as, for example, the Wolfville Sandstones and North Mountain basalts of Nova Scotia, and comparable rocks in New Jersey (Sanders, 1963).

The oldest sediments beneath the Scotian shelf are red clastics of unknown age underlying early Jurassic evaporites—the Argo Salt of Table I and Fig. 7. These are succeeded by further Jurassic sediments which show a change from restricted marine conditions to more open-water marine conditions (the

Iroquois and Mohawk formations). By contrast, the Jurassic is hardly represented at all in the coastal plain to the south, and it is clear that very different conditions must have existed there (see, e.g., Hallam, 1971). The Upper Jurassic and Cretaceous rocks may show a change from nearshore, shallow-water conditions to, possibly, a deeper water environment. For example, the Missisauga Formation of the Lower Cretaceous contains coals; the Wyandot Chalk of the uppermost Cretaceous contains large numbers of pelagic and benthonic microfauna, the pelagic forms dominating. The Wyandot Chalk formation has been described by McIver (1972) as "...perhaps the most distinctive and widely recognized lithologic unit on the shelf.... ." He says that it can be traced to the edge of the shelf, and that it may be correlated with comparable formations at the Pan Am Tors Cove well of the Grand Banks, and at Cape Hatteras. However, the well log of the Hatteras Light well No. 1 given by Swain (1952) shows that the chalk reported there is much thinner than the Wyandot Chalk in, for example, Shell well Onondaga E-84 of the Scotian shelf. No chalk was described by Swain from the North Carolina Esso well No. 2.

It may be worth commenting that some doubt has recently been expressed about the need to associate all "chalks" with relatively deep water. Scholle and Kling (1972) have described lagoonal coccolith-rich muds from waters no more than 43 m deep off British Honduras, and suggest that these are excellent modern analogs of at least some ancient chalks.

During the Upper Cretaceous and Tertiary conditions became more similar to those of the present margin. The dominant rocks are the Banquereau Formation, which are principally mudstones becoming more sandy toward the top.

IV. SURFICIAL SEDIMENTS

One might think naively that one principal source of modern sediments which would cover the shelf, slope, and rise would be sediments transported to the sea by the rivers of the eastern seaboard. This is not true (Meade, 1969, 1972; Manheim et al., 1970). These river sediments are, in fact, largely trapped in the estuaries and coastal marshes; Meade (1972) shows that the rate of rise of coastal marshes along the eastern United States, reflecting obviously the rate of rise in sea level if they are to persist, is approximately equal to the supply by rivers. This phenomenon is attributed to two causes. First, the circulation in estuaries is such that there is a near-surface seaward flow dominated by river water, but a bottom landward flow dominated by sea water. This landward flow transports fine sediments back *into* the estuary (Meade, 1969). Second, bottom currents offshore are dominantly landward (see, e.g.,

Lauzier, 1967). Consequently, much which escapes from the river mouths and estuaries is subsequently returned.

It is clear then that to understand the surficial sediments now seen on the shelf and slope we need a model with a very different starting point from that of the river mouths. A simple one appropriate to the northern area follows.

At a time during the Pleistocene the glacial ice front reached close to the edge of the Scotian shelf, the northern part of Georges bank, and Long Island (Schlee and Pratt, 1970; Stanley et al., 1972). Beneath the regions occupied periodically by ice, evidence of glacial till should be found. Beyond these regions should be found glacial outwash material. South of the ice front, because sea level was lower, what is now shelf must have been at times land, with the flora and fauna appropriate to the climate and location. Sea level rose at the close of the Pleistocene, and the ice sheets waned. The sea would rework the sediments across which it passed. As sea level rose, the agencies modifying the sediment of the open shelf would change from those of the shoreline (modified by waves and long-shore currents, for example) to those due to residual tidal currents, and to storm- and hurricane-generated waves. We might see nearshore "modern" sediment, being derived now from rivers or biogenic sources, or being derived from the reworking of Pleistocene deposits. We might see offshore "relict" sediments more characteristic of their Pleistocene origin, modified to a lesser extent by more recent agencies. Semantic difficulties might arise in describing, say, a sediment derived from a Pleistocene source, but so reworked that hardly a trace of the characteristics of the source survives. Is it modern, or is it relict? Indeed, Swift et al. (1971) have introduced the term "palimpsest" to describe sediments with petrographic attributes of both earlier and later environments.

This model would serve the student well for deducing the characteristics of the sediments of shelf and slope. Gravels dominate on the shelf only in the northern regions, in the Gulf of Maine and on the Scotian Shelf, where they occupy parts of the shallower water areas. Sands dominate the shelf to the south, and silts and muds the continental slope and the basins of the northern region. Glacially derived sediments are found in the north, fluviatile, authigenic, and biogenic sediment in the south (Emery, 1966). The relict nature of much of the sediment on the shelf is shown in a number of characteristics. Glacial end moraines are found off the coast of Nova Scotia (identified by their morphology and by the texture of their components). Silts associated with the moraines contain a fauna characteristic of brackish water (Fig. 9) (King, 1969). Peat fragments and the teeth of mastodons and mammoths are common (Whitmore et al., 1967). Iron stained material, an iron oxide coating on grains, may reflect subaerial conditions during and after the Pleistocene (Stanley, 1969). There is in fact a substantial difference in color between the regions north and south of Cape Hatteras: the sediments

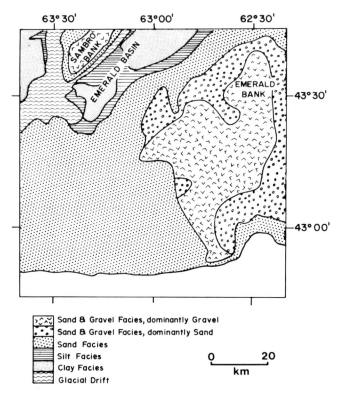

Fig. 9. Facies of the unconsolidated sediment of a part of the Scotian
shelf. Redrawn from King (1967).

on the shelf to the south are olive in color, to the north brown and yellow.
This is because the biogenic contribution is greater in the south, whereas the
glacial contribution is greater in the north.

There have been four principal sources of surficial sediment: rivers,
glaciers, underlying bedrock, and biogenic contributors. These sources led to
the characteristics of the sediment which are now observed. Biogenic sediment
—reflected in the calcium carbonate percentages—dominate south of Cape
Hatteras, which has acted as an "oceanographic barrier" (Milliman *et al.*,
1972). Waters are warm south of the Cape, but cool north of it, and as a
consequence the sediment type follows. The bedrock eroded by rivers and
glaciers has controlled the noncalcareous mineral assemblages found in the
sediments. Rivers and glaciers in the north mechanically eroded Precambrian
and Paleozoic terrain, and consequently chemically less stable minerals such
as feldspars are relatively common. The resulting sediments are arkosic. Rivers
in the south eroded coastal plain sediment, where chemical weathering had
already been important, so that feldspars are less common, and the sediments
are quartzitic. Anomalies in such generalizations occur, of course; on Georges

bank residual sediment is found, reworked coastal plain material, so that feldspars become relatively rare. Such reworking of coastal plain material has led to concentrations of glauconite and phosphorite, thought by Milliman *et al.* (1972) to be residual, derived from the bedrock and not authigenic, formed in situ. This has also been found on parts of the Scotian shelf by Stanley *et al.* (1972).

The types of assemblages of heavy minerals which are found on the shelves also depend on the source material and how it was eroded, and the degree to which the deposit containing the minerals has been reworked. For example, low-stability heavy mineral assemblages are found within the deeper waters of the Gulf of Maine (Ross, 1970). The assemblages are dominated by augite and amphiboles, and the presence of these relatively nonresistant minerals has been explained by the protection against reworking offered by Georges bank to the south. On Georges Bank itself, by contrast, the heavy mineral assemblage is dominated by garnet and staurolite. This is a more stable assemblage, which arises in part because of the greater reworking by waves and currents on the bank than within the Gulf of Maine, and by the destruction of the less stable minerals. The nature of the land surface across which the Holocene transgressive sea passed is variable. On the Scotian shelf it was predominantly glacial till. The sand of the plain between Cape Hatteras and Cape Cod was transported, however, by streams during the Pleistocene. Chenier complexes—elongate bodies of coarse sediment within a coastal marsh—are found off Virginia. The degree of unmixing of these sediments is dependent not only upon the effects of the transgression but upon the subsequent hydraulic regime. In shallow water, such as the banks of the Scotian shelf and Georges bank, tidal currents play a dominant role in unmixing, dispersing Pleistocene outwash on Sable Island bank and generating sand waves south of Sable Island, for example (James and Stanley, 1968). Tidal currents have created a complex ridge and swale topography on parts of Georges bank (Uchupi, 1968). Storm-generated currents (rather than tidal currents) maintain the ridge and swale topography of the sand plain south of Cape Cod (Uchupi, 1968).

We should draw attention here to the fact that the present shelf is a site primarily of reworking of older material. Only nearshore, and on the slope (where reworked sediment is in part deposited), do we see "modern" sediments. For these reasons the present margin is thought by some to be in disequilibrium with the present hydraulic regime, because the Holocene transgression took place such a relatively short time ago. Curray (1965) and Swift (1970) have given thought to the question whether or not the present shelf and present shelf processes are good analogs of ancient shelves and ancient processes. Clearly they are not. A schematic model Curray has developed is shown in Fig. 10. Suppose sea level rises rapidly. We will be left with a relict morphology (A in the figure), or if sediments are deposited during the transgression, with

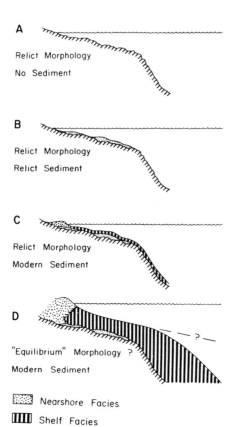

Fig. 10. Schematic representation of transition of a shelf with relict morphology (A), perhaps with relict sediments (B), to a shelf with modern sediments upon it (C), and an equilibrium morphology (D). The model assumes that there is a stable sea level and gradual regional subsidence. Reproduced from Curray (1965, Fig. 3), in: *The Quaternary of the United States,* editors, H. E. Wright, Jr. and David Frey. Copyright © 1965 by Princeton University Press. Reprinted by permission of Princeton University Press.

a relict morphology covered in part or in whole with relict sediments (B in the figure). They are relict in the sense of being unrelated to the present hydraulic regime, and will be reworked under the new regime perhaps in the way that fine sediment is now being winnowed from Georges bank into the Gulf of Maine, or from the Scotian shelf into the northern basins and onto the slope. Modern sediment—nearshore and shelf facies—progressively cover first the still-relict morphology until finally an equilibrium morphology is attained (C and D of the figure). Equilibrium morphology is relatively rare on the margins of the world today and will arise only where modern sediment influx has been very rapid.

There does not appear to be any example of an equilibrium morphology on the eastern seaboard between Florida and the Grand Banks. One example to the north may be the shelf and slope of northernmost Baffin Bay. In this region the shelf is deep, and as a consequence it seems possible that Pleistocene fluctuations in sea level were of lesser consequence than they have been on the eastern seaboard. There is no marked shelf break, and the shelf and slope merge gradually the one into the other. An example of an equilibrium morphology may be seen in seismic reflection profiles across the shelf and slope. In Fig. 11* deeper continuous reflectors grade gently from their position beneath the shelf to their position beneath the slope.

Suppose we want to analyze the transition from relict margin to equilibrium margin. Clearly, because we now generally face a situation where the margin shows features which are predominantly relict, the ancient record must be studied as well. As a result of his studies of the Cretaceous of the coastal plain, Swift (1970) put forward the following model of evolution. At first the relict sediments of the central shelf attain equilibrium in terms of the new hydraulic regime—that is, their textures, their grain size distributions, adjust. Sediment feeds the shoreline, forming the nearshore sand prism in part (the remainder of this prism being derived by fluvial transport and erosion of headlands); sediment from the central shelf also feeds the slope, the shelf edge being a boundary between a source on the one hand, a sink on the other. Eventually, estuaries at the shoreline fill, and first mud and then sand bypass the estuaries generating blankets of mud and sand which prograde across the shelf. These form the modern mud or sand blanket.

The very old-fashioned concept of a shelf showing textural grading from nearshore to offshore is found now only in the modern sand prism near to the present shoreline. Beyond this lie the relict sediments of the central shelf. Gradually this textural grading will spread as equilibrium is attained, but evidence for this comes from the ancient record, not the modern record.

It is worth asking questions about other shelves, less well known, based upon our knowledge of the Atlantic shelf. Should we expect such a degree of reworking on the shelves of parts of Baffin Bay, for example? Some of these shelves are much deeper, and because of this were probably not exposed at any time to subaerial conditions during Pleistocene low stands of sea level. Tidal currents are likely to be relatively weak in much of the area. Consequently, we might expect to find a much more homogeneous body of sediment beneath the shelf. Glacial or ice-rafted sediment will predominate during Pleistocene high stands of sea level, fluvial sediment during the extremes of Pleistocene low stands. At extreme low stands of sea level evidence of storm-generated unmixing may occur over the shallow portions of these shelves.

V. STRUCTURE

A. Ocean to Continent Transition

We need to know where the transition between oceanic and continental crust occurs so that we can be sure where the continents now bordering the Atlantic formerly fitted together; if we do not know exactly where to fit them,

* Fig. 11 faces p. 45.

apparent evidence of overlap or underlap becomes difficult to interpret. We need to know the precise nature of the change so that we can answer the question: is the "quiet magnetic zone" (described later), or indeed any portion of the ocean floor beyond the continental slope, really former continent, now foundered, with its top at or below present oceanic depths? Very nearly all the evidence to both problems—the location and the nature of the boundary—comes from geophysical observations, although in recent years wells drilled by JOIDES in the deep sea and by oil companies on the shelves have begun to help. The idea that a substantial part of the ocean floor beyond the present topographic margin is foundered continent is to some people a notion which should be rejected out of hand as fanciful, if only because it demands that a way be found to "oceanize" a continental crust, a petrological problem presumably. Yet examples of vertical changes in crustal properties with time are numerous. The oceanic crust may thicken with age away from mid-ocean ridges (Keen and Barrett, 1972). There is a suggestion that continental crusts increase in

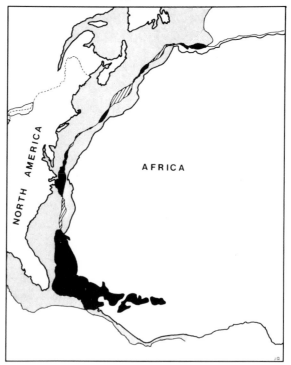

Fig. 12. Fit of Africa against North America, showing particularly the overlap of the Florida–Bahamas platform. The fit was made by Walter P. Sproll, and is reproduced from Dietz and Holden (1972, p. 112). Overlap is shown in black. From: The Breakup of Pangaea, by Dietz and Holden, Copyright © 1970 by Scientific American, Inc. All rights reserved.

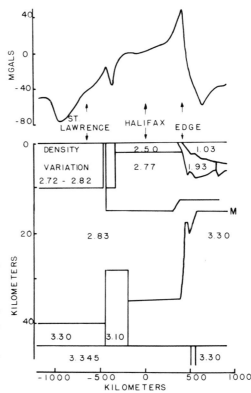

Fig. 13. A free-air gravity anomaly profile over the continental margin of eastern Canada and one model of density variations which could account for the gravity anomalies. Note the edge effect associated with the change in crustal thickness at the continent–ocean transition (from Dainty, 1967). Values of density in g/cm³. M indicates the Mohorovicic discontinuity, defined from seismic refraction lines.

thickness with age (Woollard, 1972). And, most certainly, shallow-water sediments have subsided several kilometers beneath the shelves, as we see in the wells drilled on the submerged and emerged Atlantic coastal plain.

The need to know the location of the transition from continent to ocean is apparent in reassembly of continents which are now fragmented. This can be seen clearly in the reconstruction of Pangaea by Dietz and Holden (1972). If the Bahamas' platform southeast of Florida is included as continent in the reconstruction, there is extensive overlap with Africa; if it is known that the Bahamas' platform developed upon oceanic crust, the difficulty disappears (Fig. 12).

The morphological transition from ocean basin to continental shelf is accompanied by change in crustal thickness from approximately 10 to 30 km (Fig. 13) (see, for example, Dainty et al., 1966; Keen and Loncarevic, 1966). Seismic refraction studies on the continent and ocean side of the margin suggest that the regions are in approximate isostatic balance at a depth of 40 km—that is, mass per unit area at 40 km is approximately constant. (The discrepancies are several hundred kg/cm² in about 12,000 kg/cm², as is shown in Emery et al., 1970, Fig. 31.) The compensation between the two regions does not take place near surface. Consequently, an edge effect arises at the continent–ocean transition. A large positive free-air anomaly of several tens of milligals is bounded on the ocean side by a large negative anomaly of several tens of milligals. The gravity anomaly is therefore a good guide to the continent–ocean transition even where seismic refraction measurements have not been made.

The gravity edge anomaly can be traced fairly clearly from the mouth of the Laurentian Channel to just south of Cape Hatteras (Emery *et al.*, 1970; Keen *et al.*, 1971). South of Cape Hatteras, however, the anomaly is discontinuous, reappearing only on the eastern margin of the Blake plateau and the Bahamas banks. On the evidence of gravity, at least, there is a suggestion therefore that the continent–ocean boundary would be drawn close to the Blake escarpment, not to the Florida–Hatteras slope. The gravity edge anomaly is also missing east of the Laurentian Channel, on the southern margin of the Grand Banks. This may be related, if obscurely, to the formation of the southern margin of the banks at the time of rifting.

The magnetic field over the continental margin can be divided into four characteristic zones (see Emery *et al.*, 1970; Fenwick *et al.*, 1968; Keen, 1969; Keen *et al.*, 1971; Zietz, 1970). The Appalachians show high-amplitude, short-wavelength anomalies, muted where buried beneath the coastal plain sediments. A large magnetic anomaly is associated with parts of the continental shelf and slope—the "slope" anomaly. A region 400 km wide beyond the slope anomaly possesses anomalies not larger than 200 gammas; this zone is called the "quiet zone." The quiet zone is succeeded seawards by anomalies more characteristic of the mid-ocean ridges; following Emery *et al.* (1970) we may call this the "disturbed zone" (Fig. 14).

The slope anomaly is found with few breaks from south of Cape Hatteras to the Laurentian Channel (Fig. 15). It lies to the east (seaward) of the 1000 fathom line east and north of New York, but west of this to Cape Charles, approximately. South of Cape Charles the anomaly swings westward at the northern end of the Blake plateau, and is then lost as a well-defined feature. A comparable anomaly has been described northeast of Newfoundland (Fenwick *et al.*, 1968), but south of the Grand Banks is less prominent.

No wholly satisfactory explanation has yet been given for the slope anomaly or the quiet zone. Several explanations follow.

Fenwick *et al.* (1968) pointed out that the slope anomaly ran approximately along the line of the slope northeast of Newfoundland, and in so doing truncated the general trend of the Appalachians in Newfoundland. They argued that the anomaly was an edge effect. Relatively magnetic rocks of the Appalachians were juxtaposed against relatively nonmagnetic rocks in the ocean basin, seaward. An explanation of this type does not demand universality of slope anomalies; whether or not one is found depends upon the nature of the continental and oceanic material. It is consistent with the termination of the

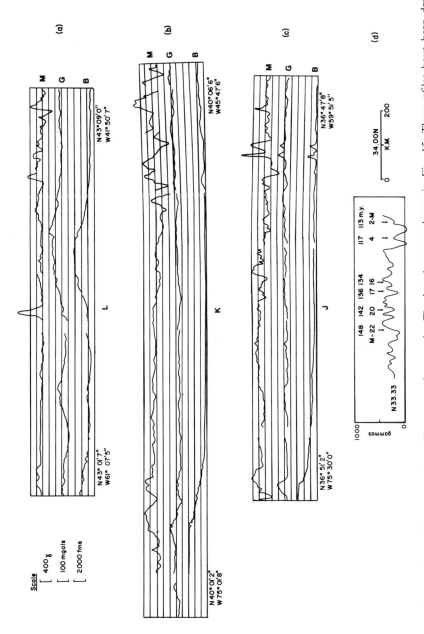

Fig. 14. Magnetic, gravity, and bathymetric profiles across the margin. The locations are shown in Fig. 15. The profiles have been drawn so that the slope anomaly on each is approximately aligned. For comparison, a magnetic profile and approximate time scale from lat. 33°N from Larson and Pitman (1972) is shown at the bottom. The data have been taken from Navado III (1967).

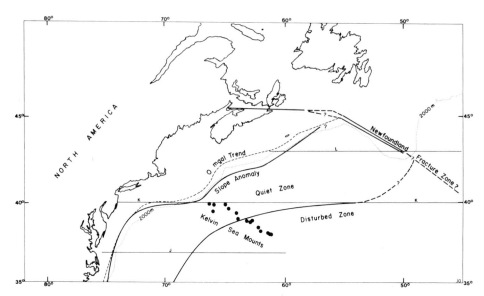

Fig. 15. The quiet zone off Canada and the U.S.A. The western boundary is defined here by the slope anomaly. The dashed line close to the 2000-m bathymetric contour is the 0-mgal free-air gravity anomaly trend. Note the suggestion of "thinning" of the quiet zone at the Kelvin seamounts. Data compiled by Renwick *et al.* (1972) from Navado III (1967) and Emery *et al.* (1970). The lines J, K, and L are the Navado profiles shown in Fig. 14.

anomaly at the northern end of the Blake plateau, for the Appalachians on land trend westward at that latitude, away from the ocean basin, as do the magnetic anomalies (Zietz, 1970).

Irving (1970) suggested that the quiet zone arises because the igneous rocks beneath were formed during initial phases of rifting. As a consequence they were emplaced subaerially, or as intrusions. The magnetic properties of rocks formed in this way are different from those extruded subaqueously (as are those at mid-ocean ridges). Their magnetic intensity is lower, and so contrasts between reversely and normally magnetized rocks would be small, and anomalies muted. The boundary between quiet zone and disturbed zone does not have to be a time line, it is process dependent.

Emery *et al.* (1970) assumed that the whole of the disturbed zone, quiet zone, and slope anomaly could be related to the Vine–Matthews model of ocean-floor spreading with normal and reversely magnetized blocks responsible for the anomalies. The quiet zone represents in their model a period in which the magnetic field was of constant (reversed) polarity—the Kiaman interval of the Permian (220–270 m.y. ago). Consequently, there are no anomalies— there are no reversals to give magnetization contrasts. The slope anomaly arises from a normally magnetized block, associated with igenous activity at the inception of rifting, 270 m.y. ago. There are several points of weakness in this explanation. The igneous activity responsible for the slope anomaly is in their model Permian in age. One might anticipate that if this is true, evidence of Permian vulcanism would be found on land. There is, however, Triassic vulcanism, but no comparable intensive Permian vulcanism. Perhaps all evidence has been eroded, or is buried. However, the oldest Mesozoic sediments beneath the outer part of the Scotian shelf are Lower Jurassic in age, resting with the Argo Salt upon Meguma (lower Paleozoic) metamorphosed sediments (McIver, 1972). The choice of a Permian age for the slope anomaly leads to an arbitrary substantial change in spreading rate for the Atlantic Ocean 100 m.y. ago. There is no other evidence for this.

Part of the problem of the quiet zone is, of course, that the anomalies in the region, being of low amplitude, are difficult to map—diurnal variation is comparable to or larger than the anomalies sought. Careful study by Vogt, Einwich, and Johnson (1972) suggested to them nevertheless that the anomalies in the quiet zone were due to reversals during the Jurassic. This implies that the reversals should be seen in paleomagnetic studies on land; unfortunately the reversal time scale for the early Mesozoic is only poorly known (Helsley, 1972).

Talwani and Eldholm (1972) by contrast suppose that part of the quiet zone west of Norway represents foundered continent, and that this may be true of part of the quiet zone off eastern North America. They suggest that seismic refraction lines of good quality within the zones show that the crust there has a significantly different compressional wave velocity from the truly oceanic crust seaward of the zone.

Part of the problem of accounting for the quiet zone and the disturbed zone is dating the floor of the western Atlantic precisely. There are relatively few JOIDES holes reaching oceanic basement, and until recently no convincing correlation has been made of magnetic anomalies in the disturbed zone with any world-wide system of anomalies whose ages are known. Some progress has, however, recently been made.

Nafe and Drake (1969) had shown that the seaward boundaries of the quiet zones on the east and west sides of the North Atlantic fitted together. They suggested that the zones represented foundered continent. Pitman and Talwani (1972) followed this idea by fitting one boundary against the other using the angular rotation appropriate to the time in the history of spreading of the North Atlantic. Because, with this rotation, the boundaries fitted, they thought that the boundaries are an isochron. To explain the quiet zones it is therefore appropriate to seek a period of very few reversals, as Heirtzler and Hayes (1967) originally did. Larson and Pitman (1972) have been able to show that the anomalies of the disturbed zone, the Keithley sequence, correlate well with anomalies in the Pacific (Fig. 14). JOIDES holes give ages, and show that the Keithley sequence in the Atlantic is Mesozoic in age, ranging from Upper Jurassic to Lower Cretaceous. This period of reversals is bounded above and below by periods of time in the Cretaceous and Jurassic of few reversals, giving rise to zones with magnetic anomalies of low amplitude both seaward and shoreward of the Keithley sequence. The seaward quiet zone boundary is approximately 150 m.y. old, if the dating is correct.

B. Mesozoic and Cenozoic Sedimentary Wedge

The shelf and slope are, as we have seen, underlain by a Mesozoic and Cenozoic sedimentary wedge. It is known from reports of wells drilled that the depth to the base of this sedimentary section is at least 5 km beneath the outer part of the shelf off Nova Scotia. Seismic refraction studies suggest that the wedge continues to thicken to a depth of 12 km beneath the slope and rise (Nafe and Drake, 1969), if the assumption is made that the base of the wedge can be identified by seismic velocity. Seismic refraction studies show also that sediment has filled deep troughs beneath the margin, although their presence has not yet been confirmed by drilling, so far as the author knows. Wells drilled beneath the Scotian shelf and the Grand Banks show that salt may lie just above known or presumed basement (Bartlett and Smith, 1971; McIver, 1972). Salt is suspected beneath the rise off the Grand Banks from reflection studies—possible salt diapirs have been detected (Pautot et al., 1970). The lowermost rocks of the sedimentary wedge were deposited at or close to sea level, sometimes under continental conditions (McIver, 1972).

The sediments of the wedge have obviously subsided (Figs. 16 and 17). The subsidence is approximately exponential with a time constant of approximately 50 m.y. (Keen and Keen, 1971; Sleep, 1971). Rates of subsidence range downwards from a few centimeters per thousand years (Keen and Keen, 1972; Rona, 1970). The salt found in the wells beneath the Scotian shelf can be related to restricted marine conditions in the early phases of continental breakup. The salt found beneath deeper water could also arise in the same way, and have been originally deposited close to sea level. Its present depth can be due to subsidence of the same sort of magnitude as is found in the wells of the coastal plain.

The Mesozoic and Cenozoic rocks are not grossly folded or faulted, except where deformed by salt diapirs; they have been affected, however, by two basement arches, the Cape Fear Arch and the Peninsular Arch, and by faulting associated with the Orpheus gravity anomaly east of Cape Breton Island, Nova Scotia. The principal phenomenon found over these basement arches is that strata of particular ages are missing or attenuated on their crests, but present on their flanks. This is the case with the Lower Cretaceous on both the arches, for example. The time of origin of the arches is not known, they have been attributed to periods between the Early Cretaceous

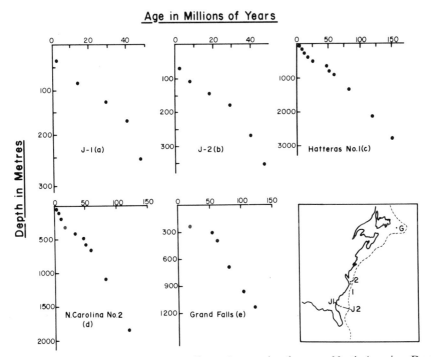

Fig. 16. Subsidence estimated from wells on the margin of eastern North America. Depth in meters is plotted against age in millions of years. Locations are shown inset, lower right. (a) JOIDES well J-1 off Florida; (b) JOIDES well J-2 off Florida; (c) Esso Standard Oil Co., Hatteras Light well No. 1, Dare County, North Carolina; (d) Esso Standard Oil Company, North Carolina Esso well No. 2, Dare County, North Carolina; (e) Amoco–Imperial Grand Falls well. Redrawn from Keen and Keen (1972), where sources of data will be found.

and Early Miocene (see Maher, 1965). Pertinent observations by Sbar and Sykes (1972) are mentioned below.

The relationship between fracture zones and faulting during the early history of opening of the Atlantic have been discussed by King and MacLean (1970), LePichon and Fox (1971), and Sbar and Sykes (1972). Four fracture zones have been suggested: the Newfoundland, Kelvin, Cape Fear, and Bahama fracture zones. Of these the best studied is the Newfoundland fracture zone and its westerly extension on the Scotian shelf (Figs. 15, 18, and 19) (King and MacLean, 1970; Auzende et al., 1971).

There is apparent continuity between faults in the Bay of Fundy, across northern Nova Scotia and the Scotian shelf, and the southern margin of the Grand Banks (Fig. 15). This system is represented on the shelf by the Orpheus gravity anomaly, caused in part by evaporites and Mesozoic sediments (Loncarevic and Ewing, 1967; McIver, 1972; King and MacLean, 1970). Cretaceous strata, but not Tertiary strata, are folded in the vicinity of this feature, although

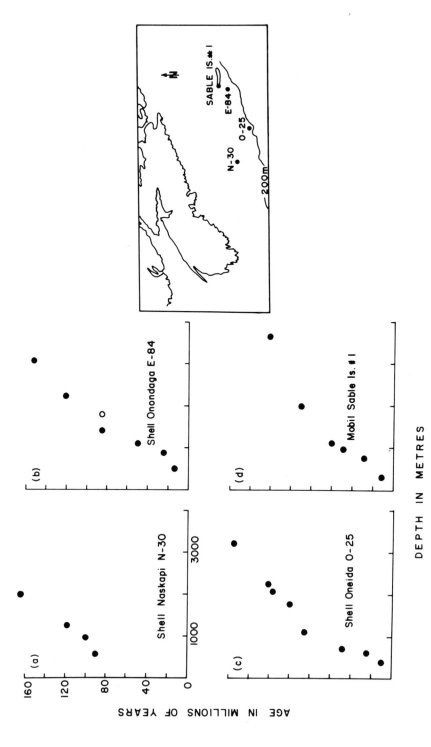

Fig. 17. Subsidence estimated from wells on the Nova Scotia shelf. Depth in meters is plotted against age in millions of years. Locations are shown inset, lower right. (a) Shell Naskapi N-30, (b) Shell Onondaga E-84, (c) Shell Oneida O-25, (d) Mobil Sable Island 1. Data compiled from *Well History Reports* by G. Renwick.

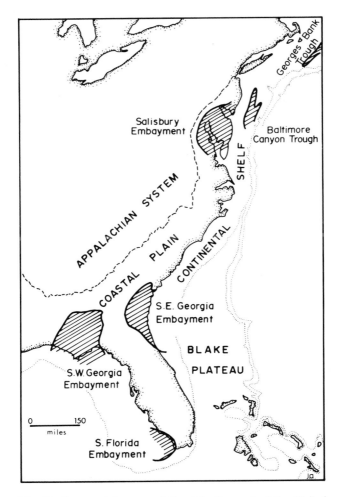

Fig. 18. Structural features of the Atlantic coast of the United
States. Redrawn from Maher (1965), by permission of the author
and the American Association of Petroleum Geologists.

there is some seismic activity at the present day. Parts of the region are known
to be associated with diapirism (Keen, 1970). King and MacLean suggest that
because of its extent the folding of the Cretaceous strata (in the Late Cretaceous
or Early Tertiary) was due to movement on the fault system, and was not
due only to diapiric deformation. This is an important point which should be
established with certainty, because it could be used to aid in timing the rifting
and spreading of the continents bordering the North Atlantic, along the lines
suggested by LePichon and Fox (1971).

The southeasterly extension along the southern margin of the Grand
Banks is marked by block faulting (Auzende et al., 1970), similar perhaps to

the structures associated with other fracture zones along continental margins
(Keen and Keen, 1972). The zone continues to the east as the Southeast Newfoundland ridge. This is a sedimentary feature formed over a rise in basement;
LePichon and Fox suggest that the other fracture zones of the eastern seaboard
have been responsible for the formation of similar sedimentary bodies in the
ocean basin. They also suggest that it terminated (so far as its oceanic expression was concerned) 80 m.y. ago. At this time the North Atlantic was
freed from constraints provided by adjacent rigid continental blocks (Africa
and North America) and the pole of rotation of these two plates shifted. This
timing is in broad agreement with King and MacLean's dating of the folding
of the Cretaceous.

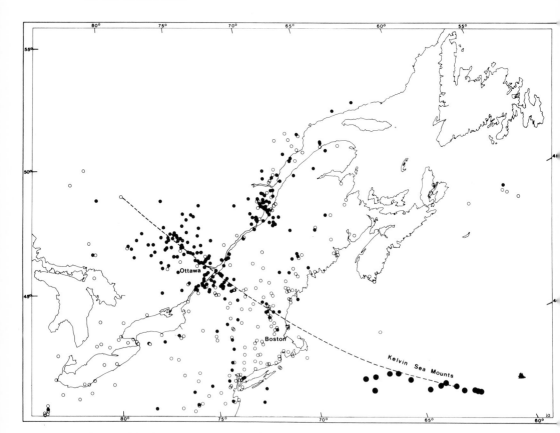

Fig. 19. Seismicity of northeastern North America from 1928 to 1959, and the possible
association with the line of the Kelvin seamounts. This is redrawn from Smith (1967),
(*Dominion Observatory Publications*) and Sbar and Sykes (1973), (*Bulletin of the Geological
Society of America*) where details will be found. All earthquakes reported by Smith are
shown; open circles show poorly located epicenters, and closed circles show well located
epicenters. No distinction has been made between magnitudes. The broken line shows the
small circle about an early pole of spreading for the North Atlantic.

The deformation of the rocks associated with the Orpheus gravity anomaly can, as we have just seen, be dated approximately. The evidence available so far does not seem adequate to date any Mesozoic or Cenozoic deformation in the other postulated fracture zones. Garrison (1970) showed that the Cretaceous and Tertiary rocks south of Martha's Vineyard and Long Island are not deformed; these strata lie along the strike of the postulated Kelvin seamount fracture zone. The Cape Fear Arch is a prominent structural feature of the central part of the Atlantic coastal plain. Maher (1965) points out that its formation has been placed variously between Early Cretaceous and Early Miocene.

Interesting observations concerning the possibility that the fracture zones just mentioned are reflected now in the present seismicity of eastern North America have been made by Sbar and Sykes (1972). They have drawn upon the compilations of seismicity made by Smith (for example, see Smith, 1967) which show that a prominent zone of earthquakes extends from northwest of Ottawa in a southeasterly direction through New England, in line with the trend of the Kelvin seamounts. This trend lies upon a small circle about one postulated pole of early opening for the North Atlantic; consequently, Sbar and Sykes, point out, if the seismic zone is located along the extension of the Kelvin seamount fracture zone, this may explain why it may be a weak zone in the mantle. Comparable observations may be made about the other fracture zones; a less prominent trend of earthquakes near Charleston, South Carolina, is also upon a small circle about the same pole of opening, and the trend is upon the extension of the Cape Fear fracture zone. The Grand Banks earthquake of 1929 lies on the extension of the Newfoundland fracture zone. One importance of the Boston–Ottawa seismic zone is that geological features associated with it geographically are known to be Mesozoic or Tertiary in age, as the Monteregian Hills are, for example.

VI. A FUTURE GEOSYNCLINE?

Drake *et al.* (1959) recognized the resemblance in form between the sedimentary wedge of the continental margin and the Appalachian geosyncline as it would be if unfolded. They equated the sediments beneath the shelf with the miogeosynclinal sediments of the Appalachians—those deposited on or adjacent to the stable platform of the continent, and the sediments of the slope and rise with the orthogeosyncline. There were several difficulties, now resolved. First, the Appalachian geosyncline, and its European equivalent, the Caledonian geosyncline, is two-sided in places. In Newfoundland and Britain, for example, Precambrian rocks bound the system to northwest and southeast. No such bilateral symmetry occurs in the modern Atlantic analog. Second, there was an implicit suggestion that the thick sediment beneath the outer part of the margin represented a trench similar to, say, the Puerto Rico trench, but filled up. Third, the Appalachian geosyncline is, of course, intensely deformed and was the site of much igneous activity. (A modern account of the Appalachians in Newfoundland has been given by Williams *et al.* (1972).

These difficulties disappear with the recognition that the Atlantic margin is possibly the site of a future geosyncline (in the sense recognized from the continental record), and that the Appalachian geosyncline may be an example of continental collision following ocean-floor spreading (see, e.g., Mitchell and Reading, 1969; Schenk, 1971). The Atlantic margin assumes importance, therefore, because of the application to the ancient record. Clearly it is not the only modern analog; Baffin Bay is an example of a future geosyncline which already has bilateral symmetry. A small ocean basin between two neighboring Precambrian cratons is filling up with sediment. If we consider then the present Atlantic margin as the early stage of the development of one type of geosyncline, there are several points of view we can take. Let us take two; we can look at modern processes occurring now—primarily a sedimentological approach—and we can look at the Mesozoic and Cenozoic stratigraphic history.

A. Modern Sedimentological Processes

We have to distinguish between sedimentological processes appropriate to times of low stands of sea level in the Pleistocene, and to times of high stands, as is the case now. During low stands the coastline will have been close to the shelf break, at least off the Scotian shelf; waves will have been unimpeded by a wide shallow shelf as they are now: the islands formed by the outer banks of the Scotian shelf will have acted as reservoirs for land-derived sediment under glacial conditions (Stanley *et al.*, 1972). It seems likely that at these times turbidites will have been generated more easily than now, and ice-rafted debris will have been common.

At present, however, the shelf is dominated not by subaerial conditions, but by fluvial and hydraulic processes, which have three principal effects. Rivers bring new sediment from the land to the coast; hydraulic processes rework Holocene sediment; sediment is transported beyond the shelf to the slope and rise. Although the processes themselves are qualitatively understood, we probably have little quantitative understanding. For example, although, as Meade has shown, we know approximately how much sediment now reaches the coast from the river systems, we do not know how much of this reaches the shelf itself, or how much bypasses it, eventually to reach the slope and rise. Or again, in northern waters part of the shelf is still covered by ice through several months of the year. We do not know the relative significance of ice-rafted material now as a source of sediment.

The slope is mantled in part by hemipelagic sediment, mainly of terrestrial origin (Fig. 20) (Pratt, 1968; Horn *et al.*, 1971). Sediment from the shelf spills

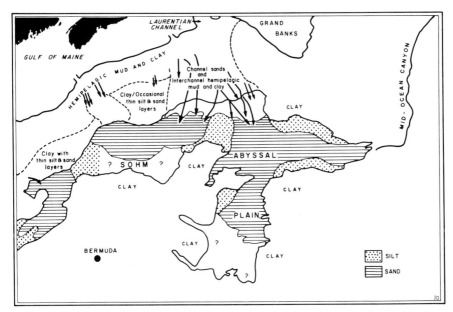

Fig. 20. Distribution of sediment on the continental margin off Georges bank, the Nova Scotia shelf, and the Grand Banks. Coarse sediment surface and near-surface sands blanket the Sohm abyssal plain, showing the ability of turbidity currents to transport large volumes of coarse sediment across the ocean floor. Redrawn from Horn *et al.* (1971, Fig. 5), from *Marine Geology*, and reproduced by permission.

over onto the upper slope and reaches the lower slope and rise through submarine canyons and in suspension. The canyons appear to be the locations of stronger current movement and of movement downslope of sediment suspension. The relative importance of the various agencies, themselves identified, is hardly known. For example, how much sediment is put into suspension by biological activity (bioturbation)? And how much is put into suspension by current action?

Turbidity currents generated the fill perhaps principally during low stands of sea level for the Hatteras and Sohm abyssal plains off eastern North America (Horn *et al.*, 1971). Some of the characteristics of these sediments are well known. Increasing distance from source is accompanied by decrease in grain size and in thickness of individual beds. Many of the sands are poorly sorted and have characteristics similar to those of at least some ancient greywackes, of flysch-type sands. We do not know, however, the relative importance either during the low stands or now of slump-generated turbidites and of rip-current-generated turbidites. We suppose that at times of lowered sea level more sediment was available on the outer shelves to be transported as turbidites. But we do not know quantitatively what this means—how many turbidity currents per century 11,000 years ago? How many now?

Part of the continental margin has been shaped by bottom currents flowing parallel to the margin (the deep western boundary currents; Heezen *et al.*, 1966). It has been suggested that such currents transport sediment southward and build, for example, structures such as the Blake–Bahamas outer ridge.

B. Mesozoic and Cenozoic Succession

Let us take the second approach and attempt to look at a stratigraphic cross section of the margin. Two sources of information can be taken profitably, the results of drilling on the Scotian shelf (McIver, 1972) (Figs. 7 and 8) and in the adjacent ocean basin (Peterson, 1970).

The wells upon the shelf show a complex of shales and mudstones, sandstones, and limestones. If a generalization can be made, the following appears to be true. Sandstones dominate nearshore (up dip); shales offshore (down dip); limestones on occasion cover much or all of the region. In terms of environment of deposition these lithologies reflect a shelf undergoing general overall subsidence, with periodic fluctuations in the position of the shoreline caused in part by changes in influx of material from the continent.

The situation offshore is very different (Figs. 21 and 22). The JOIDES

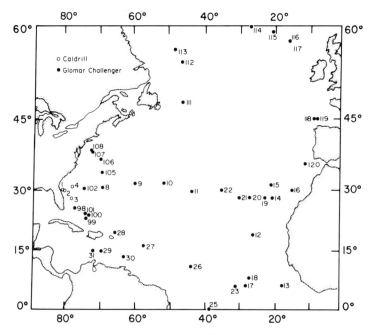

Fig. 21. Location of holes drilled by JOIDES in the North Atlantic. Open circles, holes drilled by *Caldrill*; closed circles, holes drilled by *Glomar Challenger*.

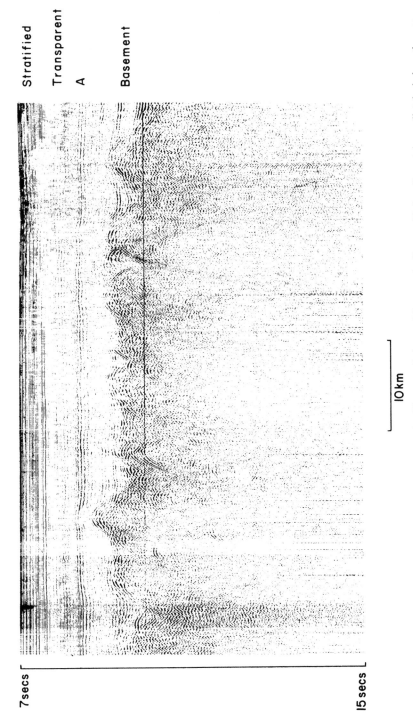

Fig. 22. Seismic reflection profile over the Sohm abyssal plain. The record shows (a) acoustically stratified sediment immediately below the ocean floor, presumably turbidites; (b) beneath these, more acoustically transparent sediment, presumably pelagic sediment; (c) a prominent reflector, perhaps Horizon A of Ewing and Ewing (1971); drilling shows this horizon to be composed of cherts, of Eocene age; (d) a rough acoustic basement, presumably basaltic rocks of the oceanic crust. This is unpublished data of the Atlantic Geosciences Centre, Bedford Institute, obtained from *C. S. S. Dawson* by D. L. Barrett and C. E. Keen.

holes across the North Atlantic (holes 8 to 12 of Leg 2 from west to east; Fig. 22) encountered an assemblage of dominantly pelagic sediment overlying in some holes basaltic basement (Peterson, 1970). The pelagic sediment varied; calcareous and radiolarian oozes, cherts, clays, and volcanic ash are examples. In hole 8, adjacent to the North American continent, channel deposits (see above) could be recognized. Two features should be mentioned. First, volcanic ash in hole 12 had been altered to the clay minerals sepiolite and palygorskite, and the suggestion was made that this had been caused by magnesium-rich brines derived from evaporites of the African continental margin. Second, pronounced hydrothermal mineralization was encountered in hole 9A, northeast of Bermuda. For example, one sample close to basaltic basement contained 11.7% hematite.

These observations lead to the development of the concept of an "Atlantic type" of geosyncline (Mitchell and Reading, 1969). The miogeosyncline is the predominantly shallow-water shelf complex of sandstones, shales, and limestones; the eugeosyncline is the predominantly deep-water complex of basic igneous rocks (of the oceanic crust), pelagic sediments (biogeneous oozes, cherts, and clays), and turbidites. (The turbidites develop on the abyssal plains when sediment can bypass the shelf, as it can more easily at times of low stands of sea level. Shelf sediment may be trapped by tectonic or diapiric dams or by delta formation; when the barriers are topped, sediment is available to the slope and rise.) As the margin progrades, different facies in turn dominate. For example, hemipelagic sediment of the rise, perhaps molded by geostrophic contour currents, will overlie the turbidites of the former abyssal plain.

ACKNOWLEDGMENTS

I should like to express my appreciation to Dianne Crouse, Ruth Jackson, Charlotte Keen, Carolyn Kennie, David Piper, and Gregory Renwick for help in preparing this chapter, and to those who have allowed diagrams to be reproduced, or who sent me manuscripts before publication. Drafting was done by Janice Aumento.

REFERENCES

Applin, P. L., 1951, Preliminary report on buried pre-Mesozoic rocks in Florida and adjacent states: U.S. Geological Survey, Circular 91.

Auzende, J.-M., Olivet, J.-L., and Bonnin, J., 1971, La marge du Grand Banc et la fracture de Terre Neuve: C. R. Acad. Sc. Paris, t. 271, Série D, p. 1063–1066.

Bartlett, G. A. and Smith, L., 1971, Mesozoic and Cenozoic history of the Grand Banks of Newfoundland: Can. J. Earth Sci., v. 8, p. 65–84.

Bunce, E. T., Emery, K. O., Gerard, R. D., Knott, S. T., Lidz, L., Saito, T., and Schlee, J., 1965, Ocean drilling on the continental margin: Science, v. 150, p. 709–716.

Charm, W. B., Nesteroff, W. D., and Valdes, S., 1969, Detailed stratigraphic description of the JOIDES cores on the continental margin off Florida: U.S. Geol. Surv. Prof. Paper 581-D.

Curray, J. C., 1965, Late Quaternary history, continental shelves of the United States, in: *The Quaternary of the United States*, H. E. Wright, Jr., and D. G. Frey, eds.: Princeton University Press, Princeton, N. J., p. 723–735.

Dainty, A. M., 1967, Crustal studies in eastern Canada: Ph.D. thesis, Dalhousie University, 144 p.

Dainty, A. M., Keen, C. E., Keen, M. J., and Blanchard, J. E., 1966, Review of geophysical evidence on crust and uppermantle structure on the eastern seaboard of Canada: *Amer. Geophys. Union*, Monograph 10, p. 349–369.

Dietz, R. S. and Holden, J. C., 1972, The breakup of Pangaea, in: *Continents adrift; readings from Scientific American*, J. Tuzo Wilson, ed.: Freeman and Co., San Francisco, p. 103–103.

Drake, C. L., Ewing, M., and Sutton, G. H., 1959, Continental margins and geosynclines: the east coast of North America north of Cape Hatteras: *Phys. and Chem. of the Earth*, v. 3, p. 110–198.

Emery, K. O., 1966, Atlantic continental shelf and slope off the United States—geologic background: U.S. Geol. Surv. Prof. Paper 529-A.

Emery, K. O., 1969, The continental shelves, in: The Ocean, A Scientific American Book: Freeman and Co., San Francisco, p. 39–52.

Emery, K. O., Uchupi, E., Phillips, J. D., Bowin, C. O., Bunce, E. T., and Knott, S. T., 1970, Continental rise off eastern North America: *Amer. Assoc. Petrol. Geol. Bull.*, v. 54, p. 44–108.

Ewing, J. and Ewing, M., 1971, Seismic reflection, in: The Sea, Vol. 4, Part 1, A. E. Maxwell, ed.: J. Wiley and Sons, New York, p. 1–52.

Fenwick, D. K. B., Keen, M. J., Keen, C. E., and Lambert, A., 1968, Geophysical studies of the continental margin northeast of Newfoundland: *Can. J. Earth Sci.*, v. 5, p. 483–500.

Garrison, L. E., 1970, Development of continental shelf south of New England: *Bull. Amer. Assoc. Petrol. Geol.*, v. 54, p. 109–124.

Grant, A. C. and Manchester, K. S., 1970, Seismic reflection profile from the continental margin off Nova Scotia, Canada: AOL Data Series (1970) 6-D, Atlantic Oceanographic Laboratory, Dartmouth, N. S.

Grant, D. R., 1970, Recent coastal submergence of the Maritime Provinces, Canada: *Can. J. Earth Sci.*, v. 7, p. 676–689.

Hallam, A., 1971, Mesozoic geology and the opening of the North Atlantic: *Jour. Geology*, v. 79, p. 129–157.

Heezen, B. C., Hollister, D. C., and Ruddiman, W. F., 1966, Shaping of the continental rise by deep geostrophic contour currents: *Science*, v. 152, p. 502–508.

Heirtzler, J. R. and Hayes, D. E., 1967, Magnetic boundaries in the North Atlantic Ocean: *Science*, v. 157, p. 185–187.

Helsley, C. E., 1972, Post Paleozoic magnetic reversals: *Trans. Amer. Geophys. Union*, v. 53, p. 363.

Horn, D. R., Ewing, M., Horn, B. M., and Delach, M. N., 1971, Turbidites of the Hatteras and Sohm abyssal plains, western North Atlantic: *Mar. Geol.*, v. 11, p. 287–324.

Inman, D. L. and Nordstrom, C. E., 1971, On the tectonic and morphologic classification of coasts, *Jour. Geology*, v. 79, p. 1–21.

Irving, E., 1970, The Mid-Atlantic Ridge at 45°N; XIV Oxidation and magnetic properties of basalt: review and discussion: *Can. J. Earth Sci.*, v. 7, p. 1528–1538.

James, N. P. and Stanley, D. J., 1968, Sable Island bank off Nova Scotia: Sediment dispersal and recent history: *Amer. Assoc. Petrol. Geol. Bull.*, v. 52, p. 2208–2230.

Keen, C. E. and Barrett, D. L., 1972, Seismic refraction studies in Baffin Bay: *Geophysical J.*, v. 30, p. 253–271.

Keen, C. E. and Loncarevic, B. D., 1966, Crustal structure on the eastern seaboard of Canada: studies on the continental margin: *Can. J. Earth Sci.*, v. 3, p. 65–76.

Keen, M. J., 1969, Magnetic anomalies off the eastern seaboard of the United States: a possible edge effect: *Nature*, v. 222, p. 72–74.

Keen, M. J., 1970, A possible diapir in the Laurentian Channel: *Can. J. Earth Sci.*, v. 7, p. 1561–1564.

Keen, M. J. and Keen, C. E., 1971, Subsidence and fracturing on the continental margin: Geol. Surv. Can. Paper 71-23, p. 23–42.

Keen, M. J., Loncarevic, B. D., and Ewing, G. N., 1971, The continental margin of eastern Canada: Nova Scotia to Nares Straits, in: *The Sea*, Vol. 4, Part 2: J. Wiley and Sons, New York, p. 251–291.

King, L. H., 1967, On the sediments and stratigraphy of the Scotian Shelf, Bedford Institute of Oceanography Report 67-2, 32 pp.

King, L. H., 1969, Submarine end moraines and associated deposits on the Scotian Shelf: *Geol. Soc. Amer. Bull.*, v. 80, p. 83–96.

King, L. H., 1970, Surficial geology of the Halifax-Sable Island map area, Marine Science Branch: Department of Energy, Mines and Resources, Ottawa, Paper 1, including Chart 4040G.

King, L. H., 1972, Relation of plate tectonics to the geomorphic evolution of the Canadian Atlantic Provinces: *Bull. Geol. Soc. Amer.*, v. 83, p. 3083–3090.

King, L. H. and MacLean, B., 1970, Continuous seismic-reflection study of the Orpheus gravity anomaly: *Amer. Assoc. Petrol. Geol. Bull.*, v. 54, p. 2007–2031.

Larson, R. L. and Pitman, W. C., 1972, World-wide correlation of Mesozoic magnetic anomalies and its implications: *Bull. Geol. Soc. Amer.*, v. 83, p. 3645–3662.

Lauzier, L. M., 1967, Bottom residual drift on the continental shelf area of the Canadian Atlantic coast: *Jour. Fish. Res. Board of Canada*, v. 24, p. 1845–1859.

LePichon, X. and Fox, P. J., 1971, Marginal offsets, fracture zones and the early opening of the North Atlantic: *J. Geophys. Res.*, v. 76, p. 6294–6308.

Loncarevic, B. D. and Ewing, G. N., 1967, Geophysical study of the Orpheus gravity anomaly: Proceedings VII World Petroleum Congress, p. 828–835.

Maher, J. C., 1965, Correlations of subsurface Mesozoic and Cenozoic rocks along the Atlantic Coast: *Amer. Assoc. Petrol. Geol.*, Tulsa, Oklahoma, U.S.A., 18 p., plates.

Manheim, F. T., Meade, R. H., and Bond, G. C., 1970, Suspended matter in surface waters of the Atlantic continental margin from Cape Cod to the Florida Keys: *Science*, v. 167, p. 371–376.

McIver, N. L., 1972, Cenozoic and Mesozoic stratigraphy of the Scotian Shelf: *Can. J. Earth Sci.*, v. 9, p. 54–70.

Meade, R. H., 1969, Landward transport of bottom sediments in estuaries of the Atlantic coastal plain: *Jour. Sed. Pet.*, v. 39, p. 222–234.

Meade, R. H., 1972, Sources and sinks of suspended matter on continental shelves, in: *Shelf Sediment Transport: Process and Pattern*, Swift, D. J. P., Duane, P. B., and Pilkey, O. H., eds.: Dowden Hutchinson and Ross, Stroudsburg, Pa.

Milliman, J. D., Pilkey, O. H., and Ross, D. A., 1972, Sediments of the continental margin off the United States: *Geol. Soc. Amer. Bull.*, v. 83, p. 1315–1334.

Mitchell, A. H. and Reading, H. G., 1969, Continental margins, geosynclines, and ocean floor spreading: *Jour. Geology*, v. 77, p. 629–646.

Nafe, J. E. and Drake, C. L., 1969, Floor of the North Atlantic—summary of geophysical data: *Amer. Assoc. Petrol. Geol. Mem.*, v. 12, p. 59–87.

Navado III, 1967, Bathymetric, magnetic and gravity investigations, H. Neith, M. S. Snellius, 1964-1965, Part 3, Hydrographic Newsletter, Special Publication Number 3, Part 3, Published by the Netherlands Hydrographer.

Pautot, G., Auzenda, J.-M., and Le Pichon, X., 1970, Continuous deep sea salt layer along North Atlantic margins related to early phase of rifting: *Nature*, v. 227, p. 351–354.

Peterson, M. N. A., 1970, Initial reports of the Deep Sea Drilling Project: U.S. Government Printing Office, Vol. II.

Pitman, W. C., III, and Talwani, M., 1972, Sea-floor spreading in the North Atlantic: *Bull. Geol. Soc. Amer.*, v. 83, p. 619–646.

Pratt, R. M., 1968, Atlantic continental shelf and slope of the United States—Physiography and sediments of the deep sea basin: U.S. Geol. Surv. Prof. Paper 629-B.

Renwick, G., Keen, C. E., and Keen, M. J., 1972, Magnetic and gravity anomalies off Nova Scotia, in preparation.

Rona, P. A., 1970, Comparison of continental margins of eastern North America at Cape Hatteras and northwestern Africa at Cap Blanc: *Amer. Assoc. Petrol. Geol.*, v. 54, p. 129–157.

Ross, D. A., 1970, Atlantic continental shelf and slope of the United States—heavy minerals of the continental margin from southern Nova Scotia to northern New Jersey: U.S. Geol. Surv. Prof. Paper 529-G.

Sanders, J., 1963, Late Triassic tectonic history of north-eastern United States: *Amer. J. Sci.*, v. 261, p. 501–514.

Sbar, M. L. and Sykes, L. R., 1973, Contemporary compressive stress and seismicity in eastern North America: *Bull. Geol. Soc. Amer.*, v. 84, p. 1861–1882.

Schenk, P. E., 1971, Southeastern Atlantic Canada, Northwestern Africa and continental drift: *Can. J. Earth Sci.*, v. 8, p. 1218–1251.

Schlee, J. and Pratt, R. M., 1970, Atlantic continental shelf and slope of the United States—gravels of the northeastern part: U.S. Geol. Surv. Prof. Paper 529-14.

Schneider, E. D., 1969, The Deep-Sea: a habitat for petroleum?: *Under Sea Technology*, v. 10(10), p. 32–57.

Scholle, P. A. and Kling, S. A., 1972, Southern British Honduras: lagoonal coccolith ooze: *J. Sed. Petrol.*, v. 42, p. 195–204.

Sleep, N. H., 1971, Thermal effects of the formation of Atlantic continental margins by continental break-up: *Geophys. J.*, v. 24, p. 325–350.

Smith, W. E. T., 1967, Basic seismology and seismicity of eastern Canada: Dom. Obs. Publ., Seis. Ser. 1966-2, 43 p.

Spangler, W. and Peterson, J., 1950, Geology of the Atlantic coastal plain in New Jersey, Delaware, Maryland, and Virginia: *Amer. Assoc. Petrol. Geol. Bull.*, v. 34, p. 1–21.

Stanley, D. J., 1969, Atlantic continental shelf and slope of the United States—color of marine sediments: U.S. Geol. Surv. Prof. Paper 529-D.

Stanley, D. J., Curray, J. R., Middleton, G. V., and Swift, D. J. P., 1969, The new concepts of continental margin sedimentation: *Short Course Lecture Notes*: American Geological Institute, Washington.

Stanley, D. J., Swift, D. J. P., Silverberg, N., James, N. P., and Sutton, R. G., 1972, Late Quaternary progradation and sand spillover on the outer continental margin off Nova Scotia, Canada: Smithsonian contributions to the earth sciences, No. 8, Smithsonian Institution Press, 88 p.

Stevenson, I. M., 1959, Shubenacadie and Kennetcook map areas, Nova Scotia: Geol. Surv. Can. Mem. 302, p. 6–41.

Swain, F. M., 1952, Ostracoda from wells in North Carolina: Part 2. Mesozoic Ostracoda: U.S. Geol. Surv. Prof. Paper 234-B, 93 p.

Swift, D. J. P., 1970, Quaternary shelves and the return to grade: *Mar. Geol.*, v. 8, p. 5–30.

Swift, D. J. P., Stanley, D. J., and Curray, J. R., 1971, Relict sediments on continental shelves: a reconsideration: *Jour. Geology*, v. 79, p. 322–346.

Talwani, M. and Eldholm, O., 1972, The continental margin off Norway and the magnetic Quiet Zone: *Trans. Amer. Geophys. Union*, v. 53, p. 406.

Uchupi, E., 1968, Atlantic continental shelf and slope of the United States—Physiography: U.S. Geol. Surv. Prof. Paper 529-C.

Vogt, P. R., 1970, Magnetized basement outcrops on the southeast Greenland continental shelf: *Nature*, v. 226, p. 743–744.

Vogt, P. R., Einwich, A., and Johnson, G. L., 1972, A preliminary Jurassic and Cretaceous reversal chronology from marine magnetic anomalies in the western North Atlantic: *Trans. Amer. Geophys. Union*, v. 53, p. 363.

Whitmore, F. C., Emery, K. O., Cooke, H. B. S., and Swift, D. J. P., 1967, Elephant teeth from the Atlantic continental shelf: *Science*, v. 156, p. 1477–1481.

Williams, H., Kennedy, M. J., and Neale, E. R. W., 1972, The Appalachian structural province, in: *Tectonic Style in Canada*: editors R. A. Price and R. J. W. Douglas, Spec. Paper Geol. Assoc. Can., Number 11, p. 181–262.

Woollard, G. P., 1972, Regional variations in gravity, in: *The Nature of the Solid Earth*, E. C. Robertson, ed.: McGraw Hill, New York, p. 463–505.

Zietz, I., 1970, Eastern continental margin of the United States: Part 1, a magnetic study, in: *The Sea*, Vol. 4, Part 2: J. Wiley and Sons, New York, p. 293–310.

Useful Charts

American Association of Petroleum Geologists, 1970, Bathymetric maps eastern continental margin, U.S.A. Three sheets. Amer. Assoc. Petrol. Geol., Tulsa, Oklahoma.

Canadian Hydrographic Service, 1969, Bay of Fundy to Gulf of St. Lawrence, Chart 801, First edition, Department of Energy, Mines and Resources, Ottawa.

Canadian Hydrographic Service, 1970, Newfoundland Shelf, Chart 802, First edition, Department of Energy, Mines and Resources, Ottawa.

Canadian Hydrographic Service, 1971, Relief diagram of the continental margin of eastern North America, Chart 810, First edition, Department of the Environment, Ottawa.

King, L. H., 1970, Surficial geology of the Halifax-Sable Island map area, Marine Sciences Branch, Department of Energy, Mines and Resources, Ottawa, Paper 1, including chart 4040G.

National Geographic Society, 1968, Atlantic ocean floor, Atlas Plate 62, Washington.

Uchupi, E., 1965, Map showing relation of land and submarine topography, Nova Scotia to Florida, U.S. Geol. Survey Misc. Geol. Inv. Map I-451, 3 sheets, scale 1:1,000,000.

Additional References

The following references which have appeared since the manuscript was written will be found useful to readers who seek up-to-date information.

Barrett, D. L. and Keen, C. E., 1973. Lineations in the magnetic quiet zone of the northwest Atlantic: *Jour. Geophys. Res.* (in press).

Emery, K. O. and Uchupi, E., 1972. Western North Atlantic Ocean: topography, rocks, structure, water, life and sediments: *Amer. Assoc. Petrol. Geol. Memoir No. 17*.

Keen, C. E., Keen, M. J., Barrett, D. L., and Heffler, D. E., 1973. Some aspects of the ocean-continent transition at the continental margin of eastern North America: *Jour. Geophys. Res.* (in press).

Swift, D. J. P., Duane, D. B., and Pilkey, O. H., 1972. (Editors), *Shelf Sediment Transport: Process and Pattern*. Dowden, Hutchinson and Ross, Inc., Stroudsburg, Pa., U.S.A.

Chapter 4

THE NORTHEASTWARD TERMINATION OF
THE APPALACHIAN OROGEN

Harold Williams, M. J. Kennedy, and E. R. W. Neale

Department of Geology
Memorial University of Newfoundland
St. Johns, Newfoundland, Canada

I. INTRODUCTION

The Appalachian orogen is a northeast-trending belt of deformed rocks that lies southeast of the Canadian Shield and Interior Platform of central North America (Fig. 1). The northernmost part of the orogen is bounded to the east by the Atlantic Ocean, and its southern part is overlain toward the east by Mesozoic sediments of the Atlantic coastal plain. The mountain belt has an exposed area of 125,000 square miles in eastern Canada and extends from Newfoundland 2000 miles southwestward along the Atlantic seaboard to Alabama in the southeast United States. The submerged offshore extensions of the orogen are covered by Mesozoic and Cenozoic sediments that form the Atlantic continental shelf. Presumably, the Appalachian orogen was once continuous with the Caledonian Mountain System which consists of similar rocks with similar structural styles in Ireland and the United Kingdom on the eastern side of the North Atlantic.

Insular Newfoundland forms the northeastern termination of the exposed Appalachian System and offers a virtually complete, well-exposed, cross section through the orogen along its northeast coastline. Probably no cross section through an old mountain belt is truly representative, but it is likely that the

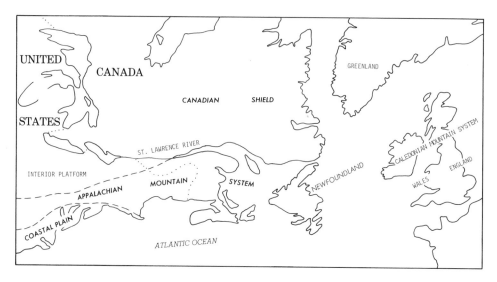

Fig. 1. The location of eastern North America with respect to western Europe before opening of the present Atlantic Ocean. After Bullard *et al.* (1965).

Newfoundland section contains more of the significant elements of northern Appalachian geology than any other. Although it has not yet received the detailed study accorded to some other parts of the Appalachians and to the Caledonides of Ireland and the United Kingdom, its superb exposures of early Paleozoic rocks and comparative lack of post-orogenic cover sequences have permitted a more searching scrutiny of the fundamental junctions among the early Paleozoic rocks of the major tectonic elements than is possible elsewhere. For these reasons, our analysis of the geology of the northeastward termination of the Appalachians will be chiefly a synthesis and interpretation of New-foundland geology, with reference to other parts of the northern Appalachian region only when required to amplify our descriptions and interpretations.

Toward this end the Newfoundland Appalachians has been subdivided into eight distinct tectonic-stratigraphic zones that are easily distinguished either by their contrasting Ordovician and earlier stratigraphy and/or their structural development. These zones are designated by geographic names in Newfoundland and by the letters A–H where they are extended southwestward across the Canadian Appalachian Region (Fig. 2). Zone I is an additional zone that is restricted to Nova Scotia and may be represented in subsurface east of zone H in Newfoundland.

The Appalachian orogen is composed mainly of late Precambrian and Paleozoic rocks that are sharply contrasted in thickness, facies, and structural style to the cover rocks of the nearby Precambrian basement of the Canadian Shield and Interior Platform (Fig. 1). The western boundary of the orogen is gradational in its southern part in the United States, where relatively thin, undisturbed carbonates and orthoquartzitic sandstones grade eastward into thicker, more deformed successions with increasingly older units at the base of the sedimentary sequence. In the northern part of the system this boundary is marked by a significant tectonic break, Logan's Line, along the St. Lawrence River (Fig. 2) that brings deformed shales and sandstones of the Appalachian Province into juxtaposition with relatively undeformed, mainly carbonate rocks of the St. Lawrence lowlands, close to the margin of the Canadian shield. A similar boundary occurs locally in western Newfoundland, where two large composite klippen have moved westward to overlie a slightly deformed Cambrian–Ordovician carbonate succession of contrasting facies. Elsewhere in western Newfoundland there is a gradual eastward progression from undeformed to deformed cover rocks analogous in most respects to the change in structural style across the Appalachian Plateau, Valley and Ridge, and Blue Ridge provinces (King, 1959) of the Appalachians in the southeastern United States.

The newly devised subdivision of the Newfoundland Appalachians forms the basis of this paper. In the following sections each subdivision or zone is briefly described, analyzed in terms of its geologic development, and correlated with similar tectonic–stratigraphic elements both in other parts of the Ap-

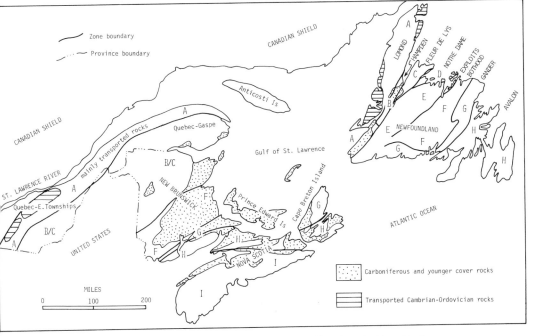

Fig. 2. Tectonic–stratigraphic zones of the Canadian Appalachian Region.

palachian System and in its continuation on the eastern side of the Atlantic as the Caledonides of Ireland and the United Kingdom. Brief sections on granites and metamorphism follow as, understandably, neither fits wholly and completely into the essentially Cambrian–Ordovician zonal pattern. A concluding section is devoted to the tectonic development of this northeastern-most part of the Appalachians, comparisons with the Caledonides, and a review of recent tectonic models that have been proposed to explain Caledonian–Appalachian evolution.

Only recent references are cited in the text. The reader is referred to the following for more complete bibliographies on northern Appalachian geology: Neale and Williams (1967), Zen *et al.* (1968), Kay (1969), Poole *et al.* (1970), Rodgers (1970), and Williams, Kennedy, and Neale (1972).

We owe a continuing debt to P. St. Julien, C. Hubert, A. Ruitenberg, W. R. Church, G. Pajari, J. Dewey, and R. Cormier who originally helped us extrapolate our zonal subdivision of Newfoundland into other parts of the Canadian Appalachian Region. A friendly running battle with three dozen colleagues and graduate students at Memorial University has produced whatever merit this work may have and will undoubtedly lead us to discard some of the ideas herein before the work is published.

II. ZONAL SUBDIVISIONS OF THE NEWFOUNDLAND APPALACHIANS

Formal names are here introduced for the eight zonal subdivisions of the Appalachians in Newfoundland that were previously designated by the letters A–H (Williams, Kennedy, and Neale, 1972). Well-known geographic localities within each are chosen as the zone names. From west to east these are as follows (Fig. 3): Lomond (A), Hampden (B), Fleur de Lys (C), Notre Dame (D), Exploits (E), Botwood (F), Gander (G), and Avalon (H).

The relationships of the distinctive lithosomes in adjacent zones are difficult to interpret in most places as the zone boundaries are faulted and it has not yet been possible to establish early Paleozoic linkages between adjacent zones everywhere. In the absence of stratigraphic or sedimentological linkages, there is no assurance that the rocks in adjacent zones did not move vast distances into their present positions.

The Lomond, Hampden, and Fleur de Lys zones in the western part of the system evolved upon a continental basement like that of the Canadian Shield and Interior Platform of North America to the west (Fig. 4). Strati-graphic analysis suggests a model related to distention with mafic dike intrusion and volcanism and the accumulation of continental rise/terrace wedge deposits (Strong and Williams, 1972; Stevens, 1970). This was followed by the develop-

Fig. 3. Tectonic–stratigraphic zones of Newfoundland and location of places referred to in the text.

ment of a carbonate bank along the newly formed continental margin (Rodgers, 1968).

The Notre Dame, Exploits, and Botwood zones in the central part of the system represent oceanic crust, at least in part, and volcanic and sedimentary rocks that were probably deposited upon oceanic crust. This inference stems from the occurrence of Early Ordovician ophiolitic suites in the Notre Dame zone that in cross section compare closely with present models of oceanic crust (Upadhyay *et al.*, 1971; Strong, 1972). Similar ophiolites, transported into the Lomond zone, are interpreted to have originated in the Notre Dame zone (Williams and Malpas, 1972). A sedimentary linkage is established between the

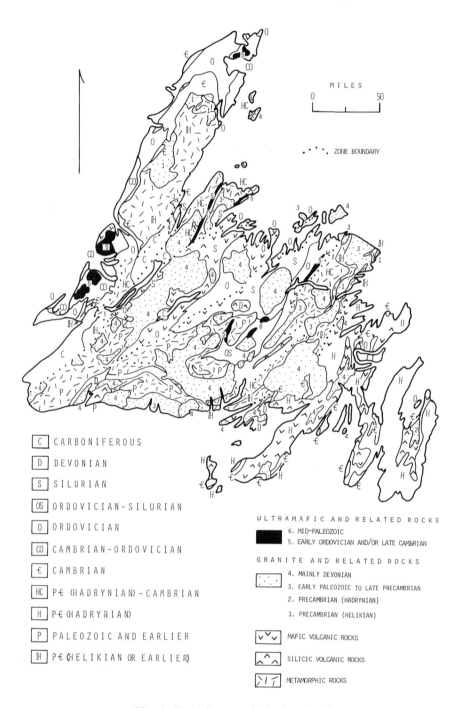

MILES
0   50

ZONE BOUNDARY

C CARBONIFEROUS

D DEVONIAN

S SILURIAN

OS ORDOVICIAN-SILURIAN

O ORDOVICIAN

CO CAMBRIAN-ORDOVICIAN

€ CAMBRIAN

HC P€ (HADRYNIAN)-CAMBRIAN

H P€ (HADRYNIAN)

P PALEOZOIC AND EARLIER

lH P€ (HELIKIAN OR EARLIER)

ULTRAMAFIC AND RELATED ROCKS
6. MID-PALEOZOIC
5. EARLY ORDOVICIAN AND/OR LATE CAMBRIAN

GRANITE AND RELATED ROCKS
4. MAINLY DEVONIAN
3. EARLY PALEOZOIC TO LATE PRECAMBRIAN
2. PRECAMBRIAN (HADRYNIAN)
1. PRECAMBRIAN (HELIKIAN)

MAFIC VOLCANIC ROCKS

SILICIC VOLCANIC ROCKS

METAMORPHIC ROCKS

Fig. 4. Geologic map of Newfoundland.

Fleur de Lys zone (west) and the Notre Dame zone (central) by the occurrence of Fleur de Lys metamorphic detritus in Early Ordovician clastic rocks above the ophiolites of the Notre Dame zone (Church, 1969). A similar linkage is established between the Botwood zone (central) and the Gander zone (east) by the occurrence of metamorphic detritus in Middle Ordovician rocks of the Botwood zone that was derived from the Gander zone (Kennedy and McGonigal, 1972). The post-Middle Ordovician history of the central part of the system (Notre Dame, Exploits, and Botwood zones) is marked by flysch infilling and then continental volcanism and red bed deposition during the Silurian and Devonian (Williams, 1967, 1969).

The Gander and Avalon zones that form the eastern part of the system contrast sharply in their stratigraphy and structural development. The Gander zone formed upon continental basement, and its development is similar in most respects to the Fleur de Lys zone in the west. It formed the eastern margin of the oceanic domain represented in the central parts of the system. Poly-deformed sedimentary rocks that overlie basement gneisses in the Gander zone may represent continental margin deposits similar to those developed in the west. There is no early Paleozoic linkage between the Gander and Avalon zones, but together they are cut by a Devonian granite batholith, and meta-morphic detritus, derived from the basement rocks of the Gander zone, occurs locally in Silurian–Devonian red beds of the Avalon zone (Williams, 1972), indicating that the two were in juxtaposition at least by Silurian–Devonian time. The Avalon zone was a platformal area during the Cambrian, and its earlier development is related to volcanism and thick red bed deposition that were either associated with a distential environment of a basin-and-range or simple rift type (Papezik, 1972a) or else volcanic islands or a volcanic island arc system built upon oceanic or continental crust (Hughes and Bruckner, 1971).

Structural trends throughout the Appalachian orogen are mainly north-easterly and parallel to the regional northeasterly trend of the mountain belt. This pattern is evident in the Lomond, Hampden, and Avalon zones, but it is much less evident in the interior zones of the orogen in Newfoundland. There, east–west structural trends are common in the Notre Dame zone, and major S-shaped configurations, as exemplified in the Gander zone (Figs. 3 and 4). This present irregular configuration of the zonal subdivision in central New-foundland and the structural trends within the zones are interpreted as the result of tectonic flexing and longitudinal faulting of originally continuous linear zones rather than a primary feature (Williams et al., 1970). The marked constriction in the system in southwest Newfoundland is marked by meta-morphic rocks and a near juxtaposing of the Lomond and Gander zones, which are separated by a wide Lower Ordovician oceanic domain in the northeast. Most likely, this marked constriction in southwestern Newfoundland represents a suture where an ancient ocean, like that represented in northeast Newfoundland, has been completely closed (Brown, 1973).

Two major deformational events are recognized in Newfoundland. The earliest and most intense is restricted to the Fleur de Lys and Gander zones, where it involves Cambrian (?) and possibly younger rocks, but does not affect adjacent Lower Ordovician rocks of the Notre Dame zone nor the Middle Ordovician of the Botwood zone (Figs. 5 and 7). The event is therefore thought to have occurred during the Late Cambrian or earliest Ordovician. It is characterized by polyphase deformation that was associated with upper greenschist to low amphibolite facies metamorphism.

The second major deformational event affected rocks up to Middle

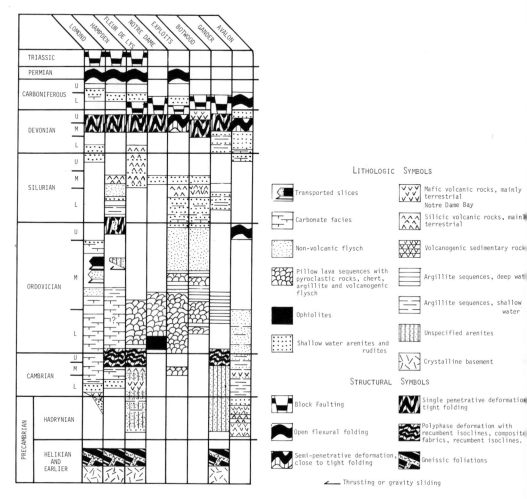

Fig. 5. Depositional and structural history

Devonian in age but did not affect Carboniferous strata. It is therefore dated as Devonian. In northeast Newfoundland isotopic ages suggest it may have commenced in Silurian time, and it is referred to as the Acadian orogeny. It is recognized across all of the Newfoundland zonal subdivisions, and it is evidenced by a single steep penetrative cleavage in most places.

Where Acadian deformation is most intense, a distinctive sequence of strains can be recognized. This sequence was initiated by regional vertical extension and northwest–southeast shortening, forming a steep penetrative foliation and tight upright folds. The sequence continued with vertical shortening forming crenulations and small folds and terminated with northeast–southwest shortening, which formed moderate-to-large, open, steeply plunging folds and kink bands. Acadian deformation is intense in the eastern part of the Lomond zone and decreases westward. It is also intense in the Hampden and Fleur de Lys zones and is locally intense in the northern and eastern parts of the Botwood zone. Acadian structures are generally more open and cleavage is less penetrative throughout the Notre Dame and Exploits zones, except for narrow belts of intense deformation.

Steep faults rarely form important tectonic features within the zones, but several important faults mark zone boundaries. The Cabot Fault, a major sinistral strike-slip fracture (Williams *et al.*, 1970) forms the eastern boundary of the Lomond and Hampden zones. Its main movement is probably of Late Devonian age. The Luke's Arm fault forms the Notre Dame–Exploits zone boundary. It is most probably a dip-slip fracture downthrown to the south in post-Silurian time but may have an older movement history. Another major fracture zone separates the Gander and Avalon zones. Its sense of displacement is unknown but it is probably either a strike-slip fault or a steeply west-dipping reverse fault.

A. Lomond Zone (Zone A)

The Lomond zone (Figs. 4 and 5) consists of a continental (Grenvillian) basement that is cut by northeast-trending mafic dikes and that is locally overlain by plateau basalts coeval with dike intrusion. All are overlain by an eastward-thickening Cambrian–Ordovician carbonate succession that passes upward into Middle Ordovician flysch deposits. The latter were derived from the east and immediately preceded the emplacement of transported clastic rocks and igneous and metamorphic complexes above the carbonates. The Lomond zone was affected by penetrative deformation (Devonian) that affects transported and autochthonous rocks alike. Many of the transported rocks, especially those in higher structural slices, were involved in several deformative events, some probably in the Fleur de Lys zone and others in an oceanic domain, prior to their emplacement into the Lomond zone.

The *Precambrian basement rocks* are gneisses and schists that are exposed in broad domes or uplifted blocks. They are complexly folded with composite gneissic fabrics and they are cut by granites that yield isotopic ages in the order of 900 m.y. Anorthosites and related intrusions, similar to those of the Grenville province of the nearby Canadian Shield, are also present in the basement complexes.

Plateau basalts locally overlie the basement gneisses (Williams and Stevens, 1969), and the flows are overlain by undated arkosic sandstones that transgress westward onto the basement rocks. Locally clastic sedimentary rocks intervene between basement and flows in the northeast part of this zone. All thicken eastward with older units represented at the base of the succession.

Archaeocyathid reef-limestones and shales conformably overlie undated arkosic sandstones and form the base of a *carbonate succession*. It is a succession of Lower Cambrian to Lower Ordovician units, all east thickening, and consists of a variety of limestones and dolomites, orthoquartzitic sandstones, and shales. An erosional disconformity separates Lower Ordovician shallow-water limestones from overlying Middle Ordovician limestones, and this is the only recognized hiatus in the succession of rock units above the basement gneisses.

Clastic sedimentary rocks conformably overlie Middle Ordovician limestones at the top of the carbonate succession and herald an abrupt and permanent change in the depositional environment that is recognized all along the western side of the Appalachian System. These clastics are interpreted in Newfoundland as the external part of a westward transgressing flysch wedge that appeared a little earlier in more easterly autochthonous exposures and earlier still in upper parts of a transported clastic succession within the Lomond zone (Stevens, 1970).

Transported rocks in the Lomond zone include a variety of sedimentary rocks and igneous and metamorphic rocks (Rodgers and Neale, 1963; Church and Stevens, 1971). All occur in a succession of separate slices (Williams and Malpas, 1972; Williams, Malpas, and Comeau, 1972) with transported sedimentary rocks at the base and igneous and metamorphic rocks in higher structural slices. The sedimentary rocks are mainly a clastic sequence, which includes a limestone-breccia and shale unit in its central part, and a condensed limestone-breccia sequence that is in places spectacularly coarse. Both sequences are largely contemporaneous with the underlying autochthonous carbonate succession. The coarse limestone-breccia sequence is interpreted as a bank-foot deposit that accumulated east of the carbonate succession (Kindle and Whittington, 1958; Rodgers, 1968). The clastic sequence is interpreted to have lain still farther east, for limestone breccias in its central unit are much finer and contain more shale interbeds than the nearshore, lithologically similar and coeval coarser breccias (Stevens, 1970). Sandstones beneath the limestone breccias of the clastic succession contain blue quartz and igneous rock fragments that suggest a westerly Precambrian source, and flysch deposits above the limestone breccia unit contain potash feldspar and ophiolite detritus that was derived from the east.

Thick *ophiolite sequences* and a variety of volcanic rocks, polydeformed greenschists and minor psammitic schists, and amphibolitic gabbros and foliated granites are represented in higher structural slices. The ophiolite sequences are stratiform plutons that consist of ultramafic rocks, gabbros, sheeted dikes, and mafic pillow lavas, from bottom to top (Williams and Malpas, 1972). Locally in northern Newfoundland, the pillow lavas are dated as Tremadocian (Williams, 1971). Amphibolitic gabbros and foliated granites in nearby slices are older, and they are cut by sheeted dikes and surrounded by pillow lavas that are similar to those of the ophiolite sequences. All recent workers (Church and Stevens, 1971; Dewey and Bird, 1971) have interpreted the ophiolite sequences as transported oceanic crust and mantle, and the older amphibolitic gabbros and granites have been interpreted as crustal remnants caught up in the ocean-spreading episode (Williams and Malpas, 1972). Deformation features in these higher slices predate emplacement into their present position (Smyth, 1971), but some may be related to displacement in the source area.

The interpretation of the transported sequences in the Lomond zone and their palinspastic restoration indicates that the highest structural slices originated farthest east and are therefore the farthest traveled.

Middle Ordovician limestones overlie transported clastic rocks at Port au Port and set an upper time limit to their emplacement (Rodgers, 1965). Subsequent deposition in the Lomond zone is recorded by a local occurrence of Siluro-Devonian red beds and a few occurrences of flat-lying Mississippian limestone and plant-bearing sandstone.

Acadian deformation is intense in the eastern part of the Lomond zone. All three characteristic strains are developed and superposed both on Cambrian–Ordovician autochthonous and allochthonous sedimentary rocks. Large-scale folding has affected the base of transported sequences in the extreme north at Hare Bay, and the intensity of foliation and tight folding decreases westward until absent at Port au Port peninsula. The age of this deformation is interpreted as Devonian (Acadian) rather than Middle Ordovician (Taconian) because of its similarities to Acadian deformation in Silurian rocks of the Hampden, Fleur de Lys, Notre Dame, and Exploits zones.

The Lomond zone with its Precambrian basement and Cambrian–Ordovician carbonate cover is similar in tectonic setting to the Lewisian basement of the Scottish Highlands in the British Caledonides with its overlying Cambrian–Ordovician carbonate succession at Durness. Unlike the Newfoundland example, the European equivalent does not have the rich profusion of transported rocks, but it has a more complete late Precambrian history represented by the thick Torridonian red beds that intervene between the basement and carbonate cover.

B. Hampden Zone (Zone B)

The Hampden zone shares characteristics of the Lomond zone to the west and the Fleur de Lys zone toward the east. Unlike either of these bounding zones its autochthonous Ordovician(?) rocks were affected by a pre-Silurian penetrative deformation (Lock, 1969). The eastern boundary of the Hampden zone is marked by the Cabot Fault and its western boundary is faulted, or interpreted as faulted, except locally on the west side of White Bay where Cambrian rocks of the Hampden zone unconformably overlie crystalline basement, and the zone boundary is drawn along the now steeply dipping unconformity.

Deformed limestones, quartzites, and shales along the western margin of the Hampden zone in the north are undated but have been traditionally assigned to the Cambrian and Ordovician and correlated with the less deformed carbonate succession of the Lomond zone. Locally the Hampden zone carbonates contain limestone-breccia beds that are similar to those that are transported in the Lomond zone, suggesting that the rocks of the Hampden zone once lay near the edge of a carbonate bank. Silurian sandstones, conglomerates, and shales with significant acid volcanic units overlie the older rocks in the north, and these are in turn overlain by Carboniferous sandstones and conglomerates.

The southern part of the Hampden zone consists of polydeformed metamorphic rocks that are similar in lithology and structural style to those of the Fleur de Lys zone to the east. These are interpreted as transported because of the contrast in structural style and metamorphic grade where they are in juxtaposition with relatively undeformed and much less metamorphosed autochthonous Cambrian and Ordovician rocks of the Lomond zone. The boundary is modified by high-angle faulting, but it may be a significant thrust that brings the polydeformed rocks directly above the carbonate succession. This structural setting is quite unlike that of the Lomond zone farther west, where transported polydeformed metamorphosed rocks are separated from underlying carbonates by unmetamorphosed transported clastic rocks that overlie autochthonous flysch at the top of the carbonate succession.

Recent work in the northern part of the Hampden zone on the west side of White Bay has shown the presence of narrow, faulted belts of migmatites and polydeformed greenschists (Lock, 1969). These are probably also transported and equivalent to similar rocks of the Fleur de Lys zone to the east.

The Hampden zone is unique among the Newfoundland tectonic–stratigraphic zones as being the only zone in which Ordovician (Taconic) penetrative deformation may have occurred. Apart from those rocks which clearly formed in the Fleur de Lys zone and which are all or in part transported, undated Cambrian–Ordovician carbonates and clastic rocks on the west side of White Bay contain two foliations whereas the overlying Silurian rocks contain a single fabric. Whether or not both or only one foliation of the pre-Silurian rocks predates the single fabric of the Silurian rocks is still uncertain. The pre-Silurian rocks, while similar to the carbonates of the Lomond zone, are also lithologically similar to pre-Lower Ordovician rocks of the Fleur de Lys zone on the east side of White Bay. They may therefore represent a westward extension of the rocks of the Fleur de Lys zone which were deformed in pre-Lower Ordovician time (Fig. 5) and which contain two penetrative foliations.

Acadian deformation, which affects Silurian rocks but predates the deposition of Carboniferous strata in the Hampden zone, has resulted in tight upright folds with an associated steep penetrative foliation and the formation of later kink bands.

The Hampden zone in Newfoundland is equivalent to parts of the northern Caledonides in Scotland where polydeformed rocks of the Moine and Dalradian sequences have been thrust northwestward to overlie less deformed rocks of the Cambrian–Ordovician carbonate succession or to lie directly upon the Lewisian basement.

C. Fleur de Lys Zone (Zone C)

The Fleur de Lys zone (Figs. 3–7) is characterized above all by a thick mixed assemblage of late Precambrian to earliest Paleozoic rocks that were complexly deformed in Late Cambrian or Early Ordovician (Church, 1969). The polydeformed rocks are underlain by a much rejuvenated gneissic basement, at least toward the west (De Wit, pers. comm., 1971), and they are faulted against Early Ordovician ophiolitic rocks along the Fleur de Lys–Notre Dame zone boundary in the east. A thin discontinuous belt of ultramafic rocks trends northeastward throughout the central portion of the Fleur de Lys zone at Baie Verte and it is bordered by polydeformed schists toward the west and a narrow belt of presumed Ordovician volcanic rocks toward the east. A chaotic conglomerate with black shaly matrix occurs locally along the ultramafic rock–volcanic rock contact. East of the ultramafic–volcanic belt the older polydeformed rocks are cut by granodiorite that shed detritus into nearby Ordovician(?) volcanic rocks, and the granodiorite is nonconformably overlain by a subaerial sequence of sediments and volcanic rocks of Silurian age (Neale and Kennedy, 1967).

Polydeformed rocks of the Fleur de Lys zone are referred to as the Fleur de Lys Supergroup and consist mainly of pelitic and psammitic schists. Marble beds and lime breccia units are conspicuous in the western part of the zone

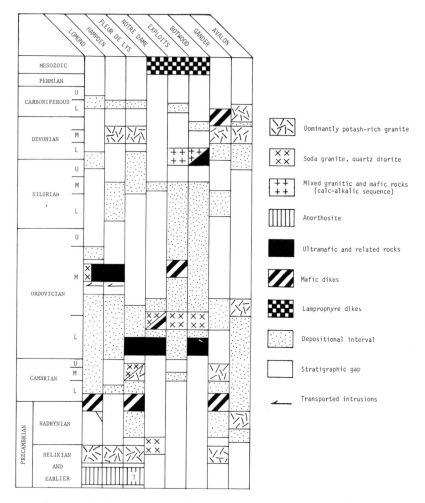

Fig. 6. Plutonic rocks of the Newfoundland tectonic–stratigraphic zones.

and conglomerates with gneissic boulders occur locally above an older metamorphic basement complex (De Wit, pers. comm., 1971). Chlorite–actinolite schist, hornblende amphibolite, intrusive porphyry, and silicic pyroclastic rocks are predominant in eastern parts of the zone.

Marble beds and lime breccias in the western part of the Fleur de Lys zone may represent easternmost occurrences of early Paleozoic bank-foot deposits along a continental margin. The thick psammitic sequence (20,000 ft) represented elsewhere is cut by pretectonic mafic dikes, and all probably relate to distention and the buildup of a continental rise prism. Mixed volcanic rocks so prevalent in eastern parts of the zone could represent offshore island volcanism.

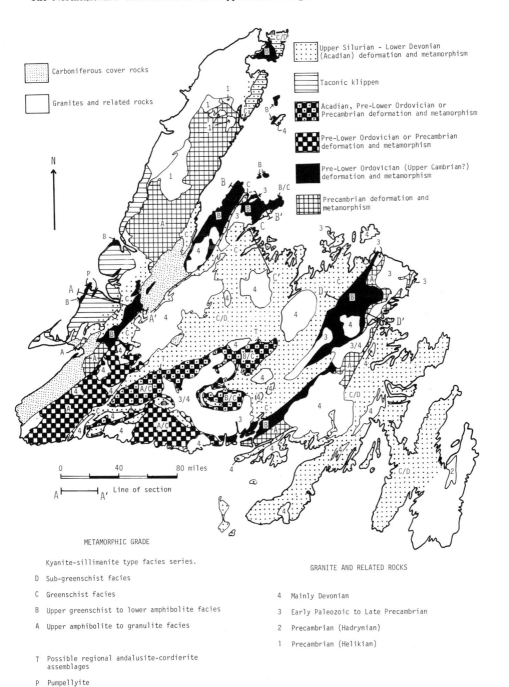

Carboniferous cover rocks

Granites and related rocks

Upper Silurian - Lower Devonian (Acadian) deformation and metamorphism

Taconic klippen

Acadian, Pre-Lower Ordovician or Precambrian deformation and metamorphism

Pre-Lower Ordovician or Precambrian deformation and metamorphism

Pre-Lower Ordovician (Upper Cambrian?) deformation and metamorphism

Precambrian deformation and metamorphism

N

0   40   80 miles

A ⊢————⊣ A'   Line of section

METAMORPHIC GRADE

Kyanite-sillimanite type facies series.

D Sub-greenschist facies

C Greenschist facies

B Upper greenschist to lower amphibolite facies

A Upper amphibolite to granulite facies

T Possible regional andalusite-cordierite assemblages

P Pumpellyite

GRANITE AND RELATED ROCKS

4 Mainly Devonian

3 Early Paleozoic to Late Precambrian

2 Precambrian (Hadrynian)

1 Precambrian (Helikian)

Fig. 7. Structural and metamorphic map of Newfoundland.

Basic schists and local occurrences of psammitic schists that are similar in metamorphic grade and structural history to those of the Fleur de Lys Supergroup form an integral part of the transported sequences in the Lomond zone. These transported rocks were metamorphosed and deformed prior to their Middle Ordovician emplacement into the Lomond zone.

The complexly deformed Fleur de Lys Supergroup displays a regional first foliation that postdates tectonic slides and that is associated with tight-to-isoclinal small-scale folds (Kennedy, 1971). This first foliation is almost completely transposed by a regional penetrative second foliation, that is associated with major recumbent isoclines. At least two pre-Acadian strain-slip foliations deformed the second fabric of the Fleur de Lys Supergroup, and these are locally associated with large-scale tight folds (Kennedy et al., 1972).

The narrow belt of Ordovician and Silurian rocks in the central part of the Fleur de Lys zone is a tight, upright synclinal structure. The rocks display a steep Acadian penetrative single foliation that is folded by flat crenulations and steeply plunging kink bands. These features indicate that all three characteristic Acadian strain regimes were developed; however, no penetrative Acadian structures have been identified in the Fleur de Lys Supergroup.

Second-phase major recumbent folds in the Fleur de Lys Supergroup face northwestward in the northwestern part of the zone at Fleur de Lys and southward in the eastern part of the zone at Mings Bight. At exposures of gneissic basement rocks near Bear Cove, the second-phase folds are upright so that this region can be interpreted as the symmetry axis from which the recumbent second-phase isoclines face outward on both sides. The gross structure therefore resembles a mantled gneiss dome.

The main (second) foliation of the Fleur de Lys Supergroup and the two later strain-slip foliations are refolded into a steep attitude by Acadian deformation along the west flank of the Ordovician–Silurian synclinal belt. The western boundary of this synclinal belt was originally interpreted as a fundamental fault that had been subsequently intruded by ultramafic rocks (Neale and Nash, 1963). However, recent work has indicated that it is unlikely that any major fault is present, and the age of the ultramafic rocks is doubtful (W. S. F. Kidd, pers. comm., 1971).

The Fleur de Lys Supergroup is similar in age, lithology, metamorphic grade, and structural style to the Moine/Dalradian metamorphic rocks that occupy a similar tectonic position in the Scottish Highlands of the British Caledonides (Philips et al., 1969; Kennedy et al., 1972). Equivalents are also present in Ireland. The Moine/Dalradian sequence overlies a metamorphic basement complex, and it is bordered by less-deformed platformal carbonates to the northwest in Scotland (Lomond zone equivalents) and Ordovician volcanic rocks (possibly Notre Dame zone equivalents) toward the southeast in the Midland Valley. The extension of the Fleur de Lys zone (zone C, Fig. 2) across the Canadian Appalachian Region is hindered by poor exposure, by younger cover rocks, and, most of all, by the uncertainty as to whether or not the polyphase deformed rocks so characteristic of this zone are in place (Fleur de Lys zone) or transported (Hampden, Lomond zones).

D. Notre Dame Zone (Zone D)

This zone is characterized by Lower and Middle Ordovician mafic pillow lava sequences with associated mafic agglomerates and tuffs (Fig. 5). Sedimentary rocks are abundant locally and are mainly greywacke, chert, siliceous argillite, and minor limestone—all of marine deposition. Ultramafic rocks that form the basal part of the Betts Cove Ophiolite Complex occur in the western part of the zone and mafic sheeted dikes underlie pillow lava sequences elsewhere (Strong, 1972). Small dioritic and granodioritic intrusions cut the mafic volcanic rocks, and in the extreme eastern part of the zone, near Twillingate, mafic sheeted dikes cut a large body of well-foliated granitic rocks and amphibolite. The foliation in these rocks predates the Ordovician pillow lava sequences.

Structural trends vary across the zone from mainly northeast in its western part to east–west in its central part to northeasterly in its eastern part. The southern boundary of the zone, separating it from the Exploits zone, is marked by a major east–west lineament known as the Luke's Arm Fault in the east and the Lobster Cove Fault in the west. Metamorphic detritus derived from the Fleur de Lys zone has been recognized in Lower Ordovician greywackes that are interlayered with mafic volcanic rocks above the Betts Cove Ophiolite Complex (Church, 1969). Tonalite and amphibolite from the Twillingate batholith and a variety of volcanic rock fragments similar to the volcanic rocks present in the Notre Dame zone occur in Silurian conglomerates south of the Luke's Arm Fault in the Exploits zone (Helwig and Sarpi, 1969).

The Betts Cove Ophiolite Complex (Upadhyay *et al.*, 1971) consists of a basal ultramafic member, transitionally overlain by a poorly developed gabbroic member, in turn overlain by a sheeted dike complex that consists of practically 100% mafic dikes. The sheeted dike complex is faulted against nearby mafic volcanic rocks, but locally the contact is gradational across a narrow zone that shows a large decrease in the percentage of dikes over a short distance. Above this the Lower Ordovician rocks are nearly 4 km thick and comprise a conformable sequence of pillow lavas, cherts, argillites, andesitic pyroclastic rocks, and immature volcanic sediments. In that the lowest pillow lava unit constitutes the upper part of the Betts Cove ophiolite complex, there is a completely conformable transition from the ultramafic member of the ophiolite complex to the top of the thick, overlying, mainly volcanic succession.

The general sequence of lithological units compares well with that of other ophiolite suites that have been interpreted to represent oceanic crust. Hence the lowest pillow lava unit of the Notre Dame zone supracrustal rocks has been interpreted as layer 2 of Early Ordovician oceanic crust, and the gabbro and sheeted dike unit as layer 3. The Moho is represented by the transitional zone between ultramafic and mafic rocks, and the basal ultramafic rocks represent mantle. According to this model, the thick sequence of immature volcanic sediments and pillow lavas of the Notre Dame zone may relate to subsequent island arc volcanism. On the other hand, if the Moho discontinuity beneath present oceans represents a phase change as opposed to a compositional change, then it is unlikely that the mafic/ultramafic rock contact at Betts Cove represents the Lower Ordovician oceanic crust/mantle interface. In the latter case, this interface may be well down in the ultramafic unit and may separate ultramafic cumulate rocks from underlying depleted mantle (John Malpas, pers. comm., 1972).

The intensity of Acadian deformation in the Notre Dame zone varies laterally and vertically. Silurian rocks commonly lack penetrative foliation, although they are tightly folded, whereas Ordovician rocks, particularly near the base of the Ordovician sections, are commonly more deformed. In places, the intensity of Acadian deformation dies out upward within Ordovician sections. It is unlikely therefore that a major angular unconformity of tectonic significance separates Ordovician from overlying Silurian rocks in this zone.

In the more foliated rocks of the Notre Dame zone all three strain regimes characteristic of the Acadian sequence of deformation are developed. In places a regional penetrative steep foliation in the Lower Ordovician mafic volcanic rocks postdates an earlier foliation that is distributed in elongate lenticular belts. These belts are interpreted as shear belts that are older than the regional deformation, but of uncertain age.

Sheeted dikes like those represented throughout the Notre Dame zone are interpreted as due to distention and to support a model of sea-floor spreading. Where the sheeted dikes cut well-foliated granitic rocks in the eastern part of the zone, it is implied that the older granitic rocks were involved in this spreading episode and that they represent older crustal remnants rafted among the evolving ophiolitic suites.

All of the relationships among the ophiolitic rocks and between sheeted dikes and earlier granitic rocks of the Notre Dame zone are equally well displayed in the higher structural slices of the transported successions in the Lomond zone at Bay of Islands. This suggests that the transported ophiolite complexes and the well-foliated granodiorites and amphibolitic gabbros of the Lomond zone originated in the Notre Dame zone.

The Notre Dame zone has not been recognized elsewhere in the Canadian Appalachian Region (zone D, Fig. 2) and it has no direct analogs of comparable size in the British Caledonides. Possibly the mafic volcanic rocks of the Ballantrae succession in the Southern Uplands of Scotland are an ophiolite succession and represent oceanic crust that is equivalent to the Newfoundland examples (Mitchell and Reading, 1971).

E. Exploits Zone (Zone E)

The Exploits zone is composed mainly of Early and Middle Ordovician sedimentary and volcanic rocks that are overlain by thick sequences of Silurian rocks (Horne and Helwig, 1969). All have been deformed by a single penetrative deformation. Ordovician sedimentary rocks are predominant, especially among higher parts of Ordovician successions, and are mainly argillites, siltstones, greywackes, cherts, and siliceous argillites. Mafic volcanic rocks, which are more abundant in basal parts of stratigraphic successions, resemble those of the Notre Dame zone except that layered pyroclastic rocks are predominant over pillow lava units. The base of the succession is nowhere exposed, so that the nature of the basement throughout this zone is unknown. A wide belt of shaly, chaotic rocks is represented in the northeast part of the zone, and it can be traced for about 20 miles southwestward. The Exploits zone is faulted against the Botwood zone, at least in its northern part, and its extension toward the southwest corner of Newfoundland is poorly defined.

Ordovician and Silurian rocks of the Exploits zone are sharply contrasted in facies (Williams, 1969). Ordovician rocks are all of marine origin with abundant graptolite faunas in Middle Ordovician shaly sequences. These are conformably and transitionally overlain by greywackes that contain Silurian brachiopods in their upper parts and that pass upward into coarse plutonic boulder conglomerates with local coralline shale interbeds. These are in turn overlain by Llandovery and Wenlock terrestrial volcanics and fluviatile, mainly red, micaceous sandstones (Williams, 1967). In the western part of this zone, unfossiliferous rocks of the Springdale Group (Neale and Nash, 1963) assigned to the Silurian consist of 10,000 ft of mixed, mainly terrestrial, volcanic rocks overlain by 5000 ft of red sandstones and conglomerates.

Acadian structures in the Exploits zone are locally the least intense of the whole system with the exception of the western extremity of the Lomond zone and the eastern parts of the Avalon zone. Although upright folds are tight, the accompanying foliation is commonly semipenetrative or absent, particularly in the north-central parts of the zone. Elsewhere a steep Acadian foliation is well developed and is affected by later crenulations and kink bands. Undeformed Carboniferous rocks unconformably overlie lower Paleozoic rocks in the central part of the zone.

The chaotic rocks in the northeast part of the Exploits zone occur in a northeast-trending fault-bounded belt from 4 to 8 miles wide. The chaotic rocks contain an extraordinary assortment of sedimentary and volcanic rock fragments and exotic blocks up to hundreds of meters long (Horne, 1969). The larger blocks are predominantly mafic lava and greywacke, and they are suspended in a shaly matrix. One such large volcanic mass contains a limestone lens that has yielded Middle Cambrian trilobites (Kay and Eldredge, 1968). Smaller clasts range from granules of quartz and lithic fragments to cobbles and boulders of varied composition. The matrix of the rock is scaly, black, pyritiferous, graphitic argillite and striped black and green argillite that is commonly homogenized into a chaotic paste devoid of primary structure. Dendroid graptolites, presumably contained in matrix material, suggest a Lower Ordovician age for the deposit in the southwest part of the belt, and the deposit is cut by a mafic intrusion dated isotopically at 480 m.y. in the northeast (Cormier, pers. comm., 1972). Acadian foliation is superposed upon the mélange in the northeast.

The origin and significance of this belt of chaotic rocks is as yet uncertain. The deposits were first interpreted as gravity-slide mélanges (Horne, 1969) and more recently as deposits typical of an oceanic trench and subduction zone (Bird and Dewey, 1970). Strikingly similar mélanges are associated with transported slices in the Lomond zone and are known elsewhere in the Fleur de Lys and Botwood zones. It is suspected that all relate to rock transport, slumping, and tectonic mixing, although the scale of the phenomenon may vary between that of a relatively small raft or klippe to that of a much larger subducting oceanic plate. If the mélange of the Exploits zone represents a subduction zone deposit, then much of this zone may have evolved upon oceanic crust like that represented in the Notre Dame zone.

Stratigraphic and sedimentological features of the lower parts of the Ordovician succession in the Exploits zone, e.g., abrupt facies changes, volcanic activity with contemporaneous erosion and sedimentation, and coeval volcanic and intrusive activity, all suggest to Mitchell and Reading (1971) that these rocks originated through island arc activity. The arcs were subsequently uplifted, unroofed, and draped by Middle Ordovician graptolitic shales during subsidence.

Sedimentary and volcanic rocks like those of the Exploits zone and with similar graptolite faunas are common throughout central parts of the Canadian Appalachian Region and the Caledonides in central Ireland, southern Scotland, and the Lake District of Britain. Silurian rocks comparable to those of the Exploits zone are also widespread throughout the Canadian Appalachian Region. In the British Caledonides, Silurian volcanic rocks are much less abundant. Red sandstones at the top of the Silurian sections in Newfoundland are in most respects similar to the Old Red sandstones of the British Isles.

F. Botwood Zone (Zone F)

The Botwood zone is composed mainly of Ordovician slates overlain by a thick sequence of Silurian conglomerates, terrestrial volcanics, and fluviatile red beds (Fig. 5). Penetrative deformation is intense in some belts, which alternate with belts of relatively undeformed rocks. Unlike the Exploits and Notre Dame zones, Ordovician mafic volcanic rocks are relatively rare except along the eastern margin of the zone in the north where a thin discontinuous belt of mafic pyroclastic rocks with associated small gabbro and ultramafic intrusions parallels the Gander River. The boundary between the Botwood and Gander zones, although geologically distinct, is poorly delineated. At Gander Lake in the north it is an unconformity modified by faulting, for Middle Ordovician greywackes there contain metamorphic detritus derived from the adjacent Gander zone (Kennedy and McGonigal, 1972). Farther north at the coast the contact is marked by a chaotic mélange that contains blocks of mafic volcanic rocks, diorite, and numerous blocks of Gander zone metamorphic rocks, including serpentinite, near the Gander zone contact. Similar mélanges that may form part of a continuous belt have been discovered recently 7 miles inland and north of Bay d'Espoir on the south coast of Newfoundland, where they also occur at or near the junction of the Botwood and Gander zones.

The Ordovician and Silurian rocks of the northern part of the zone commonly contain a penetrative steep foliation associated with tight upright folds. Locally this fabric is intense, and pebbles in conglomerates are strongly flattened in the foliation plane. Flat crenulations or chevron folds, and later steeply plunging conjugate kink bands, are the result of vertical shortening followed by northeast–southwest shortening.

The wide southern portion of the Botwood zone is comprised mainly of metamorphic rocks of questionable relationship. Some have been interpreted as the metamorphosed equivalents to Paleozoic rocks in the north, but others are probably basement to the Ordovician and Silurian rocks. Several large ultramafic bodies occur in the central portion of the zone, and these have been interpreted to cut surrounding Ordovician and Silurian rocks. One of these (at Sitdown Pond) has a well-defined ultramafic and gabbroic unit and is associated with volcanic rocks, all suggesting that it may represent an ophiolite suite. In the west-central part of the zone an extensive area of Devonian(?) silicic volcanic rocks has been interpreted to overlie folded Silurian red beds.

The Silurian and Devonian history of central Newfoundland records a period of mainly terrestrial deposition and infilling of earlier Ordovician oceanic basins. However, in New Brunswick, McKerrow and Ziegler (1971) propose the existence of a Silurian–Lower Devonian ocean trench (Fredericton

Trough) with accompanying marginal subduction zones. They suggest that Silurian and Lower Devonian andesitic volcanism in central and southern Gaspé and northern New Brunswick toward the north, and in southwestern New Brunswick and coastal Maine and Massachusetts toward the south, relates to subduction along the margins of the Fredericton trough during these periods. Its position in New Brunswick is roughly coincident with zone F (Fig. 2).

Ordovician slates similar to those of the Botwood zone contain Lower Ordovician graptolites (Charlotte Group) where they occur locally along the southeast margin of zone F in New Brunswick in the Canadian Appalachian Region (zone F, Fig. 2). Equivalent rocks in the Caledonides should be sought in central Ireland, southern Scotland, and the Lake District of England.

G. Gander Zone (Zone G)

The Gander zone (Figs. 4 and 5) consists of a basement complex that is overlain by a thick monotonous succession of polydeformed metasedimentary rocks referred to as the Gander Lake Group (Kennedy and McGonigal, 1972). This zone is continuous across Newfoundland and structural trends swing west and northwest through the Hermitage Flexure at the south coast of the island (Williams *et al.*, 1970). Basement rocks parallel the eastern margin of the zone, and its eastern boundary is a fault that brings the high-grade metamorphic basement rocks against much less deformed and metamorphosed late Precambrian rocks of the Avalon zone. Relatively unmetamorphosed sedimentary and volcanic rocks occur in southwestern parts of the Gander zone. These locally contain Devonian plant remains near La Poile Bay, but they are of presently unknown relationship with older rocks nearby.

The basement rocks of the Gander zone consist of granitic biotite gneisses and migmatites that contain xenoliths of still earlier deformed rocks. None of these rocks are dated but they may be of the same age as the Grenvillian inliers along the western part of the system (Lomond, Hampden, and Fleur de Lys zones). The overlying metasedimentary rocks are psammitic and semipelitic schists with minor graphitic schists and mafic schists, which are remarkably similar in metamorphic grade and structural style to pre-Lower Ordovician deformed rocks of the Fleur de Lys zone. Clasts of these metasedimentary rocks occur in nearby Middle Ordovician greywackes of the Botwood zone, and large schistose blocks of the metasediments with a folded second foliation are common in Middle Ordovician chaotic deposits. These relationships define a Middle Ordovician upper time limit to the stratigraphic age and time of deformation of the polydeformed metasedimentary rocks in

the Gander zone and indicate that the still older basement gneisses and migmatites are Precambrian.

The metasedimentary rocks of the Gander Lake Group have an early
penetrative foliation that has been transposed by a second dominant foliation.
This second foliation is followed by strain-slip foliations, and all are earlier
than Acadian structures in nearby Middle Ordovician rocks. In the northern
part of the Gander zone the second foliation is associated with large-scale
recumbent isoclines that face southeastward. Basement gneisses, exposed
farther southeast, apparently form the foreland to this recumbent fold complex.
In the northwest part of the zone the recumbent folds have been refolded by
Acadian deformation into upward-facing structures, but no penetrative Acadian
structures have been identified in the metasedimentary rocks of the Gander
Lake Group.

The relative positions of the Gander and Fleur de Lys zones in northern
Newfoundland and the similar lithologies, structural styles, metamorphic
grades, time of deformation, and basement relationships displayed within the
zones portray a marked symmetry to the interior part of the Appalachian
System in northern Newfoundland. In addition, the ultramafic–volcanic belt
within the Fleur de Lys zone is matched toward the east by the ultramafic–
volcanic belt along the eastern margin of the Botwood zone, near the Gander
zone boundary. It is most unlikely that this symmetrical arrangement of
tectonic–stratigraphic units is fortuitous. Since the Gander and Fleur de Lys
zones border an area that was at least in part oceanic in the Lower Ordovician,
i.e., the Notre Dame zone and possibly the Exploits zone, then the Fleur de
Lys and Gander zones represent the sialic continental margins to this early
Paleozoic ocean. Their similar pre-Ordovician sedimentary sequences and
their similar structural styles indicate that the margins evolved in a similar
fashion during ocean spreading, and then closing.

The major S-shaped configuration of the Gander zone is thought to be
tectonic rather than an original depositional trend, for similar configurations
of smaller scale resulting from Acadian northeast–southwest shortening are
common throughout central Newfoundland. This structure, the Hermitage
Flexure (Williams *et al.*, 1970), completely disrupts the symmetrical disposition
of tectonic–stratigraphic zones so well-displayed in northern Newfoundland
and brings the Gander zone metamorphic rocks close to basement rocks of the
Lomond zone in the southwest portion of the island. The resulting constriction
in the system may represent an area where an early Paleozoic ocean like that
represented toward the north was completely closed out and is represented
now by a suture where metamorphic rocks that once formed the sides of the
ocean now abut (Brown, 1973).

The Gander zone has been extended across northern Cape Breton Island
and southeast New Brunswick in the Canadian Appalachian Region (zone G,

Fig. 2). Equivalents in the Caledonides are probably rocks of the Irish Sea horst (the Mona complex of Anglesea and the Rosslare Migmatites of southeastern Ireland).

H. Avalon Zone (Zone H)

The Avalon zone consists of a thick succession of late Precambrian volcanic and sedimentary rocks in places overlain by Cambrian and Ordovician shales and sandstones. The late Precambrian–Cambrian succession is for the most part continuous and conformable, except in the Conception Bay area of eastern Newfoundland where the Precambrian rocks are locally cut by granite and where all are unconformably overlain by Cambrian shales. In most places the rocks display a steep, penetrative foliation that is locally intense but generally semipenetrative with associated tight upright folds. Where the Cambrian and Ordovician rocks locally overlie deformed Precambrian rocks and late Precambrian granite, they are undeformed. No crystalline basement rocks are known in the Avalon zone of Newfoundland. However, late Precambrian sediments in eastern Avalon Peninsula contain traces of detrital garnet and muscovite which are known to occur in basement rocks of the Gander zone but not in presediment volcanic terranes of the Avalon zone. It thus seems probable that the gneissic and granitic rocks of the Gander zone continue eastward in subsurface beneath these late Precambrian successions (Papezik, 1973a; 1973b).

Underwater sampling in the central region of the Grand Banks has revealed late Precambrian rocks like those of the Avalon zone at Virgin Rocks and Eastern Shoal (Lilly, 1965). More recently, shallow drilling at Flemish Cap, east of the Grand Banks (Pelletier, 1971) has revealed granodiorite that is approximately of the same isotopic age as granitic rocks that cut late Precambrian rocks in eastern Newfoundland. If the late Precambrian rocks are continuous in subsurface beneath the Grand Banks, then the Avalon zone is by far the most extensive in Newfoundland and is wider than all of the others combined.

The late Precambrian rocks of the Avalon zone can be broadly subdivided into three groups as follows: (a) a basal assemblage of predominantly subaerial volcanic rocks but including some sedimentary units; (b) an intermediate assemblage of siliceous slates and greywacke of marine deposition with local tuff interlayers, an important tillite horizon (Brückner and Anderson, 1971) and in places containing late Precambrian fossils (Misia, 1969); and (c) an upper assemblage of sedimentary rocks that in places includes a thick volcanic unit near its base and everywhere includes large amounts of arkose and red sandstone and conglomerate of typical shallow-water continental facies.

Volcanic rocks of the basal assemblage are intruded by the Holyrood Granite that in turn is overlain unconformably by Lower Cambrian strata. An isotopic age of 574 m.y. for the Holyrood Granite, and similar ages for late Precambrian rocks elsewhere in the Avalon zone (about 500–580 m.y.), suggest that the volcanism and intrusion might all have been related processes (McCartney et al., 1966; Hughes and Brückner, 1971).

The upper Precambrian beds are overlain in most places by a white quartzite unit, in turn followed by Lower Cambrian fossiliferous shales. The Cambrian shales are remarkably uniform throughout the zone and represent a fairly complete history of deposition. Conformably overlying Lower Ordovician beds, rocks are thicker and sandier and locally contain oolitic hematite beds. In the western part of the zone at Fortune Bay, Silurian–Devonian red sandstones and conglomerates unconformably overlie Cambrian and older rocks and contain gneissic cobbles derived from nearby basement rocks of the Gander zone. In this same area, Upper Devonian rocks are only tilted or openly folded, whereas Silurian–Devonian and Cambrian rocks are more intensely deformed (Williams, 1971).

The controlling feature of the late Precambrian development of the Avalon zone was either a series of volcanic islands probably constructed upon continental basement or tensional, continental-type block faulting. In either case, the basal late Precambrian volcanic assemblage probably represents vent facies flows and pyroclastic rocks, the intermediate marine assemblage partly represents offshore equivalents of these volcanic rocks, and the upper continental assemblage represents deposits in block-faulted basins between the earlier active volcanic centers. The late Precambrian Holyrood Granite is possibly a permissive high-level intrusion, most likely related to cauldron subsidence (Hughes and Brückner, 1971). The quartzite unit at the top of the Precambrian successions may represent reworked continental sediments along the shoreline of an advancing Cambrian sea. The uniform nature of the subsequent Cambrian deposits indicates that the Avalon zone was a gently subsiding stable element during this period.

Rocks typical of the Avalon zone are well represented in Nova Scotia and New Brunswick of the Canadian Appalachian Region (zone H, Fig. 2). In both provinces, gneisses, deformed marbles, and quartzites predate the late Precambrian volcanic and sedimentary rocks and possibly represent an older basement complex upon which the late Precambrian rocks accumulated.

In Nova Scotia the Avalon zone is bordered to the east by an extensive and thick succession of Cambrian to Early Ordovician greywackes and argillites (zone I, Fig. 2). These sediments have been interpreted as continental margin deposits that were derived from a source area that lay to the south-southeast, but that is now removed by Mesozoic continental drift and the opening of the present Atlantic Ocean (Schenk, 1971). The Avalon zone, possibly together with the Gander zone, may therefore have formed a separate continent during the early Paleozoic that was bounded by oceans both toward the west and east.

Late Precambrian rocks like those of the Avalon zone are well represented in the Caledonides by the Arvorian volcanic rocks of North Wales and the Longmyndian red beds and Uriconian volcanics of England and southwest Wales. In addition, some of the Caledonian late Precambrian rocks, e.g., at Charnwood Forest, contain almost identical seapen-like organisms to those represented in eastern Newfoundland (Misra, 1969). The Cambrian succession of Wales is almost identical to that of the Avalon zone, and both are characterized by the same European realm trilobite faunas.

I. Nova Scotia (Zone I)

The most easterly zone of the Appalachian System is recognized only in Nova Scotia (zone I, Fig. 2). Its boundary with adjacent zone H is a major transcurrent fault and no pre-Carboniferous linkages, either stratigraphic or sedimentological, are established across the zones.

Most of zone I is made up of a conformable succession of strata approximately 13 km thick that is known as the Meguma Group (Schenk, 1970). It consists of a lower greywacke unit overlain by a shaly unit that contains Lower Ordovician graptolites in its upper part. The Meguma Group is in turn overlain by mixed sedimentary and volcanic rocks and then Devonian sediments. Younger rocks represented in zone I postdate its Acadian deformation and granite intrusion and are Carboniferous sediments and Triassic basalts and red beds.

Provenance studies of the Meguma Group indicate that its thick lower greywacke unit was deposited by currents that flowed northward in southern parts of the zone, turning sharply toward the east in central parts of the zone, and flowing eastward in northeast parts of the zone. The sedimentary detritus was apparently derived from a source area that lay to the south-southeast and the transport pattern and sedimentology suggest deposition downslope along a continental margin and later transport by bottom contour currents. Similar patterns are being established in similar deposits accumulating at the present continental slope and rise along the Atlantic continental margin. This interpretation for the Meguma (Schenk, 1970) implies that at deposition it was related to a continental source area that lay to the south-southeast and that this source area has since been removed by continental drift. The model also implies that at deposition the Meguma was bounded to the west by an ocean that separated zones H and I.

III. GRANITIC ROCKS

Granitic rocks underlie about 25% of the Canadian Appalachian Region. Gravity, magnetic, and geological considerations (Garland, 1953; Weaver, 1967; G. E. Pajari, pers. comm., 1972) suggest that this relative proportion of exposed calc-alkalic plutonic material may be extended to a depth of at least 4 to 5 miles in the crust—i.e., 90,000 cubic miles of calc-alkalic material in 330,000 cubic miles of subaerial crust. However, 60% of the Canadian Appalachian Region lies beneath the sea and as granitic rocks have recently been cored from the Flemish Cap at the extreme edge of the continental shelf (Pelletier, 1971), it seems likely that granitic rocks may be equally abundant in the submerged portion of the Appalachians.

In Newfoundland granitic rocks occur within each of the tectono-stratigraphic zones (Fig. 6), the younger intrusions in places cutting across zone boundaries. It has long been known that the granites ranged in age from Precambrian to Devonian. However, there has been a tendency to consider the bulk of the granitic rocks as Silurian–Devonian (Caledonian/Acadian) based on isotopic age studies of the past two decades. This trend has undergone important reversals in the past few years as structural studies in Newfoundland (Kennedy and Phillips, 1971; Kennedy and McGonigal, 1972) have shown that granitic rocks of two or more ages are present in several of the stratigraphic zones (Figs. 4 and 6). Obviously the latest (Silurian–Devonian) period of igneous activity has masked earlier intrusions both in the field and in the geochronology laboratories.

A. Precambrian (Helikian and/or Older)

Precambrian granitic rocks are exposed within the gneissic terranes of the Lomond, Fleur de Lys, and Gander (A, C, G) zones of Newfoundland and are assumed to underlie the schists of the Hampden (B) zone.

The largest plutons are within the Grenville inlier of the Lomond zone which forms the core of the Great Northern Peninsula of Newfoundland. They consist chiefly of foliated, coarse-grained porphyritic granite which, in places, grades imperceptibly into the surrounding gneisses. At least one major pluton, the Cloud River Granite with a K/Ar age of 945 m.y., cuts discordantly across the gneisses. Granite gneisses form small inliers near Port au Port Peninsula of southwest Newfoundland and also form part of the rejuvenated basement complex of the Fleur de Lys zone. These Grenvillian gneissic granites are referred to the Helikian Era following current Canadian Precambrian time-stratigraphic nomenclature (Stockwell, 1964).

Precambrian granitic rocks of two distinct ages are distinguished in the Gander zone (Kennedy and McGonigal, 1972). The older is a biotite granite gneiss or migmatite in which the gneissic fraction has a gneissic banding formed by transposition of an earlier banding. It is tentatively classed as Helikian or older and it contains xenoliths of even older tectonites. This granite gneiss is cut by later, less deformed porphyritic biotite granite which has yielded a Rb/Sr whole-rock isochron of 600 m.y. (Fairbairn and Berger, 1969).

B. Precambrian (Hadrynian)

The 600-m.y.-old granite that cuts basement gneisses of the Gander zone is referred to the Hadrynian Era. Similar porphyritic granites, generally characterized by a single, strong penetrative fabric, are common throughout the Gander zone and are suspected to be present in northern Cape Breton and elsewhere in zone G (Fig. 2).

Hadrynian granitic rocks have long been known in the Avalon zone of Newfoundland where the 574 m.y. Holyrood Granite is overlain by fossiliferous Lower Cambrian strata. This pluton has been interpreted as a cauldron subsidence feature related to the thick pile of Precambrian (Hadrynian) Harbour Main volcanic rocks (Hughes and Brückner, 1971). A similar isochron age, 592 ± 20 m.y., has been obtained on granodiorite recovered from a drill core taken on Flemish Cap almost 400 miles to the east at the edge of the continental shelf (Pelletier, 1971). Along strike (zone H, Fig. 2), several plutons in southern Cape Breton Island, northern mainland Nova Scotia, and southern New Brunswick all resemble Hadrynian plutons in their geological settings. Recent Rb/Sr isochron age determinations (R. Cormier, pers. comm., 1972) appear to bear out this extrapolation.

C. Ordovician and Pre-Ordovician

Very few granitic plutons can be unequivocally identified as Ordovician. They are commonly small bodies apparently consanguineous with surrounding Ordovician volcanic sequences of the Notre Dame zone where they have been identified as calc-alkaline suites which range in composition from soda-rich granodiorite to diorite (Marten, 1971; Sayeed, 1970). These are the only granitic plutons that can be related with any assurance to the so-called Taconic orogeny.

Larger intrusions of somewhat similar but older soda-rich plutonic rocks are common in the Fleur de Lys and Notre Dame zones and as part of transported slices in the Lomond zone and its extension (zone A, Fig. 2) in Quebec. They include the Burlington Granodiorite which intrudes rocks of the multideformed pre-Lower Ordovician Fleur de Lys Supergroup and which contain the earliest fabrics of these rocks. The Twillingate pluton in the eastern part of the Notre Dame zone is also much more complexly deformed than surrounding Ordovician rocks. These plutons could, conceivably, be related in time and genesis to the pre-Ordovician quasi-ophiolite assemblages of greenschists and deformed ultrabasic rocks which occur within the Fleur de Lys Supergroup.

Other demonstrably pre-Ordovician granitic rocks occur within both the Fleur de Lys and the Gander zones. Fleur de Lys intrusions include the Dunamagon pluton, a granite–quartz monzonite body, and the La Scie Granite, a two-mica granite which is associated with a small body of earlier peralkaline granite. Both of these bodies were intruded early in the pre-Arenig deformation history of the Fleur de Lys Supergroup (Kennedy and Phillips, 1971; Church, 1969). Garnetiferous muscovite leucogranite which cuts the polydeformed Gander Lake Group in the Gander zone records the early fabrics of this group and also occurs as clasts in the overlying Middle Ordovician Davidsville Group (Kennedy and McGonigal, 1972).

D. Silurian-Devonian

The bulk of granitic rocks in the northeastern Appalachians apparently belong to its Caledonian/Acadian period of climactic tectonism. Geological and isotopic evidence combine to suggest that intrusive activity extended from Middle Silurian to earliest Carboniferous time: middle Paleozoic granitic rocks span the whole cross section of the northeastern Appalachians (Fig. 2) from zone A, which contains a few small plutons in Quebec, to zone I, where the Southern Nova Scotia Batholith alone underlies 2500 square miles.

Although some of these late granites cut across the Cambrian–Ordovician tectonic–stratigraphic zone boundaries of Newfoundland, there is, nevertheless, a rough correlation between the granitic rocks and the zones in which they occur. The marginal zones of the Central Mobile Belt—the Hampden and Fleur de Lys on the west and the Gander and Avalon on the east—are characterized by massive, coarse porphyritic Silurian–Devonian granitic rocks that strongly resemble the megacrystic granites of the Precambrian basement rocks, lacking only the tectonic fabrics in the margins of the Precambrian plutons. Some small granitic stocks of the Avalon zone locally show alkalic tendencies.

In contrast, the central Exploits and Botwood zones are characterized by composite calc-alkalic plutons which cut the Silurian Botwood Group and its equivalents. K/Ar ages on these rocks generally indicate an older basic phase of about 420 m.y. and a younger granitic phase of about 380 m.y. These plutons are distinguished by clearly defined, positive gravity anomalies, in contrast to diffuse gravity patterns of the porphyritic granites of the flanking Hampden, Fleur de Lys, Gander, and Avalon zones.

E. Petrogenesis

Pronouncements on the origin of the Newfoundland granitic rocks are premature at this stage when so little is known of their compositions, $^{87}Sr/^{86}Sr$ ratios, and ages. However, the following few generalizations are possibly warranted.

Many of the granitic rocks in zones underlain by Precambrian basement gneisses (zones A, B, C, G) are very coarse-grained, potash-rich porphyries. Most have little or no diorite and gabbro associated with them. This is true regardless of their ages which range from Helikian to Devonian. The only reported occurrences of peralkaline granitic rocks also occur in two of these zones; a pre-Ordovician pluton in zone C and several Devonian plutons in zone H. These granitic rocks are in areas of diffuse gravity patterns and are commonly difficult to distinguish from adjacent country rocks on the basis of Bouguer anomalies. The available evidence suggests the possibility of their formation by palingenesis within continental crust.

The soda granite–diorite suite is spatially associated with Ordovician and earlier ophiolite and quasi-ophiolite terranes both in situ (in zones C and D) and in transported sequences (zone A). Although age and other relationships are at present equivocal, origin by differentiation of oceanic crust and mantle in a volcanic arc setting is at least plausible in some cases.

The composite, calc-alkaline plutons are chiefly located in zones E and F, in the central part of the mobile belt, where they are reflected in clearly defined, positive Bouguer gravity anomalies. They are tentatively interpreted as the products of shallow differentiation of mantle-derived material and intrusion associated with climactic orogeny.

IV. METAMORPHISM

Two distinct episodes of regional metamorphism affect the late Precambrian to Paleozoic rocks of the Canadian Appalachian Region. These coincide with periods of regional deformation (Fig. 7). The first is represented in the Fleur de Lys and Gander zones where the rocks of the Fleur de Lys Supergroup and the Gander Lake Group suffered moderate-grade metamorphism accompanying the pre-Middle Ordovician polyphase deformation that affected these zones. Transported rocks that were probably metamorphosed at the same time are present in the Hampden and Lomond zones. Metamorphic rocks of the southern part of the Exploits and Botwood zones were also probably metamorphosed at this time.

The second metamorphic episode accompanied the Devonian Acadian orogeny and affected most of the central parts of the orogen. Metamorphic grade during this episode was generally low, except in the southern part of the

Exploits and Botwood zones where higher grade Acadian metamorphism may have occurred.

Basement rocks of the Lomond, Fleur de Lys, and Gander zones exhibit the effects of an earlier metamorphism of moderate to high grade that has been locally retrograded by the later pre-Middle Ordovician and Acadian episodes. These rocks are locally of granulite facies but are generally in the upper amphibolite facies. Thus, in a general way the older rocks of the Newfoundland Appalachian System exhibit the highest grade metamorphism.

The Fleur de Lys Supergroup and Gander Lake Group show considerable variation in metamorphic grade. The paragenesis belong to the kyanite-sillimanite type of facies series. Kyanite and staurolite occur locally and generally postdate the second deformation which formed the large recumbent isoclines in these rocks. Most of these terranes are in the upper greenschist or lower amphibolite facies. Garnet, biotite, hornblende, and albite or oligoclase are ubiquitous. The main metamorphic climax postdates the second deformation in the higher grade regions and predates the second deformation in the lower grade regions. Retrogression to chlorite grade has occurred locally prior to deposition of adjacent Lower or Middle Ordovician sediments which are locally rich in metamorphic detritus derived from these rocks. Rocks of similar grade in the southern part of the Exploits and Botwood zones are tentatively interpreted to have been metamorphosed in pre-Middle Ordovician time.

Acadian metamorphism is characterized by the development of chlorite, sericite, and epidote in rocks of the eastern part of the Lomond zone, the Hampden and Fleur de Lys zones, southwestern parts of the Exploits and Botwood zones, and in parts of the Avalon zone. Biotite is present locally, but in general the rocks have been only incipiently metamorphosed to low greenschist facies. The newly discovered prehnite–pumpellyite facies metamorphism in the volcanic and sedimentary rocks of eastern Avalon peninsula may also be related to an Acadian (Devonian) event (Papezik, 1972b; pers. comm.). In the Notre Dame and Exploits zones many of the Ordovician and Silurian rocks lack Acadian metamorphism or have been subjected only to subgreenschist facies conditions. Locally extensive andalusite–cordierite assemblages of the southern part of the Botwood zone may be the product of higher grade Acadian regional metamorphism, but little is known about these rocks. The Acadian metamorphism has had little effect upon the older Precambrian or pre-Middle Ordovician metamorphic rocks, except for local retrogression close to their contacts with overlying Acadian deformed sequences.

Metamorphism of an entirely different character has been identified in transported mafic igneous rocks of the Lomond zone. Pumpellyite and prehnite occur in gabbros and mafic volcanic rocks of higher structural slices, and metamorphism in the Bay of Islands ophiolite complex is apparently depth controlled (Williams and Malpas, 1972). These characteristics are absent in rocks of underlying slices, thus indicating that the high-pressure metamorphism predated Middle Ordovician transportation of the mafic igneous rocks into the Lomond zone. Similar metamorphic assemblages have not yet been recognized in the probable source terrane of the Notre Dame zone.

Deformed gabbros, granites, and amphibolites and greenschists of the most westerly transported slices at Bay of Islands (Little Port Complex) show upper greenschist to amphibolite facies metamorphism, which predates their emplacement, and predates pumpellyite-prehnite formation in nearby mafic volcanic rocks. Similar rocks occur in the Notre Dame zone (Twillingate Granite and its associated amphibolites) whose metamorphic features predate Lower Ordovician volcanism in this zone. This metamorphism may be of the same age as the early episode of the Fleur de Lys and Gander zones or it may be older.

Greenschists and amphibolites occur at the base of transported ophiolite complexes and form an integral part of the same transported slices at both Hare Bay and Bay of Islands of the Lomond zone (Smyth, 1971; Williams, 1971). The metamorphic rocks are concordantly disposed beneath the basal ultramafic component of the ophiolite complexes, and they show inverted metamorphic zonation with the highest grade rocks at the ultramafic contact and lower grade varieties below and farther away. Garnetiferous amphibolites at the contacts grade outward into black amphibolites, polydeformed amphibolites and greenschists, and then greenschists and pelitic schists. The ultramafic rocks are mylonitized at the contact and schistose nearby above the contact, suggesting that it is tectonic. However, the inverted zonation in the metamorphic rocks is clearly related to their position beneath the ultramafics, thus implying that it is the result of emplacement and early transport of the ophiolites. Some of the metamorphic rocks resemble those of the Fleur de Lys zone in both metamorphic grade and structural style, but problems remain in the comparative timing of the metamorphic episodes.

Contact metamorphism around granitic intrusions of the Newfoundland Appalachian belt varies widely. Late Precambrian to early Paleozoic granites of the Fleur de Lys and Gander zones do not generally show contact metamorphic aureoles. Aureoles may have been developed at the time of emplacement, but they have been largely destroyed by subsequent pre-Middle Ordovician deformation and upper greenschist to low amphibolite facies regional metamorphism. In contrast, Acadian batholiths display narrow but well-developed cordierite–andalusite aureoles which either postdate or are broadly syntectonic with Acadian deformation.

Pre-Middle Ordovician metamorphism in Quebec and northern New Brunswick is generally of lower grade than that of the Fleur de Lys zone of Newfoundland. Barrowvian metamorphism of the Moine/Dalradian sequence of the Caledonides is closely comparable to that of the Fleur de Lys zone, but the Dalradian locally contains high-temperature and low-pressure facies series. Possible equivalents of the Gander Lake Group show similar metamorphic grade in Cape Breton Island and eastern Massachusetts, but the Mona complex of Anglesea is unique in containing glaucophane and pumpellyite assemblages.

Acadian regional metamorphism is of similar low grade throughout the Canadian Appalachian region and the British Caledonides. Higher grade Acadian metamorphism is present in New England and possibly in the north-western part of the Meguma Group of Nova Scotia (zone I) and in northern New Brunswick.

V. STRUCTURAL DEVELOPMENT OF THE NORTHEASTERNMOST APPALACHIANS

The northeasternmost Appalachians were characterized in post-Helikian time by a distinctive two-stage orogenic history. The earliest stage was developed on both sides of the orogen, probably in latest Precambrian and in Cambrian time. Its culmination involved intense deformation in the Late Cambrian or Early Ordovician. However, synchroneity of this tectonic event has not yet been established across the width of the orogen. Its importance has been recognized only recently, and it has commonly been ascribed previously to an Ordovician (Taconic) event or to Devonian orogeny. The second event, which took place in Silurian/Devonian time, affected the entire orogen to various degrees (Fig. 7), and is generally referred to as the climactic Caledonian/Acadian orogeny.

Epi-orogenic molasse sediments accumulated in fault-bounded basins in Carboniferous to Permian time. Early Mesozoic postorogenic rifting associated with basalts and redbeds was probably a harbinger of the present stage of ocean-floor spreading and continental separation.

A. Late Precambrian/Cambrian Orogenic Cycle

Effects of the first post-Grenvillian (post-Helikian) orogenic cycle are manifest in the Fleur de Lys zone and possibly the Hampden zone in the west and in the Gander zone to the east.

In the far west, the instigation of the cycle involved subsidence, intrusion of mafic dikes, and extrusion of plateau basalts (Williams and Stevens, 1969) in the Lomond zone. The dikes have been dated at 805 m.y. (Pringle *et al.*, 1971), and the dikes and flows have been related to the inception of rifting and a spreading proto-Atlantic ocean floor (Strong and Williams, 1972). In the platformal Lomond zone deposition of a carbonate shelf sequence continued

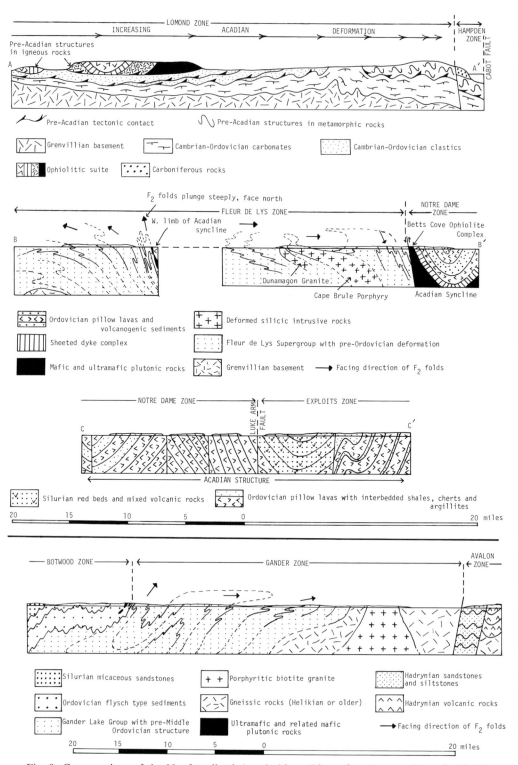

Fig. 8. Cross sections of the Newfoundland Appalachians. Lines of section are shown in Fig. 7.

uninterruptedly until Middle Ordovician time. Eastward, beyond the bank edge deposits of the Hampden zone, a thick wedge of clastic material accumulated, possibly as a continental rise deposit (Dewey, 1969). This is the Fleur de Lys Supergroup of the Fleur de Lys zone. Mafic dikes are plentiful in the rocks of this supergroup and mafic flow rocks, including pillowed varieties, are common high in the stratigraphic sequence (Kennedy, 1971). Whether or not these dikes and flows are also related to rifting and are roughly contemporaneous with the plateau basalts of the Lomond zone as first suggested by Williams and Stevens (1969) is a contentious point. The uppermost part of the Fleur de Lys sequence includes silicic volcanic rocks, small alkalic intrusions, and contemporaneous or later limestones and greywackes. In pre-Arenig, possibly Cambrian time, the rocks of the Fleur de Lys Supergroup were subjected to intense polyphase deformation; to granitic, mafic, and ultramafic plutonism; and to amphibolite facies regional metamorphism of the kyanite–sillimanite facies series. The Fleur de Lys sediments and volcanic rocks were deformed into a nappe complex over a core of Grenvillian basement rocks and then eroded to contribute detritus to Lower Ordovician flysch of the Notre Dame zone and to Middle Ordovician flysch of the Lomond zone (Fig. 8). This intense deformative event within the Fleur de Lys zone definitely preceded but may have been the first in a chain of events leading to emplacement of the transported masses in the Lomond zone in Middle Ordovician time.

The southeastward flank of the Appalachians, the Gander zone, had a somewhat similar structural history to the Fleur de Lys. Accumulation of a vast clastic sequence with a few basic volcanic horizons was followed by pre-Middle Ordovician intense polyphase deformation, granitic and basic plutonism, and amphibolite facies metamorphism which culminated in eastward-facing nappes draped over the gneissic Helikian(?) basement (Fig. 8). The similarity in sedimentational and structural history to that of the Fleur de Lys Supergroup suggests a roughly contemporaneous continental rise prism that formed on the opposing (eastward) side of a Precambrian–Cambrian ocean basin. The Avalon zone and Zone I (Meguma) to the east of the Gander zone show a virtually uninterrupted history of sedimentation from latest Precambrian through to Ordovician times.

B. Ordovician–Devonian Orogenic Cycle

The second orogenic cycle involved the continuing formation of oceanic ophiolite complexes, the building of Ordovician oceanic arc sequences, the emplacement of these oceanic assemblages into or onto continental rise and continental platform terranes, and the conversion, in Silurian–Devonian time, of an oceanic basin to a continental milieu. It terminated with Silurian–Devonian deformation and intrusion which affected the entire orogen but which was least intense on the bordering platformal zones and most intense on the inner, so-called mobile zones (Fig. 8).

There are strong lithological links with the earlier tectonic cycle, which includes highly deformed quasi-ophiolite sequences and quasi-oceanic arc sequences, which are seemingly within and part of thick clastic sequences. These links remain problematical and it must be stressed that they are not at present fully understood.

The ophiolite sequences include the Betts Cove Ophiolites (Upadhyay et al., 1971) of the Notre Dame zone which grade upward through oceanic layers 3, 2, and 1 into a thick sequence of alternating submarine lavas and volcanic clastic sediments that bear Arenig graptolites. Remnants of Early Ordovician oceanic layers occur elsewhere in the Notre Dame zone (Strong, 1972), possibly in the Exploits zone (Upadhyay and Smitheringale, 1972), and possibly in the Botwood zone. They appear to be overlain by volcanic arc assemblages (e.g., Marten, 1971), in a few places associated with limestone fringing reef deposits, some of which contain mixed shelly faunas (Neuman, 1971). Middle Ordovician deposits of the Notre Dame, Exploits, and Botwood zones consist chiefly of thick successions of chert, argillite, and greywacke. In almost all cases these Ordovician sequences are faulted against the polydeformed rocks of the Fleur de Lys and Gander zones. One possible exception is the Baie Verte belt of ultrabasic rocks and pillow lavas of the Fleur de Lys zone where the relationship is variously interpreted as a fault (Neale and Nash, 1963) or as an unconformity (Kennedy and Philips, 1971). Another exception is the Twillingate gneissic granite of zone D which obviously predates surrounding Ordovician ophiolites and appears to represent a crustal remnant incorporated within them during rifting.

Mélange zones are an integral part of several of the Cambro-Ordovician tectonic–stratigraphic zones. They occur associated with ophiolites near their contact with polydeformed rocks in the Fleur de Lys zone, at the faulted junction of the Exploits and Botwood zones, at the junction of the Botwood and Gander zones, and in the transported slices of the Lomond allochthons. The second of these, known as the Dunnage Mélange, has been interpreted as the remnants of a subduction zone (Bird and Dewey, 1970); the third has been interpreted as an accumulation related to a submarine fault scarp (Kennedy and McGonigal, 1972); and the last ones are associated with klippen transport and interpreted as at least in part tectonic (Stevens, 1970). All these chaotic deposits are very similar, and a common origin is suspected for some or all of them.

Emplacement of the transported masses in the Lomond zone and in the Hampden zone in Middle Ordovician time was not associated with penetrative deformation in the source areas. It has been suggested that emplacement was related to subduction at the eastern edge of the Exploits zone (Bird and Dewey, 1970), but, if so, it was not an orogenic event in the generally accepted use of this term—and there are no widespread and significant angular unconformities to mark this "Taconic orogeny" in the northeasternmost part of the Appalachians.

Silurian shallow-water marine to fluviatile arenaceous sediments succeeded the Middle and Upper Ordovician flysch sediments of the Notre Dame, Exploits, and Botwood zones and extended westward into the Fleur de Lys and Hampden zones. In the Lomond zone similar conditions prevailed from Middle Ordovician to Early Devonian time and in the southern part of the Avalon zone Lower Devonian shallow marine deposits transgressed onto Late Precambrian volcanic rocks. In the Exploits and Botwood zones sedimentation was apparently fairly continuous from Ordovician through Silurian time despite the distinct changes in sedimentary environment, although there are significant time gaps in other zones. Structural concordance is the rule rather than the exception, and there are few "glaring" angular unconformities on which to hang a Taconic hat.

The sole penetrative deformation to affect rocks of the Notre Dame, Exploits, and Botwood zones was the Caledonian/Acadian orogeny of Silurian–Devonian time. Its effects are recognizable, across the breadth of the orogen, weakest on the margins in the Lomond and Avalon zones, and of varying intensity in alternating northeast–southwest belts in central parts of the system. The Caledonian/Acadian folds are generally upright, tight structures. Metamorphism was low grade. Granites in the zones underlain by sialic basement gneisses are coarse, potash-rich megacrystic varieties of probable palingenetic origin, whereas those in the central zones (Notre Dame, Exploits, Botwood) are calc-alkaline composite intrusions. Paucity of fossils has hampered dating of the Caledonian/Acadian orogenic event(s). Isotopic ages of dubious value on granitic rocks range from Silurian to Middle Devonian. Fossiliferous Middle Silurian beds of the Botwood zone are folded and intruded in northern Newfoundland, and Middle Devonian beds are similarly affected in the southwest at La Poile Bay. Uppermost Devonian beds in the southwestern part of the Avalon zone are postorogenic. It is not improbable that the folding and intrusion which terminated the second orogenic cycle spanned this time interval.

C. Postorogenic Events

A Carboniferous molasse sequence occurs chiefly within a linear, graben-like basin that angles across the junction of the Lomond and Hampden zones in western Newfoundland. A small Carboniferous basin occurs within the Exploits zone at Red Indian Lake. Postorogenic, uppermost Devonian rocks of the same facies occur in the southwestern Avalon zone. These rocks are locally tightly folded, particularly where close to major faults, such as the

Cabot fault, which was reactivated in Carboniferous or post-Carboniferous time.

Triassic rocks are not known in Newfoundland but occur at the junction of zones H and I in Nova Scotia. Dated as Upper Triassic, this gently dipping sequence occupies a fault-bounded trough and consists of red beds separated by basalt flows. These deposits are probably related to rifting and onset of the current episode of Atlantic spreading.

Following the initial stages of breakup, subsidence of the Appalachian continental margin allowed accumulation of a thick Jurassic–Tertiary succession which today underlies the Grand Banks of Newfoundland and the Scotian shelf. These are discussed in another paper in this volume.

D. Extrapolations Along Strike

The two post-Grenvillian orogenic cycles can be recognized southwestward in the Appalachians (Fig. 2) and northeastward in the Caledonides. The polyphase deformed rocks of the older cycle provide the clearest basis for comparison. They form an apparently continuous belt or belts. Near the northwest flank of the orogen the Moine/Dalradian sequence of the Caledonides can be traced through the Fleur de Lys Supergroup of Newfoundland to the Macquereau and Arnold River groups of Quebec and, possibly, the Tetagouche Group of New Brunswick. The same belt of rocks appears to extend southward into the Green Mountain and Boundary Mountain anticlinoria of the northeastern United States. Polydeformed, pre-Ordovician schists of the Schickshock, Rosaire, and Caldwell groups of Quebec represent transported equivalents of this terrane similar to those in the Lomond and Hampden zones of Newfoundland. Large allochthonous masses (Taconic klippen) are recognized in New York (Zen *et al.*, 1968) and are probably present throughout Quebec on an even greater scale than in the Lomond zone of Newfoundland (C. Hubert, pers. comm., 1972). Similar allochthonous masses are not known in the British Caledonides where they might have been removed by erosion or may be covered by the surrounding seas. No exact equivalent of the early Moine thrust of the British Caledonides has been recognized west of the Great Glen/Cabot fault in the Newfoundland Appalachians.

Towards the southeast flank of the orogen, the pre-Middle Ordovician metamorphic rocks of the Gander zone can be traced southwestwards through northern Cape Breton Island and into southern New Brunswick (Fig. 2) and eastern Massachusetts. In most of these places, as in Newfoundland, the polyphase deformed cover rocks have not been distinguished from the Helikian(?) basement rocks. Northeastward, in the Caledonides, the Gander zone rocks are apparently recognizable in the Irish Sea horst where they are named the Mona complex in Anglesea and the Rosslare Migmatites in southwest Ireland. However, there are puzzling differences—for example, the Mona complex is stated to be overlain by late Precambrian (Arvonian) volcanic rocks and by Lower Ordovician sediments (Greenly, 1919; Shackleton, 1969). If so, it must be older than the age of deformation assigned to the Gander rocks or else we have misjudged this age of deformation and Gander zone deformation is actually Precambrian. A third possibility is that the Mona complex is really a composite grouping which includes both Upper Cambrian polydeformed rocks and a hitherto unrecognized, low-grade Precambrian basement which is partly overlain by the Arvonian volcanics. The platformal nature of the Avalon zone during Cambrian and Early Ordovician time is recognized in its equivalents to the southwest in southern Cape Breton Island, southern New Brunswick, and eastern Massachusetts, and to the northeast in both northeastern and northwestern Wales. Farther southeast in the orogen, the Meguma Group of zone I in Nova Scotia is probably comparable with the thick Paleozoic sequences of the Welsh basin. If so, the Meguma rocks possibly rest on Precambrian sedimentary and volcanic rocks similar to those of the Avalon zone.

The details of the younger (Ordovician–Devonian) orogenic cycle are not as closely comparable along strike in the Appalachian/Caledonian orogen. Lateral facies change and transgression are common, and it would be surprising if lithologies and events were synchronous and easily correlated. Clearly identifiable ophiolite sequences, although best developed in the Newfoundland Appalachians, are also recognizable in equivalent zones in New Brunswick, Quebec, the northeastern United States, and the Southern Uplands of Scotland. A fairly abrupt transition from Ordovician oceanic lavas and sediments to Silurian shallow-marine and terrestrial facies is common throughout the northern Appalachians and the British Caledonides. Despite scattered and sometimes equivocal evidence of angular unconformities between Ordovician and later rocks (Pavlides *et al.*, 1968), no good case has been made for widespread penetrative deformation in Ordovician time and the second climactic event was Caledonian/Acadian folding and granitic intrusion in roughly post-Middle Silurian, pre-uppermost Devonian time. In many places, e.g., zone I in Nova Scotia and in parts of the northeastern United States, this Silurian-Devonian orogenic event was much more intense than in the Newfoundland Appalachians.

VI. TECTONIC MODELS FOR APPALACHIAN DEVELOPMENT

Several plate tectonic models for the development of the Newfoundland and northern Appalachians (Dewey, 1969; Stevens, 1970; Bird and Dewey, 1970; Church and Stevens, 1971; Dewey and Bird, 1971; and Schenk, 1971) have been proposed following the original suggestion of a proto-Atlantic Ocean by Wilson (1966) and the advent of plate tectonic theory (McKenzie and Parker, 1967). These models are by far the most realistic and viable to date, and all are commendable, for they provide a mechanism for mountain building as well as endeavoring to explain the intricate geological relationships across the system. Most models rely heavily upon the relationships in western and central Newfoundland and emphasize a continental margin in the west (Lomond, Hampden, Fleur de Lys zones) and its development and interplay with an oceanic domain (Notre Dame and Exploits zones) in central parts of the system. No attempt has been made to synthesize the geology of the eastern parts of the system in the same way (Botwood, Gander, Avalon zones), except for the recent interpretation of the Meguma Group (zone I) in Nova Scotia (Schenk, 1971). This, of course, is readily understandable, for the geology of the important Gander zone has only recently been studied in any detail and its present synthesis (Kennedy and McGonigal, 1972) completely revises past interpretations. Geologic knowledge is advancing at such a rate in the northern Appalachians that it seems futile to the authors to attempt a new and sophisticated model at this point in time. Geological restraints are few, thus allowing many possible interpretations that would at most be short lived. We therefore wish to conclude by pointing out some of the changes and variations in recent plate models for the northern Appalachians and to draw attention to some remaining fundamental problems that must be incorporated in future syntheses.

The time of opening of the proto-Atlantic Ocean has been interpreted as late Precambrian–Early Cambrian (Strong and Williams, 1972), but recent isotopic ages on dikes and volcanic rocks presumably related to this event are in the order of 800 m.y. (Pringle *et al.*, 1971; Rankin *et al.*, 1969), thus implying an embarrassingly long history for this event.

The interpretation and significance of mafic and ultramafic plutonic rocks of the Lomond, Hampden, Fleur de Lys, and Notre Dame zones have changed radically since they were first depicted as essentially autochthonous intrusions. Most are now interpreted as ophiolite sequences generated by sea-floor spreading (Church and Stevens, 1971; Dewey and Bird, 1971; Upadhyay *et al.*, 1971; and Williams and Malpas, 1972). In some cases the pendulum may already have swung too far, because many of the mafic and ultramafic rocks recently included with the transported ophiolite suite in western Newfoundland (Church and Stevens, 1971; Dewey and Bird, 1971) are actually much older than previously realized and include deformed gabbros, amphibolites, and granites that predate the true ophiolite sequences. Similar rocks have recently been recognized in the Notre Dame zone where their age and significance will bear heavily upon interpretations of the central parts of the system and will require modifications of existing models. If the Notre Dame zone ophiolites are no older than Early Ordovician, then the fact that they contain rafts of older material would make it appear unlikely that a wide ocean like the present Atlantic ever existed between the Fleur de Lys and Gander zones. This reasoning in turn refutes the most attractive aspect of a wide proto-Atlantic Ocean, i.e., its closing and subduction were responsible for the intense deformation along its similar margins in the Fleur de Lys and Gander zones.

A Lower Ordovician linkage between the Fleur de Lys and Notre Dame zones and a Middle Ordovician linkage between the Gander and Botwood zones combine with the similarity and virtual continuity of Middle Ordovician rocks across the Exploits and Botwood zones to suggest that the Fleur de Lys and Gander zones were essentially linked at this time. Subsequent Middle to Late Ordovician and Silurian history represents infilling of this trough. Subduction or continental collision that might relate to Acadian orogeny is therefore not represented by a suture west of the Gander zone. Possibly the Gander–Avalon zone boundary represents such a suture or possibly it is located east of the Avalon zone, e.g., the zone H–I boundary in Nova Scotia.

The Fleur de Lys and Gander zones are sufficiently similar to suggest that the Fleur de Lys Supergroup and the Gander Lake Group are equivalents. If the Gander Lake Group is of late Precambrian–Cambrian age, it may represent a westerly offshore facies of the Cambrian of the Avalon zone in the same way that the Fleur de Lys Supergroup is interpreted as an easterly offshore facies of the Cambrian in western Newfoundland.

Lithologically similar mélange deposits occur in a variety of different geologic settings across the Newfoundland Appalachians. A few of these have been interpreted in significantly different ways and without meaningful comparative studies. Their distinctive character suggests that they conceal an important adjunct to the geological interpretation of central Newfoundland.

The interpretation of the Meguma Group (zone I, Nova Scotia) implies an ocean basin between it and zone H. The Avalon zone, possibly linked with the Gander zone, may therefore represent a continental block that intervened between North America and Africa in the remote geological past.

REFERENCES

Bird, J. M. and Dewey, J. F., 1970, Lithosphere plate–continental margin tectonics and the evolution of the Appalachian orogen: *Geol. Soc. Amer. Bull.*, v. 81, p. 1031–1060.

Brown, P. A., 1973, Possible cryptic suture in southwest Newfoundland: *Nature Physical Science*, v. 245. No. 140, p. 9–10.

Brückner, W. D. and Anderson, M. M., 1971, Late Precambrian glacial deposits in southeastern Newfoundland–A preliminary note: *Geol. Assoc. Can. Proc.*, A Nfld. Decade, v. 24, n. 1, p. 95–102.

Bullard, E., Everett, J. E., and Smith, A. G., 1965, The fit of the continents around the Atlantic: *Phil. Trans. Roy. Soc.*, v. 258, p. 41–51.

Church, W. R., 1969, Metamorphic rocks of the Burlington peninsula and adjoining areas of Newfoundland, and their bearing on continental drift in the North Atlantic, in: *North Atlantic—geology and continental drift*, M. Kay, ed.: Am. Assoc. Petrol. Geol. Mem. 12, p. 212–233.

Church, W. R. and Stevens, R. K., 1971, Early Paleozoic ophiolite complexes of the Newfoundland Appalachians as mantle-oceanic crust sequences: *J. Geophys. Res.*, v. 76, p. 1460–1466.

Dewey, J. F., 1969, Evolution of the Appalachian/Caledonian orogen: *Nature*, v. 222, p. 124–129.

Dewey, J. F. and Bird, J. M., 1971, Origin and emplacement of the ophiolite suite—Appalachian ophiolites of Newfoundland: *J. Geophys. Res.*, v. 76, p. 3170–3206.

Fairbairn, H. W. and Berger, A. R., 1969, Preliminary geochronological studies in northeast Newfoundland: 17th Ann. Progr. Rept. Mass. Inst. Tech. A.E.C. 1381-17, p. 19–20.

Garland, G. D., 1953, Gravity measurements in the Maritime provinces: *Publ. Dom. Obser.*, v. 16, n. 7, p. 185–225.

Greenly, E., 1919, *The geology of anglesea*: Mem. Geol. Surv. Great Britain.

Helwig, James and Sarpi, Ernesto, 1969, Plutonic pebble conglomerates, New World Island, Newfoundland, and History of Eugeosynclines, in: *North Atlantic—geology and continental drift*, M. Kay, ed.: Am. Assoc. Petrol. Geol. Mem. 12, p. 443–466.

Horne, G. S., 1969, Early Ordovician chaotic deposits in the central volcanic belt of northeast Newfoundland: *Geol. Soc. Amer. Bull.*, v. 80, p. 2451–2464.

Horne, G. S. and Helwig, J., 1969, Ordovician stratigraphy of Notre Dame Bay, Newfoundland, in: *North Atlantic—geology and continental drift*, M. Kay, ed.: Am. Assoc. Petrol. Geol. Mem. 12, p. 388–407.

Hughes, C. J. and Brückner, W. D., 1971, Late Precambrian rocks of eastern Avalon peninsula, Newfoundland—a volcanic island complex: *Can. J. Earth Sci.*, v. 8, p. 899–915.

Jenness, S. E., 1959, Geology of the Lower Gander River ultrabasic belt: Geol. Surv. Newfoundland, Rept. No. 14.

Kay, Marshal, ed., 1969, *North Atlantic—geology and continental drift*: Am. Assoc. Petrol. Geol. Mem. 12, 1082 p.

Kay, Marshall and Eldredge, N., 1968, Cambrian trilobites in central Newfoundland volcanic belt: *Geol. Mag.*, v. 105, p. 272–277.

Kennedy, M. J., 1971, Structure and stratigraphy of the Fleur de Lys supergroup in the Fleur de Lys area, Burlington peninsula, Newfoundland: *Geol. Assoc. Can. Proc.*, A Nfld. Decade, v. 24, n. 1, p. 59–71.

Kennedy, M. J. and McGonigal, M., 1972, The Gander Lake and Davidsville groups of northeastern Newfoundland: new data and geotectonic implications: *Can. J. Earth Sci.*, v. 9, p. 452–459.

Kennedy, M. J. and Phillips, W. E. A., 1971, Ultramafic rocks of Burlington peninsula, Newfoundland: *Geol. Assoc. Can. Proc.*, A. Nfld. Decade, v. 24, n. 1, p. 35–46.

Kennedy, M. J., Phillips, W. E. A., and Neale, E. R. W., 1972, Similarities in the early structural development of the northwestern margin in the Newfoundland Appalachians and the Irish Caledonides: Reports of 24th Int. Geol. Con., Section 3, p. 516–531.

Kindle, C. H. and Whittington, H. B., 1958, Stratigraphy of the Cow Head region, western Newfoundland: *Geol. Soc. Amer. Bull.*, v. 69, p. 315–342.

King, P. B., 1959, *The evolution of North America*: Princeton University Press, Princeton, N. J.

Lilly, H. D., 1965, Submarine exploration of the Virgin Rocks area, Grand Banks, Newfoundland: preliminary note: *Bull. Geol. Soc. Amer.*, v. 76, p. 131–132.

Lock, B. E., 1969, Silurian rocks of west White Bay area, Newfoundland, in: *North Atlantic—*

geology and continental drift, M. Kay, ed.: Am. Assoc. Petrol. Geol., Mem. 12, p. 433–442.

Marten, B. E., 1971, Stratigraphy of volcanic rocks in the Western Arm area of the central Newfoundland Appalachians: *Geol. Assoc. Can. Proc.*, A Nfld. Decade, v. 24, n. 1, p. 73–84.

McCartney, W. D., Poole, W. H., Wanless, R. K., Williams, H., and Loveridge, W. D., 1966, Rb/Sr age and geological setting of the Holyrood Granite: southeastern Newfoundland: *Can. J. Earth Sci.*, v. 3, p. 947–957.

McKenzie, D. P. and Parker, D. L., 1967, The North Pacific: an example of tectonics on a sphere: *Nature*, v. 216, p. 1276.

McKerrow, W. S. and Ziegler, A. M., 1971, The Lower Silurian paleogeography of New Brunswick and adjacent areas: *Jour. Geology*, v. 79, p. 635–646.

Misra, S. B., 1969, Late Precambrian(?) fossils from southeastern Newfoundland: *Geol. Soc. Amer. Bull.*, v. 80, p. 2133–2140.

Mitchell, A. H. and Reading, H. G., 1971, Evolution of island arcs: *Jour. Geology*, v. 79, p. 253–284.

Neale, E. R. W. and Kennedy, M. J., 1967, Relationship of the Fleur de Lys Group to younger groups of the Burlington peninsula, Newfoundland, in: *Geology of the Atlantic region*, Neale, E. R. W. and Williams, Harold, eds.: Geol. Assoc. Can., Spec. Paper No. 4, p. 139–169.

Neale, E. R. W. and Nash, W. A., 1963, Sandy Lake (East Half), Newfoundland: Geol. Surv. Canada, Paper 62-28.

Neale, E. R. W. and Williams, Harold, eds., 1967, Geology of the Atlantic region—the Lilly memorial volume: Geol. Assoc. Can., Spec. Paper No. 4, 292 p.

Neuman, R. B., 1971, An early Middle Ordovician brachiopod assemblage from Maine, New Brunswick, and northern Newfoundland: *Smithsonian Centr. Paleobiology*, n. 3, p. 113–124.

Papezik, V. S., 1970, Petrochemistry of volcanic rocks of the Harbour Main Group, Avalon peninsula, Newfoundland: *Can. J. Earth Sci.*, v. 7, p. 1485–1498.

Papezik, V. S., 1972a, Late Precambrian ignimbrites in eastern Newfoundland and their tectonic significance: Proc. 24th Int. Geol. Congress, Section 1, p. 147–152.

Papezik, V. S., 1972b, Burial metamorphism of Late Precambrian sediments near St. John's, Newfoundland: *Can. J. Earth Sci.*, v. 9, p. 1568–1572.

Papezik, V. S., 1973a, The Late Precambrian Harbour Main Group of eastern Newfoundland—an ensialic volcanic belt (abst.): Geol. Soc. Am., Northeastern Section, 8th Ann. Mtd., Abstracts, p. 205.

Papezik, V. S., 1973b, Detrital garnet and muscovite in Late Precambrian sandstone near St. John's, Newfoundland, and their significance: *Can. J. Earth Sci.*, v. 10, p. 430–432.

Pavlides, L., Boucot, A. J., and Skidmore, W. B., 1968, Stratigraphic evidence for the Taconic orogeny in the northern Appalachians, in: Studies of Appalachian geology: northern and maritime, Zen, E.-An, White, W. S., Hadley, J. B., and Thompson, J. B., Jr., eds. Wiley Interscience, New York, p. 61–82.

Pelletier, B. R., 1971, A granodiorite drill core from the Flemish Cap, eastern Canadian continental shelf: *Can. J. Earth Sci.*, v. 8, p. 1499–1503.

Phillips, W. E. A., Kennedy, M. J., and Dunlop, G. A., 1969, A geologic comparison of western Ireland and northeastern Newfoundland, in: *North Atlantic—geology and continental drift*, Kay, M., ed.: Am. Assoc. Petrol. Geol., Mem. 12, p. 194–211.

Poole, W. H., Sanford, B. V., Williams, Harold, and Kelley, D. G., 1970, Geology of southeastern Canada, in: *Geology and economic minerals of Canada*, Douglas, R. J. W., ed.: Geol. Surv. Can. Econ. Geol. Rept. No. 1, p. 227–304.

Pringle, I. R., Miller, J. A., and Warrell, D. M., 1971, Radiometric age determinations from the Long Range Mountains, Newfoundland: Can. J. Earth Sci., v. 8, p. 1325–1330.

Rankin, D. W., Stern, T. W., Reid, J. C., and Newell, M. F., 1969, Zircon ages of felsic volcanic rocks in the upper Precambrian of the Blue Ridge, Appalachian Mountains: Science, v. 166, p. 741–744.

Rodgers, J., 1965, Long Point and Clam Bank Formations, western Newfoundland: Geol. Assoc. Can. Proc., v. 16, p. 83–94.

Rodgers, J., 1968, The eastern edge of the North American continent, during the Cambrian and Early Ordovician, in: Studies of Appalachian geology: northern and maritime, Zen, E.-An., White, W. S., Hadley, J. B., and Thompson, J. B., Jr., eds.: Wiley Interscience, New York, p. 141–149.

Rodgers, J., 1970, The tectonics of the appalachians: Wiley Interscience, New York, 271 p.

Rodgers, J. and Neale, E. R. W., 1963, Possible "Taconic" klippen in western Newfoundland: Am. J. Sci., v. 261, p. 713–730.

Sayeed, U. A., 1970, The tectonic setting of the Colchester plutons, Southwest Arm, Green Bay, Newfoundland: unpublished M.Sc. thesis, Memorial University of Newfoundland.

Schenk, P. E., 1970, Regional variation of the flysch-like Meguma Group (Lower Paleozoic) of Nova Scotia, compared to recent sedimentation of the Scotian Shelf; in: Flysch sedimentology in North America, Lajoie, J., ed.: Geol. Assoc. Canada, Spec. Paper No. 7, p. 127–153.

Schenk, P. E., 1971, Southeastern Atlantic Canada, northwestern Africa, and continental drift: Can. J. Earth Sci., v. 8, p. 1218–1251.

Shackelton, R. M., 1969, The Precambrian of North Wales, in: The Precambrian and Lower Palaeozoic rocks of Wales, Wood, Alan, ed.: University of Wales Press, Cardiff, p. 1–22.

Smyth, W. R., 1971, Stratigraphy and structure of part of the Hare Bay allochthon, Newfoundland: Geol. Assoc. Can. Proc., A Nfld. Decade, v. 24, n. 1, p. 47–57.

Stevens, R. K., 1970, Cambro-Ordovician flysch sedimentation and tectonics in west Newfoundland and their possible bearing on a proto-Atlantic Ocean, in: Flysch sedimentology in North America, Lajoie, J., ed.: Geol. Assoc. Can., Spec. Paper No. 7, p. 165–177.

Stockwell, C. H., 1964, Fourth report on structural provinces, orogenies and time classification of rocks of the Canadian Precambrian shield: Geol. Surv. Canada Paper 67-17, pt. 2, p. 1–21.

Strong, D. F., 1972, Sheeted diabases of central Newfoundland: new evidence for Ordovician sea-floor spreading: Nature, v. 235, p. 102–104.

Strong, D. F., and Williams, Harold, 1972, Early Paleozoic flood basalts of northwest Newfoundland: their petrology and tectonic significance: Geol. Assoc. Can. Proc., v. 24, n. 2, p. 43–54.

Upadhyay, H. D., Dewey, J. F., and Neale, E. R. W., 1971, The Betts Cove Ophiolite Complex, Newfoundland: Appalachian oceanic crust and mantle: Geol. Assoc. Can. Proc., A Nfld. Decade, v. 24, n. 1, p. 27–34.

Upadhyay, H. D. and Smitheringale, W. G., 1972, The geology of the Gullbridge copper deposit, Newfoundland: Can. J. Earth Sci., v. 9, p. 1061–1073.

Weaver, D. F., 1967, A geological interpretation of the Bouguer anomaly field of Newfoundland: Domin. Obser., v. 35, n. 5, p. 233–251.

Williams, Harold, 1967, Silurian rocks of Newfoundland, in: Geology of the Atlantic Region, Neale, E. R. W. and Williams, Harold, eds.: Geol. Assoc. Can., Spec. Paper No. 4, p. 93–137.

Williams, Harold, 1969, Pre-Carboniferous development of Newfoundland Appalachians in: *North Atlantic—geology and continental drift*, Kay, M., ed.: Am. Assoc. Petrol. Geol. Mem. 12, p. 32–58.

Williams, Harold, 1971, Mafic–ultramafic complexes in western Newfoundland and the evidence for their transportation: *Geol. Assoc. Can. Proc.*, A Nfld. Decade, v. 24, n. 1, p. 9–25.

Williams, Harold, 1971, Belleoram map area, Newfoundland: Geol. Surv. Can., Paper 70–65, 39 p.

Williams, Harold, Kennedy, M. J., and Neale, E. R. W., 1970, The Hermitage flexure, the Cabot fault, and the disappearance of the Newfoundland central mobile belt: *Geol. Soc. Amer. Bull.*, v. 81, p. 1563–1568.

Williams, Harold, Kennedy, M. J., and Neale, E. R. W., 1972, The Appalachian structural province, in: *Tectonic styles in Canada*, Price, R. A. and Douglas, R. J. W., eds.: Geol. Assoc. Can., Spec. Paper No. 11, 1973.

Williams, Harold and Malpas, J. G., 1972, Sheeted dykes and brecciated dyke rocks within transported igneous complexes, Bay of Islands, western Newfoundland: *Can. J. Earth Sci.*, v. 9, p. 1216–1229.

Williams, Harold, Malpas, J. G., and Comeau, R. L., 1972, Bay of Islands map area, Newfoundland (12G): Geol. Surv. Canada, Paper 72-1, part A, p. 14–17.

Williams, Harold and Stevens, R. K., 1969, Geology of Belle Isle—Northern extremity of the deformed Appalachian miogeosynclinal belt: *Can. J. Earth Sci.*, v. 6, p. 1145–1157.

Wilson, J. T., 1966, Did the Atlantic close and then re-open?: *Nature*, v. 211, p. 676–681.

Zen, E.-An., White, W. S., Hadley, J. B., and Thompson, J. B., Jr., eds., 1968, *Studies of Appalachian geology: northern and maritime*: Wiley Interscience, New York.

Chapter 5

AN OUTLINE OF THE GEOLOGY OF THE ATLANTIC COAST OF GREENLAND

T. Birkelund and K. Perch-Nielsen
Institute of Historical Geology and Paleontology
Østervoldgade 10, Copenhagen, Denmark

and

D. Bridgwater and A. K. Higgins
The Geological Survey of Greenland
Østervoldgade 10, Copenhagen, Denmark

I. INTRODUCTION

Greenland is the largest island in the northern hemisphere with a length of coastline bordering the Atlantic of over 5000 km, equivalent to the eastern seaboard of North America from Labrador to the Gulf of Mexico. Greenland is made up of a marginal mountainous rim up to 250 km broad and 3000 m high surrounding a central ice-filled depression extending to 250 m below sea level. Even allowing for the isostatic effect of the 2–3 km thick Inland Ice Sheet which fills the central depression there would still be an outer rim of high ground which appears to have been lifted some 1500–2000 m and tilted slightly inward at some time before the onset of the ice age. It is tempting to suggest that this might be due to the direct effects of sea-floor spreading on a stable block bordered by the Atlantic Ocean.

Taking the bedrock beneath the Inland Ice into account, probably 75% of the total rock mass which forms Greenland is made up of Precambrian crystalline rocks which have been unaffected by major metamorphic or thermal events in the last 1600 m.y. These rocks show characters which are remarkably similar to the Canadian and northern European shields. The Greenland Precambrian shield has acted as a stable block along the margins of which sediments varying in age from the late Precambrian until the present day have been deposited (Fig. 1). Late Precambrian and early Paleozoic sediments are

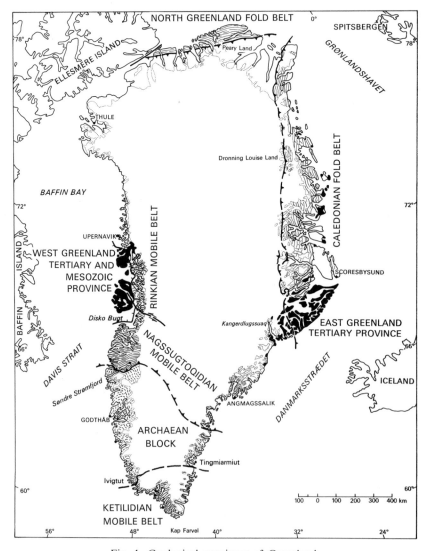

Fig. 1. Geological provinces of Greenland.

well developed in North Greenland and also in East Greenland where they form successions up to 16 km thick. Local outliers of Paleozoic sediments have been reported from the Precambrian shield of West Greenland. The major areas of late Precambrian and lower Paleozoic sedimentation together with their underlying basements were involved in metamorphism and tectonogenesis and form the mid-Paleozoic fold belts of North Greenland (Dawes, 1971) and the East Greenland Caledonian (Haller, 1971). The folded Paleozoic rocks are overlain by upper Paleozoic platform sediments in North Greenland and in East Greenland by upper Paleozoic continental and marine deposits followed by Mesozoic sediments laid down mainly under marine conditions.

A major sedimentary basin developed at least as early as the Jurassic along what is now the west coast of Greenland where it is represented onshore by approximately 2000 m of Cretaceous and Tertiary sediments. Basaltic volcanic activity began around 65–70 m.y. ago in West Greenland and possibly slightly later in East Greenland, apparently controlled by riftlike structures subparallel to the present coastline (see Chap. 11, this volume).

The basaltic lavas which form the dominant rocks of the Tertiary igneous province of East and West Greenland were followed about 50 m.y. ago by the emplacement of numerous alkali plutonic centers in East Greenland.

II. THE PRECAMBRIAN SHIELD

Four major structural provinces are recognized in the Precambrian shield of Greenland (Fig. 1).

A. Ketilidian Mobile Belt of South Greenland

An area affected by major thermal activity in the period (?)2500–1600 m.y. ago extends from the southern tip of Greenland to approximately lat. 61°30′ N on the west coast and lat. 62° N on the east coast. This mobile belt is dominated by large granitic bodies emplaced into country rocks consisting partly of reworked older gneisses and partly of metamorphosed sediments and lavas deposited on the older gneiss basement. Toward the northern margin of the belt a 5000-m-thick sequence of dolomites, shales, quartzites, graywackes, and basaltic lavas overlies an older gneissose basement. Traced southward this supracrustal sequence shows an increase in metamorphic grade and contains medium-pressure, medium-temperature assemblages of Barrovian type. The original discordance between the supracrustal cover and the under-lying gneisses is progressively destroyed southward. The gneisses and the cover rocks are truncated by the Julianehaab Granite, a complex body of intrusive rocks ranging from hypersthene gabbros to adamellites emplaced during several periods of magmatism in the period between 2500 and 1750 m.y.

ago. The Julianehaab Granite extends to the east coast and probably covers an area of at least 25,000 sq. km. The most distinctive rocks within the Juliane-haab Granite are a group of hornblende-rich basic and intermediate rocks closely resembling the appinites of the Caledonian fold belt of Scotland. South of the Julianehaab granite a second group of metamorphosed supracrustal rocks is found consisting of semipelites, basic and acid volcanics, and impure quartzites. These show considerable variation in metamorphic grade, the most abundant assemblages including biotite–cordierite–andalusite, or biotite–cordiorite–sillimanite–garnet–hypersthene, quartz, and feldspar, suggesting high-temperature, medium-to-low pressure conditions. The supracrustal rocks have been intruded, near the southern tip of Greenland and on the east coast, by a widespread group of norites, iron-rich quartz monzonites, quartz syenites, and hypersthene- or fayalite-bearing granites which locally developed rapakivi textures. The outcrops of individual intrusions cover up to 5000 sq. km. They correspond closely with a major group of adamellites found in Labrador and the rapakivi granites of the Baltic shield. The Greenland rapakivi suite has yielded Pb/U concordia ages on zircons of 1772–1784 m.y. (Gulson and Krogh, 1972).

The area affected by the Ketilidian mobile belt remained one of tectonic and magmatic activity for some 800 m.y. after the end of metamorphism in the area. East–west-trending grabens became the site for the deposition of the Gardar continental sandstones and basaltic volcanic activity. A major suite of alkali igneous bodies were emplaced in this fault-controlled zone approximately 1100–1300 m.y. ago (Upton, in press).

B. Early Precambrian Gneisses

North of the Ketilidian mobile belt an area of dominantly quartzo-feldspathic gneisses which were last affected by major thermal events some 2500–2700 m.y. ago extends northward as far as Søndre Strømfjord on the west coast and Gyldenløves Fjord (lat. 64° N) on the east coast. This old block shows complex fold patterns with no dominant structural grain and can be regarded as a single structural entity subdivided into various units including early granitic rocks, thin belts of supracrustal rocks, basic meta-igneous complexes (including chromite-bearing calciferous anorthosites and gabbro anorthosites), and several generations of intrusive granodioritic and granitic rocks; all of these predate folding and high-grade metamorphism dated at around 2800 m.y. Some of the earliest granitic rocks in the area have yielded Rb/Sr and Pb/Pb ages between 3700 and 4000 m.y. and are thus the oldest material yet dated from the crust (Oxford Isotope Geology Laboratory and McGregor, 1971). The recognizable supracrustal sequences consist dominantly of basic and ultrabasic volcanic rocks together with aluminous-rich quartzites and pelites, and some marbles and iron quartzites. Many of the rocks are interpreted as volcanic sediments. One belt at the head of Godthaabsfjord contains a major iron ore deposit. The supracrustal rocks are found as concordant layers within the quartzo-feldspathic gneisses. Most contain high-grade mineral assemblages (cordierite–garnet–sillimanite). In the southern part of the old block there is evidence to suggest that there was more than one period of formation of the supracrustal rocks and that belts of comparatively low-grade metavolcanic rocks such as the Tartoq Group (Higgins and Bondesen, 1966) overlie older higher grade gneisses containing earlier supracrustal sequences. The age of formation of these local areas of greenschist is unknown, though they yield K/Ar ages of about 2500 m.y.; whether they postdate the regional 2800-m.y. high-grade metamorphism which affected all the other major rock units mapped is uncertain. The metamorphic grade of the gneisses varies from area to area; locally cordierite granulite-facies rocks predominate, apparently developed at the expense of earlier amphibolite-facies rocks; elsewhere the gneisses are in amphibolite facies.

The early Precambrian rocks were intruded by numerous doleritic and ultramafic dike swarms which belong to at least five generations. These dikes postdate the regional metamorphism, which gives K/Ar ages of around 2500–2700 m.y. in the old block, but predate the main Proterozoic metamorphism of the Ketilidian mobile belt to the south and the Nagssugtoqidian mobile belt to the north. Similar dike swarms are widespread in Canada and Scotland. Some of the dikes near the margins of the old block are thought to have been emplaced under regional metamorphic conditions associated with the development of the younger mobile belt.

C. Nagssugtoqidian Mobile Belt

Rocks equivalent to those found in the old gneiss block were reworked in a major tectonic zone extending for at least 350 km to the north of Søndre Strømfjord (lat. 66° N: West Greenland) and Gyldenløves Fjord (lat. 64° N: East Greenland). This tectonic activity has impressed a remarkable grain on the country striking approximately east–west in West Greenland and north-west–southeast in East Greenland. Deformation can be seen to have taken place in several phases, the earliest of which appears to predate the basic dikes cutting the old gneiss block. The deformation is not developed equally throughout; islands of practically undeformed rocks varying in size from a few square meters to several hundred square kilometers have been left within the reworked zone and frequently show a much more complex fold pattern than the apparently simple gneisses surrounding them. Metamorphic mineral assemblages within the Nagssugtoqidian mobile belt vary from amphibolite to granulite facies. Pelitic rocks frequently contain large amounts of kyanite.

A group of late-tectonic to post-tectonic calc-alkaline intrusions cut the highly deformed gneisses on the east coast close to Angmagssalik. Both these intrusions and the deformed country rocks have yielded K/Ar and Rb/Sr ages of about 1650–2000 m.y. These are interpreted as representing the uplift of the area rather than the age of the major deformation which is thought possibly to have started as early as 2500 m.y. ago.

D. Northern Province (Rinkian Mobile Belt)

North of Jakobshavn Isbrae (lat. 69°15' N) on the west coast the structural pattern shown by the gneisses changes radically from the highly sheared gneisses of the Nagssugtoqidian mobile belt to an area dominated by basin and dome structures (Escher and Burri, 1967). Although the available K/Ar ages give the same 1650–1900 m.y. range as the rocks further south it is clear that there is a completely different structural province in this area separated from the Nagssugtoqidian belt by a major fault. The gneisses are overlain by major supracrustal sequences which are particularly well developed in the Umanak district (Henderson and Pulvertaft, 1967) where a succession over 5 km thick of quartzite and metagraywacke is preserved. These supracrustal rocks together with remobilized basement gneisses were metamorphosed under amphibolite facies conditions during the formation of domelike structures. To the north of Svartenhuk (lat. 71°30' N), the supracrustal sequences are cut by a major igneous body, the Prøven Granite, which covers an area of at least 10,000 sq. km. This intrusion locally contains hypersthene and mantled potash feldspar ovoids and resembles the rapakivi granites of South Greenland. To the north of the Prøven Granite the gneisses contain granulite facies minerals.

It is not known with any certainty whether the Rinkian mobile belt extends into East Greenland, since the northern part of the Nagssugtoqidian belt is overlain by Tertiary basalts.

Possible equivalent supracrustal rocks may be represented in the fragmentary exposures of the Caledonian foreland in East Greenland and also by supracrustal rocks reworked within the Caledonian fold belt; these are described below.

III. COASTAL FOLD BELT OF CENTRAL EAST AND NORTHEAST GREENLAND

The geology of East Greenland north of lat. 70° N is extremely well known in view of the remoteness and difficulties of access to the region. There remain, however, many areas known only from fragmentary observations.

An account of the history of geological exploration of the region has recently
been given by Haller (1971, p. 6–35) as part of a comprehensive treatment of
the East Greenland Caledonides. In the following pages the main geological
elements of East Greenland between lat. 70° and 82° N (Figs. 2 and 3) will be
briefly described, based partly on Haller's compilation and partly on more
recent investigations carried out by the Geological Survey of Greenland.

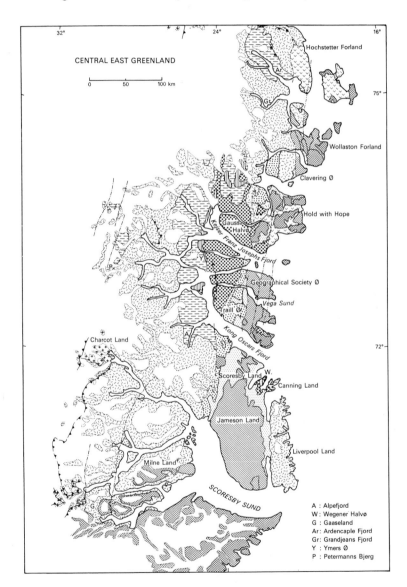

Fig. 2. General geology of central East Greenland (see Fig. 3 for legend).
Simplified after Tectonic/Geologic Map of Greenland; 1:2,500,000.

A. Geological Summary

From lat. 70–82° N East Greenland is dominated by the Caledonian fold belt whose western margin runs north–south parallel to the innermost nunataks. North of lat. 76° N the Caledonian fold belt is superimposed on the vestiges of the Carolinidian fold belt. The latter is traceable from lat. 76–82° N, mainly as partly reworked elements within the Caledonian domain, but also

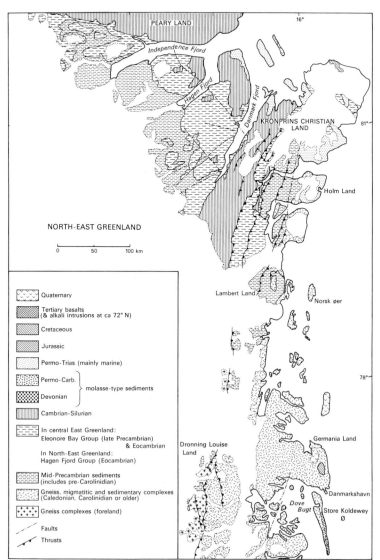

Fig. 3. General geology of northeast Greenland. Simplified after Tectonic/ Geologic Map of Greenland; 1:2,500,000.

in the Caledonian foreland. Here, middle to late Precambrian sedimentary sequences traced northward disappear underneath the cover of late Precambrian, Eocambrian, and early Paleozoic strata which crop out over much of North Greenland. The North Greenland fold belt affecting the lower Paleozoic rocks in extreme North Greenland is outside the scope of this report (see Dawes, 1971).

Intact fragments of the Precambrian shield comprising crystalline and sedimentary rocks outcrop in the Caledonian and Carolinidian forelands, and as partially or totally reworked blocks in both fold belts. Sedimentary sequences predating the Carolinidian event are known outside and within the Carolinidian fold belt in northeast Greenland. Widespread late Precambrian, Eocambrian, and early Paleozoic sedimentation preceded the development of the Caledonian fold belt with its accompanying regional metamorphism, migmatization, and granite emplacement. Late- to post-Caledonian continental deposits are found in some areas. In large areas of the outer fjord and coastal region between lat. 71° and 74° N Upper Permian and younger formations hide the Caledonian fold belt. South of lat. 70° N Tertiary basalts completely conceal its continuation.

The various reconstructions of the North Atlantic region illustrate clearly the spatial relationships between the East Greenland Caledonides and the Caledonian fold belts of Scandinavia and the British Isles. There remain, however, many uncertainties with respect to the relative placings of the various continents, and in reconstructions of the geological history of the region (Bullard *et al.*, 1965; Harland, 1969; Kay, 1969; Flinn, 1971; Pitman and Talwani, 1972).

B. Precambrian Crystalline Basement

Precambrian basement rocks unaffected by the Caledonian orogeny occur in the scattered foreland exposures along the edge of the Inland Ice. Within the Caledonian fold belt basement elements are widespread, especially in the inner fjord region of central East Greenland between lat. 70–74° N, and in northeast Greenland from lat. 76–79° N. A tentative indication of their distribution has been given by Haller and Kulp (1962, p. 20–21; see also Haller, 1971, p. 42–43). Supracrustal rocks older than the late Precambrian Eleonore Bay Group sequence occur in association with much of the basement gneisses of Central East Greenland. The pre-Carolinidian sedimentary rocks of northeast Greenland are treated separately.

1. *Gaaseland and Charcot Land* (lat. 70–72° N)

In the nunatak region between Gaaseland and Charcot Land a crystalline gneiss–granite complex and its supracrustal cover occurs beneath the marginal

Caledonian thrusts (Wenk, 1961; Steck, 1971; Henriksen, 1973). The crystalline basement has given K/Ar ages of up to 2290 m.y. (Haller and Kulp, 1962).
The cover of sedimentary and volcanic rocks more than 2000 m thick is cut in
Charcot Land by a granite which gives a zircon date of 1900 m.y. (Steiger and
Henriksen, 1972). This sequence is therefore better compared to some of the
mid-Precambrian supracrustal sequences of southern or western Greenland
rather than, as hitherto suggested, the Lower Eleonore Bay Group (Wenk,
1961; Haller, 1971; cf., Henriksen and Higgins, 1969; Henriksen, 1973).

2. *Central Metamorphic Complex* (lat. 70–74° N)

An extensive region of gneisses and granites makes up that part of the
Caledonian fold belt usually known as the central metamorphic complex.

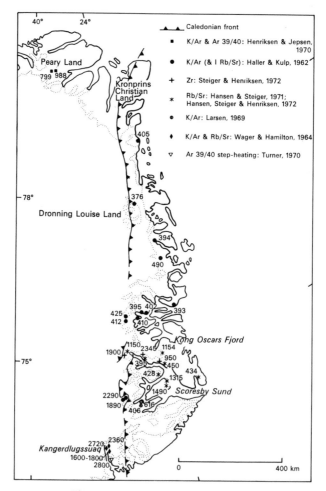

Fig. 4. Age dates of East Greenland.

Early explorers considered the region to be an Archaean massif, but most later workers have preferred to consider its characters mainly a consequence of Caledonian orogenesis. Considerable areas of crystalline basement can, however, be recognized, especially in the southwest part of the region. Here a basement of gneiss, amphibolite, and granitic rocks is overlain by a cover of high-grade metasediments. The cover series is in some areas migmatized and cut by intrusive bodies which have yielded pre-Caledonian ages (Rb/Sr mineral ages of 1100–1860 m.y.) as well as Caledonian ages (Hansen and Steiger, 1971; Hansen et al., 1972). There is a possibility that some of these meta-sediments are older than the lower Eleonore Bay Group with which they have usually been compared or correlated (Wenk, 1961; Vogt, 1965; Haller, 1971). The range of age dates now available clearly indicates that at least the southern part of the central metamorphic complex has a mixed Caledonian and pre-Caledonian origin (Fig. 4).

3. Northeast Greenland

In western Dronning Louise Land (ca. lat. 76° N) which forms part of the Caledonian foreland, Precambrian basement gneisses occur capped locally by quartzites of the Trekant Series. Traced northwards, the quartzites exhibit deformation due to Carolinidian events.

Within the Caledonian fold belt north of lat. 76° N, crystalline basement rocks make up substantial elements of thrust blocks and sheets. They have been reworked to a very varied extent during first Carolinidian and sub-sequently Caledonian orogenesis.

C. Sedimentation Preceding the Carolinidian Event

Sedimentary sequences which were dissected by basic dikes and sills prior to Carolinidian diastrophism occur over much of northeast Greenland. They were also intruded by a younger magmatic suite, peneplained during post-Carolinidian denudation, and overlain by further sediments in later Pre-cambrian and Eocambrian time.

The pre-Carolinidian sediments occur in autochthonous sequences in the Caledonian foreland of western Dronning Louise Land and the Danmark Fjord–Independence Fjord area, and are found also in the allochthonous thrust sheets of the Caledonian fold belt in the coastal region between lat. 76° and 81° N. Haller (1971) divides these sequences into three lithofacies. The lower two, which are nearly confined to the allochthonous areas, comprise quartzite–limestone and semipelitic sequences; they total ca. 3000 m in thick-ness and are considered to represent geosynclinal deposits. The uppermost psammitic lithofacies is thicker in the allochthonous sequences (ca. 3000 m:

Fränkl, 1954) than in the autochthonous sequences of either western Dronning Louise Land (0–500 m: Peacock, 1956; 1958), the Danmark Fjord area (ca. 300 m: Adams and Cowie, 1953), or southern Peary Land (>230 m: Jepsen, 1971).

In southern Peary Land dolerites cutting sandstones of the Inuiteq Sø Formation, which is correlated with the pre-Carolinidian sandstone sequences (Jepsen, 1971), have yielded an $^{39}Ar/^{40}Ar$ age of 988 m.y. (Henriksen and Jepsen, 1970). This date could reflect Carolinidian activity, but at least gives a minimum age for deposition of the sandstones.

D. Carolinidian Orogeny

The Carolinidian fold belt developed on the site of the pre-Carolinidian geosynclinal tract, and is traceable from Dove Bugt (lat. 76° N) to Kronprins Christian Land. In western Dronning Louise Land the pre-Carolinidian Trekant Series displays northwest–southeast-trending open folds which increase in intensity northward. The Trekant Series is transgressed by the unfolded post-Carolinidian Zebra Series (Peacock, 1956; 1958). Carolinidian deformation and metamorphism seem to have been superficial between Danmark Fjord and southern Peary Land.

Substantial parts of Caledonian thrust sheets north of lat. 76° N include Carolinidian complexes, mainly metasediments in the north and gneisses in the south. The main Carolinidian fold trends appear to vary between northeast–southwest to north–south, but have been partly affected by Caledonian reworking. Carolinidian metamorphism and deformation has been described as most intense between lat. 76° and 78° N (Haller, 1971); however, recent isotopic investigations at Zürich (Steiger, pers. comm., 1972) have yielded very old Precambrian ages on gneisses from this region and have not given a clear indication of a Carolinidian event.

E. Late Precambrian and Early Paleozoic Sedimentation

Major developments of late Precambrian and early Paleozoic sediments are found in two main regions: northeast Greenland, and central East Greenland between lat. 72° and 75° N. The successions in the two regions developed in very different ways, but in both regions the greatest accumulations seem to have taken place on the site of what was to arise as the Caledonian fold belt (Figs. 5 and 6).

1. Northeast Greenland

The late Precambrian and Eocambrian (post-Carolinidian) successions of northeast Greenland have been collectively termed the Hagen Fjord Group

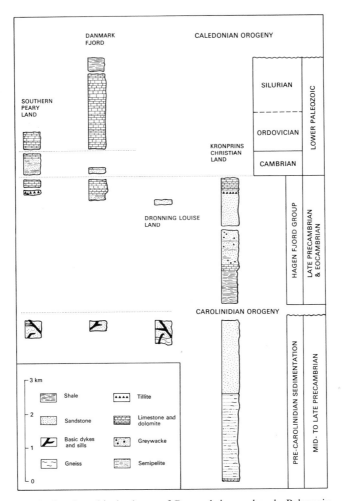

Fig. 5. Stratigraphical scheme of Precambrian and early Paleozoic strata of northeast Greenland. Data from various sources; see text for references.

(Haller, 1971). At Hagen Fjord, 2000 m are recorded and in western Kronprins Christian Land ca. 400 m; both areas are part of the present Caledonian foreland and the sequences represented in this region are considered to have accumulated in a shallow basin. The succession comprises sandstones and quartzites, followed by sandstones and dolomites, the upper beds containing stromatolitic algal reefs.

In the Caledonian thrust sheets further to the east in Kronprins Christian Land up to 5000 m of sediments occur which are considered to have accumulated in a miogeosynclinal trough sited east of the present coast. The lower part of the sequence comprises phyllites, some marbles, and graywackes (ca.

2500 m) and the middle part sandstones, conglomerates, and shales (ca. 2000 m). The upper part of the sequence (ca. 400 m) includes limestones and dolomites containing, like the uppermost strata of the foreland sequence, stromatolitic algal structures (Fränkl, 1954; Haller, 1971).

Both the autochthonous successions in the west and the miogeosynclinal successions further east contain units which have been interpreted as tillites. The significance of these rocks in the Hagen Fjord Group is a matter of debate. In southern Peary Land a well-developed tillitic sequence found at the base of the succession (Troelsen, 1949, 1956; Jepsen, 1971), has been named the Moraeneø Formation by Jepsen and is regarded as Eocambrian. A thin tillitic bed from the allochthonous succession of Kronprins Christian Land has also been considered Eocambrian (Fränkl, 1954). Haller (1971, p. 74) considers their origin as tillites and their assumed correlation with the tillites in the Mørkebjerg Formation (= Tillite Group) of central East Greenland, as possible, but not proven.

Lower Paleozoic rocks are widespread in northern Greenland, although they have not so far been recorded from allochthonous units of the Caledonian fold belt. The autochthonous succession in the Danmark Fjord region comprises more than 3000 m of mainly limestone and dolomite of Lower Cambrian to Upper Silurian age (Adams and Cowie, 1953; Cowie, 1961). Along the extreme north of Greenland lower Paleozoic rocks are involved in the North Greenland fold belt (Dawes, 1971).

2. Central East Greenland

Geosynclinal conditions were apparently established over an extensive area of central East Greenland in late Precambrian time, and sedimentation was more or less continuous up to the Middle Ordovician. Of the up to 16,000-m-thick accumulations, much the greatest part represents the Precambrian Eleonore Bay Group succession.

The main Eleonore Bay Group outcrops occur in a broad zone of the fjord region from lat. 72–75° N, with further exposures in the nunatak region between lat. 73° and 74° N, on Canning Land and Wegener Halvø (ca. lat. 71°40′ N) and in the Ardencaple Fjord area (lat. 75°30′ N). These sequences are mainly nonmetamorphic or only weakly metamorphosed and exhibit generally simple Caledonian fold structures.

The lower Eleonore Bay Group reaches its maximum development in the Alpefjord region where 8000 m of mainly unmetamorphosed pelites and quartzites are recorded (Fränkl, 1953b; Haller, 1971). Appreciable thicknesses occur also in the Petermann Bjerg (Wenk and Haller, 1953), Eleonore Sø (Katz, 1952), and Ardencaple Fjord regions (Haller, 1956; Sommer, 1957). There are wide facies variations from region to region.

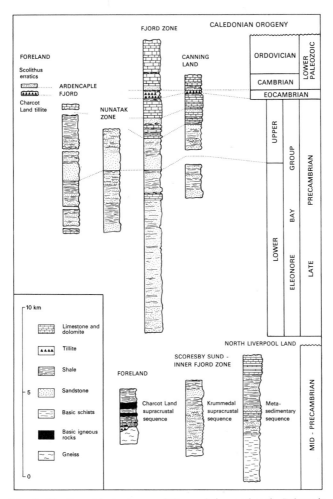

Fig. 6. Stratigraphical scheme of Precambrian and early Paleozoic strata of central East Greenland. Data from various sources; see text for references.

The lower Eleonore Bay Group is generally supposed to be represented by metasediments, migmatitic sedimentary rocks, and paragneisses over wide areas of the inner fjord zone between lat. 70° and 76° N. However, as discussed earlier, some of these sequences may be of significantly earlier date.

By contrast, the well defined sequence of quartzites, limy sandstones, limestones, and dolomites of the upper Eleonore Bay Group exhibit great uniformity of development throughout their outcrop. The southernmost outcrops are found in Canning Land (ca. lat. 71°40′ N: Bütler, 1948; Caby, 1972), with more or less continuous exposure in the fjord zone from lat. 72° to 74° N (Eha, 1953; Fränkl, 1953a, 1953b; Katz, 1952; Sommer, 1957), and further outcrops in the Ardencaple Fjord area (Sommer, 1957). Only the quartzitic part of the sequence seems to be represented in the Petermann Bjerg exposures (Wenk and Haller, 1953). The upper beds of the upper Eleonore Bay Group are rich in stromatolitic remains.

A sequence of shales, sandstone, and two composite boulder beds known as the Mørkebjerg Formation succeeds the Eleonore Bay Group. In Canning Land and throughout the fjord zone from lat. 72° to 74°30′ N it is overlain by lower Paleozoic rocks. The Mørkebjerg Formation has a uniform development.

The lower boulder bed contains mainly pebbles of the Eleonore Bay Group, and the upper boulder bed in addition contains numerous crystalline components (Fränkl, 1953a; Katz, 1954). The Mørkebjerg Formation is generally accepted as representative of the widespread Varangian glaciation.

The lower Paleozoic strata exhibit a uniform development of sandstones, silty shales, silty limestones, limestones, and dolomites ranging in age up to Middle Ordovician; they occur in the same areas as the Mørkebjerg Formation and reach a maximum of 3000 m thickness. The central East Greenland fauna shows a predominant affinity with the Paleozoic fauna of the Pacific "province."

In the foreland region between lat. 70° and 72° N there are no certain outcrops of the Eleonore Bay Group since, as indicated above, it now seems likely that the eugeosynclinal sequences here are of earlier date. A tillite occurring in Charcot Land resting directly on old crystalline rocks is possibly Eocambrian (Henriksen and Higgins, 1969). Lower Paleozoic rocks are unknown here except as numerous erratics of *Scolithus* quartzite, whose distribution pattern suggests the occurrence of lower Paleozoic rocks hidden beneath the Inland Ice between lat. 70° and 74° N (Haller, 1971, Fig. 48).

F. Caledonian Orogeny

The onset of the orogeny brought lower Paleozoic geosynclinal sedimentation to a close; the latest sediments recorded are Middle Ordovician in central East Greenland and Middle Silurian (possible Upper) in northeast Greenland. Folding, thrusting, regional metamorphism, migmatization, and emplacement of granites all played a part in the development of the Caledonian fold belt, traceable in East Greenland continuously from lat. 70° to 82° N. The main tectonic and thermal events occurred in Silurian and Early Devonian time. Magmatic activity, minor folding and thrusting continued through Devonian and into Early Carboniferous time.

The western margin of the fold belt can be traced through the nunatak zone between lat. 70° and 72° N, lat. 76° and 77°30′ N, and discontinuously from lat. 78° to 82° N. Everywhere the border zone is characterized by thrusts

with westward sense of displacement. The amount of displacement is speculative, but a figure of ca. 50 km has been suggested for Kronprins Christian
Land and a possible 50 or 100 km for Dronning Louise Land (Haller, 1971),
while in the inner Scoresby Sund region minimum figures of 60 km and 110 km
have been deduced for a lower and upper series of thrust sheets, respectively.

The dominant trends of the Caledonian fold structures are north–south
in central East Greenland and north-northeast in northeast Greenland. However, as Haller (1970) has demonstrated, these trends are superimposed on a
very varied pattern of earlier structures of widely different ages. The fold
pattern of the infracrustal regions is particularly intricate, and in central
East Greenland Haller (1955, 1971) suggests this is related to the development
of dome- and nappelike migmatitic bodies. The folding of the supracrustal
rocks has generally a simple form.

Caledonian metamorphism was apparently most intense in the southern
part of central East Greenland. The present pattern may be partly inherited
although the isograds parallel the fold belt (Henriksen, 1973). The central
metamorphic complex, a region of infracrustal gneiss, migmatite, and granite
of partly pre-Caledonian generation, suffered partial reactivation in Caledonian
time. Numerous granite bodies were emplaced mainly in zones around the
margin of the complex. They range from late tectonic to post-tectonic with
respect to the main structures.

Haller (1970, 1971) divides the development of the fold belt into three
phases. The Caledonian main orogeny (Silurian: 420–400 m.y.) includes the
main regional metamorphism, reactivation and migmatization, folding, thrust,
and nappe development. The late Caledonian spasms (Devonian: 400–350
m.y.) restricted to south of lat. 76° N include a phase of deep-seated reactivation localized in the region lat. 74–76° N, the Grandjeans Fjord complex.
Northwest-trending structures developed superimposed on the earlier northeast-trending pattern, and the grabenlike depression in which the Old Red
Sandstone accumulations are preserved was formed (see below). Post-tectonic
granites were also widespread. The third phase, the minor succeeding episodes
starting in the Middle Devonian, include warping and thrusting of the Old
Red Sandstone deposits and reactivation of Caledonian fracture systems
continuing as late as Upper Carboniferous and Lower Permian time.

G. Old Red Sandstone—Devonian

The Old Red Sandstone of central East Greenland, an accumulation of
more than 7000 m of conglomerates, breccias, arkoses, and sandstones, is
found in an extensive area of the fjord zone between ca. lat. 72°30′ and 74°30′
N. At the west margin the sediments rest unconformably on the Eleonore Bay
Group and Cambro-Ordovician sequences; the east margin is faulted. Outcrops

also occur in scattered locations further south on Wegener Halvø and Canning Land.

The sediments appear to have accumulated in an intramontane basin and comprise weathering detritus derived from the Caledonian mountain chain. The base of the sequence oversteps fold structures of the principal Caledonian folding, and throughout the sequence disconformities and, in places, angular unconformities witness to rapidly changing depositional environments. Bütler (1959) distinguishes a number of cycles of deposition. Igneous activity is significant at the northern and southern margins of the basin.

The lower part of the sequence is considered Middle Devonian. The upper part of the sequence is noted for vertebrate fossils which show it to be Upper Devonian in age (Säve-Söderbergh, 1934; Jarvik, 1950, 1961).

During deposition of the Devonian sequence the northern part of the basin area was disturbed by shallow folds and thrusts trending west-northwest to north. Granite wedges from the Caledonian intrastructure were involved in the folding. Lower Carboniferous folding affected the entire basin. A further phase of northeast- to north-northeast-trending thrusts and faults split the region into a mosaic of fault-bounded blocks. These events are considered the minor events at the end of the Caledonian orogeny. They were followed by regional uplift, and in Upper Carboniferous time a cycle of continental deposition.

IV. THE LATE PALEOZOIC AND MESOZOIC OF EAST GREENLAND

The mainly Mesozoic central East Greenland basin extends from lat. 70°20′ to 75° N. To the west it is bordered by the Caledonian mountain belt, to the east by the North Atlantic Ocean. Only in the southern part of the area is the former eastern border of the basin visible. In northeast Greenland, late Paleozoic and Mesozoic platform sediments are known from lat. 80° to 83° N. In the following pages the geological history of the central East Greenland basin will be reviewed in some detail on the basis of existing literature and more recent investigations by the Geological Survey of Greenland. The sediments in northeast Greenland will only be treated briefly.

A. Geological Summary

In the central East Greenland basin, continental Carboniferous and Lower Permian sedimentation was followed by the transgression of the Upper Permian (Zechstein) sea. The Upper Permian sequence contains limestones, gypsum, and shales and only subordinate coarser clastic layers. During the Triassic marine and continental clastic sediments, including red beds, were dominant, but the sequence also contains intercalations of gypsum, dolomite, and limestone. All the Jurassic deposits consist of sandstones and shales of marine origin except for a Rhaetic–lower Liassic deltaic sequence of plant-bearing beds. The Cretaceous sediments consist of marine conglomerates, sandstones, and shales with subordinate limestone.

In northeast Greenland continental sediments with Lower Carboniferous plant fossils (Dinantian: Witzig, 1952) were found at lat. 80–81° N. Marine Upper Carboniferous (Moscovian) and Lower Permian (Sakmarian–Wolfcampian) rocks from this area were dated by fusulinids and have been compared to the Lower Permian of Ellesmere Island (Maync, 1961). Triassic deposits are known at lat. 82°40′ N, consisting of marine shales and sandstones of upper Scythian and lower Anisian age. The fauna shows affinities with faunas from the Arctic and Cordilleran North America, Spitsbergen, and Siberia (Kummel, 1953).

B. Carboniferous and Lower Permian

Upper Carboniferous to Lower Permian continental, plant-bearing sediments outcrop in an up to 30-km-broad belt from lat. 71°15′ to 74°25′ N. To the west, these sediments are bounded by faults against Caledonian rocks. Eastwards they are generally overlain and covered discordantly by marine Upper Permian and Mesozoic sediments. They reappear on the eastern side of the basin on Wegener Halvø (ca. lat. 71°40′ N), thus indicating a basin at least 80 km broad.

The "molasse-type" Upper Carboniferous to Lower Permian sediments of central East Greenland are about 5000 to 6000 m thick and were laid down in basins bordered by the young Caledonian mountain ranges (Bütler, 1961). They consist mainly of coarse sandstones and arkoses containing, often, reddish alkali feldspars and small rock fragments. The color varies considerably and cannot be used for long-distance correlations. Polymict conglomerates occur mainly on the western and eastern borders of the basin. Limnic deposits are described from Traill Ø and in northern Scoresby Land. The observed cycles consist of pebbly arkoses overlain by fine-grained micaceous sandstones, black partly bituminous shales containing well preserved fish remains, and thin calcareous beds. No clearly marine fossils have yet been found.

Lowermost Carboniferous beds lie conformably on the uppermost Devonian. Due to tectonic activity in the Lower Carboniferous, part of the Tournaisian and probably the whole of the Visean are missing. The bulk of the continental Carboniferous was deposited in the Namurian and subsequently. From north to south, only Namurian deposits were found from Clavering Ø to Vega Sund (lat. 74°30′–72°50′ N). On Traill Ø (lat. 72°45′ N) younger beds occur, and these are well exposed along Kong Oscars Fjord. From here southward, the uppermost part of the succession has, from pollen evidence, been assigned to the Lower Permian (Kempter, 1961).

C. Upper Permian

Marine Upper Permian (Zechstein) sediments overlie the older Paleozoic formations with angular unconformity. They outcrop mainly in the same areas as the Carboniferous and Lower Permian rocks and are overlain by Lower Triassic deposits. The marine Upper Permian shows a considerable diversity of interfingering lithologies and was deposited under neritic and nearshore conditions (Fig. 7).

1. Foldvik Creek Formation

The whole Upper Permian sequence has been assigned to the Foldvik Creek Formation which consists of eight members which grade laterally into each other or may entirely replace one another and are contemporaneous (Maync, 1942, 1961) (see Fig. 7).

The conglomerate member often marks the base of the marine Upper Permian. This polygenic conglomerate attains a thickness of a few to over 100 m and contains in places large angular pebbles of Caledonian rocks, but only few marine fossils have been found.

Massive gypsum or gypsiferous shales form the gypsum member and are closely associated with the limestone–dolomite member. They can form the base of the Upper Permian, or gypsiferous intercalations may occur between the basal conglomerate and the limestone–dolomite member or form the remainder of the Upper Permian above the basal conglomerate.

Reefy limestones and dolomites of the limestone–dolomite member overlie the basal conglomerate or the gypsum member, or form the base of the Upper Permian. They vary greatly in thickness and texture, and include massive limestones and dolomites, platy limestones, coquina beds, monogenetic intra-formational breccias, and lenses of pseudo-oolithic limestone with detrital fragments. On Wegener Halvø the "reefs" are seen to thin from 150 m to a few meters from northeast to southwest towards the center of the basin.

The *Posidonia* shale member consists of grey and black, sometimes bituminous, calcareous silty shales with thin limestone intercalations and fills basins between the "reefs." It also includes layers or lenses of *Productus* limestone and *Martinia* limestone. Fish remains were found in the shales and in calcareous concretions (Nielsen, 1932, 1952; Aldinger, 1935a, 1937).

The *Productus* limestone member comprising bluish grey argillaceous and sometimes bituminous coquinas of mainly brachiopod shells, occurs either

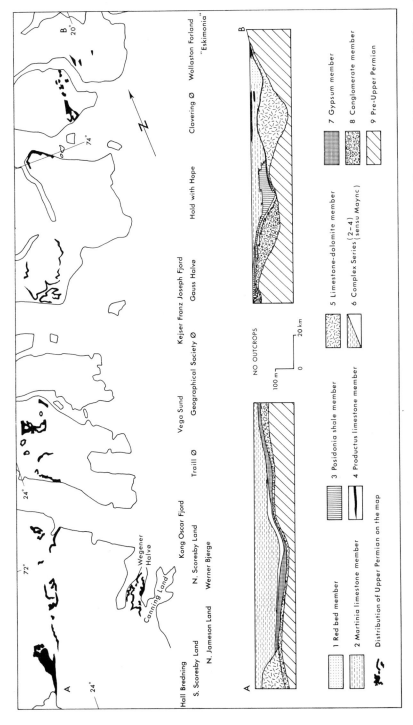

Fig. 7. Geographical distribution and facies distribution of the marine Upper Permian of central East Greenland. After Maync (1942) and own observations.

interstratified with *Posidonia* shales or *Martinia* limestone on top of the "reefs" or between tongues of the red bed member.

The *Martinia* limestone member consists of well-bedded, bluish-grey or greenish limestones, which often bear muscovite, alternating with thin marly banks. Permian ammonites (*Cyclolobus*) are found in this member. On Clavering Ø and in Scoresby Land, the boundary between this member and the overlying Lower Triassic shales is sometimes indistinct.

The red bed member contains red, ferruginous, calcareous, platy sandstones, generally medium grained with locally arkosic and conglomeratic intercalations. It occurs at the southern margin of the northern Permian borderland "Eskimonia." The benthonic fauna indicates a marine shallow-water environment. Clastic intercalations are also known from the southern area.

Maync (1961) also assigned interbedded, thin-bedded dolomitic and platy bituminous limestones, *Productus* limestone, and dark, often sandy shales that could not be ascribed to one of the above described lithological units to the Complex Series. They may occur between the basal conglomerate and the *Posidonia* shale or *Martinia* limestone, mainly on Gauss Halvø, Traill Ø, and in Scoresby Land.

2. *Paleogeography*

The fauna of the Upper Permian supports the suggestion that the central East Greenland basin at this time was part of the arctic sea which extended southward to the Zechstein Sea of northern Germany and England. The ammonites found in the *Martinia* limestone member also suggest a connection with the geosynclinal areas of the Himalayas over the arctic sea (Maync, 1942; Dunbar, 1961; Stensiö, 1961).

D. Triassic

The approximate age, the thickness of individual lithological subdivisions and the distribution of Triassic deposits in central East Greenland are summarized in Fig. 8. Most of the Triassic sequence consists of clastic deposits including red beds, conglomerates, arkoses, sandstones, siltstones, and shales. However, gypsum occurs in the Wordie Creek Formation as well as in the Gipsdalen Formation, and minor limestone intercalations occur throughout the Triassic.

1. *Scoresby Land Group*

The Triassic of central East Greenland is placed in the Scoresby Land Group (Perch-Nielsen *et al.*, in press) which outcrops from lat. 70°30′ to 74°30′ N.

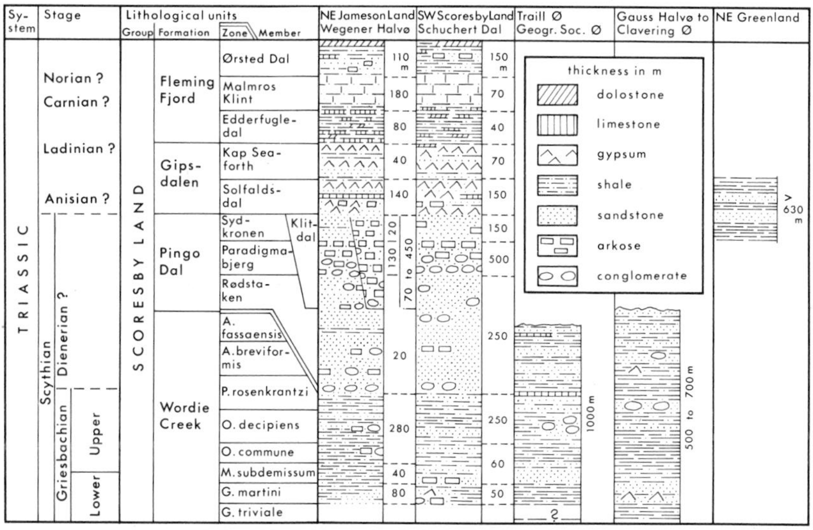

System	Stage	Lithological units Group	Formation	Zone	Member	NE Jameson Land Wegener Halvø	SW Scoresby Land Schuchert Dal	Traill Ø Geogr. Soc. Ø	Gauss Halvø to Clavering Ø	NE Greenland
TRIASSIC	Norian?	SCORESBYLAND	Fleming Fjord		Ørsted Dal	110 m	150 m	thickness in m		
	Carnian?				Malmros Klint	180	70			
	Ladinian?		Gips-dalen		Edderfugle-dal	80	40			
					Kap Sea-forth	40	70			
	Anisian?				Solfalds-dal	140	150			> 630 m
			Pingo Dal	Syd-kronen / Klit-dal		130/20	150			
				Paradigma-bjerg			500			
	Scythian (Dienerian?)			Rødsta-ken	70 to 450					
			Wordie Creek	A. fassaensis			250			
				A. brevifor-mis		20				
				P. rosenkrantzi						
	Griesbachian Upper			O. decipiens		280	250	1000 m	500 to 700 m	
				O. commune			60			
	Lower			M. subdemissum		40				
				G. martini		80	50			
				G. triviale						

Legend (thickness in m): dolostone, limestone, gypsum, shale, sandstone, arkose, conglomerate

Fig. 8. Stratigraphical scheme of selected areas of the Triassic in East Greenland. After Nielsen (1935), Grasmück and Trümpy (1969), and own observations.

A hiatus of variable importance between the Upper Permian and the Triassic Wordie Creek Formation is found in most sections, and in the southeast the Pingo Dal Formation overlies Caledonian rocks.

The Wordie Creek Formation consists mainly of marine, black to green and grey, often silty shales with intercalations of deltaic arkosic and conglomeratic layers at different levels. Thin limestone bands are also found. Gypsum is reported to occur at the bottom of the formation at Hold with Hope in the north (lat. 73°30′ N) and from southwestern Scoresby Land. Toward the top, sandstones predominate over the shales. The Wordie Creek Formation was deposited in a shallow marine basin with a paleoslope tilted northward to the open sea. The coarse clastic intercalations were supplied from the eastern and western borders of the basin; thinner and finer grained arkoses and sandstones occur toward the central part of the basin. The ammonite and vertebrate fauna of the Wordie Creek Formation is well known and allows the zonation of the sequence (see Fig. 8), and its correlation with Canadian, Spitsbergian, and Siberian sequences (Trümpy, 1969; Spath, 1935a; Nielsen, 1935, 1942, 1949).

The Pingo Dal Formation is found on Traill Ø, Scoresby Land, and Jameson Land. It overlies the Wordie Creek Formation and includes four members consisting mainly of dark red, often cross-bedded sandstones (red beds) in the lowermost and uppermost parts of the formation and red, pink, and yellowish to white arkoses and conglomerates in the middle part, and in the member represented at the eastern border of the basin. Here the Pingo

Dal Formation overlies Caledonian rocks. The Pingo Dal Formation is closely interrelated with the Wordie Creek Formation and was formed by increased supply of material from corresponding sources west and east of the slightly shifted basin. An upheaval of the borderlands together with an arid hot climate is thought to be responsible for the minimal chemical weathering and the formation of coarse, feldspar-rich sediments. The interfingering, finer grained, red sandstones in the central part of the basin are a consequence of floodplain conditions within the previously marine basin. Fossils are extremely rare in this formation, and its age is thus problematical (Defretin-Lefranc, 1969; Grasmück and Trümpy, 1969).

The Gipsdalen Formation is only known from Scoresby Land and Jameson Land. Two members can be distinguished, both containing gypsum or gypsiferous layers, sandstones, and shales. In the lower member a limestone bed with a poor marine fauna can be followed over a large area and is an indicator of marine incursions into the shallow evaporite basin. The amount of clastic material transported into the basin was minor compared to that of the Pingo Dal Formation. Pelecypods found in the limestone band imply a Middle Triassic age for this part of the sequence (Defretin-Lefranc, 1969).

Three members can be recognized in the Fleming Fjord Formation, which is found in Scoresby Land and Jameson Land. The lowermost member consists of yellowish weathering, stromatolitic dolomite and limestone, greenish and red-brown siltstones, and shales. The middle member is built of red and red-brown siltstones, mudstones, and thin-bedded sandstones. These grade into the upper member of coarse red-brown sandstones and light-colored and greenish arkoses overlain by dolomitic layers. The sedimentary environment changes locally from shallow marine to nonmarine and back to marine again in a flat basin. Fossils are scarce throughout the formation. Vertebrate remains occur sporadically and form a thin bone-bed at the top of the formation in the northern part of the area.

E. Jurassic

One of the most complete Jurassic sequences, bordering the ancient Atlantic Sea, is exposed in East Greenland between lat. 70°30′ and 77° N (Fig. 9).

All the Jurassic deposits are clastic, the earliest being nonmarine and the younger part marine, probably deposited in sublittoral environments. Clay-rich deposits are subordinate, silt being dominant in the fine clastic part of the sequence. All the sediments have a high content of mica, which may reach over 50% at certain levels. The coarse clastic deposits from the Lower Jurassic are often arkosic, while feldspar is scarce in most of the coarse clastic deposits from the Middle and Upper Jurassic.

The Rhaetic–Jurassic sequence is most complete in Jameson Land and placed in the Jameson Land Group by Surlyk *et al.* (1973).

1. *Lower Jurassic*

It can be seen in Fig. 9 that Lower Jurassic deposits are restricted to the southernmost part of the area, occurring in nearly the same area as the Upper Triassic deposits.

The lower part of the sequence consists of the Rhaetic–Liassic, non-marine, deltaic, plant-bearing Kap Stewart Formation (Harris, 1937). This increases considerably in thickness from south to north (170–400 m), and consists mainly of light-colored cross-bedded, arkosic sandstones interbedded with siltstone and shaly layers. The complete disappearance of red beds marks a climatic change to humid conditions.

The overlying Neill Klinter Formation increases also in thickness to the north (210–400 m). In the southern part the lower boundary is marked by a marine transgression giving rise to coarse arkosic sandstone containing a lower Pliensbachian fauna. This unit is overlain by more fine-grained, often ripple-bedded, sandstone and siltstone which only contains trace fossils. The formation is terminated by more coarse-grained, strongly cross-bedded, massive sandstones interbedded with shales, containing a Toarcian fauna (Rosenkrantz, 1934). The facies change toward the top suggests uplift in the source region.

2. *Middle Jurassic*

In the Middle Jurassic the sea transgressed over extensive areas to the north of Kong Oscars Fjord where the deposits rest on Triassic on Traill Ø, and on Permian or directly on the Caledonian basement further to the north. The Middle Jurassic deposits are found in two main areas separated by the Hold with Hope region. The nonoccurrence of the Middle Jurassic in the area around Hold with Hope may be due to nondeposition (Maync, 1947) or to pre-Aptian erosion (Donovan, 1957).

In the southern part of the area the Middle Jurassic Vardekløft Formation commences with uniform deposits of dark, silty shales without diagnostic fossils. These are overlain by coarse, clastic, cross-bedded sandstones, which increase rapidly in thickness to the north (10–900 m), probably a consequence of uplift of a northern source area.

In Jameson Land the sedimentation changed again to finer clastic, often glauconitic and silty deposits in the Bathonian. The change is diachronous, being earlier in southern Jameson Land than in northern Jameson Land, and in Scoresby Land the sedimentation of coarse clastic material continues through the Middle Jurassic.

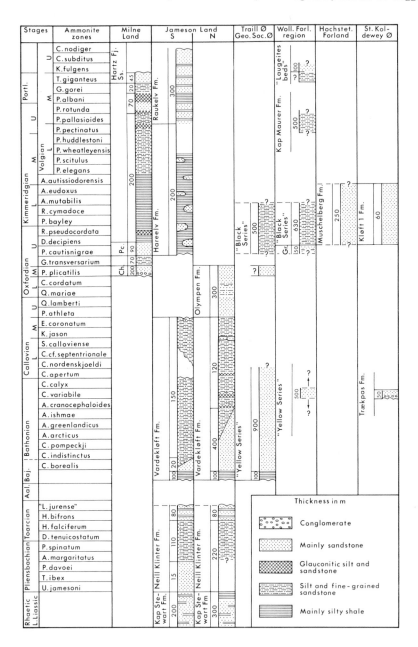

Fig. 9. Stratigraphical scheme of the Jurassic of selected areas in East Greenland. Ch., Charcot Bugt Sandstone. Pc., "Pecten Sandstone." Gr., "Grey Series." Milne Land and Jameson Land after Surlyk *et al.* (1973). Northern areas after Donovan (1957). The ammonite zonation of the Middle Jurassic is new and will be described in detail by Callomon and Birkelund.

3. *Upper Jurassic*

During the Upper Jurassic the sea transgressed further to the west in the Scoresby Sund area, where coarse, clastic middle Oxfordian deposits overlie the Caledonian basement in Milne Land (Aldinger, 1935*b*). Except for that area, the distribution of the Upper Jurassic deposits is very similar to that of the Middle Jurassic. Lower and middle Oxfordian deposits are not widely distributed (north Jameson Land, Milne Land, Traill Ø) and consist of near-shore coarse clastic sediments. Upper Oxfordian–Kimmeridgian deposits are much more widely distributed and are dominated by dark silty shales. Only in the northernmost part of the area do the sediments indicate deposition in coastal environments. In the upper Kimmeridgian (lower Volgian) the sedimen-tation changed to a more sandy facies both in southern Jameson Land and in Milne Land. In Wollaston Forland and in Milne Land middle Volgian sand-stones overlie earlier deposits slightly unconformably, indicating tectonic activity in the uppermost Jurassic (Maync, 1949; Callomon, 1961). In southern Jameson Land the tectonic activity took place slightly later, the Jurassic sequence, including upper Volgian deposits, being slightly folded before deposition of Ryazanian beds (Surlyk and Birkelund, 1972).

4. *Faunas*

Most of the Jurassic deposits are very fossiliferous (Spath, 1932, 1935*b*, 1936). Remarkable alterations in faunal realms characterize the period. The Liassic floras and faunas show close affinity to those of Europe. In the (?)upper Bajocian–Bathonian boreal ammonites occur exclusively, nearly all of which belong to *Cadoceratinae* which are only known from the northern realm. In the uppermost Bathonian–lower Callovian the fauna becomes more diverse, and from the beginning of the *S. calloviense* zone and through all the younger part of the Callovian the ammonite succession is similar to that of England. The Upper Jurassic faunas show affinities both with European faunas and with Russian Volgian faunas. The remarkable changes in faunal realms may be due to transgressions and regressions in the ancient shallow North Atlantic Sea (Callomon *et al.*, 1973).

F. Cretaceous

Cretaceous deposits are mainly distributed between Scoresby Sund and Germania Land (lat. 70–77° N). They are also known locally south of Scoresby Sund, at Kangerdlugssuaq (lat. 68°N) (Fig. 10).

All the Cretaceous sediments are marine and predominantly clastic, consisting of conglomerates, sandstones, and shales. Limestones are very

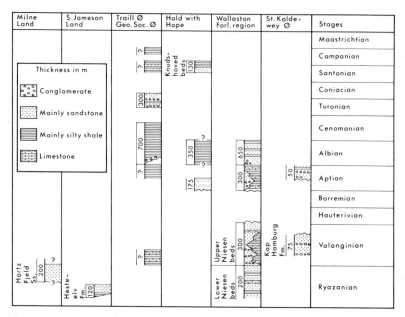

Fig. 10. Stratigraphical scheme of the Cretaceous of selected areas in East Greenland. Mainly after Donovan (1957, 1964).

subordinate, and most common in the Valanginian. The Cretaceous is less
complete than the Jurassic and large parts of the sequence are only known
from restricted areas which escaped erosion. The Valanginian, Aptian, and
Albian–Cenomanian are strongly transgressive (Donovan, 1957).

Ryazanian deposits overlie middle or upper Volgian deposits conformably
or unconformably. There is no fossil evidence of the lowermost Ryazanian in
the East Greenland sequence. In the Valanginian two principal facies realms
can be distinguished: a coarse clastic facies deposited near steep coastlines,
and a facies consisting of marls and marly shales with limestone bands, deposited under more quiet conditions. The presence of Hauterivian deposits is
doubtful, and Barremian rocks are unknown in East Greenland. Aptian sediments also show a wide range of facies variation—coarse sandstones and
conglomerates close to the ancient coastlines, and dark shales in more distant
regions. The Albian–Cenomanian deposits reach a considerable thickness.
The Albian consists in some regions of dark shales, while in other areas a
coarser clastic facies occurs. All the Cenomanian deposits consist of dark
shales. On Traill Ø a very coarse Turonian conglomerate occurs, interpreted
by Donovan (1957) as a boulder bed flanking a faulted coastline. Sandstones
and shales of Turonian age are known from the same region. Only small
outcrops of Santonian and Campanian rocks of sandy or shaly facies are
known, while rocks of Maastrichtian age have not been found.

1. *Faunas*

The Ryazanian ammonite fauna is entirely boreal and closely related to forms now known from Sandringham Sands in Norfolk (e.g., *Hectoroceras*: Casey, 1971). In the Valanginian several Tethyan faunal elements occur besides boreal elements. The Aptian fauna is rather cosmopolitan, and the Albian fauna is a mixture of boreal faunas, European and cosmopolitan forms. The Cenomanian fauna shows only affinity with European faunas. In the Santonian and Campanian forms widely distributed in the northern hemisphere predominate.

G. Structures

The sedimentary areas of central East Greenland consist of faulted and tilted blocks. To the west, the post-Devonian main fault runs from Scoresby Sund northward through all East Greenland. It was active from Carboniferous times throughout the Mesozoic and Cenozoic. The area east of the main fault acted largely as a coherent block up to the Upper Jurassic. Then an antithetic block-fault system gradually developed in the region lat. 74–75° N, the site of faulting successively shifting eastward (Vischer, 1943). The uneven subsidence of the southern part of the basin and the periodical influx of coarse

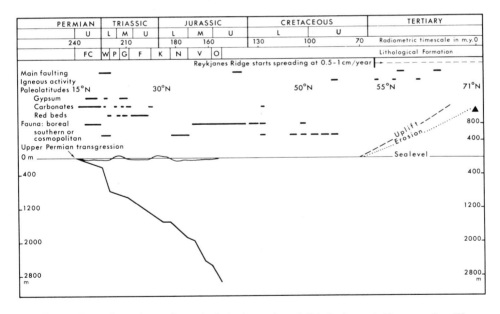

Fig. 11. Tentative scheme for paleolatitudes, selected lithologies, subsidence and uplift through time in northern Jameson Land (lat. 71°30′N, long. 23°30′W). Paleogeographical relations of the East Greenland faunas.

clastic wedges from the north through Middle and Upper Jurassic times was probably also caused by violent tectonic activity.

The general subsidence of the central East Greenland basin during late Paleozoic and Mesozoic times was followed by uplift in the early Tertiary (see Fig. 11). Erosion was active, forming the relief visible under the early Tertiary basalt flows in the Gaaseland, Milne Land, Scoresby Land, and Wollaston Forland areas. Later during the Tertiary, faulting again displaced the lava flows and sediments, especially north of Jameson Land. Gentle late Jurassic folding is known from southern Jameson Land and slight Tertiary folding occurs in northeast Greenland.

The fault pattern has been related to a crustal widening during Paleozoic and Mesozoic times, and attention has been drawn to the intensive faulting in the late Jurassic, contemporaneous with the formation of the mid-Atlantic ridge (Haller, 1971). Tertiary uplift, faulting, and igneous activity seem to be connected with the onset of spreading in the Reykjanes ridge and the opening of the Norwegian basin.

V. CONCLUDING REMARKS

The major units occurring throughout the geological column in Greenland can be well matched with rocks of similar age in North America and Europe, and both the character of rocks exposed in Greenland and their geographical distribution are consistent with the idea that Greenland represents a fragment of an earlier North Atlantic continental mass. The timing of the possible breakup between Greenland and Canada and between Greenland and Europe is a matter of debate. There is a reasonable amount of evidence from West Greenland and East Canada that the Davis Strait–Labrador Sea area has been one of tectonic activity for at least 500–1000 m.y. (see, e.g., Bridgwater et al., 1973; Fahrig et al., 1971; Henderson, 1973) and that the main movements in the late Mesozoic and Tertiary suggested by studies of magnetic anomaly plots (Pitman and Talwani, 1972) may represent the final stages of the separation of two land masses rather than the initial break.

The eastern border of Greenland has clearly been an active area for at least 600–1000 m.y., from the initiation of Caledonian geosynclinal sedimentation to the present day. Whether or not the Caledonian fold belt itself represents a late Precambrian–Paleozoic major closing of an earlier Atlantic followed by a Mesozoic opening (Wilson, 1966), or whether Greenland first became separated from Europe late in the Phanerozoic, are matters of controversy probably best resolved by studies of the rocks preserved in the Caledonian belt itself. The Eleonore Bay Group sediments are typical shelf deposits which must have been laid down at the rim of a continent, and the following

Caledonian earth movements certainly represent a major event in the development of the North Atlantic; however, the amount of pre-Caledonian material preserved in the fold belt is much larger than hitherto suggested and can be regarded as earlier continental crust.

The post-Caledonian history of East Greenland has been one of persistent crustal movement from Devonian time and onward. During the late Paleozoic and Mesozoic marine and continental sedimentation took place in shallow basins subsiding at a rate of ca. 3 cm/1000 years. The early Triassic ammonite and fish fauna includes species suggesting open seaways to northern Canada, Spitsbergen, and northern Siberia, as well as to the Tethyan Sea over western Siberia. Middle and Upper Triassic deposits including red beds and gypsum show a facies pattern similar to the germanotype Triassic of Europe. The marked change of the ammonite faunas in the beginning of the Middle Jurassic to strictly boreal faunas not known from northern Europe may be caused by an extensive regression in the North Atlantic during Bajocian–Bathonian times. In the late Mesozoic and the Tertiary uplift took place also at a rate of ca. 3 cm/1000 years.

Widespread igneous activity (intrusions, sills, and dikes) can be dated to the late Mesozoic and Early Tertiary, and early Tertiary plateau basalts cover large areas. Intensive faulting occurred during the late Paleozoic, late Jurassic, and again in the Tertiary. The late Mesozoic and Tertiary tectonic and igneous activity may be related to changes in ocean-floor-spreading patterns in the North Atlantic: the opening of the North Atlantic, the initiation and pause in the opening of the Labrador Sea, the beginning of spreading on the Reykjanes ridge, and the shift of the spreading axis in the Norwegian Sea.

ACKNOWLEDGMENTS

This paper is published with the permission of the Director of the Geological Survey of Greenland. Figure 1 was prepared by Dr. A. Escher.

REFERENCES

Adams, P. J. and Cowie, J. W., 1953, A geological reconnaissance of the region around the inner part of Danmarks Fjord, Northeast Greenland: *Meddr Grønland*, v. 111, n. 7, p. 1–24.

Aldinger, H., 1935a, Das Alter der jungpalaeozoischen Posidonomyaschiefer von Ostgrönland: *Meddr Grønland*, v. 98, n. 4, p. 1–24.

Aldinger, H., 1935b, Geologische Beobachtungen im Oberen Jura des Scoresbysundes (Ostgrönland): *Meddr Grønland*, v. 99, n. 1, p. 1–128.

Aldinger, H., 1937, Permische Ganoidfische aus Ostgrönland: *Meddr Grønland*, v. 102, n. 3, 1–392.

Bridgwater, D., Escher, A., Jackson, G. C., Taylor, F. C., and Windley, B. F., 1973, Development of Precambrian shield in West Greenland, Labrador, and Baffin Island: *Mem. Am. Ass. Petrol. Geol.*, v. 19.

Bullard, E., Everett, J. E., and Smith, A. G., 1965, The fit of the continents around the Atlantic: *Phil. Trans. Roy. Soc.*, v. A258, p. 41–75.

Bütler, H., 1948, Notes on the geological map of Canning Land (East Greenland): *Meddr Grønland*, v. 133, n. 2, p. 1–97.

Bütler, H., 1959, Das Old Red-Gebiet am Moskusoksefjord: *Meddr Grønland*, v. 160, n. 5, p. 1–188.

Bütler, H., 1961, Continental Carboniferous and Lower Permian in Central East Greenland, in: *Geology of the Arctic*, Vol. 1, Raasch, G. O., ed.: University of Toronto Press, Toronto, p. 205–213.

Caby, R., 1972, Preliminary results of mapping in the Caledonian rocks of Canning Land and Wegener Halvø. East Greenland: *Rapp. Grønlands geol. Unders*, v. 48, p. 21–38.

Callomon, J. H., 1961, The Jurassic system in East Greenland, in: *Geology of the Arctic*, Vol. 1, Raasch, G. O., ed.: University of Toronto Press, Toronto, p. 258–268.

Callomon, J. H., Donovan, D. T., and Trümpy, R., 1973, Annotated map of the Permian and Mesozoic formations of East Greenland: *Meddr Grønland*, v. 168, n. 3, p. 1–35.

Casey, R., 1971, Facies, faunas and tectonics in Late Jurassic–Early Cretaceous Britain, in: *Faunal Provinces in Space and Time*: Geol. Journ. Spec. Issue No. 4, p. 153–168, Seel House Press, Liverpool.

Cowie, J. W., 1961, The Lower Paleozoic geology of Greenland, in: *Geology of the Arctic*, Vol. 1, Raasch, G. O., ed.: University of Toronto Press, Toronto, p. 160–169.

Dawes, P. R., 1971, The North Greenland fold belt and environs: *Bull. Geol. Soc. Denmark*, v. 20, p. 197–239.

Defretin-Lefranc, S., 1969, Notes on Triassic stratigraphy and paleontology of north-eastern Jameson Land (East Greenland). III. Les Conchostracès Triassiques du Groenland Oriental: *Meddr Grønland*, v. 168, n. 2, p. 123–136.

Donovan, D. T., 1957, The Jurassic and Cretaceous systems in East Greenland: *Meddr Grønland*, v. 155, n. 4, p. 1–214.

Donovan, D. T., 1964, Stratigraphy and ammonite fauna of the Volgian and Berriasian rocks of East Greenland: *Meddr Grønland*, v. 154, n. 4, p. 1–34.

Dunbar, C. O., 1961, Permian invertebrate faunas of Central East Greenland, in: *Geology of the Arctic*, Vol. 1, Raasch, G. O., ed.: University of Toronto Press, Toronto, p. 224–230.

Eha, S., 1953, The pre-Devonian sediments on Ymers Ø, Suess Land, and Ella Ø (East Greenland) and their tectonics: *Meddr Grønland*, v. 111, n. 2, p. 1–105.

Escher, A. and Burri, M., 1967, Stratigraphy and structural development of the Precambrian rocks in the area north-east of Disko Bugt, West Greenland: *Rapp. Grønlands Geol. Unders.*, v. 13, p. 1–28.

Fahrig, W. F., Irving, E., and Jackson, G. D., 1971, Paleomagnetism of the Franklin diabases: *Can. J. Earth Sci.*, v. 8, p. 455–467.

Flinn, D., 1971, On the fit of Greenland and North-West Europe before continental drifting: *Proc. Geol. Ass.*, v. 82, p. 469–472.

Fränkl, E., 1953a, Geologische Untersuchungen in Ost-Andrées Land (NE-Grønland): *Meddr Grønland*, v. 113, n. 4, p. 1–160.

Fränkl, E., 1953b, Die geologische Karte von Nord-Scoresby Land (NE-Grønland): *Meddr Grønland*, v. 113, n. 6, p. 1–56.

Fränkl, E., 1954. Vorläufige Mitteilung über die Geologie von Kronprins Christians Land (NE-Grönland, zwischen 80°-81°N und 19°-23°W): *Meddr Grønland*, v. 116, n. 2, p. 1–85.

Grasmück, K. and Trümpy, R., 1969, Notes on Triassic stratigraphy and paleontology of north-eastern Jameson Land (East Greenland). I. Triassic stratigraphy and general geology of the country around Fleming Fjord (East Greenland): *Meddr Grønland*, v. 168, n. 2, p. 1–71.

Gulson, B. L. and Krogh, T. E., 1972, U/Pb zircon studies on the age and origin of post-tectonic intrusions from South Greenland: *Rapp. Grønlands Geol. Unders*, v. 45, p. 48–53.

Haller, J., 1955, Der "Zentrale Metamorphe Komplex" von NE-Grönland. Teil I. Die geologische Karte von Suess Land, Gletscherland und Goodenoughs Land: *Meddr Grønland*, v. 73, n. I(3), p. 1–174.

Haller, J., 1956, Die Strukturelemente Ostgrönlands zwischen 74° und 78°N: *Meddr Grønland*, v. 154, n. 2, p. 1–27.

Haller, J., 1970, Tectonic map of East Greenland (1:500,000). An account of tectonism, plutonism, and volcanism in East Greenland: *Meddr Grønland*, v. 171, n. 5, p. 1–286.

Haller, J., 1971, *Geology of the East Greenland Caledonides*: Interscience, London, 413 p.

Haller, J. and Kulp, J. L., 1962, Absolute age determinations in East Greenland: *Meddr Grønland*, v. 171, n. 1, p. 1–77.

Hansen, B. T., and Steiger, R. H., 1971, The geochronology of the Scoresby Sund area. Progress report I: Rb/Sr mineral ages: *Rapp. Grønlands geol. Unders*, v. 37, p. 55–57.

Hansen, B. T., Steiger, R. H., and Henriksen, N., 1972, The geochronology of the Scoresby Sund area. Progress Report 2: Rb/Sr mineral ages: *Rapp. Grønlands Geol. Unders*, v. 48, p. 105–107.

Harland, W. B., 1969, Contribution of Spitsbergen to understanding of tectonic evolution of North Atlantic region: *Mem. Am. Ass. Petrol. Geol.*, v. 12, p. 817–851.

Harris, T. M., 1937, The fossil flora of Scoresby Sound East Greenland. Part 5: Stratigraphic relations of the plant beds: *Meddr Grønland*, v. 112, n. 2, p. 1–114.

Henderson, G., 1973, The implications of continental drift for the petroleum prospects of West Greenland. Proc. Nato advanced study institute on continental drift, sea floor spreading and plate tectonics: implications for Earth Sciences, Newcastle 1972: Academic Press, New York.

Henderson, G. and Pulvertaft, T. C. R., 1967, The stratigraphy and structure of the Precambrian rocks of the Umanak area, West Greenland: *Meddr Dansk Geol. Foren*, v. 17, p. 1–20.

Henriksen, N., 1973, The Caledonian geology of the Scoresby Sund region, Central East Greenland: *Mem. Am. Ass. Petrol. Geol.*, v. 19.

Henriksen, N. and Higgins, A. K., 1969, Preliminary results of mapping in the crystalline complex around Nordvestfjord, Scoresby Sund, East Greenland: *Rapp. Grønlands Geol. Unders*, v. 21, p. 5–20.

Henriksen, N. and Jepsen, H. F., 1970, K/Ar age determinations on dolerites from southern Peary Land. *Rapp. Grønlands Geol. Unders*, v. 28, p. 55–58.

Higgins, A. K. and Bondesen, E., 1966, Supracrustals of pre-Ketilidian age (the Tartoq Group) and their relationships with Ketilidian supracrustals in the Ivigtut region, South-West Greenland: *Rapp. Grønlands Geol. Unders*, v. 8, p. 1–21.

Jarvik, E., 1950, Note on Middle Devonian Crossopterygians from the eastern part of Gauss Halvö, East Greenland. With an appendix: An attempt at a correlation of the Upper Old Red Sandstone of East Greenland with the marine sequence: *Meddr Grønland*, v. 149, n. 6, p. 1–20.

Jarvik, E., 1961, Devonian vertebrates, in: *Geology of the Arctic*, Vol. 1, Raasch, G. O., ed.: University of Toronto Press, Toronto, p. 197–204.

Jepsen, H. F., 1971, The Precambrian, Eocambrian, and early Palaeozoic stratigraphy of the

Jørgen Brønlund Fjord area, Peary Land, North Greenland: *Bull. Grønlands Geol. Unders*, v. 96, p. 1–42 (also *Meddr Grønland*, v. 192, n. 2).

Katz, H. R., 1952, Ein Querschnitt durch die Nunatakzone Ostgrönlands (ca. 74° n.Br.): *Meddr Grønland*, v. 144, n. 8, p. 1–65.

Katz, H. R., 1954, Einige Bemerkungen zur Lithologie und Stratigraphie der Tillitprofile im Gebiet des Kejser Franz Josephs Fjord, Ostgrönland: *Meddr Grønland*, v. 72, n. II(4), p. 1–64.

Kay, M., 1969, Continental drift in North Atlantic ocean: *Mem. Am. Ass. Petrol. Geol.*, v. 12, p. 965–973.

Kempter, E., 1961, Die jungpaläozoischen Sedimente von Süd Scoresby Land (Ostgrönland, 71½° N), mit besonderer Berücksichtigung der kontinentalen Sedimente: *Meddr Grønland*, v. 164, n. 1, p. 1–123.

Kummel, B., 1953, Middle Triassic ammonites from Peary Land: *Meddr Grønland*, v. 127, n. 1, p. 1–21.

Maync, W., 1942, Stratigraphie und Faziesverhältnisse der oberpermischen Ablagerungen Ostgrönlands (olim "Oberkarbon-Unterperm") zwischen Wollaston Forland und dem Kejser Franz Josephs Fjord: *Meddr Grønland*, v. 115, n. 2, p. 1–128.

Maync, W., 1947, Stratigraphie der Jurabildungen Ostgrönlands zwischen Hochstetterbugten (75°N) und dem Kejser Franz Joseph Fjord (73°N): *Meddr Grønland*, v. 132, n. 2, p. 1–223.

Maync, W., 1949, The Cretaceous beds between Kuhn Island and Cape Franklin (Gauss Peninsula), northern East Greenland: *Meddr Grønland*, v. 133, n. 3, p. 1–291.

Maync, W., 1961, The Permian of Greenland, in: *Geology of the Arctic*, Vol. 1, Raasch, G. O., ed.: University of Toronto Press, Toronto, p. 214–223.

Nielsen, E., 1932, Permo-Carboniferous fishes from East Greenland: *Meddr Grønland*, v. 86, n. 3, p. 1–63.

Nielsen, E., 1935, The Permian and Eotriassic vertebrate-bearing beds at Godthaab Gulf (East Greenland): *Meddr Grønland*, v. 98, n. 1, p. 1–111.

Nielsen, E., 1942, Studies on Triassic fishes from East Greenland. I. *Glaucolepis* and *Boreosomus*: *Meddr Grønland*, v. 138, p. 1–394.

Nielsen, E., 1949, Studies on Triassic fishes from East Greenland. II. *Australosomus* and *Birgeria*: *Meddr Grønland*, v. 146, n. 1, p. 1–309.

Nielsen, E., 1952, On new or little known Edestidae from the Permian and Triassic of East Greenland: *Meddr Grønland*, v. 144, n. 5, p. 1–55.

Oxford Isotope Geology Laboratory and McGregor, V. R., 1971, Isotopic dating of very early Precambrian amphibolite facies gneisses from the Godthaab district, West Greenland: *Earth Planet. Sci. Lett.*, v. 12, p. 245–259.

Peacock, J. D., 1956, The geology of Dronning Louise Land, N. E. Greenland: *Meddr Grønland*, v. 137, n. 7, p. 1–38.

Peacock, J. D., 1958, Some investigations into the geology and petrography of Dronning Louise Land, N. E. Greenland: *Meddr Grønland*, v. 157, n. 4, p. 1–139.

Perch-Nielsen, K., Birkenmajer, K., Birkelund, T., and Aellen, M., Revision of Triassic stratigraphy of the Scoresby Land and Jameson Land region, East Greenland, in press, *Bull. Grønlands Geol. Unders.* (also *Meddr Grønland*, v. 193, n. 6).

Pitman III, W. C. and Talwani, M., 1972, Sea-floor spreading in the North Atlantic: *Bull. Geol. Soc. Amer.*, v. 83, p. 619–646.

Rosenkrantz, A., 1934, The Lower Jurassic rocks of East Greenland. Part I: *Meddr Grønland*, v. 110, n. 1, p. 1–122.

Säve-Söderbergh, G., 1934. Further contributions to the Devonian stratigraphy of East Greenland. II. Investigations on Gauss Peninsula during the summer of 1933: *Meddr Grønland*, v. 96, n. 2, p. 1–74.

Sommer, M., 1957, Geologische Untersuchungen in den praekambrischen Sedimenten zwischen Grandjeans Fjord und Bessels Fjord (75°-76° n. Br.) in NE-Grönland: Meddr Grønland, v. 160, n. 2, p. 1–56.

Spath, L. F., 1932, The invertebrate faunas of the Bathonian-Callovian deposits of Jameson Land (East Greenland): Meddr Grønland, v. 87, n. 7, p. 1–158.

Spath, L. F., 1935a, Additions to the Eo-Triassic invertebrate fauna of East Greenland: Meddr Grønland, v. 98, n. 2, p. 1–115.

Spath, L. F., 1935b, The Upper Jurassic invertebrate faunas of Cape Leslie, Milne Land. I. Oxfordian and Lower Kimmeridgian: Meddr Grønland, v. 99, n. 2, p. 1–82.

Spath, L. F., 1936, The Upper Jurassic invertebrate faunas of Cape Leslie, Milne Land. II. Upper Kimmeridgian and Portlandian: Meddr Grønland, v. 99, n. 3, p. 1–180.

Steck, A., 1971, Kaledonische metamorphose der praekambrischen Charcot Land serie, Scoresby Sund, Ost-Grönland: Bull. Grønlands Geol. Unders., v. 97, p. 1–69 (also Meddr Grønland, v. 192, n. 3).

Steiger, R. H. and Henriksen, N., 1972, The geochronology of the Scoresby Sund area. Progress report 3: Zircon ages: Rapp. Grønlands Geol. Unders, v. 48, p. 109–114.

Stensiö, E., 1961, Permian Vertebrates, in: Geology of the Arctic, Vol. 1, Raasch, G. O., ed.: University of Toronto Press, Toronto, p. 231–247.

Surlyk, F. and Birkelund, T., 1972, The geology of southern Jameson Land: Rapp. Grønlands Geol. Unders, v. 48, p. 61–74.

Surlyk, F., Callomon, J. H., Bromley, R., and Birkelund, T., 1973, Stratigraphy of the Jurassic–Lower Cretaceous sediments of Jameson Land and Scoresby Land, East Greenland: Bull. Grønlands Geol. Unders., v. 105, p. 1–76 (also Meddr Grønland v. 193, n. 5).

Troelsen, J. C., 1949, Contributions to the geology of the area round Jørgen Brønlunds Fjord, Peary Land, North Greenland: Meddr Grønland, v. 149, n. 2, p. 1–29.

Troelsen, J. C., 1956, The Cambrian of North Greenland and Ellesmere Island: Int. Geol. Congr., 20th, Mexico, Symp. 3, v. 1, p. 71–90.

Trümpy, R., 1969, Notes on Triassic stratigraphy and paleontology of north-eastern Jameson Land (East Greenland). II. Lower Triassic ammonites from Jameson Land (East Greenland): Meddr Grønland, v. 168, n. 2, p. 77–116.

Turner, G., 1970, Thermal histories of meteorites, in: Palaeogeophysics, Runcorn, S. K., ed.: Academic Press, London and New York, p. 491–502.

Upton, B. G. J., in press, The alkaline province of South West Greenland, in: The alkali rocks, Sørensen, H., ed.: Interscience, London and New York.

Vischer, A., 1943, Die postdevonische Tektonik von Ostgrönland zwischen 74° und 75° N. Br. Kuhn Ø, Wollaston Forland, Clavering Ø und angrenzende Gebiete: Meddr Grønland, v. 133, n. 1, p. 1–194.

Vogt, P., 1965, Zur Geologie von Südwest-Hinks Land (Ostgrönland 71°30'N): Meddr Grønland, v. 154, n. 5, p. 1–24.

Wager, L. R. and Hamilton, E. I., 1964, Some radiometric rock ages and the problem of the southward continuation of the East Greenland Caledonian orogeny: Nature, v. 204, p. 1079–1080.

Wenk, E., 1961, On the crystalline basement and the basal part of the pre-Cambrian Eleonore Bay Group in the southwestern part of Scoresby Sund: Meddr Grønland, v. 168, n. 1, p. 1–54.

Wenk, E. and Haller, J., 1953, Geological explorations in the Petermann region, western part of Fraenkels Land, East Greenland: Meddr Grønland, v. 111, n. 3, p. 1–48.

Wilson, J. T., 1966, Did the Atlantic close and then reopen? Nature, v. 211, p. 676–681.

Witzig, E., 1952, Neues zur Stratigraphie des grönlandischen Karbons: Ecl. Geol. Helv., v. 44, p. 347–352.

Chapter 6

THE SCANDINAVIAN CALEDONIDES

Robin Nicholson

Department of Geology
The University
Manchester, England

I. INTRODUCTION

Almost the whole of the western edge of the landmass of Scandinavia is occupied by a 1700-km belt of deformed and sometimes metamorphosed rocks on which in the west lie a few isolated masses of relatively undeformed and completely unmetamorphosed Old Red Sandstone. The belt, shown in Fig. 1, where it contrasts with the undivided and unornamented autochthon on its east side, is subdivided in two ways. Firstly and fundamentally there is lateral division into elements running great lengths of the chain. These elements, the five zones of Fig. 1, are distinguished from one another by different rock assemblages as well as different levels of deformation and metamorphism. Boundaries between zones are not steep but gently dipping known, or presumed, thrust faults. Secondly, there is a division into Southern, Central, and Northern Caledonides over the late culminations of Grong and Rombak, which although very convenient for description are not as fundamental since the major zones appear to pass over them.

The Scandinavian Caledonides have a sharp east margin where allochthonous rocks overlie an autochthonous sequence of late Precambrian to Silurian sedimentary rocks themselves unconformable on rocks of the Baltic shield. This marginal autochthonous sequence which is thin except in the Oslo

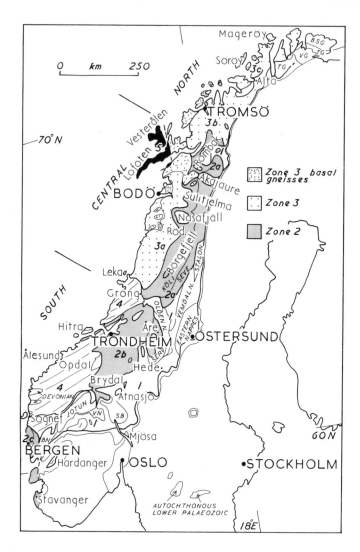

Fig. 1. Map of the Scandinavian Caledonides showing position on the western margin of the Baltic shield, and division into five zones and three elements over the Grong and Rombak culminations. Zone 1, eastern nonmetamorphic zone rich in late Precambrian and Eocambrian arkoses, including also the Jotun Nappe. Zone 2, the eastern metamorphic belt: zone 2a, the type area for the Seve–Köli including (in the Köli) metamorphosed but sometimes fossiliferous lower Paleozoic rocks; zone 2b, the Trondheim depression, a supposed continuation of the Köli unit; zone 2c, the Bergen arcs and adjacent areas. Zone 3, the western metamorphic belt: zone 3a, the Central Caledonides; zone 3b, the Northern Caledonides of Troms; zone 3c, the Northern Caledonides of Finnmark. Zone 4, the western gneiss area of the Southern Caledonides. Zone 5, the Lofoten region. Abreviations: Röd, Rödingfjell; VN, Valdres Nappe; J, Jotun Nappe; VG, Vestertana Group; SB, Sparagmite basin; BN, Bergsdalen Nappes; TG, Tanafjord Group; BSG, Barents Sea Group.

graben in the Southern Caledonides and Finnmark in the Northern Caledonides, is mainly restricted to a narrow zone along the edge of the allochthon.
Autochthonous Palaeozoic only becomes widespread when the South Baltic
Coast of Sweden is reached. According to Asklund (1960, p. 130) in part,
at least, this absence of a widespread cover reflects original distribution and
not merely post-Caledonian erosion.

Since the allochthon contains sure evidence that some of its metamorphic
rocks are Silurian and involved in the peak episodes of metamorphism and
deformation, a Late Silurian or Early Devonian climax of both must be
accepted for much of it. However, there are various indications, particularly
in the Northern Caledonides, of an earlier climactic phase comparable with
the Early Ordovician one of the Caledonides of the British Isles. At present
the definition of such areas is vague, but it seems they must constitute the
lesser part of the Scandinavian mountain system. Thus comparison of the
time of climactic deformation with the British Caledonian relates the mass
of the Scandinavian Caledonian with its high content of metamorphic rocks
to the generally nonmetamorphic southeastern sector of the British Caledonides
rather than with the earlier deformed Moinian and Dalradian metamorphic
core (Roberts, 1971). (See, however, Phillips, 1973 on the extent of the
early event in the British Isles.)

Knowledge of the clear thrust structure of the eastern edge of the Scandinavian Caledonides has dominated thinking in Sweden both in the initial stages
of geological enquiry some 100 years ago (Högbom, 1909) and in the last 40
years; and the now widely recognized major thrusting of the belt was first
described from the southern Central Caledonides of Sweden. In parts of the
north Central Caledonides, too, it is obvious that there are very large thrust
nappes since central windows like Rombak (Fig. 1) reveal gneissic Baltic
basement and an autochthonous thin Caledonian sedimentary cover without
Caledonian metamorphism or structure under Caledonian metamorphic rocks.
Relations between basal gneissic rocks and metasedimentary cover well within
the Scandinavian Caledonides and largely in Norway, however, are more
obscure since structural and metamorphic conformity of cover and basal
gneisses often is developed. Consequently, nappe interpretations of relationships are less than pressing, unless urged by wider considerations—cir-
cumstances that explain the relative reluctance in the past to accept nappe
interpretation in Norway.

II. THE AUTOCHTHON

The Baltic basement east of the Caledonides is extensive and varied in
age and lithology ranging from very old pre-Karelian in the north with ages

of 2800 m.y. or more to the 950–1150-m.y.-old Sveco-Norwegian in the south (Wilson and Nicholson, 1973). So far the distribution of Sveco-Norwegian ages shows no Caledonian pattern, although ages get close to the time of possible commencement of deposition of the dominantly arenaceous younger Precambrian/Eocambrian sediments (Föyn, 1967*b*, p. 7) which form such an important part both of the Scandinavian Caledonides and of the rest of the North Atlantic Caledonian. Those first sediments which lie with marked unconformity on often plutonic or metamorphic older Precambrian rocks principally occur, however, in the deformation belt, while outside it on the Baltic shield lower Paleozoic rocks generally lie directly and unconformably on the older Precambrian. Thus the younger Precambrian/Eocambrian distribution fits the later Caledonian pattern of deformation and together with the succeeding lower Paleozoic is distinguished here as the Caledonian sequence.

The stratigraphic relation between lower Paleozoic and older members of the Caledonian sequence beneath is best seen in the autochthonous–parautochthonous development of the Northern Caledonides (Banks *et al.*, 1971, p. 201) (Figs. 1 and 4), details of which appear in Tables I and II. A slight unconformity divides the older sedimentary succession. Above it lie Eocambrian tillites which presumably correlate with the Boulder bed of the Dalradian (Dewey, this volume) and below lies the Tanafjord Group of Siedlecki and Siedlecki (1971, p. 261) which can be compared with the upper part of the authochthonous Torridonian and thus perhaps part of the Moine also. A Rb/Sr whole-rock isochron (Pringle, 1973) from shales in the Northern Caledonian tillite sequence of about 700 m.y. dates the glacial episode, and another from shales in the lowermost part of the Tanafjord Group of about 800 m.y. fits upper Torridonian results from Scotland. The Barents Sea Group (Figs. 1 and 4) may underlie the Tanafjord Group (Siedlecki and Siedlecki, 1971, p. 294) and then be equivalent to the Eleonore Bay Group of northeast Greenland.

In the North Caledonides the postglacial shallow-water regime above the tillites (Banks *et al.*, 1971, p. 220–230) eventually includes trace fossils indicative of Cambrian age: the sequence then continues up into levels of Tremadocian age where it is cut off by a major thrust. Elsewhere the autochthonous marginal sequence usually reaches up only into Cambrian rocks, although Ordovician and Silurian are present in the Oslo area and Ordovician in Jamtland, near Östersund in the central Caledonides. In the North Caledonides autochthonous Eocambrian tillites are found well inside the fold belt at the south of the southern Alta window (Figs. 1 and 4), and similar internal occurrences are known from the Atnasjö and Brydal windows in the Southern Caledonides (Figs. 1 and 2). In all of them, however, the young Precambrian element is thin and the Eocambrian tillites lie almost directly on the old basement.

TABLE I

Selected Sequences in the Eastern Autochthon. Jämtland Is the Area Broadly to the West of Östersund (Fig. 3)

Autochthon	Southern Caledonides		Central Caledonides Jämtland	Northern Caledonides
	Oslo	Mjösa		
Downton	Sandstone			
		(Missing)		
Ludlow	Shale			
	Limestone	Sandstone		
Wenlock	Shale	Shale	(Missing)	
	Limestone	Limestone		
Llandovery	Sandstone and shale			
Ashgill	Sandstone (Missing)	(Missing)		(Missing)
Caradoc	Limestone and shale	Limestone	Limestone shale conglomerate	
Llandeilo	Shale	Shale	(Missing)	
Llanvirn	Limestone and shale		Limestone	
		Limestone and shale		
Arenig	Limestone		Shale	
			Sandstone	
Tremadoc	Alum		(Missing)	
		Shale		Sandstone and shale
U. Cambrian	Shale		Shale conglomerate	
M. Cambrian			(Missing)	
		Limestone	Sandstone conglomerate	Sandstone and siltstone of Vestertana group
L. Cambrian	(Missing)	Shale and siltstone		
	(Old Precambrian)	Vemdal Formation etc. below (see Table II)	(Old Precambrian)	etc. below
Eocambrian				(see Table II)
	Strand and Kulling, 1972	Strand and Kulling, 1972	Strand and Kulling, 1972	Banks et al., 1971

Note. In order to keep tables simple and to emphasize lithology rather than nomenclature, a minimum of formation names etc. is used in them. Age assignments are made in tables only where certainly identified.

TABLE II

Selected Sequences in Zone 1 (Fig. 1)

Zone 1	Southern Caledonides		Central Caledonides Jämtland		N. Caledonides (Gaissa Nappe etc.)
	Valdres Nappe	West Sparagmite Basin	Olden Nappe	Vemdal Nappe	
Ludlow					
Wenlock			(Missing)	Greywacke Shale Lst Sandstone	
Llandovery	(Missing)			(Missing)	
Ashgill			Shale	Shale	
Caradoc			Greywacke	Lst greywacke Shale	(Missing)
Llandeilo	Graptolitic (Mellsenn Formation)		Shale	Shale	
Llanvirn	Shale	Grey (Graptolites)	Conglomerate	Limestone (Missing)	
Arenig	Shale with limestone	Phyllites	(Old Precambrian)	Limestone Shale	
Tremadoc	Quartzite	Alum (Phyllite Formation)	(Thrust base)	(Missing) Alum shale	Sandstone and shale
U. Cambrian				(Missing)	
M. Cambrian	Shale	Shale		Shale	
L. Cambrian		Shale Siltstone Sandstone		Sandstone Conglomerate	Sandstone
Eocambrian	Quartzite (Vemdal Formation) Green-grey Shale Tillite Arkose Conglomerate Arkose Conglomerate Arkose	Quartzite (Vemdal Formation) Green-red shale Tillite Arkose Unconformity Shale Limestone Arkose		Quartzite (Vemdal Formation) (Thrust base)	Sandstone (Vestertana group) and Siltstone Tillite Dolomite Tillite Unconformity Sandstone (Tanafjord group) Siltstone
Precambrian	Precambrian Jotun-type rocks	Old Precambrian			Old Pre-cambrian
	Nickelsen, 1967	Strand and Kulling, 1972	Strand and Kulling, 1972	Strand and Kulling, 1972	Banks et al., 1971

166

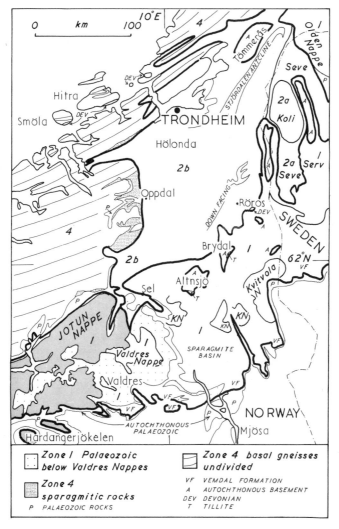

Fig. 2. The eastern part of the Southern Caledonides: KN, Kvitvola
Nappe.

In the autochthon of the Östersund area of the county of Jämtland there
is no late Precambrian or Eocambrian sequence, and the Middle Cambrian
is partly littoral: there are indications of tilting in the lower Caradoc (Thorslund, 1960, p. 101). In the Oslo area also, for the most part, no younger Precambrian or Eocambrian sedimentary sequence intervenes between lower
Paleozoic rocks and the older Precambrian basement (Henningsmoen, 1960,
p. 133). To the north of Oslo, however, along the north part of Lake Mjösa
(Fig. 2), there is the edge of an early Caledonian sedimentary development

equivalent to that of the North Caledonides (Strand, in Strand and Kulling, 1972, p. 17); Table II shows the division of these rocks and others above them in the so-called Western Sparagmite basin. The edge of the area of thicker sediments approximates to the edge of the allochthon so that the transition between the presumed troughs of accumulation of thicker western sequences and the shelf to the east is not well known. It was from these earlier defined western sedimentation areas or troughs, however, that transgression spread east and south in the Eocambrian and Cambrian, the whole Oslo region being inundated by the Middle Cambrian. This Oslo sequence generally is one of limestones and shales, indicating shallow-water sedimentation with a considerable break between about Ordovician and Silurian. In the Silurian, a sandier facies develops in the west (Henningsmoen, 1960, p. 147).

Both the Paleozoic of the Oslo graben and that of the North Caledonides are folded, but in the latter area no rocks younger than Tremadocian are present. The Oslo structures, however, involve the Ludlovian/Downtonian Ringerike Sandstone and so demonstrably are late. Folding occurs only in the northern part of the long Oslo region: folds are disharmonic, dying away to the basement surface (Strand, in Strand and Kulling, 1972, p. 13). As clearly expressed by Strand (in Strand and Kulling, 1972, p. 15, 16), there is some difficulty in dividing allochthon from autochthon at the north end of Lake Mjösa, since the disharmonic folds south of the obvious thrusts probably involve southern transport over basement so that there is a transition between allochthon and autochthon.

The involvement of the Downtonian Ringerike Sandstone in folds at the north end of the authochthon and immediately south of the obvious thrusts may set a minimum age for the most marginal nappe movements in the region. It is probable that the site of marginal thrusting was fixed by the edge of internal sequences with thick arenaceous Eocambrian, the sandstones sliding south over the shales of the thinner shelf sequence and above autochthonous basement (this chapter and Gee, 1972, p. 12, for a similar situation in the Central Caledonides in Sweden).

III. ZONE 1: THE NONMETAMORPHIC BELT

A. Introduction

Zone 1, the most easterly of the Scandinavian Caledonian zones distinguished here, broadly lies with the eastern marginal thrust or thrusts of the deformation belt below it and the thrust base of the metamorphic allochthon of zone 2 above. However, sequences assigned to it in part appear to be continuous into autochthonous ones in the Northern Caledonides (Fig. 1) where the marginal thrust dies out eastwards. Elsewhere, for example north of Oslo, much of zone 1 appears to be no more than parautochthonous.

Zone 1 is composed largely of the young Precambrian–Eocambrian arenaceous rocks mentioned above, but includes also some of the lower Paleozoic rocks that stratigraphically succeeded them as well as some allochthonous basement rocks. The Paleozoic element is very like that on the basement of the autochthon to the east and is best regarded as once continuous with it. Ross and Ingham (1970, p. 398), in an analysis of some fossil provincial affinities, seem to consider the two as capable of being in different Caledonian plates (see p. 198); on structural grounds, this view seems mistaken (Nicholson, 1971a). In the region east of Ostersund in Jämtland in the Central Caledonides (Fig. 3) Silurian occurs in the allochthon of zone 1 although not in the autochthon (Table II here and Kulling, in Strand and Kulling, 1972, p. 159, 186); the youngest Silurian of zone 1 is Wenlockian.

The sequences of zone 1 generally are little affected by Caledonian metamorphism and are less deformed than the rocks of more westerly zones. Some of the uppermost structural elements of zone 1, however, are transitional in structure and metamorphism to the metamorphic allochthon above. These higher units, the Serv Nappe (Strömberg, 1961) and the Kvitvola Nappe (Strand, in Strand Kulling, 1972, p. 26) of the Southern Caledonides also differ in containing dolerite dikes. In the Northern Caledonides, where the whole metamorphic allochthon is far richer in basic intrusives than the rest of deformation belt, dolerites occur in the parautochthonous Gaissa Nappe (Table II) as well as in the Laksefjord Nappe (which lies about at the same structural level as the Serv Nappe). On the whole, however, the character of zone 1 suggests a role of passive reaction to the emplacement of the metamorphic allochthon from westerly and active zones of the belt.

The Paleozoic rocks of zone 1 often are described as miogeosynclinal: they are almost entirely nonvolcanic, relatively thin, probably of fairly shallow-water origin, and hardly metamorphosed, although they may be considerably deformed into both thrust and fold nappes (Strand, 1960, p. 128). In the south part of the Hardanger region, southwest of the *H* of Hardanger of Fig. 1 (Sörbye, in Strand and Kulling, 1972, p. 48), there is an upward increase of metamorphism described from a phyllite–schist sequence above basement said to grade up into gneissic rocks presumed to belong to the Jotun Nappe. Müller and Wurm (1970) report similar metamorphic features from the Stavanger area where volcanic rocks are also reported. The Stavanger rocks also have strong resemblances to the lower Paleozoics of the Bergen district, here distinguished as zone 2c.

Eocambrian tillites (p. 164) are not known outside zone 1 except as autochthonous cover revealed in windows down to the autochthon cut through zones 1, 2, and 3. There the sequence below the tillites is thin, and the sequence

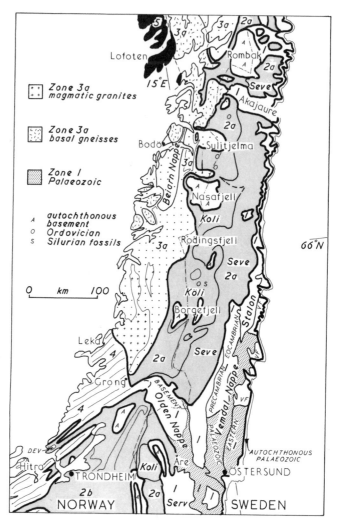

Fig. 3. The Central Caledonides, Trondheim to Lofoten. Blank right-hand part of the map is the Baltic shield: VF, Vemdal Formation; DEV, Devonian.

might be viewed as lying on the west side of the initial troughs of sedimentation or perhaps as on elevations within them. However, arenaceous rocks of similar lithology and structural position to the upper parts of the young Precambrian–Eocambrian do occur in the metamorphic sequences of other zones.

Zone 1 is completed in the Southern Caledonides by the Jotun Nappe (Strand, in Strand and Kulling, 1972, p. 38), composed of distinctive charnockitic and anorthositic rocks. It is arguable whether or not it ought to be included in zone 1 or merged with the Trondheim structure and put in zone 2.

The key area is in Sel and Vågå (Strand, 1951) where the Jotun and Trondheim regions meet. There it is reported by Gjelsvik (1948) and Strand (1951; and in Strand and Kulling, 1972, Fig. 22) that Trondheim facies sediments may well rest unconformably on a gneissic complex containing anorthosites of Jotun type and origin. Unfortunately, it is unclear still whether or not these gneisses are directly continuous *now* into the Jotun rocks proper, and this doubt together with the apparently negligible Caledonian metamorphism and deformation of the Jotunheim rocks has lead to their placing here in zone 1. This division follows Kulling (in Strand and Kulling, 1972, p. 279). It conforms also to the place in the Valdres Nappe of Jotun rocks unconformably under late Precambrian and Eocambrian sequences distinguished by Loeschke (1967, p. 60) as typical of what is here designated zone 1. In contrast, Jotun-type rocks are included in zone 2c where they allochthonously overlie rocks of Trondheim affinities.

B. Regional Developments

1. Valdres–Hede Region—Southern Caledonides

At the eastern edge of the allochthon in this tract (Figs. 1, 2, and 3), as well as further north, there often are thrust sheets of the arenaceous Eocambrian Vemdal Formation lying on more or less autochthonous Cambro-Ordovician. North of Oslo and northwest of such thrust Vemdal rocks, there is the wide so-called West Sparagmite basin mentioned above (Sparagmite after Esmark's term for the pre-Paleozoic but young Precambrian and Eocambrian arenaceous rocks; Holtedahl, 1960, p. 110; Strand, in Strand and Kulling, 1972, p. 7), now an anticlinal zone showing lower Paleozoic down to the base of the late Precambrian (Skjeseth, 1963). This basin lies to the west of a basement ridge beyond which there is a second and eastern basin which itself has a continuation into Sweden narrowing toward Hede.

The Swedish extension of the Sparagmite basin has been considered to be autochtonous, an anticlinal or horst zone, over which to the west the frontal nappes of the Vemdal Formation must find their roots (Asklund, 1955). Here we have taken Kulling's view that the Vemdal Formation roots within the Sparagmite basin, and not west and over it (Kulling, in Strand and Kulling, 1972, p. 185). Recent detailed work in the southern Central Caledonides of Sweden (Gee, 1972, p. 11, 208–222) has confirmed Kulling's opinion.

At the eastern thrust edge of the Southern Caledonides from the Oslo region northward to beyond Östersund (Fig. 2) the Vemdal Formation is succeeded stratigraphically by a conformable, thin Paleozoic sequence. This sequence is called the Phyllite Formation in the region southwest of the West Sparagmite basin, where it is overlain by the Valdres Nappe of Kulling (initial definition, 1961) and Nickelson (detailed description, 1967). The latter is made up of folded rocks lying on a clean-cut thrust which may more or less follow the middle limb of a recumbent anticline–syncline pair. These higher rocks now are known to belong to the young Precambrian/Eocambrian sequence typical of zone 1 (the Sparagmite sequence of looser terminology) although first described by Goldschmidt (1916) as Taconic flysch stratigraphically as well as structurally overlying the Ordovician Mellsen Formation (Table II). Further, that formation is upside-down and above the thrust and on top of right-way-up Ordovician of the Phyllite Formation cover of the Vemdal Formation further below. The oldest Valdres sediments lie unconformably on lithologies typical of the Jotun Nappe structurally above (Table II).

The other structural unit recognized in the sparagmite area of the southern Caledonides is the Kvitvola Nappe (Strand, in Strand and Kulling, 1972, p. 26), largely defined on content of a different facies of arenaceous rock from the southern basins. This is the light sparagmite, a well-layered, quartzofeldspathic rock, which is the typical sparagmite outside the marginal basins and which, apparently younger than the tillites, is Eocambrian in the sense of Føyn (1967b, p. 7). The Nappe has its type area in the Southern Caledonides about halfway between Valdres and Hede on the Norwegian–Swedish border where it reaches to the front of the deformation belt. Extensive areas of the light sparagmite occur further north, however, toward the overlying Trondheim rocks and also are put in the Kvitvola Nappe. The light sparagmite is a metamorphic rock of low greenschist facies.

In a broad way the Kvitvola Nappe lies below the level of the Jotun Nappe and at about the level of the Valdres Nappe, but a straightforward reading of the 1:1,000,000 map of Norway (Holtedahl and Dons, 1960; Strand, in Strand and Kulling, 1972, Fig. 6) seems to make the northwest edge of the Kvitvola Nappe fit under the lower Paleozoic of the Phyllite Formation which has the Valdres Nappes on top. The Phyllite Formation, however, is said to be in stratigraphic order structurally on top of the Sparagmite basin sequence, thus leaving no room for the Kvitvola Nappe. Strand (in Strand and Kulling, 1972, p. 38) has had doubts on the detailed correlation of Kvitvola and Valdres nappes.

The Serv Nappe of the Swedish side of the border north of Hede (Figs. 2 and 3) is at the top of its local zone 1 sequence, presumably below the Jotun Nappe level, and thus at about the same level as the Kvitvola Nappe. The Serv Nappe, like the latter, also is constructed of quartzofeldspathic rocks with a distinct if low-grade metamorphic fabric (Strömberg, 1961); some of them are greywackes, which are not known from the Kvitvola Nappe but which resemble rocks in the Valdres Nappes (Strand, in Strand and Kulling, 1972, p. 38).

The highest unit of the structural sequence of the south Caledonides of zone 1 is the Jotun Nappe of the Jotunheim, much of it composed of distinctive

charnockitic and anorthositic rocks. As we have noted, its structural independence of zone 2 is not certain. On the south side of the main mass of the
Jotun Nappe there are apparent topographic outliers or klippe of the nappe
more or less directly on the autochthon. To the north of them the main mass
of the Jotun Nappe dips down into a trough that runs northeast–southwest
and from which it rises to the northwest to again lie flatly on basement. It has
been regarded by some as rooting in the trough (Goldschmidt, 1912; Smithson
and Ramberg, 1970); the southeastern marginal klippen would then not be
representative of the Jotun structure as a whole. Here, on the contrary, the
klippen are taken as resembling the main nappe in rooting northward of
their present position.

Jotun-type rocks in the Bergen area may be a one-time continuation of the
Jotun Nappe. At the west end of the Jotun Nappe, east of the Bergen area
and lying between the nappe and the autochthon, lie the Bergsdalen nappes
(Fig. 1) of Kvale (1955) composed of basement and an apparently marginal
lower Palaeozoic sequence. The Bergsdalen nappes thus have a similar structural position to the Valdres Nappe. However, their Precambrian rocks are
very like nearby autochthonous rocks to the southeast and not exotic like
the Jotun rocks of the Valdres Nappe.

No radiometric ages are available from the Jotun Nappe proper, but
acceptance of correlations between its rocks and typical Jotun types unconformably beneath the late Precambrian of the Valdres Nappe defines its
rocks as Precambrian. Hossack (1968) has reported only retrograde Caledonian
metamorphic effects from the Jotun Nappe of the Bygdin area of northeast
Valdres.

2. North of Hede—Southern Central and Northern Caledonides

Some 60 km north of Hede (Figs. 1 and 3) in Jämtland and west of Östersund in Sweden there is a region roughly comparable with that around Valdres:
once again there is a wide tract of probably allochthonous lower Paleozoic
which here, crossing over the national border to the northwest and into Norway, thins like the Serv Nappe above it (Fig. 3). The major difference between
interpretations of the structure of this Jämtland Paleozoic by Asklund (1960,
p. 131–134, 144–148) and Thorslund (1960, p. 101–110) on the one hand, and
Kulling (in Strand and Kulling, 1972, p. 178–200) on the other, centers on the
supposed character of the extreme marginal nappes of the Vemdal Formation.
Here we adopt, for regions south of Hede, Kulling's interpretation of them as
rooting together with some of the allochthonous Paleozoic of the region on
top of the allochthonous Eocambrian and late Precambrian on their west
side (see Gee, 1972). To Asklund, however, these latter rocks form another
anticlinal or horst zone like that he proposed south of Hede and over which

thrust Vemdal rocks have to root. Not all the allochthonous Paleozoic lies above the nappes of the Vemdal Formation, however; some lies below them in what Kulling has distinguished as the East Jämtland Nappes (Kulling, in Strand and Kulling, 1972, p. 178–180). (Jämtland is the county in which Östersund lies.)

All the Jämtland nappes except the highest are cover nappes, that is, they include no basement and with the same exception begin their Paleozoic sequences with the Cambrian. There are breaks especially at the top of the Cambrian and at the top of the Ordovician, but sequences go up into the Silurian. However, the highest Jämtland nappe, the Olden Nappe, which lies directly under the Serv Nappe, has Arenig rocks as its oldest Paleozoic (Table II) deposited on old basement also in the nappe (Asklund, 1955), and contains Caradoc greywackes which thin eastwards, in primary fashion. Greywackes also are found lower in the structural sequence at the top of the lower nappe that contains the Vemdal Formation, the Vemdal Nappe of Kulling (in Strand and Kulling, 1972, p. 18); these greywackes are Wenlockian (Table II), and they are said to thin eastwards similarly (Kulling, in Strand and Kulling, 1972, Fig. 138). The gneisses at the base of the Olden Nappe (Fig. 3) appear to be continuous with the basal gneisses of the Grong culmination: a relation that taken at its simplest has led Asklund (1955) and Kulling (in Strand and Kulling, 1972, p. 278) to suggest that the western basal gneisses (Figs. 1 and 3) also all belong to the Olden Nappe.

Above and northeast of the Jämtland allochthonous Paleozoic, and below rocks of zone 2, is a thrust unit of Eocambrian and late Precambrian rocks (Fig. 3) comparable with the Serv Nappe but lacking dolerites. This is Kulling's Stalon complex (Kulling, 1955, p. 122; in Strand and Kulling, 1972, p. 192). Similar rocks occur at this structural level for a considerable distance further north but at about lat. 67° N, about that of Sulitjelma of Fig. 3, are replaced by gneissic rocks of presumed Precambrian basement origin. First descriptions were made by Hamburg (1910) who defined a Syenite Nappe in the southern Akajaure area (Fig. 1) which Kulling (1964) has suggested is continuous below ground with the basal gneisses on the Norwegian side of the border which therefore also are allochthonous. Kautsky (1953, p. 169) has offered an alternative explanation suggesting that the supposed coherent basement nappe is an imbricate zone of locally derived scales of basement overlying autochthonous basement common to regions on both sides of the Norwegian–Swedish border. This view better fits the proposal that the autochthonous and generally un-Caledonized basement of the Rombak window to the north of the Akajaure area merges in depth with the Caledonized basal gneisses to the west in Norway (this chapter and Nicholson and Rutland, 1969, p. 75).

In the eastern marginal regions of the Caledonian allochthon north of the Akajaure area and east of the Rombak window (Fig. 3), there is a nappe apparently comparable with Hamburg's Syenite Nappe, the Abisko Nappe (Kulling, 1964, p. 82). Its Precambrian basement rocks are without Caledonian metamorphism and lie with strong structural discordance on an imbricate zone of Eocambrian and younger Precambrian rocks which die out westward allowing the higher nappe to lie directly upon the autochthon. The junction with zone 2 rocks above is said by Kulling (Strand and Kulling, 1972, p. 203) to be much less abrupt than with the rest of zone 1 below, the particular upper junction rocks here being the so-called "hardskiffer," an often flinty, well-banded, and well-lineated quartzofeldspathic rock with some resemblances to members of the arenaceous Eocambrian sequence. This is the lowest level at which a penetrative Caledonian lineation is prominent in this region.

North of the Rombak window possible equivalents of the lower and little metamorphosed nappes of further south may be defined. These are the arenaceous Storfjell Group (Gustavson, 1966, p. 15) and the similar lower grade rocks of Oleson (1971) plus the unit of sparagmitic rock of the Birtavarre area (Fig. 4; Padget, 1955). In the Northern Caledonides there is a lower nappe composed of the Tanafjord Group (Siedlecki and Siedlecki, 1971, p. 261) and an upper one of the presumed Precambrian Laksefjord Group (Fig. 4), both under the metamorphic Kolvik Nappe (Gayer and Roberts, 1971, p.

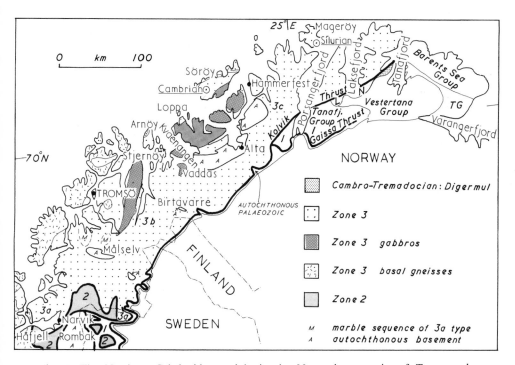

Fig. 4. The Northern Caledonides, mainly in the Norwegian counties of Troms and Finnmark: TG, Tanafjord Group; LN, Laksefjord Nappe.

28–42). Banks *et al.* (1971, p. 233) have suggested that if the rocks of the Laksefjord Nappe are young Precambrian, as their general arkosic character might suggest, then the very considerable differences between their thickness and sedimentary development and those of the young Precambrian of the autochthon could indicate an original separation of several hundred kilometers.

IV. ZONE 2: THE EASTERN METAMORPHIC BELT

Zone 2, separated from the higher zone 3 by a presumed thrust fault, is distinguishable from it in two ways. Firstly, its metamorphic grade is not uniformly high as is that of zone 3. Secondly, and apparently as a consequence, fossils are still found in its rocks so that stratigraphic work is possible in a way it is not in zone 3.

Zone 2 is divisible into three parts as yet only loosely correlated. One of these, zone 2a, in the Central Caledonides of Sweden, is the type area for the overthrusting of metamorphic rocks on to unmetamorphosed Paleozoic and seems to have a similar stratigraphy through its length of over 500 km. The second, the Trondheim region of Norway, may be more complex in structure and metamorphism. Zone 2c, the Bergen district of Norway, is less well known than either 2a or 2b but, as noted below, may display early Paleozoic deformation comparable with that of the Dalradian of Scotland as well as the later and probably Silurian deformation so typical of the Scandinavian Caledonides.

It can be argued that much of zone 1 cannot have been too distantly transported since it includes a bulk of unmetamorphosed Caledonian sediments whose stratigraphy has close links with that of the autochthon. The metamorphic rocks of zones 2 and 3 which lie above them offer no such easy argument. They have better developed and more obscuring deformation and metamorphism than the rocks of zone 1 and no obvious stratigraphic likeness to the autochthon except for the common presence of sparagmitic rocks often in basal situations. Also they have an ambiguous relation with the basal gneissic rocks under them in the west. In the east zone 2 rocks lie on zone 1, usually sharply but sometimes with something of a transitional character.

A. Zone 2a: The Seve–Köli

Zone 2a ranges from the north end of the southern Caledonides to the region of the Rombak window (Fig. 3). In the south it includes the Åre area with metamorphic rocks clearly overlying the allochthonous fossiliferous Paleozoic of the Jämtland nappes. The zone 2a metamorphic sequence is

broadly divisible into two subunits with long-established names of Seve below
and Köli above (Kulling, in Strand and Kulling, 1972, p. 161). The Seve,
generally regarded as Precambrian, has not yielded much of stratigraphic
value, although results which are beginning to come in may in part disturb
that assumption (Zwart, 1972). Essentially, the name Seve is given to higher
grade unfossiliferous metasediments at the base or front of a complex of
metamorphic rocks whose upper part, the Köli, has yielded more stratigraphic
information.

Zachrisson (1969, p. 17) has reported a tectonic contact between Seve
and Köli, while Trouw (1972) has emphasized lithological and metamorphic
continuity, the Seve having a higher grade core than margins, metamorphism
being Caledonian. According to Glass (1972) granulite facies grades are
reached in part of the Seve southeast of Borgefjell, an interesting comparison
with the Jotun Nappe at about this structural level whose granulite metamorphism probably is Precambrian. It has recently been shown (Nicholson
and Rutland, 1969; Zachrisson, 1969, p. 27) that, like other stratigraphical/
structural units in the Central Caledonides, at least, the Seve thins out westward, being absent in the Norwegian Central Caledonides and apparently
unknown in the Southern and Northern Caledonides too (unless we chose
to regard the Jötun Nappe of the former as an extension of the Seve).

The Köli has been the subject of extended treatment recently (Zachrisson,
1969; Kulling, in Strand and Kulling, 1972); here only outlines of its stratigraphy and structure are given. Firstly, the type region east of Borgefjell
(Figs. 1 and 3) has two developements, east and west of an anticline revealing
sparagmitian rocks of zone 1 (Table III). The eastern depression contains
useful fossils while the western does not. No Cambrian fossils are known and
there is not much room for them in the Köli between the lowest known
Ordovician and the Seve: upward the Köli sequence reaches upper Llandovery
with further so far unfossiliferous rocks above. Isotope work at Sulitjelma
north of Nasafjell in the Köli confirms the view of Silurian tectonism (Wilson,
1971). There Upper Ordovician fossils are known from rocks similar to those
further south: again no Cambrian fossils are yet known (Kulling, in Strand
and Kulling, 1972, p. 231–242).

One distinctive feature of the zone 2a Köli are the serpentinite conglomerates of the Ordovician, all at a well-defined stratigraphic level. It has been
suggested that they represent contemporaneously eroded sea-floor extrusives
(Kulling, in Strand and Kulling, 1972, p. 245). At Otta at the southwest end
of the Trondheim region, serpentinite conglomerates have a rich Lower–
Middle Ordovician fauna (Strand, in Strand and Kulling, 1972, p. 36).

Zachrisson (1969, p. 16) has divided the Swedish Köli directly east of
Börgefjell (Fig. 3) into several nappes with the higher occurring only in the
west by the Norwegian border and extending over it so that the Köli of the

TABLE III

Sequences in Zone 2a East and West of the Anticline East of Borgefjell Revealing Zone 1 Rocks Under Those of Zone 2a (Fig. 3)

Zone 2a	Central Caledonides	
	Eastern Köli (Remdalen)	Western Köli (Bjorkvattnet)
	Basic volcanics	
	Dark graphitic phyllites	
	Calcareous phyllites	
	Quartzite conglomerate	Quartzites in calcareous phyllites
	Calcareous phyllites many gabbros	Calc-quartzite, greywacke, slate
Llandovery	Basic and acid volcanics	Slate with graptolites
		Quartzite with fossils
Ashgill	Thin impersistent limestone and quartzitic conglomerate or quartzite	Limestone with fossils, calcareous quartzite
		Marl, limestone conglomerate
		Quartzitic conglomerate
		Quartzite unconformity
	Graphitic phyllite quartzite	Black slates
	Conglomerate, limestone	Greywackes, basic volcanics
	Basic and acid volcanics	Basic volcanics, slates
	Grey and graphitic phyllite	
	Serpentine conglomerate	Serpentine conglomerate
	Varied sediments, tuffs	
	Other volcanics	
	Seve	Seve
	Zachrisson, 1969	Strand and Kulling, 1972

Swedish side merges with the lower grade metamorphic rocks of the eastern Norwegian Central Caledonides (Strand, 1955). The latter are overlain to the west by high-grade rocks without fossils or way up structures. The Köli in this broad sense underlies such high-grade rocks for the whole of its western limit, the type locality for the relation being in north Västerbotten on Rödingsfjell from which Kulling (1955, p. 215) defined a new western unit, the Rödingsfjell Nappe (Fig. 3), to contain the whole tract of higher grade metasediments and granitic rocks that extend so widely in the western part of the Central Caledonides. In fact, this precise type of arrangement had been described earlier by Kautsky (1953, p. 95) from further north in the Central Caledonides in the Sulitjelma area whose west side shows, without room for doubt, the way in which the whole Köli unit of Sulitjelma thins, like the Seve beneath it, so that both disappear westward in the cover of the Nasafjäll basement window (Nicholson and Rutland, 1969, p. 26).

Nicholson and Rutland (1969, Fig. 17) have speculated on the scale of Köli extension north of Sulitjelma from their own work and that of Kautsky (1955), Gustavson (1966), and Foslie (1941), and these results and those of Kulling (1964) suggest that high-grade rocks of the Rödingfjell Nappe in the broad sense of Kulling's definition reach north and east of the Rombak window almost to the eastern edge of the allochthon (Fig. 4) so that no Seve or Köli is present there.

The metamorphism of the Köli and Seve rocks, except for possible Caledonian granulite facies development in the latter, everywhere is Barrovian. Often near the top of the Köli higher grade types lie above lower. At least at Sulitjelma isograds in the Köli are not coincident with supposed major thrusts (Henley, 1970, p. 121). Nicholson and Rutland (1969, p. 71) suggest that increase of grade upward may indicate hotter rocks thrust in above; Mason (1967) and Henley (1970) prefer to regard it as a result of intrusion above. Elsewhere in the Köli (Zachrisson, 1969) the not uncommon position of higher grade rocks structurally above lower has been explained by the over-thrusting of the higher grade.

B. Zone 2b: The Trondheim Depression

West of the Åre area of Jamtland (Figs. 1 and 2) low-grade rocks belonging to the Köli dip west under the edge of the wide area of Paleozoic of the Trondheim region which Törnebohm (1896) made the anticlinal core of the Scandinavian Caledonian with thrusting of fold nappes to east and west from it. However, broadly (Fig. 2) the region lies with basement of the central domes of the fold belt in the east and broad areas of basal gneiss in the west. Its north end is crossed by the basement of the tract of culmination west from Grong into Sweden, while at the south end only a couple of kilometers of Trondheim rocks separate inwardly dipping sparagmite sequences above basement from one another. Thus the Trondheim region occupies a structural depression.

In the view of Wolff (Wolff et al., 1967, p. 126–127) and Roberts et al. (1970, p. 133) the northern part, at least, of the Trondheim depression has a somewhat mushroomlike anticlinal spine (the Stjördalen anticline of Fig. 2) flanked by relatively old but still Late Silurian fold nappes. This is an explanation not unlike the very early one of Törnebohm (1896) noted above. The interpretation fits the presence of generally younger rocks on the edge of the Trondheim depression and the greater age and higher grade metamorphism of the rocks of its core and is an alternative to the general explanation of similar metamorphic change upward in the Köli in terms of thrust sheets of higher grade rocks above low-grade ones. Rui (1972, p. 19–20), from the southeast of the Trondheim region, has described regional inversion of rock sequences from the lower Paleozoic compatible with the postulated fold nappes of the north (Fig. 2), the structures proposed having some resemblance to the Tay (fold) Nappe of the Dalradian of Scotland (Dewey, this volume, Chap 7).

In a broad way the Trondheim depression seems continuous into the trough that contains most of the Jotun Nappe. Neither there nor north of the Trondheim region in the Köli with which the Trondheim Paleozoic usually is stratigraphically and structurally correlated, however, is there any trace of the early-formed central anticlinal spine which, according to Roberts (in Wolff et al., 1967, p. 88), was present before the generally assumed large-scale thrusting eastwards of the Trondheim sequences.

Table IV shows the most recent view of the stratigraphy of the barely metamorphosed rocks to the southwest side of the spine of the northern Trondheim region, not far southwest of Trondheim itself (Chaloupsky, 1970). This stratigraphy represents considerable changes from that of Vogt (1945) and will necessitate consequent changes in extrapolations based in part on Vogt's work (e.g., Kautsky, 1948). Among the oldest rocks of the region are the Stören greenstones, a thick sequence of basaltic volcanics rich in pillow lavas. Chaloupsky (1970, p. 279) judged their age to be Tremadoc–Arenig, although that is not too well fixed, depending upon views of the age of grapto-lites of the Bogo Shale, a part of the succeeding sequence.

An important stratigraphic break occurs above the Stören greenstones, but none has been suggested between it or its equivalents elsewhere in zone 2b and the schists beneath. And since these rocks below the greenstones have not yielded fossils older than Tremadocian, the importance of Cambrian rocks is obscure in zone 2b (Rui, 1972, p. 19) as it is in zone 2a.

The movement phase represented by the post-Stören unconformity, Vogt's Trondheim disturbance (Vogt, 1945), has been compared with the Grampian event of the metamorphic Scottish and Irish Caledonides (Skeving-ton and Sturt, 1967), but while the two may be roughly contemporary they certainly are not similar (Roberts, 1971). The field evidence of the Trondheim region is probably of a Silurian major deformation and metamorphism, a view confirmed by recent isotope studies (M. R. Wilson, pers. comm.). On the island of Hitra and islands southeast of Smöla (Fig. 2), northwest of Trondheim fjord, there is Old Red Sandstone unconformable on a western extension of the rocks of the Trondheim area (Strand, in Strand and Kulling, 1972, p. 58): fossils (Störmer, 1935) indicate a Downtonian age for some of the higher Old

TABLE IV

Sequence in the Holönda-Hulsjöen District, Southwest of Trondheim, Zone 2b (Fig. 3)

Zone 2b	Southern Caledonides Hölonda-Hulsjöen (Trondheim District)
	Sandstone shale
	Conglomerate
	(Missing)
Ashgill	Dark slates
	Sandstone, limestone (fossils)
Caradoc	Conglomerate
	(Missing)
	Acid volcanics
	Sandstone, conglomerate
	Slate
	Basic volcanics
Arenig	Slate, limestone (fossils)
	Conglomerate
	(Missing)
	Basic volcanics
	(Stören)

Chaloupsky, 1970

Red rocks. In the Trondheim area itself, in Old Red Sandstone near Röros, there are plants of a Lower Devonian age (Strand, in Strand and Kulling, 1972, p. 128).

C. Zone 2c: The Bergen Arcs

The Bergen arcs (Fig. 1) are said to be formed of a compound synform later bent so as to be convex eastward in plan and overturned to the west. The core is occupied by Jotun-type rocks lying on lower Paleozoic sequences exposed in two concentric arcs, one on either side of the Jotun-type rocks. The Paleozoic sequences, which appear to be the same, have some resemblance to that of the Trondheim region (Strand, in Strand and Kulling, 1972, p. 70) containing volcanics and intrusives. Both, in turn, lie to both concave and convex sides of the arcs on top of basement rocks typical of the western gneiss area of the Southern Caledonides.

Perhaps the most interesting aspect of the Bergen arcs district for the fold belt as a whole is the indication of early Ordovician deformation and metamorphism there as in the Northern Caledonides. There is an important break and presumed unconformity in the Bergen Cambro-Silurian between upper rocks with no intrusives and relatively little deformation and a more deformed sequence below which includes intrusive rocks. The upper rocks contain Upper Ordovician and younger fossils. No fossils have yet been discovered below (Kvale, reported by Strand, in Strand and Kulling, 1972, p. 228).

The lower Paleozoic sequence of the Bergen arcs extends north to regions where it is overlain by Devonian conglomerates. A similar sequence occurs in the Stavanger district.

The Jotun-type rocks above Trondheim-type Cambro-Silurian themselves are overlain unconformably by Devonian and form part of a nappe that once may have been continuous with the extensive Jotun Nappe of zone 1. However, these Jotun-type rocks not only lie above Trondheim sequences but also have been involved in postemplacement deformation that formed the curving pattern of the Bergen district and to which there is no analogy in the Jotunheim to the east but which might be correlated with the gneiss dome tectonics found in the western parts of the deformation belt. In some ways, therefore, the sequences and structures of the Bergen region are intermediate between Jotunheim and Trondheim developments.

V. ZONE 3: THE WESTERN METAMORPHIC BELT

A. Introduction

Zone 3 (Fig. 3) in the Central Caledonides is composed of high-grade schists often granite veined and migmatitic which in the east lie on lower grade rocks of zone 2a. The junction between the two is known from the north side of the Grong culmination (Oftedahl, 1956); west of Borgefjell (Strand, 1955); south of Nasafjell at the type locality of the Rödingsfjell Nappe (Kulling, 1955, p. 215); and in the Sulitjelma region north of Nasafjell (Kautsky, 1953, p. 95; Henley, 1969, p. 113; Nicholson and Rutland, 1969, p. 58; Nicholson, 1971b), the type locality of the partly comparable Gasak Nappe of Kautsky.

Zone 3 is definable on its east side with fair precision in the Central Caledonides lying mostly in the broadly folded western half of the fold belt but with lobes extending east between the only partly Caledonized basement of the central domes and into the regionally flat-lying eastern sector of the belt. North of Rombak the deformation belt also can be divided into a broadly folded western sector and a flat-lying eastern one most of the way to Mageröy. However, it is not possible yet to be sure that the division into continuations of zones 2a and 3a is applicable to the extreme north, and so a transitional region (3b) is defined between 3a and the latter.

No obvious equivalent to zone 3 is known from the Southern Caledonides, but it is possible that rocks of equivalent structural position to those of zone 3 occur in the gneiss area north of the Jotunheim—zone 4 of this account. As pointed out below, much has yet to be learned of zone 4.

B. Zone 3a: The Central Caledonides

Although Kautsky's definition at Sulitjelma of the Gasak Nappe has priority over Kulling's definition of essentially the same relation between high-grade rocks above and lower grade Köli beneath at Rödingsfjell further south, the latter has achieved wider use. Kulling's Rödingsfjell Nappe (1955, p. 215), unlike Kautsky's Gasak Nappe, was defined to include the higher grade rocks known to stretch to the Norwegian coast and shown on many maps as rich in Caledonian granites. Work in the northern coastal Central Caledonides (Rutland and Nicholson, 1965, p. 79) has shown that at least there the so-called Caledonian intrusives are either domes of Caledonian basal gneisses of Precambrian aspect or, generally at higher structural levels, vein complexes in metasediments, the basal gneisses often occurring in late steep-sided or overturned domes with fold nappes developing from them (Nicholson and Walton, 1963).

The gneisses of the Caledonian coastal domes appear to lie at the same level in the regional structural sequence as the broader area of Caledonized basal gneisses west of the Rombak window and the un-Caledonized gneisses of the window itself [first shown clearly in Foslie's map (1941) of the Tysfjord area], a relation pointing to the probable Precambrian age of the former. These relations may be taken more controversially to indicate the likely autochthonous character of the western rocks so that, as put forward by Rutland and Nicholson (1965, p. 80), the gneisses of the coastal domes, perhaps with their common immediate quartzofeldspathic cover of sparamitian aspect, form an autochthonous–parautochthonous unit within the area of the Rödingsfjell Nappe as originally defined. Isotope work has subsequently confirmed the Precambrian origin of the basal gneisses of the north coastal district of the Central Caledonides (Wilson and Nicholson, 1973).

Although a strong contrast in degree of metamorphism and igneous intrusion exists in east Sulitjelma between the Gasak Nappe above and Köli rocks below them, the location of the thrust between is not too obvious. On the north side of Sulitjelma there is an interesting conglomerate of gneiss of basement type lying on a kilometer slab of parent gneiss, with Köli below and the higher grade assemblage above, which may represent a slice of basement and a covering conglomerate at the bottom of the Gasak Nappe (Kautsky, 1953, p. 97–98; Nicholson, 1971*b*). Isograds in the Köli beneath the nappe, at any rate at Sulitjelma itself, cut through the Köli so that higher grade Köli rocks overlie lower grade ones. The limit high/low grade which is not at the supposed Gasak thrust at the top of the Köli but somewhat below it (Vogt, 1927; Henley, 1970, p. 133) does not seem to correspond with any thrust in the Köli.

Nicholson and Rutland (1969, p. 26) have shown that the distinctive granite-veined and high-grade Gasak Nappe of Kautsky rapidly thins westward like the Köli and Seve beneath: where thin it shows no metamorphic contrast with the Köli. In the west its thinned edge eventually is covered by marble-rich sequences (containing little or no intrusive material) belonging to another structural unit so far unnamed but here called the Fauske Nappe. The Fauske Nappe itself thins west and is covered itself by the vein-rich Beiarn Nappe also rich in marbles. And the Beiarn Nappe appears to rest directly in the west by the coast on the possibly autochthonous complex containing the basal gneisses (Rutland and Nicholson, 1965, p. 96).

Correlations and extensions of Sulitjelma lithologies and structure are easy to the west of the Rombak window and into the Håfjell syncline of Foslie (1941) immediately west of the *B* of Rombak, Fig. 3, but not clear yet to the southern Central Caledonides. There (Fig. 3), northeast of Leka, many granites of magmatic origin are shown on existing maps: if this description is correct then that region is the one part of the whole fold belt in which magmatic granites occur on any scale. Kollung (1967) has recently described part of this tract.

At the south end of the Central Caledonides the rocks of the Rödingsfjell Nappe, interpreted in Kulling's original broad sense, overlie the basement of the Grong culmination. Off the coast, the island of Leka, composed of a serpentine massif and some adjacent low-grade sediments, may belong to a level beneath the high-grade rocks of the mainland (Strand, in Strand and Kulling, 1972, p. 91). Ultramafic rocks are found not uncommonly in Rödingsfjell high-grade rocks but most usually as small, much-altered masses strung out at particular structural levels rather than in masses the size of that on Leka.

It is common to describe the west part of the Central Caledonides as containing geosynclinal or eugeosynclinal rocks, but while basic rocks of probable volcanic origin are not uncommon the most obvious characteristic of the metasediments is richness in carbonate, both in marble and calcareous schist. Around the possible autochthonous coastal gneiss massifs well-layered psammites occur next to basal gneisses and are like sparagmitic rocks found in zone 1. Some of the marble sequences contain thick dolomites, marble conglomerates, and sedimentary iron ore deposits (Vogt, 1897, p. 200; Bugge, 1948; Nicholson and Rutland, 1969, p. 31).

C. Zone 3b: The Northern Caledonides of Troms

The southern limit of zone 3b, the Rombak window, is crossed by the major elements of zone 3a (Gustavson, 1966; Nicholson and Rutland, 1969), and the general character of the tract north of Rombak suggests extension of the area of marble-rich sequences of 3a to Tromsö (Fig. 4). To the west, basal gneisses are exposed that correspond to those of the west side of the Hafjell syncline and the coastal gneisses of zone 3 in general (Figs. 3 and 4).

Both the work of Gustavson (1966) and Oleson (1971) can be interpreted as implying that no Seve–Köli is present around the Målselv window southeast of Tromsö (Fig. 4), with low-grade parautochthonous rocks directly below Rödingsfjell Nappe equivalents. Further north there are long windows in the northwest-trending valleys cut through the flat-lying Caledonian metamorphic allochthon (Binns, 1968, for example) revealing unmetamorphosed lower Paleozoic rocks unconformable on autochthonous basement and themselves covered by phyllonites and quartz-rich rocks with higher grade and more lithologically varied rocks in turn above them. Extensive areas of deformed sparagmitian lithologies have been reported at Birtavarre (Fig. 4) (Padget, 1955) which, like the lower grade rocks of Olesen and Gustavson, may correspond to levels of zone 1 further south. Kulling (in Strand and Kulling, 1972, p. 217) has correlated the overlying higher grade rocks at Birtavarre (Fig. 4) with the Seve–Köli and not with the Rödingsfjell Nappe.

Little detailed structural work has been reported from zone 3b but Oleson (1971, p. 358) has described recumbent folds in his presumed Rödingsfjell sequence and has attributed to transversely trending folds a strong transverse lineation apparently like that of other more easterly regions of the fold belt (although usually described from levels below the Rödingsfjell Nappe).

D. Zone 3c: The Northern Caledonides of Finnmark

The higher grade rocks of zone 3c, although apparently continuous into zone 3b and to 3a and 2 (Fig. 4), possess two characteristics not general in zone 3. Neither, however, seems to demand any break between 3c and the rest of zone 3: firstly, the vastly greater amount of basic plutonic activity there and secondly the preservation on an unusually wide scale in the high-grade rocks of sedimentation structures allowing facing determinations. A third feature of 3c, however, may suggest a need in the high-grade rocks of some break between the early structures of 3c and of zone 2a and any northern extension of it: later structures may well be common to the whole belt. This feature is the evidence of climactic deformation and metamorphism in the Early Ordovician and not in the Late Silurian climax recognized widely elsewhere and especially in the Köli of zone 2.

1. *Structural Features*

The broad structure of 3c obviously is a continuation of that of 3b (Figs. 1 and 4). In Finnmark the un-Caledonized basement appears by the coast in the Alta windows (Reitan, 1960, p. 67) so that the eastern sector of the fold belt with obvious thrust superposition of high-grade rocks on rock either of low grade or no Caledonian metamorphism at all, occupies most of the main-land development of the fold belt. According to Gayer and Roberts (1971) the easternmost part of the metamorphic allochthon, of the northeast edge of 3c, the Kolvik Nappe, has a cleancut basal thrust and itself is of upper green-schist to lower amphibolite facies metamorphism. Locally, instead of lying on the autochthonous parautochthon rocks of zone 1, the Laksefjord Nappe of probable late Precambrian lies between the two, as the Serv Nappe lies between zones 3a and 2 in the Central Caledonides.

There are roughly syntectonic dolerite dikes in the Kolvik Nappe that may correlate with the apparently similar dolerites of the Laksefjord Nappe and even with dolerites in the parautochthon (Gayer and Roberts, 1971, p. 63). No tectonic break has yet been reported between the Kolvik thrust beside Porsangerfjord and the high-grade regions further west on Söröy (Fig. 4).

On Söröy, detailed stratigraphic and structural studies have defined a complex history notable for folds arcuate on a large scale, so that single but long folds can be successively transverse and Caledonoid in trend (Ramsay, 1971a, p. 16). The island is notable too for the presence of lower Cambrian archaeocyathids in a relatively undeformed raft of limestone in gabbro involved in the early deformation (Holland and Sturt, 1970; Ramsay, 1971b, p. 34), the only fossil occurrence known from the high-grade rocks of zone 3. Finally, a Rb/Sr whole-rock isochron (Pringle and Sturt, 1969) on specimens from an anatectic vein produced by an early syntectonic gabbro points to a time 525 ± 45 m.y. ago for intrusion, thus indicating an early deformation phase. There is supplementary isotope evidence in Sturt et al. (1967) and Pringle (1971).

In 1960 (Föyn, 1967a) fossils were found in Mageröy (Fig. 4) to the north-east of Söröy, where a sequence of low-grade fossiliferous Middle Ordovician to Late Silurian strata "is juxtaposed" (Sturt, 1971) against a high-grade and possibly Eocambrian/Cambrian sequence. Conglomerates in the possibly younger low-grade series contain boulders that may come from the higher grade and possibly earlier rocks. The latter, additionally, may itself be continuous with the high-grade rocks of Söröy. The low-grade Mageröy rocks demonstrate a later deformation episode than those of Söröy, since the fossiliferous rocks are quite strongly deformed although only weakly metamorphosed. Nothing, however, is yet available of any details of the structural relation between high- and low-grade rocks on Mageröy or of the probable presence of the Late Silurian deformation in the western rocks of Söröy itself. In terms of age of sediments and time of deformation, the low-grade Mageröy rocks display Köli characteristics.

2. *Igneous Geology*

Igneous activity in northern and western Finnmark is marked by a wealth of basic and ultrabasic intrusions together with carbonatite types unknown in the rest of the Scandinavian Caledonian (Sturt and Ramsay, 1965). It has been suggested by Brooks (1970) that the 100-mgal positive Bouguer anomaly centered over Söröy, itself a culmination on an elongate positive anomaly along the outer edge of the coastal region of the North Caledonides, may be associated with massive enrichment of the upper crust in basic and ultrabasic material and a shallowing of the Conrad discontinuity. The anomaly is similar in shape, orientation, and position to that of the Vesterålen–Lofoten 300 km southwest and may be continuous with it; there, however, there is no trace of Caledonian intrusive activity. In the Vesterålen–Lofoten area Brooks' alternative explanation for the Söröy anomaly, stressing granulitic country rocks there, provides a better fit.

3. *Regional Correlation*

The North Finnmark (3c) indications of an early Caledonian orogenic event are clear, although there are major gaps in areal cover. The history of Late Silurian deformation and metamorphism based on work in the classic Köli territory in the Central Caledonides is at least as clear. Unfortunately, these are fairly distantly separated regions. However, it seems that the recognition of the Köli unit can be extended to the north side of Rombak, where it has the usual relation with Rödingsfjell Nappe sequences, the mass of 3a. These Rödingsfjell sequences seem to persist in turn into the center of 3b, where, too, it is at least arguable that typical 3a coastal granitic gneisses extend from the Rombak region to within 100 km of Söröy itself.

These seems little doubt that the northern high-grade Söröy region has a stratigraphy in common with the region from Söröy south to Loppa (Ramsay, 1971*b*, p. 31). It also seems that the stratigraphy may be used in the Vaddas area (see Fig. 4 and Table V) south of Kvaenangenfjord and no more than 100 km in turn from the carbonate-rich Rödingsfjell-type sequences near

TABLE V

Sequences in the Metamorphic Allochthon of Zone 3c, in the County of Finnmark in the Northernmost Northern Caledonides

Zone 3c	Northern Caledonides			
	Söröy	Loppa	Skjervöy	Vaddas
	Schists-presumed turbidites			
	Graphitic			
	Phyllite			
	Quartzite			Quartz biotite schist
	Graphitic			
	Phyllite			
		Pelite		
Lower Middle Cambrian	Marble	Marble	Marble	Marble
			Augen	
	Calc-silicate rocks		Gneiss	
	Pelites and psammites	Graphitic pelite		
	Pelites	Garnet mica schist	Semipelite	Pelites
Eocambrian?	Well-bedded white	Calcareous		
		Psammite	Psammite	Psammites
	Psammite	Psammite		
	Ramsay, 1971	Ramsay, 1971	Ramsay, 1971	Armitage *et al.*, 1971

Tromsö (Armitage *et al.*, 1971). Thus while zone 3c is different in many respects from the rest of zone 3 it has an overall structure like the 2–3a combination further south and clearly in some ways is continuous with it. The general area of Troms between the well-described 3c areas and 3a and 2 presumably contains the key to problems of the extent of the 3c early phase.

VI. ZONE 4: THE WESTERN GNEISS AREA OF THE SOUTHERN CALEDONIDES

Immediately under the mass of the Jotun Nappe and on its north and west side lie allochthonous rocks of the Valdres Nappes and then lower Palaeozoic in its stratigraphic place above Precambrian gneisses (for example, the head of Sognfjord; Banham and Elliot, 1965). The basement is affected only by subsidiary mylonitization and shearing during the Caledonian. To the northeast and about 150 km away in the Oppdal area (Fig. 2), however, there is a thoroughly deformed sequence of Eocambrian and generally granitic gneisses below them both lying beneath deformed and metamorphosed allochthonous lower Paleozoic of the Trondheim area, zone 2b (Holtedahl, 1938). In addition, it has been suggested that there are big fold nappes in the lower sequence and that some of the basal gneisses were produced from Eocambrian rocks by Caledonian metasomatism (Strand, in Strand and Kulling, 1972, p. 77).

Finally, in the gneiss area north of Bergen two principal members have been distinguished (Brynhi and Grimstad, 1970): an upper varied one perhaps of metamorphosed lower Paleozoic and other Caledonian supracrustals, and a lower homogeneous granitic member of probably older Precambrian origin. There is no certain evidence that the supracrustals are not Precambrian themselves, although there are petrographic similarities between some of them and the early Caledonian sparagmitic rocks known elsewhere.

These three situations exemplify some of the variety of geology and interpretation offered by the so-called gneiss region north and west of the Trondheim–Jotun complexes. The major problem is that of the origin of the gneissic rocks; a second question is that of the possible allochthonous nature of the gneiss complex or its members. Brueckner (1969) has suggested that the widespread distribution of mafics and ultramafics in the upper supracrustal part of the complex north of Bergen and their absence in the lower granitic one implies the transport of the upper from the site of emplacement of the intrusives to its present position over the granitic gneisses. The view that the western basal gneiss complex is merely an extension of thrust sheets of basement gneisses in zone 1 in the east has already been raised. In the north Central Caledonides it seems more likely that the western basal gneisses are an extension of the autochthon than of such higher thrust sheets.

Between Bergen and Ålesund, the gneissic basal complex is overlain by Devonian Old Red Sandstone facies, which is a little deformed, presumably dating some of the structure of the lower complex as Devonian. So far isotope studies and field work in the gneisses are not sufficiently comprehensive to provide a solution of their problems, although it has been confirmed that while most of its rocks show Caledonian influence (Caledonian K/Ar mineral ages) there are rocks which provide whole-rock Rb/Sr isochron ages of 1000 m.y. and more (Brueckner, 1972; Wilson and Nicholson, 1973).

Among the mafic rocks present in the coastal areas of the western gneiss complex, for example, around Ålesund, there are eclogites which have figured prominently in discussions of eclogite genesis (O'Hara and Mercy, 1963; Brynhi *et al.*, 1970). It has been suggested that these and other intrusives of the Ålesund region define a steep and complex shear zone with an extension of some 100 km along the chain (Lappin, 1966). These are the rocks that Brueckner has decided have a sheetlike occurrence. It seems unlikely that there is any steep discontinuity in the fold belt (Nicholson, 1971*b*) which rather is characterized by gently inclined junctions.

VII. ZONE 5: THE LOFOTEN REGION

Zone 5 is composed of the gneissic rocks of the Lofoten Islands and the western Vesterålen nearer the mainland (Figs. 2 and 3). A traverse west from the Caledonian metasediments of the core of the Håfjell syncline of the north Central Caledonides (immediately west of the *R* of Rombak, Fig. 4) soon reaches the thoroughly Caledonized western basal granitic gneisses of its western limb, which are structurally conformable with their metasedimentary cover. It is these gneisses that apparently are continuous north into the coastal side of zone 3b in the Northern Caledonides. They also appear to be continuous in depth with the Precambrian gneisses of the Rombak window on their east.

Foslie (1941) and Kautsky (1946), in contrast, considered the granitic gneisses under the western metasediments of the Håfjell syncline to be Caledonian intrusives derived from the basement by melting. With them in this Caledonian category were usually also included the varied but generally gneissic rocks further west on western Vesterålen and Lofoten, rocks rather like some of the southwestern basal rocks of the Southern Caledonides as well as the Jotun Nappe (Heier and Compston, 1969) but not particularly close in lithology to the Rombak basement not far east. Isotope work (Heier and Compston, 1969) has shown, however, that both the Caledonized granitic rocks and the western Lofoten gneisses have whole-rock Rb/Sr isochrons defining Precambrian ages and that some of the latter may be very old, perhaps pre-Karelian.

While there is no published detailed work on the junction between Caledonized and un-Caledonized western rocks, according to Professor R. W. R. Rutland (pers. comm.) it is a narrow zone of very strong linear structure which seems to dip gently west beneath the Lofotens. Romey (1971) has suggested that the Lofoten gneisses are exposed in a window comparable to that of Rombak. Alternatively it has been proposed that the Lofoten gneisses

were tectonically high during the Caledonian, so escaping Caledonian deformation (Brueckner, 1971). A third possibility is that they were not part of the
sometimes postulated lower Paleozoic Scandinavian plate (Dewey, 1969)
at all but belonged to a western or American one which converged on the
Baltic plate during the Caledonian orogeny (Rutland, pers. comm.).

VIII. DISCUSSION

A. Introduction

The Scandinavian Caledonide region displays thrust tectonics on a grand
scale. In that respect it resembles the East Alpine belt of Austria: like it,
there are nappes composed essentially of either nonmetamorphic or of metamorphic rock. Unlike the Eastern Alps, however, most of the metamorphic
rocks involved in Scandinavian nappes achieved their minerals and fabric
in the Caledonian itself and not in tectonic episodes much earlier than
thrusting.

As distinguished in the preceding account, there are major unanswered
questions concerning the mutual relation of penetrative and thrust deformations as well as problems of correlation of metamorphism and deformation
from one part of the belt to another. However, these questions seem more
likely to be answerable than others about the place of origin of the rock masses
involved in the fold belt and, even more remotely, the paleographies concerned.
We shall begin with the former category of question and end with consideration of questions such as the latter, whose formulation may require the use of
some coordinating hypothesis.

B. Lineation and Thrusting

There has been much speculation (Kvale, 1953, 1955; Lindström, 1961;
for example) on the relation between movements on the obvious thrusts of the
fold belt and the often strong transverse lineation of nappes at the top of zone 1
and the metamorphic allochthon (especially the Köli).

Before deciding on a relation between the two, a distinction needs to be
made between the final en masse emplacement of the rock sheets coherently
over the eastern basement and the earlier more penetrative movements in
which some of the thrust sequences were built. Production of linear fabrics
during late eastward thrusting may be the only mode of production for the
lineation of rocks not involved in Caledonian deformation until the arrival of
the higher metamorphic nappes, but not for rock of zones 2 or 3 which had a
complex structural history before thrusting. Even in zone 1, however, Hossack

(1968) has attributed the strong deformation of the Bygdin Conglomerate of the Valdres Nappe to flattening later than the superposition of the Jotun Nappe on the latter rather than to thrust emplacement itself. In addition, Nickelsen (1967) has shown that the penetrative elongation lineation of the rocks of the Valdres Nappe in fact is parallel to the axes of arcuate recumbent folds within the thrust nappe and not to the likely direction of transport of the whole.

C. Recognition of the Early Metamorphic/Structural Event

There seems no doubt that an early Caledonian peak to deformation and metamorphism comparable with that of the metamorphic Caledonides of the British Isles has been established in the Northern Caledonides of Scandinavia. However, the question of the extent of the region affected by the early episode remains, since there is no doubt that for much of the fold belt the climax was late in the Silurian. The problem is not one merely of the existence of different deformation and metamorphic phases in the stratigraphically different regions of the zones 2 and 3. It also is one of the extent and significance of the early phase in the western metamorphic allochthon itself, since there seems to be a transition of metamorphic characteristics over the zone 2–3 boundary at Sulitjelma in the Central Caledonides so that some zone 3 metamorphism also is Silurian. Furthermore, west of this boundary and in zone 3a, not even a transitional metamorphic division of the kind recognized further east can be made even though the same structural/lithological level is revealed again round the Rishaugfjell dome (Nicholson and Rutland, 1969, p. 29–30). If the early Caledonian phase is present in the western rocks of the Central Caledonides, then it has been well obscured by the late one. The western rocks do not seem to have a more complex metamorphic history than eastern, which cannot have been affected by the early phase of activity since they are not old enough.

As we have noted, there has been an attempt to fix the early event in zone 2a, the Trondheim depression (Skevington and Sturt, 1967). However, as Roberts (1971) has described, while there may well be an unconformity of a suitable age present above the Stören volcanics there is no good evidence that the preceding deformation or metamorphism was comparable to that in the Scottish Caledonides: the major deformation is Silurian.

In the Bergen area, a break is recognized that may correspond to the one being sought in zone 3a. There the older rocks are distinguished from younger by a greater richness in intrusives as well as greater deformation. In a parallel way it might be that zone 3a is a mixture of older and younger structural/ metamorphic elements, the older being vein-rich and intrusive-rich bodies like the Gasak and Beiarn nappes. So far there is no structural or isotope evidence in favor of this proposal (Wilson, 1971). In any case there are resemblances of lithology and sequence between the marble-rich Beiarn Nappe and vein-free complexes below. Better definition and then extension of the Mageröy distinction is one way to get a solution to the problem of distribution of tectonic episodes; another is the wide extension of isotope studies.

D. The Problem of the Western Gneisses

As discussed below, there have been attempts to divide the Scandinavian Caledonides, with the western metamorphic sector belonging to an American plate (Wilson, 1966). This raises the secondary question of the whereabouts of the dividing zone or suture (Dewey, 1969), a problem itself concerned with the Baltic or American origin of constituent parts.

Asklund (1955) has suggested that the basal gneisses of the Olden Nappe, lying below a cover of little metamorphosed sediments of Baltic affinities, are directly continuous into the basal gneisses of the Grong area between and below zones 2a and 2b (Fig. 5). A similar proposal by Kulling (1964) for eastern thrust nappes of Precambrian gneiss and basal gneisses to their west in the area immediately south of Rombak (Fig. 5a) has been described. In each case the proposed continuity has been supposed to imply an extension westward also of the thrust separating gneissic bodies from the Baltic shield below.

If such correlations are accepted, then, while the western basal gneisses may lie on a late thrust or thrusts, they may at the same time have lain outside the American plate. Any suture could not lie lower than the top of the marginal gneissic bodies or their supposed continuation to the west.

However (this chapter and Nicholson and Rutland, 1969, p. 75), the western basal gneisses of the northern Central Caledonides may be regarded as joining directly underground with autochthonous gneisses appearing, for example, in the Rombak window and from which they are locally separated by only 15 km of synformal schists. Then the eastern thrust nappes of Baltic basement might be fragments separated from their source in the autochthon and carried east during the movement of the higher metamorphic allochthon rather than the thin edge of a very large nappe of basement.

Both at Rombak and in the Nasafjell window further south (Fig. 5a), basal gneissic rocks seem first to show a development and then an increase in Caledonian penetrative deformation from east to west without any sharp break. Such a transition may be closer to relations in the west between units separated at the margin by thrusts than the model of straightforward extension westward of thrust surfaces. Then the definition in the west of allochthonous character becomes difficult. Similar conclusions might be drawn from relations between the Baltic shield and the gneiss area of zone 4 in the Southern Caledonides.

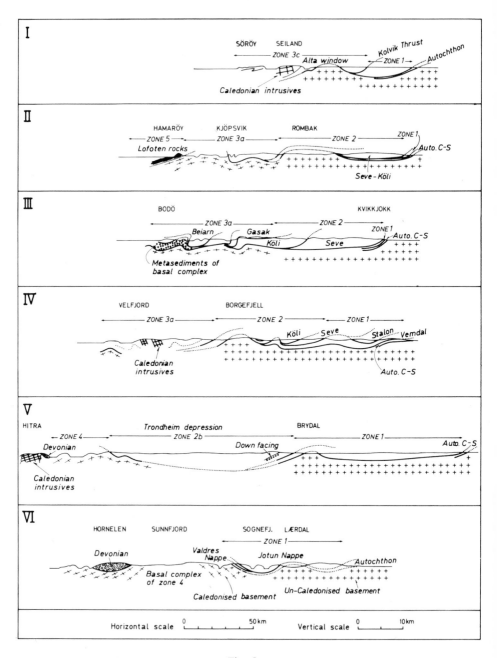

Fig. 5a.

Fig. 5. (a) Cross sections of the Scandinavian Caledonides showing in particular relations between eastern and western basal gneisses and units of the Caledonian metasedimentary cover. (b) Outline map of the Scandinavian Caledonides showing division into zones and the position of the cross sections of (a).

A gradual change in the type of strain westward from sharp break to ductile transition would also help to explain the apparent complete continuity in the northern Central Caledonides of metamorphism from the leading edges of obvious metamorphic nappes such as the possibly American Seve–Köli westward to zone 3, where there is a continuity of metamorphism as well as deformation from basal gneisses of possible Baltic origin up into the metasedimentary cover. Such continuity of structure will make difficult plate divisions in the cover.

It may be suggested that the emplacement of eastern thrust nappes of Caledonian metamorphic rock onto the eastern basement was contemporary with the relatively late and in any case postmetamorphic gneiss domes developed in the western sector of the chain. Then the western gneiss domes of the northern Central Caledonides are, in a restricted sense, autochthonous, while their equivalents in the east, the nappes, have to be described as allochthonous. In this case the base of the Seve–Köli Nappe of zone 2a may not correspond to a plate boundary having formed after plate conjunction although possibly separating rocks largely belonging to one plate or the other.

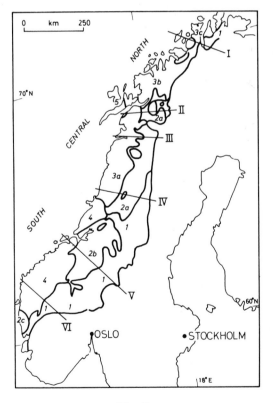

Fig. 5b.

Kautsky's ingenious proposal for the north Central Caledonides of thrusting of already metamorphosed rock over inactive Baltic basement followed
by further metamorphism and deformation was aimed at explaining problems
of metamorphic and structural continuity outlined above (Kautsky, 1946).
However, the metamorphic zones 2a and 3a cannot be divided simply into a
compound west zone and a simpler east zone. More fundamentally the metamorphism common to east and west has to be the younger, since the east,
zone 2b, can only show that phase.

Finally, returning to the question of a possible suture, it is worth recalling
that there seems to be an absence in the fold belt of those features hitherto
assigned to sutures (Nicholson, 1971a). Certainly none of the thrusts so far
distinguished in it has them. The primary tectonics at the plate margin are
more likely perhaps to involve the major strain features of the belt rather than
possibly late thrusts. One of these features in the Scandinavian Caledonides
is the curious wedge shape of components of the metamorphic belt and the
plastic imbrication to which it has been attributed (Nicholson and Rutland,
1969).

E. Comparison with the Caledonian of the British Isles

Some comparisons between the Caledonian of Scotland and Scandinavia
have already been provided. One of the most detailed concerns the glacial
deposits of the two regions. A more general correlation can be made of the
older Caledonian arenaceous sequences and the Moine and other North
Atlantic Caledonian developments of the same kind. So far not even broad
correlation can be offered for the lower Paleozoic, and it is doubtful that the
necessary similarities of sequence rather than lithology exist.

Comparison has been made also of the age of deformation and metamorphism in the British Isles and in Scandinavia, a subject not resumed here.
Nothing has been written, however, of the related comparison of the kind of
metamorphism and its relation to deformation in the two regions, nor of
structural or igneous comparisons. These now follow.

1. *Metamorphism*

The regional metamorphism of the Scandinavian Caledonides is not varied
like that of Scotland. Only the Barrovian sequence is known: there are no
equivalents of either the Buchan andalusite developments of the Scottish
Highlands or of the low-temperature, high pressure metamorphism of the
Girvan district of South Scotland. The relation of metamorphism and deformation, however, is more complex in Scandinavia than in Scotland: while in
Scotland isograds cross structure, in Scandinavia they generally appear to run

parallel to known or presumed major structural junctions. As we have seen, however, this description of the relation may be too simple to fit detailed observations and detailed work is rare enough to make it uncertain how often such complexity is the rule.

2. *Intrusion*

The principal difference between the igneous activity of Scandinavia and the British Isles is the apparent almost complete absence in the former of the posttectonic granites so widely developed in Scotland, a difference that is the more striking when it is noted that there is some minor acid volcanic activity in the Norwegian Devonian north of Bergen. In part the contrast might be explained by allowing a greater allochthonous character to the Scandinavian Caledonides so that much of the presently observable metamorphic allochthon came to lie on an inactive basement unable to provide the granitic magma necessary for late intrusions. Instead of diapiric intrusions of melt, the Scandinavian Caledonides developed the "concordant" intrusions of gneiss-dome type.

3. *Structure*

The structure of the Scandinavian Caledonides is dominated by large thrust nappes, which in the Central Caledonides, at least, thin successively westward. The belt may be divided into two with nappes of Caledonian metamorphic rock lying in the eastern sector on un-Caledonized rocks but in the western with structural and metamorphic conformity on basal gneisses. A set of late domes lies about on the junction of east and west sectors, and in them, for example at Rombak, a transition may be observed between the two types of arrangement.

Outside the fold belt the area of remaining autochthonous Caledonian sediments is much less than in the British Isles. Vulcanism is unknown in them. Clearly, the active crust of the Caledonian of Scandinavian did not extend much outside the present area of the deformation belt, while in the British Isles there is much vulcanicity and development of sedimentary basins outside the core of the deformation belt.

Structural comparisons between Scotland and Scandinavia are easiest when the thrust edges are considered. Then, for example, equivalents to zone 1 may be found below the Moine Thrust. Above the thrust some sort of structural (but not stratigraphic) comparison might be made between zone 2 and the lower Moine nappe in which sedimentation structures are not infrequent, with zone 3 roughly equivalent in those terms with the higher Moine nappe in which stratigraphy is difficult to establish.

Only the Paleozoic of the Trondheim depression seems to have any structural comparison with the Dalradian. There large fold nappes have been postulated that may have some resemblance to the Tay Nappe of the Dalradian.

F. Paleogeographic Reconstructions

The only Caledonian entity sufficiently unmodified to allow reconstruction of something of its predeformational pattern without recourse to some hypothetical model of the tectonic system seems to be zone 1 which apparently grew from an early trough or sets of troughs founded on the Baltic shield and from which the Eocambrian/Cambrian transgression spread east over much of the rest of the shield. This sedimentary assemblage seems to have been overwhelmed by the metamorphic nappes which may have reached their final eastern position by sliding out over the eastern basement on wet shale sequences. The early Caledonian volcanic-free sequences seem to have largely lain outside any active zone itself until the western nappes impinged on them.

The organization of the complex western metamorphic sector of the mountain system is a mystery. In the past it has been the subject of comparison with geosynclinal models largely based on the Western Alps—not with much success, for synthesis was hampered then by an even greater ignorance of the characters of the belt than today. More recently the plate tectonic hypothesis has been applied, and the first of such attempts by Wilson (1966), proposing a closing Atlantic during the Caledonian, was rather in line with much earlier proposals by Holtedahl (Bailey and Holtedahl, 1938) that continental drift had played a part in the construction of the Scandinavian Caledonides. However, it allowed, in addition, in an intriguing way for the division of the belt into two by making the metamorphic sector "American" and separated by a steep division or suture from eastern Baltic elements (Dewey, 1969).

Dewey (1969), in the extension of a synthesis for Britain, included the Scandinavian Caledonides in a regime of ocean opening and continental rifting prior to the closure of which the late Scandinavian nappes might be witness. It might be that this opening regime was the environment of accumulation of zone 1 rocks. The whole North Atlantic Caledonian has this type of arenaceous early assemblage.

Since faunal surveys so far published are suggestive rather than definitive and require consideration of stratigraphic and structural work for their interpretation [see comments on Ross and Ingham (1970)], the proposal that some of the rocks of the deformation belt may once have belonged to a distant western mass and arrived to impinge on a Baltic plate in the Late Silurian is not settled let alone definable in any detail. Also, because of ignorance of provenance it is not clear whether we have at any time before the postulated plate collision to regard the western edge of the Baltic plate as tectonically active on its own, above a region of subduction in plate terms, or only involved in activity when the western plate impinged upon it.

At its simplest, the plate hypothesis, employing the hints of provinciality so far obtained, requires deposition in the American plate of the rocks of the Trondheim–Köli belt beyond or actually on top of earlier formed Caledonian metamorphic rocks perhaps of varied grade and deformation. Then this complex assemblage has to combine with a less active or possibly inactive Baltic Caledonian sequence so as to deform and metamorphose for the first time its own Köli–Trondheim element and some of the Baltic plate. In addition, it has to metamorphose again some of its older Caledonian rocks, thus making obscure their relation to the younger. Then, finally, there has to be a push eastward of the complex metamorphic assemblage over still undeformed Baltic sequences now in zone 1. And the latter part of their story, for Köli metamorphism onwards, has to take place between Middle and Late Silurian. A more complex Baltic story arises if, alternatively, it is argued that both older and younger Caledonian deformation took place in the eastern plate. This requires, however, reconsideration of the significance of provinciality as hitherto defined.

REFERENCES

Armitage, A. H., Hooper, P. R., Lewis, D., and Pearson, D. E., 1971, Stratigraphic correlation in the Caledonian rocks of S. W. Finnmark and North Troms: *Norges Geol. Unders.*, v. 269, p. 318–322.

Asklund, B., 1955, Norges geologi och fjellkedje problem: *Geol. Fören. Stockholm Förh.*, v. 77, p. 185–203.

Asklund, B., 1960, The geology of the Caledonian Mountain Chain and of Adjacent Area in Sweden: *Sveriges Geol. Undersökn.*, Ser Ba, N:O, v. 16, p. 126–149.

Bailey, E. B. and Holtedahl, O., 1938, *Regionale Geologie der Erde, Bd. 11, North Western Europe, Caledonides*: Leipzig, Akademische Verlag, 76 pp.

Banham, P. H. and Elliot, R. E., 1965, Geology of the Hestbrepiggen area: *Norsk Geol. Tidsskr.*, v. 45, p. 189–198.

Banks, N. L., Edwards, M. B., Geddes, W. P., Hobday, D. K., and Reading, H. G., 1971, Late Precambrian and Cambro-Ordovician sedimentation in east Finnmark: *Norges Geol. Unders.*, v. 269, p. 197–236.

Binns, R. E., 1968, A preliminary account of the geology of the Signaldalen–upper Skibotndalen area, Inner Troms, North Norway: *Norges Geol. Unders.*, v. 247, p. 231–251.

Brooks, M., 1970, A gravity survey of coastal areas of West Finnmark, northern Norway: *Quart. J. Geol. Soc. Lond.*, v. 125, p. 171–192.

Brueckner, H. K., 1969, Timing of ultramafic intrusion in the core zone of the Caledonides of southern Norway: *Amer. J. Sci.*, v. 267, p. 1195–1212.

Brueckner, H. K., 1971, Age of the Torset Granite, Langöy: *Norsk Geol. Tidsskr.*, v. 51, p. 85–87.

Brueckner, H. K., 1972, Interpretation of Rb–Sr ages from the Precambrian and Paleozoic rocks of southern Norway: *Amer. J. Sci.*, v. 272, p. 334–358.

Brynhi, I., Fyfe, W. S., Green, D. H., and Heier, K. S., 1970, On the occurrence of eclogite in western Norway: *Contr. Mineralogy and Petrology*, v. 26, p. 12–19.

Brynhi, I. and Grimstad, E., 1970, Supracrustal and infracrustal rocks in the gneiss region of the Caledonides west of Breimsvatn: *Norges Geol. Unders.*, v. 266, p. 105–140.

Bugge, J. A. W., 1948, Rana gruber: *Norges Geol. Unders.*, v. 171, p. 149.

Chaloupsky, J., 1970, Geology of the Holonda-Hulsjoen area, Trondheim region: *Norges Geol. Unders.*, v. 266, p. 277–304.

Dewey, J. F., 1969, Evolution of the Appalachian/Caledonian orogen: *Nature*, v. 222, p. 124–129.

Foslie, S., 1941, Tysfjords geologi: *Norges Geol. Unders.*, v. 149, 289 pp.

Føyn, S., 1967a, Stratigraphic consequences of the discovery of Silurian fossils on Mageröy, the island of North Cape: *Norges Geol. Unders.*, v. 247, p. 208–222.

Føyn, S., 1967b, Dividal gruppen ("Hyolithus-sonen) i Finnmark og dens forhold til de eocambrisk-cambrisk formasjoner: *Norges Geol. Unders.*, v. 249, 483–485.

Gayer, R. A. and Roberts, J. D., 1971, The structural relationships of the Caledonian nappe of Porsangerfjord, West Finnmark, N. Norway: *Norges Geol. Unders.*, v. 269, p. 21–67.

Gee, D. C., 1972, The regional geological context of the Tåsjö Uranium project, Caledonian front, Central Sweden: *Sveriges Geol. Undersökn.*, Ser C, Nr. 671, 36 pp.

Gjelsvik, T., 1948, Anorthositkomplexet i Heidal: *Norsk Geol. Tidssk.*, v. 26, p. 1–58.

Glass, J., 1972, Some observations on the petology and metamorphic history of the Marsfjallet migmatite–gneiss complex (Västerbotten): Resume of lectures of the X Nordiske Geologiske Vintermote, Oslo, 1972, p. 19.

Goldschmidt, V. M., 1912, Die Kaledonische Deformation der sud-norwegischen Urgebirgstafel: *Skr. VidenskSelsk. Christ.*, v. 19, 11 pp.

Goldschmidt, V. M., 1916, Konglomeraterne inden Hoifjeldskvartsen: *Norges Geol. Unders.*, v. 77, 61 pp.

Gustavson, M., 1966, The Caledonian mountain chain of the southern Troms and Ofote areas, Part 1, Basement rocks and Caledonian metasediments: *Norges Geol. Unders.*, v. 239, 162 pp.

Hamburg, A., 1910, Gesteine und Tektonik des Sarekgebirges nebst einen Überblick der Skandinavischen Gebirgskette: *Geol. Foren. Stockolm Forh.*, v. 32, p. 681–724.

Henley, K. J., 1970, The structural and metamorphic history of the Sulitjelma region, Norway with special reference to the nappe hypothesis: *Norsk Geol. Tidssk.*, v. 50, p. 97–136.

Henningsmoen, G., 1960, Cambro-Silurian deposits of the Oslo Region: *Norges Geol. Unders.*, v. 208, p. 130–150.

Heier, K. and Compston, W., 1969, Interpretation of Rb–Sr age in high-grade metamorphic rocks, North Norway: *Norsk Geol. Tidsskr.*, v. 49, p. 257–283.

Högbom, A. G., 1909, Studies in the post-Silurian thrust region of Jamtland: *Geol. Foren. Stockholm Forh.*, v. 31, p. 330–349.

Holland, C. H. and Sturt, B. A., 1970, On the occurrence of archaeocyathids in the Caledonian metamorphic rocks of Söröy and their stratigraphical significance: *Norsk Geol. Tidsskr.*, v. 50, p. 341–355.

Holtedahl, O., 1938, Geological Observations in the Opdal–Sunndal–Trollheim district: *Norsk Geol. Tidsskr.*, v. 18, p. 29–41.

Holtedahl, O., 1960, Stratigraphy of the Sparagmite Group including the sandstone divisions of Finnmark: *Norges Geol. Unders.*, v. 208, p. 351–357.

Holtedahl O. and Dons J., 1960, Geologisk kart over Norge, berggrunskart: *Norges Geol. Unders.*

Hossack, J. R., 1968, Structural history of the Bygdin area, Oppland: *Norges Geol. Unders.*, v. 247, p. 78–107.

Kautsky, G., 1946, Neue Gesichtspunkte zu einigen nordskandinavische Gebirgsprobleme: *Geol. Foren. Stockholm Forh.*, v. 68, p. 589–602.

Kautsky G., 1948, Stratigraphische Grundzuge im westlichen Kambrosilur in den skandinavischen Kaledoniden: *Geol. Foren. Stockholm Forh.*, v. 71, p. 253–284.

Kautsky, G., 1953, Der geologische Bau des Sulitjelma-Salojauregebietes: *Sveriges Geol. Undersökn*, Ser. C, N:O, 528, 228 pp.

Kollung, S., 1967, Geologiske undersökelser i sörlige Helgeland og nordlige Namdal: *Norges Geol. Unders.*, v. 254, p. 95.

Kulling, O., 1955, Den kaledoniska fjallkedjans berggrund inom Vasterbottens Ian. Beskrivning till bergsgrundskarta over Västerbottens Ian, 2, Sveriges Geol. Undersökn, Ser. Ca, v. 37, p. 101–296.

Kulling, O., 1961, On the age and tectonic position of the Valdres Sparagmite: *Geol. Foren. Stockholm Forh.*, v. 83, p. 210–214.

Kulling, O., 1964, Oversikt over Norra Norrbottensfjallens Kaledonberggrund: *Sveriges Geol. Undersökn*, Ser. Ba, Nr. 19, 166 pp.

Kvale, A., 1953, Linear structures and their relation to movement in the Caledonides of Scandinavia and Scotland: *Quart. J. geol. Soc. Lond.*, v. 109, p. 51–73.

Kvale, A., 1955, Dekkestrukturer mellom Bergensbuene og Sogn: *Norsk Geol. Tidsskr.*, v. 35, p. 221–237.

Lappin, M. A., 1966, The field relations of basic and ultrabasic masses in the basal gneiss complex of Stadlandet and Almlovdalen, Nordfjord, S. W. Norway: *Norsk Geol. Tidsskr.*, v. 46, p. 429–495.

Lindström, M., 1961, Tectonic fabric of a sequence of areas in the Scandinavian Caledonides, Geol. Fören. Stockholm Forh.: v. 83, p. 15–64.

Loeschke, 1967, Zur Stratigraphie und Petrographie des Valdres-Sparagmit und der Mellsenn-Gruppe bei Mellane/Valdres (Sud-Norwegen): *Norges Geol. Unders.*, v. 243A, p. 1–63.

Mason, R., 1967, The field relations of the Sulitjelma gabbro, Nordland: *Norsk Geol. Tidsskr.*, v. 47, p. 237–248.

Müller, G. and Wurm, F., 1970, Die Gesteine der Halbinsel Strand, Beitrage zur Metamorphose und zum Aufbau der kambro-siluris Gesteine des Stavanger-Gebietes, II und III: *Norges Geol. Unders.*, v. 267, 90 pp.

Nicholson, R., 1971a, Faunal provinces and ancient continents in the Scandinavian Caledonides: *Bull. Geol. Soc. Amer.*, v. 82, p. 2349–2355.

Nicholson, R., 1971b, The sedimentary breccias of the Sorjusvann region on the Norwegian–Swedish border north of Sulitjelma: *Norsk Geol. Tidsskr.*, v. 51, p. 149–160.

Nicholson, R. and Rutland, R. W. R., 1969, A section across the Norwegian Caledonides; Bodø to Sulitjelma: *Norges Geol. Unders.*, v. 260, 86 pp.

Nicholson, R. and Walton, B. J., 1963, The structure of the Navervatn–Storglomvatn area, Glomfjord, North Norway: *Norsk Geol. Tidsskr.*, v. 43, p. 1–58.

Nickelsen, R. P., 1967, The structure of Mellene and Heggeberg, Valdres: *Norges Geol. Unders.*, v. 243C, 65–120.

Oftedahl C., 1956, Om Grongkulminasjonen og Grongfeltets skyvedekker: *Norges Geol. Unders.*, v. 195, p. 57–64.

O'Hara, M. J. and Mercy, E. L., 1963, Petrology and petrogenesis of some garnetiferous peridotites: *Royal Soc. Edinburgh Trans.*, v. 65, p. 251–314.

Oleson, N. Ö., 1971, The relative chronology of fold phases, metamorphism and thrust movements in the Caledonides of Troms, North Norway: *Norsk Geol. Tidsskr.*, v. 51, p. 355–377.

Padget, P., 1955, The geology of the Caledonides of the Birtavarre region, Troms, northern Norway: *Norges Geol. Unders.*, v. 192, 107 pp.

Phillips, W. E. A., 1973, The pre-Silurian rocks of Clare Island, Co. Mayo, Ireland and the age of the metamorphism of the Dalradian in Ireland: *J. Geol. Soc. Lond.*, v. 129, p. 585–606.

Pringle, I. R., 1971, A review of radiometric age determinations from the Caledonides of West Finnmark: *Norges Geol. Unders.*, v. 269, p. 191–196.

Pringle, I. R., 1973, Rb–Sr determinations on shale horizons associated with the Varanger Ice Age: *Geol. Mag.*, v. 109, p. 465–472.

Pringle, I. R. and Sturt, B. A., 1969, The age of the peak of the Caledonian orogeny in West Finnmark, North Norway: *Norsk Geol. Tidsskr.*, v. 49, p. 435–436.

Ramsay, D. M., 1971a, The structure of northwest Söröy: *Norges Geol. Unders.*, v. 269, p. 15–20.

Ramsay, D. M., 1971b, Stratigraphy of Söröy: *Norges Geol. Unders.*, v. 269, p. 314–317.

Ramsay, D. M. and Sturt, B. A., 1970, Polyphase deformation of a polymict Silurian conglomerate from Mageröy, Norway: *J. Geol.*, v. 78, p. 264–280.

Reitan, P. H., 1960, Precambrian of northern Norway, Windows in the Caledonides in the Ofoten district, Troms and Finnmark, Raipas windows: *Norges Geol. Unders.*, v. 208, p. 68–72.

Roberts, D., 1971, Timing of Caledonian Evorogenic Activity in the Scandinavian Caledonides: *Nature*, v. 232, p. 22–23.

Roberts, D., Springer, J., and Wolff, F. C., 1970, Evolution of the Caledonides in the northern Trondheim region, Central Norway: a review: *Geol. Mag.*, v. 107, p. 133–145.

Romey, W. D., 1971, Basic igneous complex, mangerite, and high grade gneisses of Flakstadöy, Lofoten, Northern Norway: 1. Field relations and speculations on origin: *Norsk Geol. Tidsskr.*, v. 51, p. 33–61.

Ross, R. J. and Ingham, J. K., 1970, Distribution of the Toquima–Table Head (Middle Ordovician Whiterock) faunal realm in the Northern Hemisphere: *Geol. Soc. Amer. Bull.*, v. 81, p. 393–408.

Rui, I. J., 1972, Geology of the Rörös district, southeastern Trondheim region with a special study of the Kjöliskarvene–Holtsjöen area: *Norsk Geol. Tidsskr.*, v. 52, p. 1–21.

Rutland, R. W. R. and Nicholson, R., 1965, Tectonics of the Caledonides of part of Nordland, Norway: *Quart. J. Geol. Soc. Lond.*, v. 121, p. 73–109.

Siedlecki, A. and Siedlecki, S., 1971, Late Precambrian sedimentary rocks of the Tanafjord–Varangerfjord region of Varanger peninsula, northern Norway: *Norges Geol. Unders.*, v. 269, p. 246–294.

Skevington, D. and Sturt, B. A., 1967, Faunal evidence bearing on the age of the Late Cambrian–Early Ordovician metamorphism in Britain and Norway: *Nature*, v. 215, p. 608–609.

Skjeseth, S., 1963, Contributions to the geology of the Mjösa district and the classical Sparagmite area in southern Norway: *Norges Geol. Tidsskr.*, v. 220, 126 pp.

Smithson, S. B. and Ramberg, I. B., 1970, Geophysical profile bearing on the origin of the Jotun Nappe in the Norwegian Caledonides: *Bull. Geol. Soc. Amer.*, v. 81, p. 1571–1576.

Störmer, L., 1935, *Dictyocaris*, Salter, a large crustacean from the upper Silurian and Downtonian: *Norsk Geol. Tidsskr.*, v. 15, p. 265–298.

Strand, T., 1951, The Sel and Vågå map areas: *Norges Geol. Unders.*, v. 178, 116 pp.

Strand, T., 1955, Sydöstligste Helgelands geologi: *Norges Geol. Unders.*, v. 191, p. 56–70.

Strand, T., 1960, Cambro-Silurian stratigraphy, introduction: *Norges. Geol. Unders.*, v. 208, p. 128–130.

Strand, T., and Kulling, O., 1972, Scandinavian Caledonides; Part I, the Norwegian Caledonides, T. Strand, p. 1–145, and Part II, the Swedish Caledonides, O. Kulling, p. 147–285: Wiley Interscience, New York.

Strömberg, A. G. B., 1961, On the tectonics of the Caledonides in the southwestern part of the county of Jämtland, Sweden: Bull. Geol. Inst. Univ. Upsala, v. 39, p. 1–92.

Sturt, B. A., 1971, The timing of orogenic metamorphism in the Norwegian Caledonides, Abstracts, Edinburgh and Glasgow Geological Societies Meeting, Dating events in the metamorphic Caledonides, Symposium, Edinburgh, September, 1971.

Sturt, B. A., Miller, J. A., and Fitch, F. J., 1967, The age of alkaline rocks from West Finnmark, Northern Norway, and their bearing on the dating of the Caledonian orogeny: Norsk Geol. Tidsskr., v. 47, p. 255–273.

Sturt, B. A. and Ramsay, D. M., 1965, The Alkaline complex of the Breivikbotn area, Söröy, northern Norway: Norges Geol. Unders., v. 231.

Thorslund, P., 1960, The Cambro-Silurian: Sveriges Geol. Undersökn., Ser. Ba, N:0 16, p. 67–110.

Törnebohm, A. E., 1896, Grunddragen af det centrala Skandinaviens bergbyggned: Geol. Fören. Stockholm Förh., v. 10, p. 328–336.

Trouw, R. A. J., 1972, The structural and metamorphic transition from the Köli to the Seve units in south Västerbotten: Lecture resume, X Nordiske Geologiske Vintermöte, Oslo, 1972, p. 23–24.

Vogt, J. H. L., 1897, Norsk Marmor: Norges Geol. Unders., v. 22, 364 pp.

Vogt, T., 1927, Sulitelmafeltets Geologi og Petrographi: Norges Geol. Unders., v. 121, 560 pp.

Vogt, T., 1945, The geology of the Hölonda-Horg district, a type area in the Trondheim region: Norsk Geol. Tidsskr., v. 25, p. 449–528.

Wilson, J. T., 1966, Did the Atlantic close and then re-open?: Nature, v. 211, p. 676–681.

Wilson, M. R., 1971, The timing of orogenic activity in the Bödo-Sulitjelma tract: Norges Geol. Unders., v. 269, p. 184–190.

Wilson, M. R., and Nicholson, R., 1973, The structural setting and geochronology of basal granitic gneisses in the Caledonides of part of Nordland, Norway: J. Geol. Soc. Lond., v. 129, p. 365–388.

Wolff, F. C., Chaloupsky, J., Fediuk, F., Siedlecki, A., Siedlecki, S., and Roberts, D., 1967, Studies in the Trondheim region, central Norwegian Caledonides II: Norges Geol. Unders., v. 245, 146 pp.

Zachrisson, E., 1969, Caledonian geology of Northern Jämtland-southern Västerbotten, Köli stratigraphy and main tectonic outlines: Sveriges Geol. Unders., Ser C, Nr. 644, 33 pp.

Zwart, H. J., 1972, Some results of structural and petrological investigation of the Trondheim area and Vasterbotten: Lecture resumé, X Nordisk Geologiske Vintermöte, Oslo, 1972, p. 27.

Chapter 7

THE GEOLOGY OF THE SOUTHERN TERMINATION
OF THE CALEDONIDES

John F. Dewey

Department of Geological Sciences
State University of New York at Albany
Albany, New York

I. INTRODUCTION

Caledonian earth movements are manifest in the Caledonides. This truism raises the partly semantic, yet vexing, question of a proper definition of the terms Caledonides and Caledonian. If one adheres rigidly to definitions based upon Scottish mountains, the type example, the Caledonides involve the Northwest and Grampian Highlands and the Southern Uplands. The term Caledonian would then apply to earth movements from Precambrian to Middle Devonian. The terms, however, are currently, and sensibly, used to embrace orogenic zones, orogenic stratigraphic sequences, and orogeny in a distinct and unified, albeit complex, belt from Spitzbergen and Finmark through Scandinavia to the British Islands. When viewed on a pre-Early Jurassic sea-floor spreading reconstruction of the continents around the North Atlantic, the Caledonides can be interpreted as forming part of an orogenic belt continuous into the Northern Appalachian System of eastern North America and probably involves the Paleozoic orogenic belt of East Greenland (Fig. 1). Thus, on this model, it is proper to regard the Northern Appalachians and the Caledonides as a single mountain belt if we are to fully evaluate and

understand the evolution of either continent. The problem is further compounded by the fact that certain distinctive zones are common to the Caledonides, the Northern Appalachians, the Southern Appalachians, the Mauritanides and the Hercynides (Fig. 1). This essentially means that, when viewed
on a sufficiently large geographic scale, classically defined orogenic belts are
not discrete separate units but form parts of larger systems of crustal mobility.
Nor is the problem alleviated by choosing upper and lower time limits to
define an orogenic system. For example, the evolution of the Scottish Caledonides, from time of deposition of the first sediments to final stabilization,
occurred during an interval from about 900 m.y. to about 360 m.y. ago.
During this time the Celtic Orogenic Cycle (Wright, 1969) occurred in the
southern part of the Caledonides and in the Hercynides, but not in the Mauritanides. Thus, rocks of a certain age that are basement to one part of an
orogenic belt are intrinsic parts of the orogenic sequence in another. The
concepts of basement, orogenic system, and cover, although highly significant

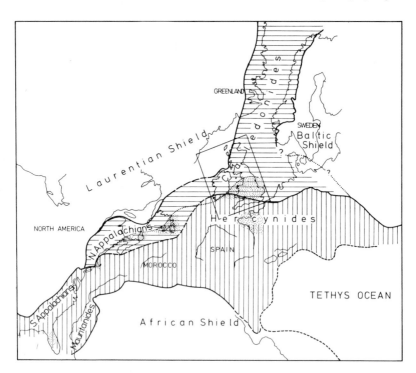

Fig. 1. Outline of the spatial relationship of the Appalachian, Caledonian, and
Hercynian (Variscan) orogenic belts on a Permian reconstruction of the continents
around the North Atlantic. Horizontal ornament: regions largely stabilized by the
Upper Devonian; vertical ornament: regions largely stabilized by Permian time;
dotted ornament: areas affected by late Precambrian deformation (Celtic event).
Area inside rectangle shown in more detail in Fig. 2.

for a particular zone, cannot be applied to a whole system, since the larger the area that is chosen in which to apply basement/cover time limits the larger the time span that is involved. However, in order to define Caledonide/Caledonian limits for the purpose of describing the southwestern termination of the European Caledonides in the British Islands the following definitions are adopted. The Caledonides comprise those rock sequences, in the British Islands, that accumulated and were deformed to varying degrees, and at different times, during a Caledonian time interval defined as from the beginning of Moine Series sedimentation about 900 m.y. ago to the base of the Upper Devonian about 350 m.y. ago. In this definition the youngest pre-Caledonian basement is Pentevrian, probably represented by the Rosslare complex in southeast Ireland (Fig. 2). Post-Caledonian cover consists of Upper Devonian and younger rocks (Fig. 2).

The Caledonides in the British Islands occur in a series of inliers of varying size (Fig. 2), isolated by post-Caledonian cover sequences. Some of these inliers are sufficiently large to enable a fully coherent pattern of stratigraphic and structural development to be worked out; others are very small and merely provide tantalizing glimpses. It has long been recognized that different parts of the British Caledonides are characterized by different development paths— stratigraphic, structural, and metamorphic sequences. Read (1961), for example, distinguished the northern metamorphic from the southern nonmetamorphic Caledonides. The incisive syntheses of the stratigraphic development of the Caledonides by George (e.g., 1960, 1963, 1965) have emphasized the great complexity of lateral thickness and facies variation. In this chapter, 11 zones (A–K, Fig. 2) are recognized. In Section II the main stratigraphic, structural, and metamorphic characters of each of these zones are described. Each zone is divided into stratigraphic/structural elements (1, 2, etc.) that record distinct episodes in the sequential evolution of the zone (summarized in Fig. 3). The elements are bounded by unconformities and/or major facies changes. The writer wishes to emphasize that the zones and elements chosen could certainly be subdivided. Although they accord with geologic data, they are chosen to highlight what the writer believes to be events of greatest significance for the evolution of the belt. Thus, Section II lays a basic outline of the stratigraphic/structural framework of the British Caledonides in as factual a way as consistent with incomplete data. In Section III some regional problems of the British Caledonides are discussed, such as the basement problem and the age of major deformational events.

Our knowledge of the British Caledonides stems from data accumulated by intensive and detailed work by hundreds of researchers over a period of more than 150 years. The sheer volume of literature is formidable, and for this reason it is impossible to individually reference every paper that contributes to the pool of knowledge. The reader is referred to the following publications

with their accompanying bibliographies on which the writer has drawn heavily,
in addition to those cited specifically in the text.

Bassett (1963, 1969); Cummins (1969); Dewey (1969a, 1971); Dewey and
Pankhurst (1970); Dunning (1966); Eastwood (1953); Edmunds and
Oakley (1947); Friend (1969); George (1960, 1963, 1965); Harper (1948);
Johnson (1965a, 1965b); Johnstone, Smith, and Harris (1969); Kennedy

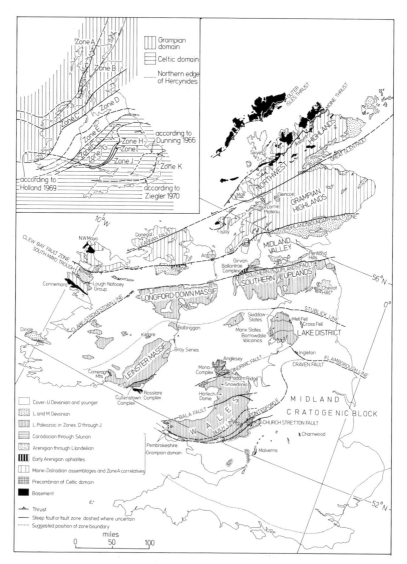

Fig. 2. Outline geologic map of the British Islands showing the distribution of
pre-Upper Devonian rocks and some major Caledonian structures. Inset shows
schematically the zones referred to in the text.

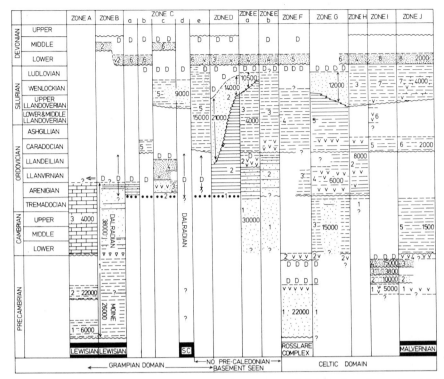

Fig. 3. Schematic illustration of the ages, thicknesses, and lithologies of stratigraphic/ structural elements in the various zones of the British Caledonides referred to in the text. Zone C subdivisions: (a) Highland Boundary fault; (b) Tyrone; (c) South Mayo trough; (d) Connemara; (e) Ballantrae. Zone E subdivisions: (a) Lake District/Isle of Man; (b) Leinster. Thicknesses given in feet. Key to ornament: double-barred arrow, nappe emplacement; D, deformation; SC, pre-Caledonian basement of Connemara; bricks, shallow marine carbonates; dashed horizontal lines, shallow marine clastics; horizontal lines, deep marine lutites and cherts; wide spaced dots, flysch; close-spaced dots, molasse; v, acid/intermediate volcanics; triangles, basic volcanics; black circles, ophiolites.

(1958); Pocock and Whitehead (1948); Pringle and George (1948); Rast (1963, 1969); Rast and Crimes (1969); Read (1961); Shackleton (1969); Stewart (1969); Swett (1969); Walton (1965a, 1965b, 1969); Williams (1969); Wood (1969); Wright (1969); Ziegler (1970).

II. ZONES AND ELEMENTS

A. Zone A

In this zone, the basement consists of the Lewisian complex. The Lewisian contains the oldest rocks known in the British Islands and has been divided (Sutton and Watson, 1951) into an older, Scourian, assemblage of granulite facies gneisses with pegmatites yielding ages from 2500 to 2000 m.y. and a younger, Laxfordian, assemblage of almandine–amphibolite facies gneisses permeated by granites yielding ages from 1650 to 1200 m.y. The Stoer Group (element 1) (Fig. 3) consists of gneiss breccias and conglomerates, red arkosic sandstones and siltstones and occasional limestones, filling irregularities in, and burying a strong topography cut into the Lewisian. The environment of deposition seems to have been one of bajada fans and playa lake cycles of desiccation interrupting a fluvial red sand facies. The Torridonian sequence unconformably overlies the Lewisian and has been shown by Stewart (1969) to consist of two major groups (Stoer and Torridon groups) separated by an unconformity. Near the top of the Stoer Group, volcanic debris is common. The Stoer Group wedges out eastward beneath the Torridon Group and was deposited by west-flowing currents. The Torridon Group (element 2) is regionally more extensive than the Stoer Group, overstepping onto the Lewisian. The lowest division, the Diabaig Formation, consists of locally derived breccias, red flaggy sandstones, and grey siltstones and sandstones burying an irregular Lewisian topography. The succeeding Applecross Formation comprises cross-bedded red fluvial arkoses with pebbles of locally derived Lewisian rocks and "exotic" pebbles of tourmaline aplite, quartz–fuchsite schist, metaquartzite, chert, porphyritic rhyolite, and acid tuff. The overlying Aultbea Formation is similar but is pebble-free. The topmost division, the Cailleach Head Formation, consists of numerous cycles of grey shale (possibly shallow marine) succeeded by planar cross-bedded, and then trough cross-bedded, red sandstone. The Torridon Group becomes finer eastward and was deposited mainly by eastward-flowing currents. In Skye, the Sleat Group, consisting of grey current-bedded fluvial sandstones and greenish-grey shales (probably shallow marine), comes in conformably below the Torridonian and thickens eastward, suggesting a basin of deposition in that direction. There is now some evidence that indicates a considerable time interval between the deposition of the Stoer and Torridon groups in that shales in the lower group have yielded 935 ± 24 m.y. and shales in the upper group 751 ± 24 m.y. by the Rb/Sr whole rock method. The paleomagnetic break recorded by a $50°$ shift in paleomagnetic pole position (Irving and Runcorn, 1957) coincides with the Stoer/Torridon unconformity.

During the late Precambrian, the Torridon Group was tilted westward and eroded before a lower Cambrian marine transgression that laid down the shallow marine sandstone sequence of the Eriboll Formation. The Eriboll Formation is divided into a lower current-bedded feldspathic/orthoquartzite member and an upper flat-bedded orthoquartzite member (Swett, 1969). The overlying An t-Sron Formation comprises a lower member, the Fucoid Beds, consisting mainly of dolomitic siltstones and bearing *Olenellus*, and an upper member, the Serpulite Grit, consisting principally of orthoquartzites. The

An t-Sron Formation is succeeded by dolostones, limestones, and cherts of the Durness Formation. The Eriboll, An t-Sron, and Durness formations, comprise element 3 and record the progressive marine inundation of zone A and the establishment of a carbonate shelf environment. The highest unit of the Durness Formation, the Durine Member, contains a conodont fauna of upper Canadian age (*Didymograptus protobifidus* zone of North America; Higgins, 1967). The correlation of the North American and British Lower Ordovician has been a subject of considerable controversy (Berry, 1968; Dewey, Rickards, and Skevington, 1970). Berry's scheme would place the Durine Member in the early part of the Llanvirnian Series; that of Skevington, 1968, would mean that the Durine Member is of Arenigian age (*Didymograptus nitidus* zone). Sometime after the deposition of the Durness Formation, the eastern part of zone A was severely deformed during the emplacement of the Moine Nappe of zone B. Large folds, such as the Lochalsh fold, were formed with penetrative cleavages below the Moine Nappe in Skye and a complex history of thrust imbrication (e.g., in the Assynt Bulge) affecting a narrow strip beneath the Moine Nappe. The Outer Isles thrust was probably developed at this time.

B. Zone B

Zone B consists of the Northern and Grampian Highlands of Scotland, bounded by the Moine thrust zone and Highland Boundary fault zone (Fig. 2). The zone continues westward into Ireland where it appears as inliers in Antrim, Donegal, and northwest Mayo. The northwest margin of the zone in Ireland is not seen and the southern margin is taken as the Clew Bay fault zone in Mayo, a probable continuation of the Highland Boundary fault zone of Scotland. The main element of zone B is the Moine/Dalradian sequence (element 1) whose composite thickness probably exceeds 60,000 ft. The Moine assemblage consists of some 26,000 ft of quartzofelspathic psammite, semi-pelite, pelitic schist, calc-silicate, and rare marble. Lewisian basement inliers are widespread within the Moine assemblage (Fig. 2). These are in various stages of Caledonian "reworking" and indicate with certainty that at least the northwestern part of zone B is underlain by pre-Caledonian continental crust. The Dalradian assemblage consists of two separate successions whose temporal relationships are not clear. The Ballappel succession conformably overlies the Moine assemblage and occurs in a narrow tract along the northwestern part of the Grampian Highland. It consists of mature quartzite, pelite, limestone, and dolostone. The Iltay succession also conformably overlies the Moine assemblage but is tectonically juxtaposed against the Moine by the Iltay Boundary Slide along much of its lower contact. The Iltay Dalradian succession may be divided into two facies assemblages. The lower and middle

Dalradian comprise the first assemblage, one of some 26,000 ft of flaggy quartzite, pelite, and carbonate with abundant evidence of shallow marine deposition. The base of the middle Dalradian is marked by the Portaskaig Boulder Bed lying between a marble (Islay Limestone) below and an orthoquartzite (Islay Quartzite) above. This distinctive conglomerate occurs in several localities in the Grampian Highlands and in the Dalradian of Donegal, northwest Mayo, and Connemara and is probably of glacial origin (Kilburn, Pitcher, and Shackleton, 1965). The upper Dalradian of the Iltay succession is composed of about 12,000 ft of mainly immature sediments (graded grits, greywackes, and slates) many of which were deposited by turbidity currents. Basic vulcanicity was important during the early stages of upper Dalradian deposition.

There is now abundant paleontologic evidence for the Cambro-Ordovician age of the higher parts of the Dalradian assemblage (Downie *et al.*, 1971). An unequivocal topmost Lower Cambrian age for the Leny Limestone is given by *Pagetides*. The Upper Dolomitic Group of the middle Dalradian Islay Quartzite and the overlying Easdale Slate has yielded spheaeromorphs of probable uppermost Precambrian age. The Tayvallich Limestone at the base of the upper Dalradian has abundant acritarchs of probable Lower Cambrian age (Downie *et al.*, 1971). Thus the Precambrian/Cambrian boundary appears to lie in a conformable sequence between the Easdale Slate and the Tayvallich Limestone. A minimum stratigraphic age for the Dalradian assemblage is given by the McDuff Slates in which highly altered chitinozoa indicate at least an Arenigian and probably a Llanvirnian age (Downie *et al.*, 1971). The time of onset of Moine sedimentation on a Lewisian basement and correlations between elements 1, 2, and 3 of zone A and element 1 of zone B are equivocal. Rb/Sr isotopic data suggest a *maximum* age for the onset of Moine sedimentation of about 1000 m.y. (Long and Lambert, 1963). Lambert (1969) has shown that the western part of the Moine assemblage gives an Rb/Sr isochron of about 800 m.y. and argues that this represents a distinct Precambrian metamorphic episode, the Morarian event. Supporting evidence comes from Rb/Sr ages of 765 $\pm$ 15 m.y. from micas in a pegmatite from Knoydart (Giletti, Moorbath, and Lambert, 1961). On this basis the Stoer/Torridon Group unconformity of zone A could be an external foreland manifestation of the Morarian event. However, there is at present no structural data in the Moine assemblage supporting this view. The Moine and Dalradian assemblages appear to be a coherent conformable stratigraphic entity whose first major deformation and metamorphism occurred during the Early Ordovician. In broad terms it seems likely that Stoer Group sedimentation in zone A and Moine sedimentation in zone B began about 950 m.y. ago, that the Stoer Group is equivalent to the lower part of the Moine assemblage, the Torridon Group is equivalent to the upper part of the Moine assemblage and the lower part of the Dalradian assemblage and that the upper Dalradian equates with the Eriboll/Durness sequence (Fig. 3).

Element 1 is affected by a complex polyphase structural and metamorphic history involving D1/D2 nappe formation followed by an episode of basic magmatism and migmatite "permeation" at which time the major metamorphic imprint occurred. This was succeeded by large-scale D3 antiforms and synforms and, generally, retrogressive metamorphism. A minimum stratigraphic age for this deformation and metamorphism (Grampian Event) is given by relatively flat-lying Lower Devonian sediments and volcanics in the Lorne plateau, at Glencoe, and along the southern border of the Grampian Highlands. A full discussion of the age and regional relationships of the Grampian event is given in Section III.

C. Zone C

Zone C is defined as that region bounded by the Clew Bay–Highland Boundary fault zone on the north and the southern Uplands fault, and its westward Irish extension, on the south. Zone C is stratigraphically and structurally complex, and is beset with unsolved major regional problems. Element 1 comprises a lower/middle Dalradian metamorphic assemblage exposed in Connemara and Tyrone. In Connemara recently recognized, preelement 1, basement inliers (Dr. M. Max, pers. comm.) occur in the Connemara migmatite belt. Potash feldspar gneisses in this belt yielded an Rb/Sr isochron of 725 ± 175 m.y. (Leggo, Compston, and Leake, 1966). Whether this age refers to a basement age, or an intraelement 1 metamorphic event, is not clear, but this age is close to those in the western Moine Series of zone B in Scotland that Lambert regards as signifying the latter interpretation. Element 2 consists of variously developed members of an ultramafic/mafic suite of intrusive and volcanic rocks that, where paleontologically dated, are of late Tremadocian/ early Arenigian age. In Connemara, element 2 pillow lavas and red cherts with minor acid/intermediate rocks (Lough Nafooey Group) comprise the lowest part of the South Mayo trough, lower Paleozoic sequence. However, the base of the group is not seen and the group has a slightly unconformable relationship with overlying Arenigian strata of element 3. Serpentinites along the Clew Bay fault zone and serpentinites–gabbros, pillow lavas, and sediments (Highland Border Series) along the Highland Boundary fault probably form an integral part of element 2, as probably do gabbros and pillow lavas of the Tyrone igneous complex (Cobbing, Manning, and Griffith, 1965). Element 2 is best developed in the Ballantrae complex of Ayrshire (Fig. 2), where it consists of an assemblage of serpentinites, gabbros, trondjemites, pillow lavas, and cherts. The Ballantrae complex is locally affected by blueschist (crossite) metamorphism and suffered this metamorphism and accompanying deformation prior to the Caradocian. In the South Mayo trough of western Ireland a 36,000-ft-thick section of Ordovician sediments and volcanics shows a striking southward thinning and facies change to a 7000-ft-thick section (Dewey, Rickards, and Skevington, 1970). The thick northern section consists of 21,000 ft of pyroturbidites, turbidites, and argillites (element 3) below, conformably overlain by 15,000 ft of red cross-bedded sandstones with green siltite tongues, and ignimbrite horizons (element 4). Element 3 is a "flysch" sequence and consists of a lower group of argillites, coarse turbidites, and boulder slides (2000 ft), a middle group of pyroturbidites (10,000 ft), and an upper group of turbidites (5000 ft) below and slates (4000 ft) above. Element 3 thins southward to a shallow marine (? partly fluviatile) sequence (5000 ft) with conglomerates and ignimbrites. Locally, the threefold division characteristic of the northern sequence is recognized, but both the middle volcanic and the upper nonvolcanic groups overstep onto element 2 basic volcanics. Element 4 is a dominantly fluviatile molasse sequence that shows a similar southward thinning and facies change to a 2000-ft-thick sandstone/fanglomerate sequence. Elements 3 and 4 in the South Mayo trough are of Arenigian/Llanvirnian age, and their relationships to the Dalradian rocks of Connemara pose some difficult problems involving the age of the deformation and metamorphism of the Dalradian rocks of element 1. These problems, together with the allied problems of zone A, B, and C relationships, will be fully discussed in Section III.

Element 5 consists of a varied assemblage of mainly shallow marine sediments of Caradocian to Ludlovian age lying unconformably on older rocks. In regional zone C terms, element 5 thins northwards, the base becomes younger northward, and the depositional environment shallowed northward. Deformation and low-grade metamorphism affected most of zone C in Late Silurian times, but locally, in the Midland Valley of Scotland, sedimentation continued into the Lower Devonian. The Lower Devonian (element 6) is a molasse sequence resting mainly unconformably on older rocks, derived from both zone B and zone D source areas. Faulting and open folding was widespread during the Middle Devonian and Upper Devonian, and younger strata form a relatively flat-lying cover.

In southwestern Ireland the zone affinities of Silurian and Devonian rocks on the Dingle peninsula are obscure. Over 5000 ft of shallow marine Wenlockian/Ludlovian sediments with conspicuous acid/intermediate volcanic intercalations are conformably overlain by the Dingle Group, a thick "molasse" sequence of fluviatile purple/red sandstones and conglomerates (Holland, 1969). These Silurian/Devonian strata were folded prior to the deposition of an Upper Devonian cover. Holland (1969a) has argued that the Dingle rocks belong to the southern margin of the British Caledonian mobile belt (zone H or I of this chapter), whereas Rast and Crimes (1969) and Dewey (1969b) have suggested a zone C affinity arguing that the Clare/Strokestown Line

(Fig. 2) passes south of Dingle as the westward continuation of the Southern Upland fault.

D. Zone D

The lowest element (1) in zone D is a suite of early Arenigian pillow lavas and cherts forming inliers in the northern part of the Southern Uplands and rare occurrences in the Longford/Down massif. These are the oldest rocks exposed in Zone D, and there is no evidence for a pre-Caledonian continental basement as there is in zone A and parts of zones B and C. Element 1 pillow lavas are conformably overlain by element 2 sediments that range essentially conformably from Arenigian to, locally in Ireland, early Ludlovian. Two major and one minor facies comprise element 2. A thin chert/argillite facies below is overlain by a very thick turbidite (flysch) facies; the boundary is strongly diachronous from Llanvirn/Llandeillian in the north to upper Llandoverian in the south. Ziegler (1970) and Dewey (1971) have compiled stratigraphic data indicating the gradual southward-spread of the flysch wedge across the chert/argillite environment. The provenance of the flysch wedge until upper Llandoverian times was to the north of the Southern Uplands fault. Sediments in the flysch wedge contain detritus indicating that igneous and metamorphic rocks of the Ballantrae complex were exposed in zone C, that biotite-grade metamorphic rocks were exposed in zone B by Llandeilian times, and that garnet- and andulusite-bearing metamorphics were being eroded in zone B by Caradocian times. Caradocian acid/intermediate volcanics form a minor constituent of element 2. "Cockburnland," an important positive paleogeographic element, appeared along the northern edge of zone D during the upper Llandoverian. This appears to have been a growing swell or ridge that rose above sea level, growing by the deformation of the flysch wedge (Ziegler, 1970). From upper Llandoverian times onward it formed a rising and southward-growing barrier, feeding the flysch wedge to the south and a northward-spreading flysch/molasse assemblage along the southern edge of zone C. Elements 1 and 2 were deformed and invaded by granite in Late Silurian times. During the Lower Devonian, zone D low-grade metamorphic lands shed molasse sediments into zone C, and acid/intermediate vulcanicity developed in the Cheviots. A milder deformation occurred during the Middle Devonian, and the Caledonian "cycle" was closed by the progressive onlap of an Upper Devonian/Carboniferous cover sequence.

E. Zone E

This zone incorporates the Caledonides of the Lake District, the Cross Fell inlier, the Isle of Man, and most of the Leinster massif of southeast

Ireland (Fig. 2). Clear zone E boundary relations are only seen in the southeast corner of Ireland. The zone D/zone E boundary is entirely concealed by the cover sequence, and therefore its position and nature are uncertain. Zone E is characterized by large thicknesses of Ordovician acid/intermediate/basic volcanic rocks and a prominent sub-Caradocian unconformity. No pre-Caledonian continental basement is exposed within the zone. The zone does not maintain a constant linear unity of stratigraphic development, and for this reason it is transversely divided into subzones E_a and E_b (Fig. 3). Element 1 in zubzone E_a comprises the Manx Slate Series, some 25,000 ft of argillites, turbidites, and slide breccias, and the Skiddaw Slate "Series" about 7000 ft of cleaved argillites and siltites. The age of the Manx Slate Series is unknown, but the Skiddaw Slate Series is Arenigian/Llanvirnian. Possibly the two series form a conformable Cambrian/Early Ordovician sequence. Fifteen thousand feet of andesites with minor basalts and rhyolites (element 2), the Borrowdale Volcanic Series, probably of Llandeilian age, overlie the Skiddow Slate Series. There is some controversy concerning the nature of the sub-Borrowdale contact, Simpson (1967) and Helm (1970) claiming a major unconformity, Soper (1970) arguing for a conformable relationship. There is considerable evidence to suggest mild local sub-Borrowdale unconformity, although it seems clear that the complex polyphase deformation of the Skiddaw Slate Series was largely end-Silurian. Element 3 comprises a conformable Caradocian to Ludlovian sequence resting unconformably on older rocks. Pre-Caradocian folding of the Borrowdale volcanics produced open north-northeast-trending folds. Element 3 consists of a Caradocian assemblage of calcareous shales and ashy sediments from 150 to 1200 ft thick, overlain by 100 ft of Ashgillian limestone and shale, and 1250 ft of Llandoverian/Wenlockian shales and mudstones capped by 10,500 ft of Ludlovian turbidites. This turbidite sequence could be the frontal portion of the flysch wedge that spread southward across zone D from the advancing margin of Cockburnland. Lower Devonian molasse sediments are probably represented by the Mell Fell Conglomerate (element 4) resting unconformably on deformed lower Paleozoic rocks.

Element 1 in subzone E_b is the Bray Series, an 8500-ft-thick assemblage of green and red turbidites and shales with thick, pure-white quartzite intercalations. The presence of *Oldhamia* (*antiqua* and *radiata*) suggests Cambrian age. Element 2, the Bannow Group (Crimes and Dhonau, 1967), is a thick argillite, silitite, greywacke, orthoquartzite sequence, dominantly grey, green, and black, but with red "Bray-like" intercalations. The group is partly of early Ordovician age, but its relationships with the underlying Bray Series are equivocal. Element 3 is a thick pile of Caradocian andesitic/rhyolitic/basaltic volcanics with locally a thin basal limestone resting unconformably on the Bannow Group. Little-studied Silurian argillites west of Waterford have uncertain relationships to the volcanic sequence but are here included in element 3 by comparison with the Ordovician/Silurian sequence of the Balbriggan inlier. The Comeragh Group (element 4), a coarse red molasse facies, is probably of Lower Devonian age (Capewell, 1956). The Comeragh Group rests unconformably on element 3, indicating that the major deformation of subzone E_b was accomplished prior to Lower Devonian times, although around most of the Leinster massif the cover sequence consists of progressively onlapping Upper Devonian/Carboniferous sediments.

F. Zone F

The rocks of zone F are exposed in County Wexford in the southeast corner of Ireland, in Anglesey, and possibly at Ingleton (Fig. 2). Crimes and Dhonau (1967) have shown that the Precambrian rocks of zone F in Wexford comprise two discrete metamorphic complexes, an older Rosslare complex consisting of gneisses, schists, and migmatites and a younger Cullenstown complex consisting of ortho- and para-quartzites, greywackes, and red to green albite–chlorite–sericite schists. The Rosslare complex clearly forms a pre-Caledonian metamorphic basement possibly of Pentevrian age. The Cullenstown Group (element 1) is insufficiently exposed to allow a coherent stratigraphy to be established. However, it is almost certainly equivalent to the Mona complex of Anglesey in which Shackleton (1969) has established a very comprehensive geologic history. The Monian succession (element 1) consists of over 20,000 ft of apparently conformable metasediments and metavolcanics; graded greywackes and quartzites at the base becoming more pelitic upward, with the incoming of pillow lavas, red cherts, limestones, boulder conglomerates, and a spectacular slide breccia; and finally at the top a dominantly acidic volcanic group. Element 1 has a protracted polyphase history of deformation and metamorphism. Of particular note is the rapid cross-strike change metamorphic grade from lawsonite-bearing blueschists to sillimanite–almandine amphibolites. Some of the high-grade gneissic rocks of Anglesey were regarded by Greenly (1919) as part of a pre-Monian basement complex, but Shackleton (1969) has challenged this view showing that large areas of the gneisses developed from parts of the Monian sequence.

Element 2, the Arvonian Series, is an acidic volcanic sequence of ignimbrites and tuffs resting unconformably on the Mona Complex. Element 3 consists of about 3000 ft of Arenigian conglomerates and sandstones resting unconformably on a deeply eroded surface on the Mona complex overlain by over 1000 ft of finer Arenigian and Llanvirnian sediments. The Cambrian is possibly represented by small, locally derived pockets of conglomerate beneath the transgressive base of element 3. Strongly discordant Caradocian conglomerates overstepping the Lower Ordovician northwestward onto the Mona complex form the base of element 4, the rest of which is composed of Caradocian shales and about 1400 ft of Llandoverian and Wenlockian shales. Strong Late Silurian penetrative deformation of elements 3 and 4 and non-penetrative deformation of older rocks was followed by the deposition of element 5, a Lower Devonian molasse sequence of basal conglomerates overlain by red and purple sandstones, marls, and cornstone. Element 5 deposition was followed by renewed earth movements (?Middle Devonian) when the south-ward-moving Carmel Head thrust sheet was emplaced.

Southeast of the Lake District in the Ingleton region, Caradocian sediments rest unconformably upon strongly deformed but weakly metamorphosed greywackes and pelites. These rocks, the Ingletonian "Series" are of uncertain, but probable late Precambrian age. It is not clear to which zone the Ingleton inlier belongs, but the writer adopts the view that the Ingletonian rocks are part of the Monian sequence and that they belong to zone F.

G. Zone G

This zone comprises the bulk of the classic Welsh geosyncline (Jones, 1938). It is a zone of considerable stratigraphic and structural complexity with great thickness and facies contrasts from one part of the zone to another. Monian (element 1) and Arvonian rocks (element 2) are overlain by a sequence of Cambrian strata (element 3, 5000 ft) consisting of conglomerates, grits, and ignimbrites below, passing conformably upwards into a slate sequence. Wood (1969) has shown that the Arvonian and Cambrian rocks are conformable to the southeast but that northwestwards the Cambrian unconformably overlies the Arvonian. To the southeast in the Harlech dome, element 3 is a 15,000-ft-thick turbidite succession of poorly fossiliferous Cambrian greywackes and shales (Harlech Series) whose base is not seen. In South Wales element 1 consists of about 4000 ft of Cambrian shallow-water sandstones, mudstones, and shales resting unconformably upon Precambrian acidic/intermediate volcanic (Pebidian complex) and intrusive (Dimetian) rocks. Element 4 comprises strata of Arenigian, Llanvirnian, and Llandeillian age either resting unconformably upon older rocks (notably in the northwest and southwest) or showing a marked facies contract to them. The basal member is usually a shallow marine conglomeratic grit or ortho-quartzite and element 4 sediments generally become progressively finer upward. Acidic/intermediate/basic volcanic rocks are prominent immediately to the southeast of the Harlech dome and in southwest Wales. Rast (1969) has cogently argued a relationship between the distribution of Arenigian/Llanvirnian volcanic rocks and a possible southward extension of the Bala fault (Fig. 2). Widespread uplift occurred during the Llandeilian. In South Wales 2500 ft of sandy calcareous flags and limestones accumulated, but in North Wales Llandeilian rocks appear to be absent, and early Caradocian mudstones and shales rest disconformably on similar Llanvirnian rocks.

Element 5 is a Caradocian to Ludlovian sequence with minor disconformity between Caradocian and Ashgillian strata. Spectacular developments of acid/intermediate/basic volcanics occur in the Caradocian, notably in Snowdonia (Fig. 2). Element 5 Ordovician sediments are mainly of shallow-water origin and commonly have evidence of contemporaneous vulcanism. Element 5 Silurian sediments, however, except in South Wales, show a marked change to nonvolcanic "basinal" conditions. Turbidite sedimentation began in upper Llandoverian times and shifted northward during the Wenlockian and Ludlovian (Cummins, 1969). During Late Silurian times, the rocks of zone G were deformed, and in South Wales Lower Devonian molasse sediments (element 6) rest unconformably upon them.

H. Zones H and I

Zone H is defined as a narrow belt including the Shelve and Builth inliers, bounded to the east by the Pontesford Hill fault and to the west by the Builth/Welshpool Line. Its main distinguishing feature is a strong sub upper Llandoverian unconformity and shallow marine Silurian sediments overlapping southeastward (Ziegler, 1970). In zone I this overstep is continued, Llandoverian andesites and rhyolites are developed (in South Wales), and the Caradocian oversteps the Cambrian and Early Ordovician rocks to rest unconformably upon the Precambrian. Thus during Ordovician times the Pontesford Hill fault was the eastern boundary of the Welsh basin, whereas during Silurian times the shelf edge moved westwards to the Builth/Welshpool Line. Zone I may have formed, during the Ordovician and Early Silurian, a ridge between a deeper water basin to the west and a shallow-water platform (zone J) to the east. Precambrian rocks are extensively developed in zone I. Element 1, the Uriconian, is a dominantly acidic volcanic complex. Element 2, the Stretton Series, consists of 10,000 ft of red–purple and green shales, siltstones, and sandstones with thin andesites near the top. The Portway Group (element 3) comprises up to 3800 ft of conglomeratic sandstone and purple and green shales. Element 4, the Wentnor Series, consists of over 15,000 ft of mainly purple conglomerates, sandstones, and shales. All these Precambrian rocks were folded prior to the Cambrian and lie in a broad but tight syncline. The Stretton Series have a "flyschlike" aspect, whereas the Portway Group and the Wentnor Series are molasse-type sediments. Unconformities are developed below both the Portway Group and Wentnor Series, and the latter oversteps westward onto the Uriconian.

The Late Silurian deformation, widespread in zone G, appears to have had little representation in zones H and I, where molasse-type Lower Devonian sediments (element 4 in H, element 8 in I) lie without angular discordance on shallow marine Silurian sediments.

I. Zone J

Zone J is the classic Midland Cratogenic block or Midland shelf characterized by a relatively thin, little-deformed, lower Paleozoic section lying unconformably on Precambrian rocks. The oldest Precambrian rocks are a gneissic pre-Caledonian basement complex, the Malvernian. Uriconian (element 1), Strettonian (element 2), and Wentnorian (element 3) rocks are developed along the northwestern margin of the zone. In the Charnwood Forest region (Fig. 2) pyroclastics overlain by conglomerates and slates may correlate with the Uriconian and Strettonian, respectively (Wright, 1969). In the Malverns, the Warren House volcanics (acid/intermediate) rest unconformably on the Malvernian and may be equivalent to the Uriconian, or to the Arvonian of zone G. Small enigmatic outcrops of metamorphic rocks such as the Primrose Hill Gneiss and the Rushton Schists could belong to the Malvernian basement complex, may be equivalent to the Monian, or could be more strongly deformed equivalents of the Uriconian or Strettonian.

Element 5 is a 1500-ft-thick Cambrian sequence of shallow marine transgressive quartzites followed by glauconitic sands and shales. Element 6 consists of a Caradocian shallow marine sandstone shale sequence some 2000 ft thick. Forty thousand feet of shallow marine sandstones, shales, and limestones with, south of Bristol, a local development of Llandoverian/Wenlockian andesites, comprise element 7. Element 8 consists of 2000 ft of Lower Devonian cyclothemic molasse sediments lying conformably on element 7.

J. Zone K

This is a zone whose northwest boundary is poorly defined and whose existence as a region, discrete from the Midland shelf, has been recognized entirely from borehole information. Lower Paleozoic rocks in boreholes south and east of the postulated zone J/zone K boundary in Fig. 2 are of "geosynclinal" aspect similar to those of the Ardennes massif in Belgium.

III. DISCUSSION

It will be clear from the foregoing zone descriptions, even as presented in barest outline, that the stratigraphic/structural development of the British Caledonides was a long and complex history of zone differentiation and evolution. Minor unconformities, disconformities, and nonsequences, too numerous to mention in this brief account, occur throughout the Caledonides in addition to the major deformational and metamorphic events. George (1963) summarized and analyzed the distribution and significance of sediment thickness variation in relation to unconformity development. He has clearly shown the difficulties of establishing ages of diastrophism from unconformities, illustrating how many unconformities are inherited from much earlier events and how some are the product of cumulative movement and protracted progressive onlap. McKerrow (1962) plotted the distribution and ages of unconformity in the British Caledonides and cogently emphasized that the term Caledonian as applied to diastrophism is not particularly meaningful because Caledonian deformation was manifested in a variety of ways in different places and at different times. It is clearly of great importance to distinguish between merely recognizing the existence of an unconformity, being able to date the major deformation of rocks below it, and recognizing protracted events such as vertical uplift before the final onlap. This problem is enhanced by the difficulty of relating flysch and molasse sequences to associated "orogenic events" manifested by deformation and metamorphism elsewhere. If we take fluviatile cyclothemic molasse sequences to represent the rapid uplift of adjacent, just-deformed terranes, we can recognize three *major* deformational events. Wright (1969) has argued that the Wentnor Series of zone I is a late orogenic molasse sequence to the deformation of the Monian complex. He has further suggested that the Stretton Series represents an exogeosynclinal flysch sequence contemporaneous with much of the deformation and metamorphism of the Mona complex. This late Precambrian deformation has been termed the Celtic event (Wright, 1969) and the Cadomian event (Rast and Crimes, 1969). It was widespread from zone F southward into continental Europe where the Brioverian sequence of France is the probable partial equivalent of the Mona complex. The effect of the Celtic event was to terminate the Celtic Cycle (Wright, 1969) of crustal mobility and to produce a wide lower Paleozoic shelf area of shallow marine sedimentation (zone J) and a bounding northwest outer ridge (zone F). Regions of lower Paleozoic crustal mobility within the Celtic domain, with thick sedimentary sequences and vulcanism, such as the Welsh basin (zone G) and the southeast England/Brabant trough (zone K), may represent areas of *continued* mobility, or basins initiated after the Celtic event. Zone F, the outer ridge, formed an important paleogeographic element, the Irish Sea Landmass, during Cambrian times and was progressively inundated during the Early and Middle Ordovician. The writer accepts the basic premise of Wright (1969) that the Longmyndian sequence is an external manifestation of deformation and metamorphism in zone F. In this view zone F is the core of a Celtic metamorphic protolith that suffered polyphase deformation and metamorphism from about 680 m.y. to about 590 m.y., with a Longmyndian exogeosynclinal sediment wedge recording the progressive deformation, uplift, and unroofing of the protolith. The exogeosyncline was deformed, stabilized, and eroded before the basal Cambrian transgression and, although the Monian protolith was stabilized by early Cambrian times, it formed a positive element that was not finally buried until Caradoc times.

The next major deformational event in Caledonian development was the Grampian event that terminated the Grampian Cycle and established the Grampian metamorphic protolith of zones B and C (Dewey, 1969*b*; Rast and Crimes, 1969). The age of the Grampian event, that is the age of the deformation and metamorphism of the Dalradian and Monian assemblages, is enigmatic and riddled with apparent paradoxes. Apart from the Stoer Group/ Torridon Group unconformity and its possible manifestation in the western Moines (Lambert, 1969), the Moine and Dalradian assemblages form a coherent stratigraphic element with essentially continuous sedimentation from the Precambrian (? 900 m.y.) to at least the Early Ordovician. The late Precambrian Celtic event did not affect zones B and C. It is now apparent that the Grampian event is not simply represented by instantaneous deformation and metamorphism in uplifted terranes, followed by widespread erosion and subsequent transgression. It has been the search for a narrowly and well-defined unconformity, representing the age of Grampian deformation and metamorphism, that has led to many difficulties and altercations. If we adopt the precedent of starting with the recognition of a molasse sequence, as we did with the Celtic event, such a molasse sequence is the Partry Supergroup in western Ireland. The Partry Supergroup is of Llanvirnian/?Llandeilian age, thins rapidly southward across the South Mayo trough, and contains detritus that indicates the progressive unroofing of a rising Dalradian metamorphic terrane in Connemara to the south (Dewey, 1963). Thus, at least in parts of zone C, the main climactic metamorphism (M3) that postdated the large-scale early deformations involving massive recumbent folds occurred prior to part of the Llanvirnian. Although the upper part of the Partry Supergroup (Maumtrasna Group) oversteps the lower part (Mweelrea Group) to rest unconformably on earlier Ordovician rocks on the southern edge of the South Mayo trough, the base of the Partry cannot be regarded as providing a clear younger limit to the Grampian event, since in the northern part of the Ordovician outcrop the Partry Group rests conformably on 21,000 ft of Arenigian/Llanvirnian flysch. This Early Ordovician flysch sequence was clearly framed by the same paleogeographic elements as the Partry Supergroup since it shows the same rapid southward thinning and facies change to a shallow marine conglomerate/volcanic sequence. The oldest fossils recovered from this thin sequence are of lower *Didymograptus nitidus* zone age (Dewey, Rickards, and Skevington, 1970). Thus, the South Mayo trough was in existence during Arenigian times and was bounded to the south by a rising cordillera in Connemara. There is suggestion but not positive proof that *Dalradian* metamorphic rocks were exposed in this cordillera during Arenigian times. The Mt. Partry Group contains boulders of schistose semipelite and psammite very similar to Connemara Dalradian rocks. The Leenane Group contains boulders of an altered oligoclase–microcline granite very similar to the Oughterard Granite of Connemara that intruded the Dalradian rocks at a late stage in their structural history and has yielded a Rb/Sr isochron of 510 ± 36 m.y. (Leggo, Compstom, and Leake, 1966). The Arenigian flysch sequence is full of detrital chlorite and mica suggesting a low-rank metamorphic source, and detrital staurolite and garnet are present in the Derrylea Group. The framing of the South Mayo trough by a southerly bounding cordillera contributing low-rank, then high-rank metamorphic detritus to the flysch/molasse sequence, suggests to the writer that the major nappe-forming events in the western Irish Dalradians were completed by, or during, *D. nitidus* times. If they were not, the South Mayo trough was initiated as a rootless structure carried along on the backs of moving nappes. It seems to the writer that there is a strong analogy between the Ordovician flysch/molasse to Dalradian relationship in Connemara and the Longmyndian flysch/molasse to Monian relationship of the Celtic domain in that the flysch/molasse sequences may record the late stages of deformation of adjacent metamorphic terranes. The Celtic event was finally terminated by folding of the Longmyndian prior to the Cambrian transgression; in Connemara the Grampian event is bracketed by the northward onlap of transgressive upper Llandoverian sediments that postdated large-amplitude folding of the Ordovician flysch/molasse sequence. In Tyrone an even older age limit for the post-Grampian onlapping sequence is provided by Caradocian sediments resting unconformably on the Tyrone complex, Dalradian rocks, and probably early Ordovician basic and ultrabasic rocks. Thus, as Kennedy (1958) argued, it is only the sub-Caradocian unconformity of zone C that gives an unequivocal upper age limit to the Grampian event. This unconformity is expressed by the northward onlap of shallow marine Caradocian conglomerates and limestones across the Ballantrae complex (Fig. 2). The Ballantrae complex is an ophiolite suite, affected locally by a blueschist facies metamorphism, in which argillites contain an early Arenigian graptolite fauna. Recently, Prof. D. Skevington (pers. comm.) has recovered an earliest Arenigian graptolite fauna from the Lough Nafooey Group that is unconformably overlain by *D. nitidus* bearing sediments of the Glensaul Group. Conceivably, therefore, the Ballantrae complex, the Lough Nafooey Group, and parts of the Tyrone complex form part of a pre- or syn-Grampian ophiolite suite. The ultrabasic screens along the Clew Bay fault zone and Highland Boundary fault zone, together with the Highland Border Series, could be part of this ophiolite assemblage, since Johnson and Harris (1967) have put forward structural data that convincingly supports coeval deformation of the Highland Border Series and adjacent low-grade Dalradian rocks. The Highland Boundary/Clew Bay fault zone, surely initiated as a fundamental fracture after the formation of Grampian nappes, probably started moving during the Arenigian to delimit the zone B/zone C boundary. Thus data in zone C suggest that the polyphase Grampian deformation and metamorphism occurred between earliest Arenigian and Caradocian times. D1 nappe-forming events were probably completed by, or during, *D. nitidus* zone times, and the major metamorphic climax (M3) was completed during, or by, Llanvirnian times. In zone B the evidence is not in accord with this picture. Recently, Downie *et al.* (1971) have shown that the highest Dalradian rocks, the MacDuff Slates, affected by regional Grampian deformation and metamorphic events contain probable chitinozoa that are not older than Arenigian and are probably Llanvirnian or Llandeilian. Similarly, the Moine nappe was emplaced during the earliest Grampian structural phase onto the edge of zone A after some part of the *D. protobifidus* zone. An extreme younger age limit for the Grampian event in zone B is given by Lower Devonian sediments and volcanics resting unconformably on Dalradian metamorphic rocks. The absence of older post-Grampian unconformable sequences in zone B appears to be due to the progressive uplift, unroofing, and cooling of zone B throughout Ordovician and Silurian times, unlike zone C that was progressively buried by Caradocian and younger sequences. The progressive uplift and stripping of zone B is recorded by a K/Ar radiometric "age" spread in the zone B metamorphic protolith and by the southerly advance of a flysch wedge, containing Grampian metamorphic detritus, into zone D from Llandeilian times onward.

The writer's earlier view (Dewey, 1961) that Grampian deformation and metamorphism was *completed* prior to the Arenigian is no longer tenable, but the history of the South Mayo trough/Connemara cordillera couple remains a major problem. The simplest view, not directly countered by paleontologic evidence, that could be adopted, is that the Grampian D1 nappe-forming events, including the initial emplacement of the Moine nappe, occurred synchronously during *D. protobifidus* zone times throughout zones B and C. This is the oldest possible age limit for D1, postdating a regional zone C ophiolite suite and predating the deposition of the Owenmore and Mt. Partry groups during the initiation of the South Mayo trough. In this model the M3 metamorphic climax could have occurred at any time before or during the earliest Partry Supergroup molasse deposition, and the large northeast-trending D3 folds postdating M3 in the Dalradian could be synchronous with the post Partry/?pre-Caradocian folding of the South Mayo trough. Alternatively, the D1 nappes could have formed synchronously during the late Arenigian and/or early Llanvirnian. In this model, MacDuff Slate and Durness Limestone deposition up to early Llanvirnian age could be "accommodated." The pre-Partry flysch sequence would be part of the Dalradian assemblage, and D1 nappe formation would be represented by the conformable flysch–

molasse transition in the South Mayo trough. On balance, the writer favors a model for the establishment of a "Grampian" metamorphic protolith during an early Arenigian/pre-Caradocian interval *possibly* involving diachronous structural and metamorphic phases. D1 and D2 were probably over by the beginning of Mt. Partry deposition but may entirely postdate the Lough Nafooey Group/Ballantrae complex basic volcanic episode. Whole-rock K/Ar ages, between 500 and 510 m.y., in Dalradian rocks along the Highland Border (Harper, 1967) probably reflect the beginning of post D1/D2 zone B uplift and the initiation of the Clew Bay/Highland Boundary fault zone. The gradual rise of isotherms that culminated in the M3 metamorphic climax about 486 m.y. ago may have occurred progressively during Arenigian and early Llanvirnian times accompanied by uplift of zone B and the local growth of positive elements such as the Connemara cordillera in zone C. During Llanvirnian and Llandeilian times zone B and local zone C uplift was fast, contributing to molasse sedimentation in the South Mayo trough and Llandeilian conglomeratic flysch deposition in zone D. However, the only truly unequivocal statement one can make about the age of the Grampian event is that it was completed prior to Caradocian times, as witnessed by the sub-Caradocian unconformity in Tyrone and at Ballantrae. The histories of zones B and C from Caradocian times onwards were divergent, zone B rising and zone C sinking with a faulted hinge zone along their boundary.

The Celtic and Grampian major metamorphic and deformational events affected separate and distinct regions, the former establishing, at the beginning of the Cambrian, a zone F ridge, a zone G/zone H basin, and a zone I/zone J stable shelf domain, then the latter establishing, by Caradocian times, metamorphic lands in zone B (Eocaledonia: Ziegler, 1970) and a subsiding platform in zone C. An important sub-Caradoc unconformity, however, developed over most of the Celtic domain and in zone E. Rast and Crimes have labeled the mid-Ordovician movements represented by this unconformity the Lakelandian event from its strongest expression in the English Lake District. The significance and, indeed, magnitude of this event are difficult to assess. In Leinster and North Wales it is associated with instability marking the onset of powerful Caradocian acid/intermediate vulcanicity. In the Lake District it postdates Llandeilian vulcanicity. In zone J it is marked by a mild break and in South Wales and the Shelve area by little if any expression.

The last major deformational episode to affect the British Caledonides was a Late Silurian deformation whose effects were widespread between zones B and J. Rast and Crimes (1969) have termed this the Cymrian event from its strong expression in Wales. The widespread Lower Devonian cyclothemic red bed sequences clearly bear a molasse relationship to this deformation. In areas that suffered the strongest deformation, such as the South Mayo trough and zones D through G, Lower Devonian molasse sequences are absent or thin and unconformable on older rocks. In these areas the Cymrian deformation is announced by a harbinger of the gradual spread of Silurian flysch sequences marking increasing crustal mobility from upper Llandoverian times onwards. In parts of the Midland Valley of Scotland, and in zones I and J, Silurian and Lower Devonian molasse sequences are conformable and the molasse is thick, reflecting the accumulation of debris shed from adjacent metamorphic lands. In these areas the final Caledonian episode of deformation is recorded—a post-Cymrian Middle Devonian episode of faulting, open fold-ing, and warping prior to the gradual onlap of Upper Devonian and younger cover sequence.

In this chapter the writer has attempted to present those data which appear to him to be the most critical to understanding and evaluating the large-scale evolution of the British Caledonides in as factual a way as is con-sistent with his prejudices. Many aspects have been superficially treated, and details have been omitted that may be of great importance. The reader is referred, for a wealth of additional information, to those references quoted at the end of Section I.

In conclusion, it seems appropriate to comment briefly and in general terms on a model that seems, to the writer, to explain at least the gross aspects, and many of the fine details, of Caledonian evolution. J. Tuzo Wilson (1966) was the first to suggest that the Appalachian/Caledonian orogen was the result of the contraction of an old ocean. Since this suggestion other workers (Bird and Dewey, 1970; Bird, Dewey, and Kidd, 1971; Church and Stevens, 1971; Dewey, 1969, 1971; Dewey and Bird, 1971; Fitton and Hughes, 1970; Mc-Kerrow and Ziegler, 1971; Stevens, 1970; Upadhyay, Dewey, and Neale, 1971) have attempted to explain various facets of Appalachian/Caledonian evolution as geologic corollaries of lithosphere plate tectonics. The various models all differ in details and often in gross aspects, but all invoke the genera-tion and subsequent destruction of oceanic crust and mantle at accreting and consuming plate margins, respectively. The *raison d'être* and justification for viewing the evolution of the British Caledonides in a plate tectonic regime is many sided. The main reason for adopting models involving oceans analogous, or even homologous, to those of the present day is the development of Ap-palachian/Caledonian ophiolite suites that conform in great detail to the sequence, structure, and petrology of the present-day oceanic crust and mantle (Church and Stevens, 1971; Dewey and Bird, 1971; Upadhyay, Dewey, and Neale, 1971). Additional reasons involve facies changes such as that between the Cambrian rocks of zones A and B that seem to be best explained by a continental shelf edge (Rodgers, 1968), paired metamorphic belts (Miyashiro, 1967) comparable to those in young island arcs and orogens around the Pacific Ocean, and geochemical polarity of volcanic sequences (Fitton and Hughes, 1970) similar to those of island arcs. There are, however, additional reasons why Caledonian evolution is best explained in terms of large horizontal motions involving ocean-floor destruction. The Celtic and Grampian domains have utterly different tectonostratigraphic developmental histories that would be difficult to reconcile with evolution in their present relative positions and a history of purely vertical movements. Although the Celtic, and Grampian, domains are underlain, certainly in large part, by a pre-Caledonian continental basement, that basement appears to belong to different Precambrian provinces. The Grampian domain is underlain by Lewisian basement (Archean locally reworked by a mid-Proterozoic deformation and metamorphism), whereas the Celtic domain is underlain by a basement similar to the younger Pentevrian basement province of Brittany that may be of Grenville age (Rast and Crimes, 1969). If *continuity* of pre-Caledonian basement or of a single pre-Late Silurian element between zones C and G could be demonstrated, an ocean-based plate tectonics model would be nonviable. Such continuity has not been demonstrated. Pre-Caledonian basement rocks are unknown in zones D and E, and no Caledonian element can be traced continuously across them from zone C into zone G. The oldest rocks in zone E are Cambrian sediments, and those in zone D are early Arenigian pillow lavas. Zone D is the obvious site for a lower Paleozoic ocean whose contraction brought Grampian and Celtic domains into juxtaposition during Late Silurian times. Other oceans may have been involved in Caledonian evolution if the premise that orogeny is the result of either subduction at continental margins or island arc, or due to continent/continent or continent/island arc collision (Dewey and Bird, 1970) is correct. The Celtic event may have been associated with the loss of a Precambrian ocean or oceans. If blueschist metamorphism is one of the hallmarks of subduction, the boundary between zone F and zone G may have originated along the site of a subduction zone in late Precambrian times. The Welsh basin (zone G) may itself have been a small ocean, possibly of marginal, Sea of Japan type that became filled by over 40,000 ft of lower Paleozoic sediments and volcanics. Bird, Dewey, and Kidd (1971) and Upadhyay, Dewey, and Neale (1971) have argued that ophiolite complexes in the Grampian domain originated as the oceanic crust and mantle of such marginal basins. Whether such models are correct or not is of secondary importance to the fundamental importance of continued detailed mapping projects, geophysical studies, and large-scale integrated syntheses of the Caledonides in relation to the Appalachian and Hercynian orogenic belts. Only by the tenacious pursual of such studies will we even begin to understand Caledonian evolution.

REFERENCES

Bassett, D. A., 1963, The Welsh lower Palaeozoic geosyncline: a review of recent work on stratigraphy and sedimentation, in: *The British Caledonides*, Johnson, M. R. W. and Stewart, F. H., eds.: Oliver and Boyd, Edinburgh, p. 35–69.

Bassett, D. A., 1969, Some of the major structures of early Palaeozoic age in Wales and the Welsh Borderland: an historical essay, in: *The Precambrian and Lower Palaeozoic rocks of Wales*, Wood, A., ed.: University of Wales Press, Cardiff, p. 67–107.

Berry, W. B. N., 1968, Age of Bogo Shale and western Ireland graptolite faunas and their bearing on dating early Ordovician deformation and metamorphism in Norway and Britain: *Norsk. Geol. Tidsskr.*, v. 48, p. 217–230.

Bird, J. M. and Dewey, J. F., 1970, Lithosphere plate–continental margin tectonics and the evolution of the Appalachian orogen: *Geol. Soc. Amer. Bull.*, v. 81, p. 1031–1060.

Bird, J. M., Dewey, J. F., and Kidd, W. S. F., 1971, Proto-Atlantic oceanic crust and mantle: Appalachian/Caledonian ophiolites: *Nature*, v. 231, p. 28–31.

Capewell, J. G., 1956, The stratigraphy, structure and sedimentation of the Old Red Sandstone of the Comeragh Mts. and adjacent areas, Co. Waterford, Ireland: *Geol. Soc. Lond. Quart. Jour.*, v. 112, p. 393–412.

Church, W. R. and Stevens, R. K., 1971, Early Paleozoic ophiolite complexes of the Newfoundland Appalachians as mantle–oceanic crust sequences: *Jour. Geophys. Res.*, v. 76, p. 1460–1466.

Cobbing, E. J., Manning, P. I., and Griffith, A. E., 1965, Ordovician–Dalradian unconformity in Tyrone: *Nature*, v. 206, p. 1132–1135.

Crimes, T. P. and Dhonau, N. B., 1967, The Precambrian and lower Palaeozoic rocks of southeastern Co. Wexford, Eire: *Geol. Mag.*, v. 104, p. 213–221.

Cummins, W. A., 1969, Patterns of sedimentation in the Silurian rocks of Wales, in: *The Precambrian and Lower Palaeozoic rocks of Wales*, Wood, A., ed.: University of Wales Press, Cardiff, p. 214–238.

Dewey, J. F., 1961, A note concerning the age of the metamorphism of the Dalradian rocks of western Ireland: *Geol. Mag.*, v. 98, p. 399–405.

Dewey, J. F., 1963, The lower Palaeozoic stratigraphy of central Murrisk, Co. Mayo, Ireland, and the evolution of the South Mayo trough: *Geol. Soc. Lond. Quart. Jour.*, v. 119, p. 313–344.

Dewey, J. F., 1969*a*, Structure and sequence in paratectonic British Caledonides, in: *North Atlantic—Geology and Continental Drift*, Kay, Marshall, ed.: Amer. Assoc. Petrol. Geol. Mem. 12, p. 309–335.

Dewey, J. F., 1969*b*, Evolution of the Appalachian/Caledonian orogen: *Nature*, v. 222, p. 124–129.

Dewey, J. F., 1971, A model for the lower Palaeozoic evolution of the southern margin of the early Caledonides of Scotland and Ireland: *Scott. Jour. Geol.*, v. 7, p. 219–240.

Dewey, J. F. and Bird, J. M., 1970, Mountain belts and the new global tectonics: *Jour. Geophys. Res.*, v. 75, p. 2625–2647.

Dewey, J. F. and Bird, J. M., 1971, Origin and emplacement of the ophiolite suite: Appalachian ophiolites in Newfoundland: *Jour. Geophys. Res.*, v. 76, p. 3179–3206.

Dewey, J. F. and Pankhurst, R. J., 1970, The evolution of the Scottish Highlands in relation to their isotopic age pattern: *Roy. Soc. Edinb. Trans.*, v. 68, p. 361–389.

Dewey, J. F., Rickards, R. B., and Skevington, D., 1970, New light on the age of Dalradian deformation and metamorphism in western Ireland: *Norsk. Geol. Tidsskr.*, v. 50, p. 19–44.

Downie, C., Lister, T. R., Harris, A. L., and Fettes, D. J., 1971, A palynological investigation of the Dalradian rocks of Scotland: Inst. Geol. Sci. Rep. 71/9, London, 29 p.

Dunning, F. W., 1966, Tectonic map of Great Britain and Northern Ireland: Inst. of Geol. Sci., London.

Eastwood, T., 1953, British regional geology: northern England: Inst. Geol. Sci., London, 72 p.

Edmunds, F. H. and Oakley, K. P., 1947, British regional geology: the central England district: Inst. Geol. Sci., London, 80 p.

Fitton, J. G. and Hughes, D. J., 1970, Volcanism and plate tectonics in the British Ordovician: *Earth Planet. Sci. Lett.*, v. 8, p. 223–228.

Friend, P. F., 1969, Tectonic features of Old Red sedimentation in North Atlantic Borders, in: *North Atlantic—Geology and Continental Drift*, Kay, Marshall, ed.: Amer. Assoc. Petrol. Geol. Mem. 13, p. 703–710.

George, T. N., 1960, The stratigraphical evolution of the Midland Valley: *Geol. Soc. Glasgow Trans.*, v. 24, p. 32–107.

George, T. N., 1963, Palaeozoic growth of the British Caledonides, in: *The British Caledonides*, Johnson, M. R. W. and Stewart, F. H., eds.: Oliver and Boyd, Edinburgh, p. 1–33.

George, T. N., 1965, The geological growth of Scotland, in: *The Geology of Scotland*, Craig, G. Y., ed.: Oliver and Boyd, Edinburgh, p. 1–48.

Giletti, B. J., Moorbath, S., and Lambert, R. St. J., 1961, A geochronological study of the metamorphic complexes of the Scottish Highlands: *Geol. Soc. Lond. Quart. Jour.*, v. 117, p. 233–272.

Greenly, E., 1919, Geology of Anglesey: Geol. Surv. G. B. Mem., 2 vols.

Harper, C. T., 1967, The geological interpretation of potassium–argon ages of metamorphic rocks from the Scottish Caledonides: *Scott. Jour. Geol.*, v. 3, p. 46–66.

Harper, J. C., 1948, The Ordovician and Silurian rocks of Ireland: *Liverpool Geol. Soc. Prov.*, v. 20, p. 48–67.

Helm, D. G., 1970, Stratigraphy and structure in the Black Cambe Inlier, English Lake District, Yorks.: *Geol. Soc. Proc.*, v. 38, p. 105–125.

Higgins, A. C., 1967, The age of the Durine Member of the Durness Limestone Formation at Durness: *Scott. Jour. Geol.*, v. 3, p. 382–388.

Holland, C. H., 1969a, The Welsh Silurian geosyncline in its regional context, in: *The Precambrian and lower Palaeozoic rocks of Wales*, Wood, A., ed.: University of Wales Press, Cardiff, p. 203–217.

Holland, C. H., 1969b, Irish counterpart of Silurian in Newfoundland, in: *North Atlantic—Geology and Continental Drift*, Kay, Marshall, ed.: Amer. Assoc. Petrol. Geol. Mem. 13, p. 298–308.

Irving, E., and Runcorn, S. K., 1957, Analysis of the palaeomagnetism of the Torridonian Sandstone Series of northwest Scotland: *Roy. Soc. Lond. Phil. Trans. Ser. A.*, v. 250, p. 83–99.

Johnson, M. R. W., 1965a, Torridonian and Moinian, in: *The Geology of Scotland*, Craig, G. Y., ed.: Oliver and Boyd, Edinburgh, p. 80–113.

Johnson, M. R. W., 1965b, Dalradian, in: *The Geology of Scotland*, Craig, G. Y., ed.: Oliver and Boyd, Edinburgh, p. 115–160.

Johnson, M. R. W. and Harris, A. L. 1967, Dalradian–(?) Arenig relations in parts of the Highland Border, Scotland, and their significance in the chronology of the Caledonian orogeny: *Scott. Jour. Geol.*, v. 3, p. 1–16.

Johnstone, G. S., Smith, D. I., and Harris, A. L., 1969, Monian assemblage of Scotland, in: *North Atlantic—Geology and Continental Drift*, Kay, Marshall, ed.: Amer. Assoc. Petrol. Geol. Mem. 13, p. 159–180.

Jones, O. T., 1938, On the evolution of a geosyncline: *Geol. Soc. Lond. Quart. Jour.*, v. 94, p. 60–110.

Kennedy, W. Q., 1958, Tectonic evolution of the Midland Valley of Scotland: *Geol. Soc. Glasgow Trans.*, v. 23, p. 107–133.

Kilburn, C., Pitcher, W. S., and Shackleton, R. M., 1965, The stratigraphy and origin of the Portaskaig Boulder Bed Series (Dalradian): *Geol. Jour.*, v. 4, p. 343–360.

Lambert, R. St. J., 1969, Isotopic age studies relating to the Precambrian history of the Moinian of Scotland: *Geol. Soc. Lond. Proc.*, n. 1652, p. 243–243.

Leggo, P. J., Compston, W., and Leake, B. E., 1966, The geochronology of the Connemara granites and its bearing on the antiquity of the Dalradian Series: *Geol. Soc. Lond. Quart. Jour.*, v. 122, p. 91–188.

Long, L. E. and Lambert, R. St. J., 1963, Rb–Sr isotopic ages from the Moine Series, in: *The British Caledonides*, Johnson, M. R. W. and Stewart, F. H., eds.: Oliver and Boyd, Edinburgh, p. 217–247.

McKerrow, W. S., 1962, The chronology of Caledonian folding in the British Isles: *Natl. Acad. Sci. Proc.*, v. 68, p. 1905–1913.

McKerrow, W. S., and Ziegler, A. M., 1971, The Lower Silurian paleogeography of New Brunswick and adjacent areas: *Journal Geology*, v. 79, p. 635–646.

Miyashiro, A., 1967, Orogeny, regional metamorphism and magmatism in the Japanese Islands: *Medd. dansk. geol. Foren.*, v. 17, p. 390–446.

Pocock, R. W. and Whitehead, T. H., 1948, British regional geology: the Welsh borderland: Inst. Geol. Sci. London, 79 p.

Pringle, J., and George, T. N., 1948, British regional geology: South Wales: Inst. Geol. Sci. London, 100 p.

Rast, N., 1963, Structure and metamorphism of the Dalradian rocks of Scotland, in: *The British Caledonides*, Johnson, M. R. W. and Stewart, F. H., eds.: Oliver and Boyd, Edinburgh, p. 123–142.

Rast, N., 1969, The relationship between Ordovician structure and volcanicity in Wales, in: *The Precambrian and lower Palaeozoic rocks of Wales*, Wood, A., ed.: University of Wales Press, Cardiff, p. 305–335.

Rast, N. and Crimes, T. P., 1969, Caledonian orogenic episodes in the British Isles and Northwestern France and their tectonic and chronological interpretation: *Tectonophysics*, v. 7, p. 277–307.

Read, H. H., 1961, Aspects of Caledonian magmatism—Britain: *Liverpool Manchester Geol. Jour.*, v. 2, p. 653–683.

Rodgers, J., 1968, The eastern edge of the North American continent during the Cambrian and Early Ordovician, in: *Studies of Appalachian Geology—Northern and Maritime*, Zen, E-an, White, W. S., Hadley, J. B., and Thompson, J. B., Jr., eds.: Interscience, New York, p. 141–149.

Shackleton, R. M., 1969, The Precambrian of North Wales, in: *The Precambrian and lower Palaeozoic rocks of Wales*, Wood, A., ed.: University of Wales Press, Cardiff, p. 1–22.

Simpson, A., 1963, The stratigraphy and tectonics of the Manx Slate Series, Isle of Man: *Geol. Soc. Lond. Quart. Jour.*, v. 119, p. 367–400.

Simpson, A., 1967, The stratigraphy and tectonics of the Skiddow Slates and the relationship of the overlying Borrowdale Volcanic Series in part of the Lake District: *Geol. Jour.*, v. 5, p. 391–418.

Skevington, D., 1968, British and North American Lower Ordovician correlation: discussion: *Geol. Soc. Amer. Bull.*, v. 79, p. 1259–1264.

Smith, B. and George, T. N., 1961, British regional geology: North Wales: Inst. Geol. Sci. London, 97 p.

Soper, N. J., 1970, Three critical localities on the junction of the Borrowdale Volcanic Rocks with the Skiddow Slates in the Lake District, Yorks: *Geol. Soc. Proc.*, v. 37, p. 486–498.

Stevens, R. K., 1970, Cambro-Ordovician flysch sedimentation and tectonics in West Newfoundland and their possible bearing on a proto-Atlantic ocean: Geol. Assoc. Can. Spec. Paper No. 7, p. 165–177.

Stewart, A. D., 1969. Torridonian rocks of Scotland reviewed, in: *North Atlantic—Geology and Continental Drift*, Kay, Marshall, ed.: Amer. Assoc. Petrol. Geol. Mem. 13, p. 595–608.

Sutton, J. and Watson, J., 1951, The pre-Torridonian metamorphic history of the Loch Torridon and Scourie areas in the northwest Highlands and its bearing on the classification of the Lewisian: *Geol. Soc. Lond. Quart. Jour.*, v. 106, p. 241–307.

Swett, K., 1969, Interpretation of deposition and diagenetic history of Cambrian–Ordovician succession of northwest Scotland, in: *North Atlantic—Geology and Continental Drift*, Kay, Marshall, ed.: Amer. Assoc. Petrol. Geol. Mem. 13, p. 630–646.

Upadhyay, H. D., Dewey, J. F., and Neale, E. R. W., 1971, The Betts Cove Ophiolite Complex, Newfoundland: Appalachian oceanic crust and mantle: *Geol. Assoc. Can. Proc.*, v. 24, p. 27–34.

Walton, E. K., 1965a, Lower Palaeozoic rocks—stratigraphy, in: *The Geology of Scotland*, Craig, G. Y., ed.: Oliver and Boyd, Edinburgh, p. 161–200.

Walton, E. K., 1965b, Lower Palaeozoic rocks—palaeogeography and structure, in: *The Geology of Scotland*, Craig, G. Y., ed.: Oliver and Boyd, Edinburgh, p. 201–227.

Walton, E. K., 1969, Lower Paleozoic rocks in southern Scotland, in: *North Atlantic—Geology and Continental Drift*, Kay, Marshall, ed.: Amer. Assoc. Petrol. Geol. Mem. 13, p. 265–266.

Williams, A., 1969, Ordovician of British Isles, in: *North Atlantic—Geology and Continental Drift*, Kay, Marshall, ed.: Amer. Assoc. Petrol. Geol. Mem. 13, p. 237–264.

Wilson, J. T., 1966, Did the Atlantic close and then re-open?: *Nature*, v. 211, p. 676–681.

Wood, D. S., 1969, The base and correlation of the Cambrian rocks of North Wales, in: *The Precambrian and lower Palaeozoic rocks of Wales*, Wood, A., ed.: University of Wales Press, Cardiff, p. 47–66.

Wright, A. E., 1969, Precambrian rocks of England, Wales, and southeast Ireland, in: *North Atlantic—Geology and Continental Drift*, Kay, Marshall, ed.: Amer. Assoc. Petrol. Geol. Mem. 13, p. 93–109.

Ziegler, A. M., 1970, Geosynclinal development of the British Isles during the Silurian Period: *Jour. Geol.*, v. 78, p. 445–479.

Chapter 8

THE GEOLOGY OF THE WESTERN APPROACHES

Thomas Richard Owen

Department of Geology
University College
Swansea, Great Britain

I. INTRODUCTION

The term "Western Approaches" has different meanings for different authors, and in its widest sense can cover both the northwestern and southwestern waterways into the British Isles. In this chapter the area will be restricted to the southwestern flanks of Great Britain and Ireland, but (for convenience of description) the northwest tip of France—Brittany—will also be included. It is also meaningful to include Brittany because the region described forms the segment in which the Armorican mountain chains (the "Variscides") run westward to the continental margin (Fig. 1). The area has been subdivided into eight portions (four land areas and four water areas). This description of the geology and structure of each portion is then followed by an attempt to synthesize the paleogeographical and structural setting of the complete region.

The geology of the land area has been known for some considerable time but that of the sea floor has not been studied in detail until very recently. In this respect, the work carried out by the universities of Bristol and Cambridge, University College, London, and the officers of the Institute of Geological Sciences deserves particular mention.

The land areas rise ruggedly from the sea. The west and southwest coasts cut across the structural grain of the Armorican folds, and the differential

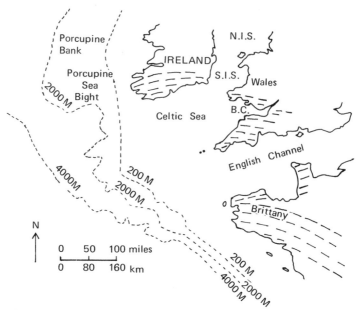

Fig. 1. The western approaches. Key: BC, Bristol Channel; NIS, northern
Irish Sea; SIS, southern Irish Sea.

coastal erosion into rocks of varying hardness has resulted in a deeply indented
coastal form, especially in southwest Ireland and Brittany. The land areas
frequently rise to heights above 300 m and in some cases to over 700 m above
sea level (as in southwest Ireland and South Wales). Geologically, they expose
rocks ranging in age from Precambrian (especially in Brittany) to Cenozoic,
with Paleozoic strata covering the greatest area.

The continental shelf (0–200 m depth) is wide, particularly to the west
of Cornwall (225 miles). Off the southwestern flank of Brittany the shelf is
100 miles wide, followed seaward by a narrow and steep continental slope
of only 25 miles width. Southwest of Ireland the slope is much wider (up to
200 miles) while to the west of southern Ireland the marked Porcupine bank
(depth only 200–300 m) breaks the continuity of the slope over a width of
some 60 miles (Fig. 1).

Tertiary sediments form the (solid) sea floor over much of the shelf and
slope but in places on the latter, Upper Cretaceous and even pre-Mesozoic
rocks are exposed. Upper Cretaceous strata form extensive parts of the floor
of the Celtic Sea and English Channel, but are covered (along downwarps)
by Tertiary sediments. The most complex structure, as far as the sea areas are
concerned, occurs in the English Channel.

Quaternary ice sheets spread southward down the Irish Sea reaching on
one occasion as far south as the tip of southwest England. Variable thicknesses

of glacial deposits, therefore, cover the solid rocks in the Irish Sea and over parts of the Celtic Sea. The western portion of the English Channel, however, is remarkably free of superficial deposits.

II. REGIONAL SURVEY

A. Southern Ireland

This land area lies south of the east–west line from Galway Bay to Dublin (Fig. 2). Lower Carboniferous carbonates occupy much of the surface of the central areas and parts of the south and give rise to relatively low, rolling country, the limestones being largely covered by extensive glacial and peat deposits. Lower Paleozoic slates, grits, and greywackes, giving rise to higher hills, come to the surface in a number of anticlinal inliers in central Ireland (Fig. 2) and occupy a more extensive upland area in southeastern Ireland (counties Wicklow and Wexford), where they were intruded by the large (Caledonian) Leinster Granite, 70 miles long. Precambrian rocks, slates, quartzites, and greywackes are restricted in southern Ireland to the Bray area (south of Dublin) and to the Rosslare region in the extreme southeast of Ireland where an older (possibly even Lewisian) Rosslare complex of horn-

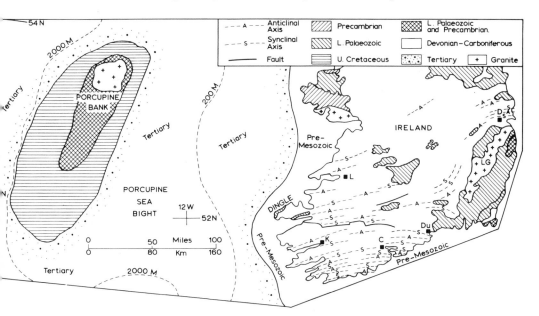

Fig. 2. The geology of the Porcupine bank and southern Ireland. Key: C, Cork; D, Dublin; Du, Dungarvan; K, Kenmare; L, Limerick; LG, Leinster Granite. Note that the igneous complex at the northern end of the Porcupine bank includes gabbro and other basic intrusives.

blende and biotite mylonite gneisses has suffered at least two periods of Pre-cambrian deformation. The complex is overlain by a group of schistose quartz-ites, greywackes, and greenschists (the Cullenstown Group), which have been compared to the Monian of Anglesey (Baker, 1969, p. 257). The Rosslare complex is intruded by the Carnsore Granite, which has yielded a radiometric age of 429 $\mp$ 7 m.y. The fairly extensive tract of Cambrian–Ordovician rocks east of the Leinster Granite are folded along northeast–southwest lines, the main flexures being the Arklow and Bray anticlines and the Wicklow and Campile synclines (Gardiner, 1970, p. 47). The Leinster Granite is probably a batholith but could be partly laccolithic (Gardiner, 1970, p. 50). Radiometric dates suggest intrusion at the beginning of Devonian times. Crimes and Crossley (1968), in their studies of the lower Paleozoic rocks of County Wexford, have shown the Cambrian to be some 2600 m thick and have traced three phases of Caledonian folding of the lower Paleozoics (p. 185). These sediments were formed in a northeast–southwest-trending trough bounded to the southeast by an Irish Sea landmass. Crimes and Crossley have demon-strated repeated eastward upfaulting of regional blocks from Cambrian to Permian times along this Wexford coastal area of Ireland (p. 214). The central blocks of this complex Paleozoic horst, situated in what is today the southern Irish Sea, could have supplied some lower Paleozoic sediment to the Welsh depositional area.

The other lower Paleozoic areas of central and southern Ireland are smaller and are restricted to rocks of Ordovician but more particularly of Silurian age. These inliers include Slieve Bernagh, the Galtee Mountains, Slieve Bloom, Slievenaman, the Comeragh Mountains, and the Dingle penin-sula (on the west coast). Holland (1969) has described the Silurian rocks of the Dingle peninsula, which include a volcanic formation of rhyolites, agglom-erates, and tuffs with ignimbrites. This Silurian sequence, some 1700 m thick and ranging from Wenlock to Ludlow, probably passes up conformably into the Siluro-Devonian Dingle Group, 2700 m of grey to purple sandstones, mudstones, and locally thick conglomerates, all probably of fluvio-lacustrine origin. The Silurian and the Dingle Group were folded during a late episode of the Caledonian orogeny and are seen to be inverted within many gigantic overfolds, e.g., on Clogher Head. Much of the steepening and inversion of the limbs of isoclinal folds occurred during later (Armorican) earth movements, which also caused extensive faulting and sporadically developed zones of cleavage (Holland, 1969, p. 306).

A prolonged interval of uplift and erosion followed the Lower Devonian folding of the Silurian and Dingle Group of the Dingle peninsula, and deposi-tion was probably not resumed until Upper Devonian times over most parts of southern and central Ireland. These Upper Devonian rocks therefore rest unconformably on the lower Paleozoics and Dingle Group of the Dingle peninsula. The postorogenic Devonian sediments here comprise 1600 m of red, brown, and yellow-grey sandstones and conglomerates of aeolian and fluvio-lacustrine origin. The basal Inch Formation is a thick immature conglomerate or breccia of metamorphic provenànce, with an imbrication indicating a southerly source (Horne, 1970, p. 57). This basal member appears to have been transported rapidly northwards from an elevated ridge which occupied the present site of Dingle Bay and which separated a Dingle basin from a Munster basin, further south.

The Devonian succession thickens appreciably south of Dingle Bay, being 4000 m thick in the McGillicuddy's Reeks (Walsh, 1968, p. 11) and as much as 7000 m thick in the Caha Mountain region (between Dingle Bay and Kenmore River). These thick piles of purple, red, grey, and green sandstones, siltstones, slates, cornstones, and conglomerates are largely of fluvio or fluvio-lacustrine origin. Their pre-Devonian base is not seen, and the sparse fossil (plant) evidence suggests that only the Upper Devonian comes to the surface over these extensive areas of southwest Ireland. It is obvious that the basin of their thick deposition was one of geosynclinal magnitude but that a marked thinning or wedging out takes place northward across southwest Ireland. The significance of this northward wedging with respect to the position of the major "Armorican front" of southern Ireland will be discussed later.

The Carboniferous succession of central Ireland can be divided into a lower carbonate unit and an (upper) shale–turbidite–paralic Coal Measure sequence. The lower division—the Carboniferous Limestone, about 1000–1200 m thick—includes a sheet reef up to 600 m in thickness and occurring over an area of nearly 3000 square miles. Contemporaneous volcanics occur in the Limerick basin. The Upper Carboniferous rocks outcrop widely in County Clare and comprise 1100 m of shales, siltstones, and sandstones, displaying slumps and sand volcanoes (Brindley and Gill, 1958). Still higher Carboniferous strata (the Coal Measures) are restricted to the Castlecomer (west of the Leinster Granite), Ardagh, and Kanturk coalfields. Most of the coals are anthracites.

Two features of the Carboniferous areas merit further attention. Firstly, a marked change occurs in the Lower Carboniferous sequence when traced southward across County Cork and the extreme south of County Kerry, the typical Irish carbonate succession giving way southwards to a much thicker pile (the so-called "Carboniferous Slate") of cream, white, and grey sandstones and dark to black slates. This Lower Carboniferous sequence of southernmost Ireland is over 2500 m thick but probably transcends the Lower/Upper Carboniferous boundary. The second feature of note is the lateral thinning which takes place in both a northerly and southerly direction from the Shannon estuary region within the "Millstone Grit" (Namurian) sequence. Hodson and Lewarne (1961) have shown that to the south of this Namurian trough there existed a positive area of restricted deposition running near to the Dingle peninsula, and which could be a western continuation of the Leinster massif, a paleogeographic element of Carboniferous times.

While intervening earth movements must have occurred in central Ireland in Middle Ordovician, Late Ordovician, and Middle Carboniferous times, the major structural effects occurred at the climaxes of the Caledonian and Hercynian (Armorican) orogenies. The former occurred somewhere within a Late Silurian–Upper Devonian interval. McKerrow (1962, Fig. 1) favored a Middle Devonian age for the Caledonian climax in central Ireland. The peak of the Hercynian orogeny probably occurred toward the close of Carboniferous times—according to Cole (1922) during the Stephanian age.

Structurally, the southern half of Ireland can be divided into two portions, the dividing line being the so-called "Armorican front" which runs in an east–west direction from Dingle Bay to Dungarvan. North of this line, the lower Paleozoic rocks seen in the inliers were folded along northeast–southwest or (in the west) ENE–WSW axes by the Caledonian movements into anticlinoria and synclinoria. Strike faulting is common, but it is often difficult to separate the amount of Caledonian and later (Hercynian) fracturing. The upper Paleozoic strata of this northern portion are also folded along caledonoid or subcaledonoid lines so that the older trends and cratons must have modified the north–south Hercynian compression. The influence of the Leinster Granite accounts further for the marked northward swing of the Hercynian fold on the western margin of the pluton (Fig. 2). The Hercynian folds are of concentric type, but there is a striking difference between the simple fold style of the large anticlines in Devonian (with exposed lower paleozoic cores) and the more complex small-scale fold frill of the intervening Carboniferous areas (Gill, in Coe, 1962, p. 52). Middle Carboniferous shales particularly suffered strong crenulation and faulting with some "abschierung" at these horizons. Thicker Lower Carboniferous limestones, especially reefs, have a bolder box-type fold style.

South of the Dingle Bay–Dungarvan line, the thick Devonian and Carboniferous rocks are much more intensely folded into large-scale anticlinoria and synclinoria of cleavage or parallel-type flexuring. The area lay on the northernmost fringe of the great "Armorican Arc" that curved westward through southwest England and Brittany, turning slightly south of west in southwest Ireland. The complex anticlinoria frequently bring up the thick Devonian successions to the surface and include the Mangerton, Clashmore, and Kilcrohane (amplitude 4000 m) anticlines. The downfolds preserve the Carboniferous successions and include the Bantry Bay, Blarney, Ardmore, Cork, and Minane synclines. The ria inlets of southwest Ireland invariably mark the downfolds while the headlands are dominated by the Devonian cores of the major upfolds. Superimposed on all the major upfolds and downfolds are a very large number of minor flexures ranging from hundreds of feet to mere inches in amplitude. Structural zones, characterized by different fold types, have been delineated in several areas, e.g., in the Sneem region (Capewell, 1957) and in the Beara peninsula (Coe and Selwood, 1963). More than one phase of deformation has been detected, e.g., in the Sheep's Head peninsula (Coe and Selwood, 1964). Coe (1969) has detected a complex pattern of deformation and igneous intrusion in southwest Cork. The intrusions were controlled by the structure occurring mostly where dips were steep or vertical and along steep strike faulting. Coe thinks the tectonic style suggests movements on block faults in an underlying rigid basement and the igneous activity was probably connected with fault block movement at depth rather than fold development at high crustal level. Coe and Selwood (1963, p. 58) in their study of the Beara peninsula, County Cork, suggest that the final uplift of the central zone of this complex diapiric fold could be related to the emplacement at great depth of a granite pluton. Moreover, "a granite here would be the logical conclusion to H. H. Read's granite series developed from central France through Brittany and southwest England."

A complex fault pattern exists in southernmost Ireland and includes ENE–WSW thrusts, powerful wrench faults trending east–west (see Capewell, 1957), and a large number of fractures ranging from northwest–southeast through north–south to NNE–SSW. Gill (in Coe, 1962) has noted how northwest–southeast wrench movement accompanied the growth of the folds. In the Ardmore district, north–south faults appear to postdate the east–west folds (Dawson-Grove, 1955). North–south faults in the Reeks must have a large accumulative easterly downthrow in order to cancel out the persistent westerly plunge (of about 15–20 deg) of the folds in this area (Walsh, 1968, p. 19).

The "Armorican front," i.e., the Dingle Bay–Dungarvan line across southernmost Ireland, has been described by Walsh (1968) in his study of the area west of Killarney as a line marking a sharp contrast in structural elevation and fold style. The position of the "front" may well have been decided before the climax of the Hercynian orogeny. The line is near to the Upper Devonian barrier (along Dingle Bay) noted by Horne (1970). It lies also near to the southern positive area of Middle Carboniferous times, noted by Hodson and Lewarne (1961). Moreover, it lies along a zone where the excessively thick Devonian and Carboniferous sedimentary piles were rapidly thinning northwards (see Coe and Selwood, 1968; Walsh, 1968, p. 20). Such northward restriction could result in (a) a marked piling upward of the southern folds and (b) eventual "giving way" northward of slices along powerful thrusts.

The character of the Armorican front changes from east to west across southern Ireland. In the vicinity of Loch Leane (near Killarney), powerful faults with large northerly downthrows such as the Muckross–Millstreet fault

(3000 m throw), the Benson's Point fault (1600 m throw), and the Black Lake fault (3300 m throw) may together mark a wide zone of upthrusting stretching all the way to the west coast at Rossbeigh. Walsh (1968) has described these fractures and noted fault plane inclinations of between 60 and 90 deg. The central portion of the front in Ireland may also be largely of thrust character. Gill (in Coe, 1962) states that in front of the Mangerton Range the dislocation is for the most part along a single thrust dipping at 45 deg southwards, and which in places faults Devonian against Coal Measures. Near Mallow, a zone of lower angled "schuppen" occurs, involving slices of Devonian and Carboniferous rocks all essentially the right-way up (Gill, in Coe, 1962, p. 54). North of Mallow, Philcox (1963) has detected differential "compartment" deformation with decollement at the base of the Namurian and of the Lower Carboniferous.

Philcox draws attention to the marked change in the frontal zone which takes place from Mallow eastwards to Dungarvan. In this eastern section, the marked structural elevation of the southern area dies away and no major thrusting can be traced. Philcox believes that the front splits up east of Mallow with echelon zones and swells in the vicinity of Mitchelstown, and along the Comeragh and Slievenaman hills. This "fanning out" may in some way be related to caledonoid influences in this eastern area, and/or to the presence of the Leinster pluton.

B. South Wales

Like southern Ireland, South Wales (Fig. 3) displays rocks of wide geological age. Precambrian acid volcanics, intruded (and perhaps also underlain) by Precambrian granites, granophyres, and diorites, come to the surface in Pembrokeshire at St. Davids, Trefgarn, and Johnston. North Pembrokeshire is noted for its classic Cambrian succession, about 1400 m thick, of shales, grits, greywackes, and flags. These are overlain unconformably by an Ordovician sequence that is noted for its contrasting sandy calcareous, nearshore facies (with a trilobite–brachiopod–coral fauna) and an open-water graptolitic facies of cleaved mudstone with occasional greywacke intercalations. The shoreline of Ordovician times lay somewhere over southeast Wales so that the "shelly" facies is best developed in that direction. In North Pembrokeshire, and again in the Builth area of central Wales, vulcanicity was intense at times, and thick piles of volcanics, ranging from acid keratophyres and ignimbrites to basic spilites, swell the total Ordovician sequence to some 3500 m in the Strumble–Fishguard–Prescelly areas of North Pembrokeshire.

The facies contrast of Ordovician deposition continued into the Silurian, the major line of facies demarcation continuing to lie near the line of the Towy Anticlinorium (also an important Caledonoid structure). Contemporane-

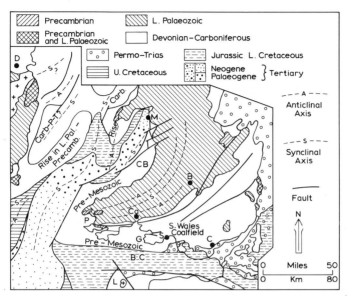

Fig. 3. The geology of southern Wales, the Bristol Channel, and the southern Irish Sea (geology of the sea areas based on the work of various authors mentioned in the text). Key: B, Brecon; BC, Bristol Channel; C, Cardiff; Ca, Carmarthen; CB, Cardigan Bay; D, Dublin; G, Gower; L, Lundy Granite; M, Mochras borehole; S, Swansea.

ous earth movements from Late Ordovician through Silurian times appreciably break the continuity of the Silurian sequence along the southeast flank of that upfold. Appreciable uplift occurred along fault blocks in South and Central Pembrokeshire prior to the deposition of upper Llandovery sands and muds, which rest unconformably on Precambrian igneous rocks in the vicinity of the Johnston and Benton faults (see George, in Stewart and Johnson, 1963). These long-established basement fractures probably curved northeastward into east-central Wales and the Welsh borderland, and caused similar Silurian–Precambrian unconformable relationships in Radnorshire and over the S. Longmynd of Shropshire.

Two further features of the Silurian of South Wales are worth noting. First, Ziegler *et al.* (1969, p. 409) have reallocated the thick (1000 m) pile of basaltic volcanics on Skomer Island to the Lower Silurian. This vulcanicity probably connects with the (Wenlock?) activity in the Dingle peninsula of Ireland. Secondly, Walmsley (in Owen *et al.*, 1971, p. 47) has shown that the Ludlovian fauna of South Pembrokeshire has greater affinities with North America than with the type Ludlovian of Ludlow, Shropshire.

Before the commencement of Devonian deposition, the Caledonian orogeny deformed the lower Paleozoic rocks of South Wales. The main Caledonian northeast–southwest structures of central Wales swing round into a more east–west direction when traced southwestward into Pembrokeshire. Powerful strike fractures accompany the fold axes. In North Pembrokeshire some strike faults replace fold limbs and have stratigraphical displacements of up to 2000 m. Intrusions of gabbro and dolerite pierce the Cambro-Ordovician rocks of North Pembrokeshire, though these probably date back to later Ordovician times.

The Devonian strata of South Wales are up to 2000–2500 m thick and were deposited in a fluviatile environment on the northern edge of a sea which lay over southwest England. Thick red marls with cornstones pass upward into micaceous flags, tough red sandstones, and conglomerates. Traced westward across Carmarthenshire, the Devonian transgresses the Silurian on to the Ordovician and is in turn overstepped by the Lower Carboniferous on the northern flanks of the Pembrokeshire coalfield. South of that coalfield, however, the Devonian thickens rapidly and may exceed 5000 m on the northern side of Milford Haven. South of that inlet, wedges of coarse conglomerate characterize the middle portion of the Devonian sequence. These rudites were derived from a southerly source and contain pebbles of lower Paleozoic rocks with fossils of North French affinities. The Upper Devonian Skrinkle Sandstone of South Pembrokeshire is a largely marine sequence, heralding the major marine invasion that was to come in Lower Carboniferous times.

The Carboniferous rocks of South Wales are divisible into

 (iii) The Coal Measures   3000 m maximum thickness
 (ii) The Millstone Grit   600 m maximum thickness
 (i) The Carboniferous Limestone 1500 m maximum thickness

The Lower Carboniferous carbonates are best exposed in the southern peninsulas of Pembrokeshire and Gower and include bioclastic limestones, oolites, calcilutites, and dolomites. This shelf-deposited sequence thins markedly where it reappears along the northern and eastern rim of the major downfold which is the South Wales Coalfield, and gaps appear in the sequence. The carbonate sea was bordered to the north by a relatively low-lying landmass, called "St. George's Land." A rejuvenation of this land area flushed spreads of sand, gravel, and mud to the South Wales area in Middle Carboniferous times, forming the so-called Millstone Grit. The silting-up of the Carboniferous area of deposition led ultimately to the swampy paralic environment of Coal Measure times.

The Coal Measure sequence falls into a lower, shaley portion with many workable coal seams and an upper sandy unit (the "Pennant Measures") with fewer coals. These upper sandstones are of subgreywacke type and represent the "molasse" of newly emergent mountain chains (Kelling, in Van Straaten, 1964) somewhere to the south of South Wales (probably over southwest England). The coals of South Wales change with depth from bituminous to anthracite and also show this change when traced at any one horizon across the coalfield from southeast to northwest.

The highest (Stephanian) Coal Measures are missing—probably through erosion—in South Wales. Permian rocks are also completely absent and were probably never deposited over the area. This great erosional episode followed the climax of the Hercynian (Armorican) orogeny in South Wales, and in the Vale of Glamorgan (south of the main coalfield) flat-lying Upper Triassic and Lower Jurassic strata rest with marked unconformity on folded and fractured Carboniferous, Devonian, and even Silurian rocks.

The Hercynian structures of South Wales are dominated by the east–west downfold of the coalfield, broken by the waters of Carmarthen Bay into the main portion and the Pembrokeshire Coalfield. The fold has a main east–west axis but is closed on the eastern side by the NNE–SSW-trending Usk anticline which exposes the Silurian. The Pembrokeshire continuation is a much narrower belt due to the greater compression and internal thrusting in this western region.

The Carboniferous rocks of the South Wales Coalfield are much affected by faulting, which follows three main trends: (i) NNW–SSE or northwest–southeast; (ii) east–west, and (iii) northeast–southwest. The first type are particularly numerous but are of complex origin, involving both lateral and vertical movements. Many of them separate areas of differing fold pattern and obviously grew with the folding. Some of these fractures have throws approaching 400 m. The second type are particularly prevalent on the "South Crop" of the coalfield and comprise both low-angled thrusts and "lag faults." The third type include marked belts of disturbance which enter the coalfield from the Devonian tract to the north. They comprise narrow but intense belts of caledonoid folding and faulting, the latter being of sinistral wrench type along the Neath and Swansea Valley disturbances but of thrust type along the Careg Cennen (or Church Stretton) disturbance which lies outside the northern limit of the main coalfield (Fig. 3). All three disturbances probably turn westward into Pembrokeshire where they have behaved as thrusts, one of which has pushed Precambrian volcanics and intrusives (with a Silurian capping) northward on to Millstone Grit and Coal Measures. These northeast–southwest disturbances appear to lie on the sites of ancient fractures in the basement and the north–south Armorican compression was appreciably modified along their length (Owen, 1971a, p. 1310). Carboniferous (and to some extent even Devonian) deposition was influenced by contemporaneous growing movements along these basement-controlled belts. Owen (1971a) has shown that the coalfield downfold was also to some extent growing during Carboniferous times.

It is, however, in two of the southern peninsulas of South Wales, i.e., South Pembrokeshire and Gower, that the greatest Hercynian deformation occurred. In Gower and Pembrokeshire (especially), the upper Paleozoic rocks are tightly folded and dips are steep. In Gower, Devonian rocks occupy the cores of upfolds, but in South Pembrokeshire, Ordovician and Silurian strata are exposed along the anticlinal axes (as at Freshwater East and Orielton). Many strike thrusts occur, and there are numerous cross-faults, which in Gower frequently separate blocks of differing tectonic pattern (George, 1940). Dextral shifts of fold axes occur along many NNW–SSE fractures in South Pembrokeshire (Dixon, 1921, Fig. 13).

The line marking the southern edge of the South Wales (and Pembrokeshire) Coalfield is often referred to as the position of the Armorican front in South Wales. The writer is, however, hesitant to place the Armorican front into South Wales and thinks that it might well lie more along the north coast of Devon. (Ramsbottom, 1970, Fig. 1, also suggests this more southerly position.) The relatively moderate folding of the sub-Mesozoic rocks in the Vale of Glamorgan certainly does not warrant their positioning along the main Armorican front. As stated below, the North Devon (or Cannington Park) thrust brings Devonian strata on the south against probable Coal Measures on the north, a structural relationship very reminiscent of the front in southern Ireland.

The post-Hercynian history of South Wales can only in part be reconstructed. Triassic marls and breccias are restricted to the Vale of Glamorgan and to very small patches in Gower and South Pembrokeshire. Lower Jurassic (Lias) shales and limestones occur only in the Vale of Glamorgan. Cretaceous strata are missing, but there is a small remnant of Tertiary (Oligocene) in Pembrokeshire. Mid-Tertiary movements affected South Wales, reactivating older faults and producing new fractures. High Pliocene sea levels beveled flat surfaces across the rocks of South Wales producing a falling "staircase" of topographic levels from at least 200 m (perhaps even from 800 m) to 60 m (this lower surface probably being of Pleistocene age). The last major geological event in South Wales was, of course, the Pleistocene Ice Age with ice advancing down over South Wales from the interior and also down the Irish Sea.

C. The Southern Irish Sea

The Irish Sea separates Ireland from England and Wales and is divisible into two parts by a line from Anglesey (northwest Wales) to Dublin. The southern Irish Sea is alternatively known as St. George's Channel with a broad indent into the West Wales coastline known as Cardigan Bay.

The rocks bordering both sides of the southern Irish Sea are mostly old and are dominated by the lower Paleozoic. Precambrian rocks occur in Anglesey, on the tip of the Lleyn peninsula, in Pembrokeshire, southeast Wexford, and in Bray. The upper Paleozoic is restricted and post-Paleozoic rocks are absent on these (southern) Irish and Welsh coasts.

The pioneer views of O. T. Jones (1951, 1956) have been upheld by the deep borehole drilled at Mochras, on the coast near Harlech, Merionethshire, between 1967 and 1970 (Wood and Woodland, 1968). This borehole reached a depth of nearly 2000 m and proved 609 m of Tertiary (Oligocene?) and Quaternary deposits unconformably overlying 1219 m of Lias mudstones and limestones (the thickest Lias succession in Britain), which in turn rested on 39 m of Rhaetic and Keuper (Triassic) marls and dolomites. This extremely thick Lower Jurassic marine sequence occurred in what many paleogeographers had claimed was a Jurassic land area! The borehole supported earlier geophysical work by Griffiths, King, and Wilson (1961) in Tremadoc Bay to the north and particularly upheld their view of a major westward-downthrowing fracture along this Harlech coast. Dobson (1971) has drawn attention to the great throw (over 4570 m) of this fracture. He believes (1971, Fig. 3) this fault to be displaced by northeast–southwest fractures running out from the Welsh mainland into the southeast portions of Cardigan Bay. The Bala fault would appear to be one of these transcurrent faults. According to Dobson these caledonoid fractures pass to the northwest of the North Pembrokeshire coast (Fig. 3) and help to define the southeastern edge of an important Mesozoic–Tertiary graben. The nature of the Mochras sequence, however, demands more than one phase of westward downfaulting along the Harlech coast with pre-Oligocene and post-Oligocene movements.

Geophysical work in the main portion of the southern Irish Sea has been carried out by the geophysical sections of the Institute of Geological Sciences, Birmingham University and the University College of Wales, Aberystwith. The picture (still not complete) emerging from the published work of Bullerwell and McQuillin (1968), Blundell, Davey, and Graves (1970), Al-Shaikh (1969, 1970), and Bamford, Blundell, and Davey (1970) is of a complex, downfaulted, downwarped, sedimentary basin, over 6000 m deep extending from the north of Cardigan Bay southwestward between Pembrokeshire and southeast Ireland (see Fig. 3). The Cardigan Bay portion is divided from the major St. George's Channel basin by an upwarp running north–south from the Lleyn peninsula toward northeast Pembrokeshire. The main basin to the west of this upfold is believed to have a sedimentary fill of 6096 m of which over 2000 m in the center of the basin are of gently folded marine sediments interpreted as being of Paleogene age (Blundell, Davey, and Graves, 1970). It is suggested that much of the underlying sediment is of Permo-Trias and Jurassic age. Bamford, Blundell, and Davey (1970, p. 79) explain the regional gravity field across Ireland and Wales as due to the presence beneath the Irish Sea of a layer between 25 and 30 km deep having a seismic velocity of 7.3 km/sec and a density of 3.1 g/cm^3. "There appears to be, moreover, a direct link between the presence of this layer, the subsidence of basins like those beneath Cardigan Bay and Tremadoc Bay during the Tertiary, and basaltic vulcanicity such as the Antrim (northern Ireland) plateau basalts. These same features have been observed together in certain other parts of the world, for example, in the Rhine rift of Germany."

Blundell, Davey, and Graves (1970) believe the St. George's Channel caledonoid basin to be attenuated sharply to the southwest at a line approximating to the position of the Armorican front and that it then appears to divide to form two shallower synclinal structures (involving Tertiary sediments) separated by an extensive area of chalk.

A positive belt or rise appears to run from Carnsore Point (southeast Ireland) towards the Lleyn peninsula of North Wales with a faulted southeastern edge. Rast (in Wilson, 1968) has suggested that the Aber–Dinlle fault in North Wales could continue to the south of Carnsore Point. Such a fracture could be one of great age but reactivated as a monocline in post-Carboniferous times.

Details about the more northern portions of the southern Irish Sea have been given by Dobson (1971). A Caernarvon basin, northwest of the Lleyn peninsula, has a fill of 1370 m, probably mainly of Carboniferous and Permo-Trias. So also has a central Irish Sea basin midway between Lleyn and Dublin, the thicker fill here being up to 4570 m. This basin is separated by another rise from the Kish bank basin, east of Dublin.

Summing up, the ancient rocks of southeast Ireland and Wales are separated by deep basins with faulted margins, preserving sedimentary fills of Mesozoic and Tertiary sediments together with, in some cases, underlying uppermost Paleozoic strata. The trend of the basins is essentially northeast–southwest, that is, reflecting the caledonoid grain of the Irish and Welsh margins. The fractures bounding the basins are probably, in some cases at least, of great antiquity, and a major rise extending from the southeast tip of Ireland toward northwest Wales was a major paleogeographical element of lower Paleozoic times ("the Irish Sea geanticline" of O. T. Jones, 1938). The Coal Measures of South Wales may have extended to Cardigan Bay (Owen, 1971a, p. 1314). Subsidiary folds to the northwest of Pembrokeshire may be associated with salt bodies (Bullerwell and McQuillin, 1968, p. 4). They could indicate the presence of Permian deposits.

D. Southwest England

Southwest England (Fig. 4) is an area of irregular coastal outline and surface topography. The latter varies from the high rugged moors of the granite masses (650 m O.D. on Dartmoor) to low-lying mudflats in the Bridgewater and Bristol areas. A number of well developed land surfaces occur in

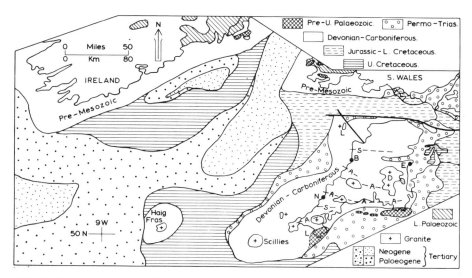

Fig. 4. The geology of southwest England and the Celtic Sea (geology of the sea areas based particularly on the work of Curry, Hamilton, and Smith, 1970, and of Hamilton and Blundell, 1971). Key: A, major anticlinal axis; B, Bude; D, Dartmoor Granite; E, Exeter; L, (southern) Lizard; L, (northern) Lundy; N, Newquay; S, major synclinal axis.

southwest England at heights of 330, 230, and 130 m O.D. Spectacular cliffs occur on the Cornish coasts with the Lands End Granite and the Lizard metamorphics providing prominent headlands.

Devon and Cornwall, like southern Ireland, effectively display the eroded cores of the Hercynian (Armorican) mountain chains. Moreover, the orogenic belt was intruded by a great ENE–WSW-trending granite batholith whose upper surface cupolas are today the granite masses of Dartmoor, Bodmin, St. Austell, Land's End, etc. Extensive mineralization accompanied the igneous intrusion.

Much of the surface of Devon and Cornwall is formed of folded Devonian and Carboniferous rocks, folded into a main central synclinorium preserving the Carboniferous or "Culm" of North Cornwall and central Devon with the Devonian outcropping to the north on Exmoor and on the south in South Cornwall and South Devon. Apart from the granite masses the only other variants to this Devonian–Carboniferous tract are the metamorphosed areas of the Lizard and the Start Point peninsulas.

East of a north–south line running through Exeter, the Cornubian orogen passes beneath a Permian–Mesozoic cover, sediments which dip generally eastward at gentle angles. The Permo-Trias rocks, predominantly red, are sandstones and marls with local breccias and pebble beds. Lower Jurassic clays and calcilutites give way eastward to Middle Jurassic calcarenites. The Upper Cretaceous is markedly transgressive and rests on Trias in easternmost

Devon and on Permian near Exeter. In the Bristol–Mendips district, however, Hercynian folding is again revealed through the erosion of this Mesozoic blanket, the Carboniferous and Devonian being sharply folded along east–west lines in the Mendip Hills but along very variable directions in the vicinity of Bristol.

Apart from small inliers of Tremadocian well north of Bristol, the oldest rocks in southwest England appear to be the metamorphic–igneous complexes which occur in the Lizard peninsula at the southernmost tip of Cornwall, and in the extreme south of Devon, between Start Point and Bolt Head. These two areas have been thrust northward on to the Devonian rocks of southwest England. The Start region is formed of mica schists but the Lizard mass is much more complex, involving mica and hornblende schists, invaded by peridotite, gabbro, and troctolite; basaltic dikes and sills; and material producing the Kennack Gneiss. All these intrusions accompanied a period of regional metamorphism. The gneiss appears to have been metamorphosed 350–390 m.y. ago (Dodson, 1961), and the higher figure would make the Lizard metamorphism Caledonian sensu-stricto, i.e., the end of the Silurian Period. On the other hand, the younger dating would bring the deformation forward to the earliest Carboniferous, a structural episode supported by Hendriks (1959), who believes that great nappe formation and serpentine intrusion occurred at about the Devonian–Carboniferous time junction. That some even pre-Devonian rocks and structural events are preserved in these Lizard–Start areas is, however, shown by the date of 442 $\pm$ 24 m.y. obtained from the Lizard Schists (Miller and Green, 1961). It may be that severely deformed lower Paleozoic or Precambrian (Brittany type) masses have been pushed from more southerly positions on to these southern extremities of southwest England.

Lower Paleozoic rocks are of very restricted occurrence in Cornwall, being confined to exotic masses of Normandy-type Ordovician quartzite (at Carne and other areas near Dodman), Wenlock limestone (at Porthluney), and mixed horizons in the Meneage crush zone on the northern (thrust) border of the Lizard mass. At Nare Head in Roseland, between the Lizard and Dodman, "foreign" blocks of Ordovician and Silurian—of southern aspect— occur within black slates that may be Middle or Upper Devonian age. It is rather peculiar that Ordovician blocks occur lower down in the slate sequence than those of Silurian age. The Meneage crush zone also has its problems, as it involves not only tectonic crushing but also subaqueous sliding of almost unconsolidated sediments (Lambert, 1962, 1964).

The Devonian rocks of southwest England occur in two main belts on either side of the Central Devon Synclinorium in Carboniferous (Culm) sediments. The northern area extends from North Devon into West Somerset. In this Exmoor region, the Devonian succession (a mixed one of continental and marine facies) is about 5000 m thick, but the base is not seen. The oldest rocks occur along the north coast. The continental units are red, purple, and grey sandstones, yielding plants or fish, while the marine representatives include slates, sandstones, and limestones. A gradation occurs, within the Pilton beds, up into the Carboniferous.

The Devonian rocks of South Devon and Cornwall, though almost entirely marine (the Dartmouth Slates near the base yield mainly fish remains), present far greater stratigraphical and structural problems, being severely deformed in South Cornwall and moreover frequently displaying lateral changes in sediment and volcanic lithologies over the whole outcrop. It is extremely difficult to give a total thickness to this southern Devonian succession, but in the vicinity of Padstow, on the north Cornish coast, the upper Eifelian to lower Famennian sequence (involving slates, turbiditic limestones, agglomerates and tuffs, and goniatite-bearing slates of grey green and purple color) totals at least 1500 m (Gauss and House, 1971). In southernmost Cornwall, the Lower Devonian appears to be represented by the Mylor beds followed northwards by the Middle Devonian Gramscatho Series (grits, sandstones, and slates of flysch origin and probably deposited in a geosynclinal trough bordered to the south by an actively rising cordillera). From Newquay to Looe and eastward to Dartmouth, the main outcrop of Lower Devonian occupies the axial zone of the major Watergate Anticlinorium. The Middle Devonian, with thick lenticular areas of limestone, is well exposed in the Torquay and Brixham areas. Volcanics, mainly spilitic, characterize the Upper Devonian near here and again on the North Cornwall coast.

On the southern side of the Central Devon Synclinorium, the Lower Carboniferous is formed of shales, quartzites, spilitic tuffs, limestones, cherts, and black calcareous slates. The overlying Namurian consists of shales passing upwards into greywacke sequences. Thicknesses are again difficult to estimate because of the folding and thrusting, and the Namurian Crackington Measures of the North Cornwall coastal area have been claimed to have a great thickness, though this is now considered doubtful. Into the center of the main synclinorium in northwest Cornwall the (Westphalian) Bude Sandstone Series (1200 m thick) and the overlying Welcombe beds (650 m) complete the Upper Carboniferous succession, though it should be noted that these beds reach no higher than a lower Westphalian position.

Still further north into the Bideford area of West Devon, the Carboniferous sequences become more involved. The bulk of Lower Carboniferous time here falls within the upper Pilton beds, followed by cherts and shales in part representing a very condensed lower and middle Namurian sequence. There then follow turbidites, paralic sequences, and further turbidites taking the sequence from the higher Namurian up into the lower Westphalian. Correlation is extremely difficult for the tectonic blocks along this coastal area of West Devon–North Cornwall, and great changes have been periodically made to these Upper Carboniferous successions. Reading (1965) has suggested that some succession blocks are in an exotic position—in a depositional sense—because of thrusting and overriding.

One further point of interest concerning the Upper Carboniferous of southwest England has been highlighted by Ramsbottom (1970). In his studies of the Carboniferous faunas of the area, he states that a low Namurian fauna near Drewsteignton, near Exeter, appears to have migrated from northern Spain. Higher in the Carboniferous sequence, goniatites near Bude and at nearby localities in Devon (and also in South Wales and Bristol) are unlike forms in other British areas but closely resemble American species.

The dating of the main Hercynian (Armorican) orogeny in southwest England is uncertain. No proved horizon higher than the lower Coal Measures has been yet detected in the Culm. The great intrusive batholith of Dartmoor–Lands End yields dates of about 290 m.y., and this intrusion postdates the Cornubian folding and thrusting. The wide spread of Pennant molasse into South Wales, the Bristol area, and Kent suggests that the main Armorican climax could have been reached in southwest England by the beginning of upper Coal Measures times (Owen, 1971a). That the Hercynian orogeny was complex and polyphase in southwest England has been clearly demonstrated by Dearman (1969, 1970) and by Dearman, Leveridge, and Turner (1970). Zones of differing radiometric dates (Dodson and Rex, 1971) across southwest England range over almost 100 m.y., while the phases of folding appeared to have steadily moved northward (Dearman, 1969, Fig. 4).

Dearman (1969, 1970) has subdivided the area into four tectonic zones (excluding the Lizard–Dodman areas). From north to south they comprise: (i) a zone of upright folds in Westphalian strata (near Bude); (ii) a zone of sliced overturned folds in Namurian strata between Boscastle and Bude; (iii) a zone of thrust slices involving Upper Devonian and Carboniferous near Tintagel and Launceston; and (iv) a wide zone of recumbent folds in the main Devonian tract south of the Central Devon Synclinorium (1970, Fig. 1). He has demonstrated how differences in the suprastructure and infrastructure (with the development of slaty, second, and crenulation cleavages in the latter) arose with the continuation of the earth movements. Late effects were reversals to normal type movement northward along existing thrusts and eventually the rise of the great granite batholith, the contact metamorphism, now seen around the various cupolas, and (in later Tertiary times) dextral wrench movements along northwest–southwest faults (Dearman, 1963).

Superimposed on all the complex Hercynian structures of southwest England are a number of major Hercynian "master folds," such as the Truro, Watergate, Kit Hill, North Exmoor upfolds and the Central Devon Synclinorium. Major thrusting probably marks the north coast of Exmoor. On geophysical grounds, Bott, Day, and Masson-Smith (1958) postulate a low-angled fracture beneath the Exmoor Devonian, with Carboniferous beneath. This thrust could be a westward continuation of the Cannington Park thrust, near Bridgewater. This line appears to separate two distinct Lower Carboniferous environments, the succession to the north (in the Mendips and Bristol area) being a shelf carbonate one. In the Mendips, this Carboniferous Limestone, together with Devonian and Upper Carboniferous, has been folded into four main east–west anticlines arranged "en echelon" and affected by powerful east–west thrusting. At Vobster, this thrusting has pushed Limestone over high Coal Measures.

Important mineralization accompanied the granite intrusion in southwest England, and to a lesser extent the Hercynian deformation in the Mendips. In the vicinity of the granites there occur numerous granitic "elvan" dikes, tourmaline, and aplite veins and mineral veins all along ENE–WSW or northeast–southwest lines. Cassiterite, wolframite, galena, sphalerite, and a little uranium occur. Hill and Vine (1965) believe that both upthrust of the metamorphic basement and the batholith intrusion are accommodations of a change in structural trend between the western portion of the English Channel and southwest England.

As far as the Mesozoic–Cenozoic history of southwest England is concerned, only two points need to be stressed in this account. Firstly, the base of the Upper Cretaceous is increasingly transgressive westwards across West Dorset and East Devon. Secondly, small, but important, areas of downfaulted Oligocene pipeclays and sands occur at Bovey Tracey (near Torquay) and at Petrockstow (in central Devon). They appear to lie near to the important Sticklepath fault which trends out to sea in Bideford Bay.

E. The Celtic Sea and Bristol Channel

The Celtic Sea is that part of the shelf which is bordered by the south coast of Ireland, southwest Wales, and the north coast of Cornwall. Its southern boundary may be considered to lie at lat. 50° N, i.e., roughly at the position of the Scilly Isles. The Bristol Channel is the waterway which separates Wales and southwest England. Much of our knowledge of the sea floor of these seas is largely due to the work of university groups at Bristol, Birmingham, and Swansea but aided also by Prof. Curry and Dr. A. J. Smith. A recent synopsis of the work has been given by Hamilton and Blundell (1971, p. 297–300), and their main conclusions are incorporated in Figs. 3 and 4.

The Paleozoic rocks of southern Ireland extend beyond the south coast for about 12 to 15 miles and then suddenly give way to Upper Cretaceous or Tertiary. The straightness of this change strongly suggests a fault which could be a continuation of that forming the southeastern border of the Irish Sea Geanticline. Chalk of Campanian to Maestrichtian age occurs widely south of this line (Curry *et al.*, 1967), but there is a narrow downfold of Tertiary sediments immediately south of the Paleozoic boundary between long. 8° and 6°30′ W. Neogene to Recent sediments along the axis of this trough unconformably overlie gently dipping Paleogene. Large lenticular masses within the youngest sediments may be infillings of glacial scour (Hamilton and Blundell, 1971, 298).

Another (more major) downfold involving Neogene to Recent sediments occurs midway between North Cornwall and Ireland. Seismic profile records show this to be a broad fold, which probably forms the southwestern continuation of the main basin in St. George's Channel. The Neogene to Recent layer appears to be at least 300 m thick. Broad open folds occur but faults are rare. On the southeastern flank of this downfold, Upper Cretaceous sediments appear to transgress the Jurassic on to Permo-Trias deposits which in turn abut against the Devonian–Carboniferous fringe of North Cornwall. On the southern margin of the Celtic Sea, Cenomanian to Turonian Chalk rests directly on the upper Paleozoic rocks which are pierced by the granite mass of Haig Fras (at long. 8° W). This granite belongs to the Scillies–Cornwall suite and has given a radiometric age of 277 ± 10 m.y. (Curry, Hamilton, and Smith, 1970, p. 9). The Chalk appears also to directly fringe the Devonian on the southern side of the Scillies.

Between the Haig Fras Paleozoic Granite "rise" and southwest Ireland, Eocene sediments occupy the main downfold, though these lower Tertiary sediments probably give way to higher Tertiary before the edge of the continental shelf. This main downfold continues that of St. George's Channel and the mid-Celtic Sea but with the caledonoid northeast–southwest trend turning into a more WSW–ENE direction, reflecting the influence of an older (Hercynian) trend.

Along the Bristol Channel, the post-Hercynian downfolding follows an east–west trend. A Jurassic sequence (from Lias to almost Portland) has been proved in the Bristol Channel, and is probably 1600 m thick. The underlying Triassic (of Keuper Marl facies) has a narrow outcrop close to the North Devon coast and is believed to be about 230 m thick (Donovan, Lloyd, and Stride, 1971, p. 294). The fully marine character of the Jurassic, now so close to the Devonian slates and limestones of the North Devon mainland (height over 350 m above sea level), implies major east–west faulting of the Mesozoic margin very close to the North Devon coast and with an accumulative northerly downthrow of about 2000 m. These post-Jurassic faults closely follow the Hercynian thrust, postulated by Bott, Day, and Masson-Smith (1958) along the North Devon coastline.

At the western end of the Bristol Channel, a westward dipping sequence of Middle Jurassic to Lower Cretaceous rocks is believed to occur (Hamilton and Blundell, 1971, p. 298). This suggests that the east–west Bristol Channel downfold gives way to the northeast–southwest downfold of the Celtic Sea. A further complication at this western extremity of the Bristol Channel is the Lundy Granite. This Eocene intrusion rises to over 130 m above sea level and is intruded into Devonian slates. It appears to be sharply cut off on its eastern side by the Sticklepath fault of southwest England (see Owen, 1971b, Fig. 1). The underwater extension of this intrusive complex is not yet fully known. An elongated negative magnetic anomaly extending northwestward from Lundy is interpreted by Cornwell (1971, p. 288) as being due to a Tertiary dike swarm—perhaps along the Sticklepath fracture. The Lundy Granite may border a larger igneous complex of basic composition (Brooks, 1972). The significance of this Tertiary pluton will be discussed later.

Two features of the geology of the Celtic Sea and Bristol Channel must be further stressed. Firstly, the marked caledonoid and armoricanoid trends present in the area. Secondly, the important transgression by the Upper Cretaceous deposits (mainly Chalk) across older formations. Important Middle Cretaceous earth movements occurred in southern England prior to the deposition of the Albian stage. These movements appear to have been widespread also in the sea area west of Cornwall. Owen (1971b) has suggested that the main downfolding of the Jurassic rocks in the Bristol Channel occurred in Middle Cretaceous times, whereas east–west and northwest–southeast fracturing occurred later, in Tertiary times.

F. Brittany ("The Armorican Massif")

The type area for the Armorican fold belt is the "Armorican massif." This embraces Brittany, the western portion of Normandy (known as the Cherbourg or Cotentin peninsula), and, further south, the area known as the Vendée. The Channel Islands, west of the Cotentin, are also to be included. The eastern and southern limits of the massif are determined by the unconformable Mesozoic blanket which, in the Straits of Poitou (in the extreme southeast) separate this northwestern massif of France from another—the Central massif.

The rocks of the Armorican massif range from Precambrian to Upper Carboniferous (Stephanian) with very small outliers of Tertiary sands, gravels, and clays. By far the greatest part of the surface is formed of the Precambrian, known as the "Brioverian" with even older Precambrian ("Sarnian") claimed in northeast Brittany near St. Brieuc and St. Malo. Gneisses and schists in western Brittany also could be this older Precambrian or more metamorphosed Brioverian.

The Armorican folds dominate the "grain" of the massif, the Precambrian dominating major upfolds such as the Leon, Mancellia, Rennes, and Cornouailles anticlines ("anticlinoria" would be a better term) while the Paleozoic rocks are preserved in the major downfolds which include the Morlaix, Chateaulin, and Ancennis synclines (the only downfolds preserving Carboniferous strata) and, in the Cotentin peninsula, the synclines of Nehou, St. Sauveur le Vicomte, Coutances, and Granville (Fig. 5). An outstanding feature of the fold trend is the change from a ENE–WSW trend in the north of the Armorican massif to a WNW–ESE trend in the south. This is virtually the mirror image of the change in southwest England from the WNW–ESE trend in the main part of the area to the WSW–ENE trend in the extreme south of Cornwall and especially along the Lizard–Dodman–Start line (Dearman, 1969).

The Precambrian (Brioverian) rocks of Brittany are frequently highly folded and metamorphosed (up to the sillimanite grade) and have been intruded by granite. Relatively unmetamorphosed Brioverian occurs in places, e.g.,

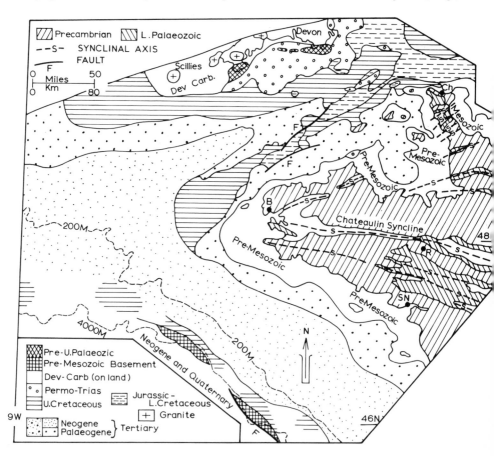

Fig. 5. The geology of Brittany, the western English Channel, and the adjacent continental shelf and slope (geology of the sea areas based mainly on the work of Curry, Hamilton, and Smith, 1970, and of Stride *et al.*, 1969). Key: B, Brest; Ch, Cherbourg; R, Rennes; SN, St. Nazaier.

east of Brest and in the Bay of Douarnenez, and include shales, graded grey-wackes, sandstones, and spilitic volcanics with pillow structure. White quartz-ites occurring "within" the Brioverian may be contemporaneous, though they may represent downfolded outliers of lower Paleozoics. Conglomerates in central Brittany, containing Brioverian pebbles, are probably still Brioverian in age but must have followed mid-Brioverian earth movements. Particularly interesting are the occurrence in Normandy and northeast Brittany of glacial and periglacial deposits—thick tillites—reminding one of the Dalradian of Scotland or the Eocambrian of Scandinavia. These tillites too contain eroded Brioverian products and point to glaciations following mid-Brioverian earth movements.

Pre-Brioverian orogeny is evidenced by the unconformity seen near St. Brieuc between the Brioverian and the underlying Sarnian (or Pentevrian) granitic and granodioritic gneisses, folded along north–south axes. This earlier Pentevrian orogeny dates back to 1100–1300 m.y. ago. Large parts of southern Brittany, and also the extreme northwest (Leon), are formed of metamorphosed rocks. In the Leon anticline the belts of metamorphics are orientated along ENE–WSW lines. Mica schists, graphite schists, garnetiferous mica schists, granite gneiss (e.g., the Gneiss de Brest) lenses, and veins of granitic material and pyroxenite characterize this northern area. Further south, in the Cornouailles anticline, micaschists dominate with belts of granitic gneisses. Opinions differ as to whether (doubtfully) these metamorphics represent rocks earlier than the Brioverian or whether, as Barrois considers, they are more metamorphosed Brioverian.

Precambrian deposition was brought to a close by the Cadomian orogeny dated by granite intrusion at about 580 m.y. This orogeny, not unexpectedly, was a polyphase one and preceded, as mentioned above, by earlier pulses. Bishop *et al.* (1969) have distinguished at least three Cadomian phases. Cado-mian complexity is not rendered any easier by the reactivating and reintrusion of Cadomian granites during the later Hercynian (Armorican) orogeny.

Cambrian sedimentation followed a long period of post-Cadomian erosion, and in Normandy, the basal Cambrian conglomerates clearly rest unconform-ably on the Brioverian. In South Brittany and the Vendée, however, there is no clear boundary but various conglomerates (perhaps within the higher Brioverian) are correlated with the post-Cadomian rudites further north. The lower Ordovician, on the other hand, is a more readily recognized unit, being mainly formed of the well-known "Grès Armoricain" of Brittany, under-lain by basal conglomerates and quartzites on the southern side of the Cha-teaulin syncline and south of Brest in the Crozon peninsula. The Grès Ar-moricain is a beach facies deposit and is formed of pure white or liver colored well-sorted quartzite. The deposit reaches a thickness of 1000 m in the region of the Rade de Brest but thins toward Morlaix. It is known to occur in northern

Spain and even further south. The British Triassic pebble beds contain pebbles of it so that it could also occur beneath the Permo-Trias blanket of the English Channel or even the southern English mainland. Ordovician transgression appears to have been from the southeast through the "Fosse Armoricaine" of Pruvost (1949) and a southerly source is indicated for many of the Ordovician units of Brittany. In western Brittany, the remainder of the Ordovician sequence (up to Caradocian) is made up of carbonaceous shales, sandstones (often micaceous), calcareous shales, and interbedded volcanic tuffs. In the May downfold of the Cherbourg peninsula, shales ("with *Calymene*") containing iron ore deposits are followed by another liver colored arenite—the Gres de May—and thence by black graptolitic shales, with no break into the Silurian. Silurian deposition was mostly quiet water with slow black shale accumulation, but some occurrence of basal Silurian conglomerate points to movements at the Ordovician–Silurian boundary in the Camaret area (southwest of Brest).

The main Devonian areas in the Armorican massif are the northern and southern flanks of the Chateaulin syncline (being especially well exposed in the Rade de Brest coasts) and in the Laval syncline further east. Elsewhere outcrops are small. The quiet deposition of Silurian times was broken by a great influx of sediment in Devonian times into the Armorican trough. On the north side of the Rade de Brest, quartzites show slumping toward the south and pass up into widespread reddish fossiliferous sandstones. Higher in the Devonian succession, lenticular and nodular limestones occur within shales and Eifelian to Famennian faunas compare with those of the German Rhine sections.

The Armorican (Hercynian) orogeny was also polyphase and protracted in time over the massif. George (in Coe, 1962) has drawn attention to the numerous breaks in the Bretonic Paleozoic successions with unconformities at the base of the Devonian, within the Devonian, at the base of the Carboniferous, the Namurian (on Givetian Devonian in the Vendée), Westphalian, and Stephanian (see his Fig. 2/111, p. 28). That one of the major climaxes of the Hercynian orogeny took place in the Devonian–Carboniferous interval is shown by the strong unconformity developed at the base of the Tournaisian (Lower Carboniferous) which transgresses across older Paleozoics on to Brioverian. These Tournaisian sediments are coarse grits and conglomerates often of deltaic origin and probably deposited in a number of independent basins or "fosses." The overlying Viséan, on the other hand, is more marine and of Culm (Devon) type but with some bioclastic limestones. On the Cotentin they transgress beyond the Devonian. Viséan granitic intrusion (unroofed and transgressed by the Upper Carboniferous) points further to the repeated buildup of the Armorican "earth storm" in the massif. It is a matter of opinion therefore where one places the major Armorican climax in Brittany—at the

Early Carboniferous "Bretonic phase," at the "Sudetic phase" (pre-Namurian), or even at the post-Westphalian–pre-Stephanian interval, because the Stephanian deposits are so strikingly different from their earlier Carboniferous counterparts, being coarse conglomerates, arkoses, shales, and poor, thin coals of isolated "limnic" environments (probably small intermontaine basins of upland aspect). These Stephanian outliers are much faulted but are not folded or metamorphosed. On the other hand, the conglomerates contain much metamorphosed and granitic material.

The Hercynian folds are open type folds grouped into major anticlinoria and synclinoria with asymmetrical flexures facing southward. How much the varying trend from west-southwest in the north to west-northwest in the south is due to (a) control by basement trends (Cadomian) and (b) some twisting of the earlier fold influences is not clear. In some cases the Paleozoic cover folded while the underlying basement responded by faulting to the later compressions. Later Hercynian phases involved widespread vertical movements and migmatite injection followed by granite intrusion, both largely along growing master fractures. Surface vulcanicity occurred locally in Upper Carboniferous times. Large sinistral east–west wrench fractures were fairly late features of the orogeny and invite comparison with those in southwest Ireland. Of the Hercynian intrusions, large binary granites (e.g., Quintin) were intruded into the Devonian–Carboniferous cover of the Chateaulin syncline. Comparison with southwest England is invited by the pegmatite-aplite and mineral veins. Other granite bodies in the Precambrian belts are often difficult to differentiate from ancient granitic gneisses.

The post-Hercynian history of the Armorican massif began with a long period of erosion during Permo-Triassic times with the eroded products being carried northward and northeastward. Jurassic marine transgressions then followed, but over a relatively stable area which did not appreciably sink until Upper Cretaceous times, only to be uplifted again by the Maestrichtian. In the Cenozoic, marine invasions by the Paris and Aquitaine seas of the massif were localized and often restricted to narrow fault-controlled gulfs or valleys. Mid-Tertiary earth movements produced shallow but irregular folds and appreciable faulting.

G. The Western Portion of the English Channel

Unlike the southern Irish Sea, the coasts of the English Channel show rocks ranging in age from basement Precambrian through Paleozoic to Mesozoic and Cenozoic. It was to be expected, therefore, that the sea-floor geology of the channel would be just as varied and this is now known to be the case. It has, however, been found that lower Paleozoic rocks are absent. The results of early work was published by Hill and King (1953) and a masterly review of the geological history of the English Channel was presented by King (1954). King showed that the older (upper Paleozoic and Precambrian) floor was covered by "New Red Sandstone" (Permo-Trias), Jurassic, and Cretaceous; some thin Tertiary veneers in the western part of the channel thickened and increased in area in the eastern portion. He further showed that the Lower Cretaceous and Jurassic thinned markedly in the western channel and that to some extent this was true also of the Chalk (his Fig. 2). One other important result of the work by Hill and King was to show the irregularity of the sub-Permo-Trias floor south of Devon and Cornwall, leading King to comment that "much of the present upland areas of the west of Britain... was blocked out before the end of the Trias and is now merely showing a redeveloped and uplifted topography" (King, 1954, p. 100).

Much further work has now been carried out in the English Channel, and the most up to date report on the results, as far as the western approaches to the English Channel are concerned, has been given by Curry, Hamilton, and Smith (1970). Their geological map, covering some 24,000 square nautical miles (80,000 sq. km), is based on the results of systematic sampling and on continuous seismic surveys. About 1600 gravity core stations were manned, and almost 2400 miles (4400 km) of sparker track covered. It has been possible to sample the solid rock floor extensively, because most of this western part of the channel has only a very thin veneer of superficial sediments. This sea floor is remarkably flat with an average slope of 1 in 1000.

A wide range of geological horizons has been demonstrated. The basement is composed of granite and gneiss, together with metamorphosed and un-metamorphosed Paleozoic sediments. Granitoid gneiss occurs in localities south of Plymouth, and at one place it breaks sea surface to give the Eddystone Rocks (isotope age of 375 ± 17 m.y.). Around the Cornubian peninsula, the Paleozoic rocks are dark grey slates of Devonian to Carboniferous age. The Permo-Trias is up to 1000 m thick, but is very uneven in thickness off Devon and Cornwall because of the irregular floor to these brick red, occasionally pale green, sandstones, marls, and siltstones.

The Jurassic shows a continuity of deposition with southern England and totals up to 1000 m thick. Clays and calcilutites in the Lower Jurassic give way to pale clays and calcarenites in the Middle Jurassic. Upper Jurassic rocks were not detected. Lower Cretaceous sediments were restricted to inliers particularly southwest of the Hurd Deep (Fig. 5). These sandstones, arkoses, and micaceous clays seemed to be of terrestrial type, and one sample yielded a rich Wealden microflora. French workers have recorded similar inliers northwest of Ushant. The Lower Cretaceous appears to be about 150 m thick.

The extensive Upper Cretaceous cover consists almost entirely of Chalk, exceptions being sandstones and dolomites near the Scillies. West of these islands, lower and middle Chalk predominates but higher Chalk horizons are characteristic further east, with Campanian or Maestrichtian Chalk dominating areas on either side of the Hurd Deep. The base of the Upper Cretaceous is strongly transgressive everywhere and the Chalk is up to 500 m thick. A similar maximum thickness is recorded for the Eocene beds which are predominately carbonates, in striking contrast to the sands and clays of the land basins. Eastward overstep occurs within the Eocene, but there is little Oligocene present. Miocene buff, sandy, calcareous rocks are of marine origin, foramini-fera-rich, and with an abundance of globigerinids; 120 m of the Miocene is present, the formation resting with strong unconformity on the Eocene and Cretaceous. Plio-Pleistocene sediments are restricted and less than 30 m thick. Igneous rocks in the western English Channel include granites around the Scillies and Seven Stones (radiometric age 281 ± 9 m.y.), north of Brittany, and at Haig Fras. Phonolite occurs in two places south of Lands End. Their age has ranged from 113 ± 7 to 262 m.y.

Curry, Hamilton, and Smith have rightly stressed the presence in this channel sequence of four important unconformities. These occur at the base of the Permo-Trias, Upper Cretaceous, Eocene, and Miocene, and particularly at the first two mentioned positions. (This marked transgression at the base of the Upper Cretaceous has been noted in the account on the Celtic Sea and Bristol Channel.) Whereas the Permo-Trias, Jurassic, and Lower Cretaceous rocks are commonly faulted and folded, the Upper Cretaceous and Eocene are only very broadly flexured (1970, p. 5). Curry, Hamilton, and Smith suggest a "structural high" in the Cretaceous and Eocene in the area between Devon and the Cherbourg peninsula with thinning or even absence of Upper Creta-ceous in this area.

Structurally, the western part of the English Channel extends seaward to that part of the shelf which, as shown by Stride *et al.* (1969), had a long period of subsidence in Mesozoic and Cenozoic times. The main structure in the English Channel is shown by Curry, Hamilton, and Smith to be a single, large syncline occupying the central part of the western approaches to the waterway. The downfold trends WSW–ENE and plunges at a low angle westward. The outermost flanks of the fold (in Cornubia and Brittany) are composed of Paleozoic or basement rocks. Pre-Upper Cretaceous rocks are fairly tightly folded and faulted in places, especially at the northeastern end of the Hurd Deep. Asymmetrical folds are associated with low-angled reverse faults, dipping southward. The strongest northeast–southwest fracture occurs near the south western end of the Hurd Deep and has a throw of 500 m. An earthquake occurred here in 1925. The Lizard–Dodman–Start fault along the southern edge of Cornubia is believed to be responsible for the sudden southward thickening of the Permo-Triassic.

H. "The Edge of the Continent"

The main features of the British–French edge of the European continent are (a) a wide continental shelf west of southwest England and the English Channel with a relatively wide continental slope (becoming even wider northward toward lat. 50° N) but with prominent spurs—the Goban and Austell spurs; (b) a wide slope west of Ireland (wide because of the large north–south-aligned Porcupine bank at long. 14° W) but broken into from the south by the relatively deep (2000 m +) Porcupine Sea Bight; (c) a relatively wide shelf west of France but with a much narrower, steeper continental slope.

A major change in the continental margin occurs moreover at about lat. 49° N (Laughton, in Day and Williams, 1970). North of this latitude, the shelf is growing seaward with prograding sediments accumulating while south of it active canyon erosion is cutting back the shelf.

Sampling and sparker records of the sea bed along the continental margin has been carried out by several British and French workers. Upper Cretaceous has been proved on the southwest side of the Porcupine bank, near the Goban and Austell spurs, and on the continental slope off France. Eocene and Miocene sediments have been proved in several places, especially high on the continental slope, but on the northern portion of the Porcupine bank, igneous and metamorphic pebbles have yielded K/Ar dates of lower Paleozoic to Precambrian.

The most extensive survey of this portion of the European margin has however been carried out by Stride *et al.* (1969), their reconnaissance being by means of a 60-kJ continuous reflection profiler, supplemented by Boomer profiles and by extensive Asdic coverage of the continental shelf. Fourteen reflection profiles were carried out between the Faeroes and northwest Spain. Of these, profiles 4 to 9 are relevant to that part of the continental edge described in this chapter. Profiles 4 and 5 crossed from southwest Ireland to the northern and southern parts of the Porcupine bank, respectively. These traverses showed that Tertiary (flat-lying Neogene on more tilted Paleogene) sediments reached more than 1000 m in thickness in the Porcupine Sea Bight, and appeared to rest on somewhat more contorted Cretaceous. Over the northern part of the Porcupine bank, the Cenozoic–Mesozoic film disappears and basement metamorphics with igneous intrusions, particularly gabbro, occur on the sea bed (Fig. 2). The west side of the bank shows post-Cretaceous faulting, while at the foot of the continental slope, flat (late?) Tertiary turbidites rest on westward-dipping Cretaceous. Faulting affects the Tertiary sediments on the east flank of the main Porcupine Sea Bight which appears to be a downwarped, downrifted structure. At the southern end of the Porcupine bank there are only at most about 200 m of Tertiary sediment.

The main Celtic Sea and English Channel downfolds (ENE–WSW to east–west) appear to extend in a broadened fashion to the edge of the shelf, but that they remain separate downwarps is suggested by the extensive Eocene occurring west of Haig Fras and by the Cretaceous on the Goban Spur (Fig. 6). Day and Williams (1970) have shown that a ridge of high gravity extends from Cornwall to beyond the shelf edge, caused (they say) by a rise in the metamorphic basement which probably extends beyond the shelf edge forming the Goban Spur. Associated with this basement rise and also extending to the shelf edge is a line of gravity lows (their Fig. 1) interpreted as evidence of an extension of the granite batholith of Cornubia. The metalliferous mineral band associated with the granites may also extend to the shelf edge beneath the younger sediments. According to Day and Williams, a further ridge of high gravity (another rise in basement) extends from Brittany to the shelf edge

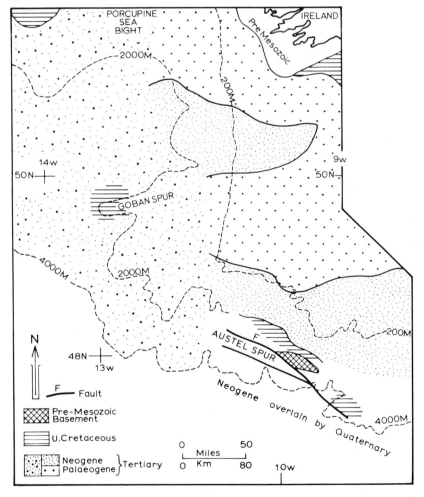

Fig. 6. The geology of the continental shelf and slope west of the English Channel and Celtic Sea. Based mainly on the work of Stride et al. (1969).

(the Austell Spur?). Hill and Vine (1965, Fig. 4) suggested that the metamorphic basement exposed in the Lizard–Start belt extended out to the shelf edge at about lat. 48°45′ N.

Profiles 6 and 7 of Stride *et al.* (1969) were carried out near to each other, east of the Austell Spur, and on the slope due west of northwest Brittany. Paleogene and Neogene Tertiary sediments are much deformed and slumped on this continental slope and rest on Cretaceous rocks which in turn rest with irregular contact on the basement. This basement is exposed in one place low on the slope, while major faults upthrow (oceanward) at least two portions, with throws of 500 m.

Profile 8, from the shelf to the slope southwest of Brest, showed the basement passes seaward into crumpled Chalk and in turn through Paleogene into flat-bedded Neogene which thickens oceanward. Out on the slope, slumped Eocene and Miocene occur with basement appearing at about 3000 m depth but with thick (over 800 m) Neogene–Quaternary, covering Paleogene and Cretaceous, present at the foot of the slope (Fig. 6).

Off the River Loire at long. 9° W, profile 9 showed Eocene, resting on basement between the islands of Belle Ile and Ile d'Yeu, followed seaward by Miocene sediments. On the continental slope here the Tertiary sediments are thinner and are turned landward by slump-block rotation. They rest on Cretaceous which in turn rests on basement which appears to outcrop at depths of 3000–4000 m. The slope falls abruptly here and a fault with an oceanward throw of about 1500 m is indicated.

Stride *et al.* (1969) see five main development stages to this West European continental margin. After (i) its origin in the Mesozoic, there was (ii) deposition and downwarping in the Cretaceous; (iii) Late Cretaceous to early Tertiary times was an erosional interlude, followed by (iv) downwarping and further deposition in Tertiary times (with a pause in the Paleogene–Neogene interval). During (v) late Tertiary and Quaternary times there occurred a second erosional phase which included faulting, slumping, and canyon cutting.

Turning their attentions to the problem of continental rifting, Stride *et al.* stress the dramatic change from the older rocks (Paleozoic downward) to the younger deposits laid on the shelf edge and on the continental slope. They draw attention to the fact that the outermost Paleozoic rocks appear to have no structural relationship with the older basement that would favor any proposed thesis for a Paleozoic Atlantic Ocean. Stride *et al.* suggest that the originally deeper Porcupine Sea Bight was initiated by lateral northward and westward translation of a "few tens of kilometers" to the Porcupine bank relative to the continental margin west of Ireland "where there is a scar of suitable size and shape." "Such a replacement would avoid the apparent overlap of the bank with the west side of the Atlantic in reconstructions of the supposed pre-continental drift land."

Further details of the Porcupine Sea Bight have been given by Clarke and Bailey (1970) who show that at least 1.3 km of sediment occur in this north–south-elongated trough. Vertical or subvertical dislocation planes can be detected around the trough's margins. Sparker records show diapiric structures toward the northern end of the trough. A regional nonsequence across the trough suggested that a tectonic trough filled to shelf depths had been rejuvenated by the onset of more active subsidence. Stacey (p. 86) in the discussion of Clarke and Bailey (1970) pointed out that a large Bouguer anomaly occurred along the north–south axis of the sea bight between lat. 51° and 53° N. This anomaly could be interpreted as either high-density material in the lower crust or as a thin crust. Seismic refraction work suggested that little lateral extension of the bight had occurred during the deposition of the topmost (and consistent) 2 km of sediment layer. Stacey believed there was no good evidence for major northward or westward movement of the Porcupine bank relative to Ireland and that the formation of the sea bight by a downwarping of the crust still remained a distinct possibility.

Stride *et al.* stress that the oldest continental slope deposits on the West European margin are of Cretaceous age. About 4 km of "postrift" sediments (Cretaceous–Recent) occur in the western continuation of the English Channel and in the Aquitaine basin, and a similar thickness must be present on the adjacent continental slope. "This thickness is sufficient to mask the original shape of the edge of Europe, supposedly resulting from continental drift, so lessening the perfection of its fit with North America and Greenland. The original Bay of Biscay may well have extended eastward beneath a part or the whole of the Aquitaine basin, almost to the Mediterranean so that its two sides were originally almost parallel." Late Cenozoic erosion of the continental slope south of lat. 49° N is impressive with canyons 1000 m deep.

Summing up, Stride *et al.* claim that the episodic evolution of the West European continental margin may be an indication that the Atlantic Ocean was formed during *two* periods of continental drift—(a) up to Late Cretaceous times and (b) during Eocene–Miocene times—and that drift was unimportant during Late Cretaceous and during Pliocene–Quaternary times.

III. PALEOGEOGRAPHIC–TECTONIC SYNTHESIS

Reconstructions of Precambrian paleogeography must be treated with caution because of the difficulties of correlation from area to area and from sedimentary to igneous environments. The oldest Precambrian rocks in the western approaches probably occur in the Rosslare complex of Southeast Ireland and as the Sarnian of Brittany. The latter gneisses have yielded a date of 1100–1300 m.y., which would place them at the upper limit of the Laxfordian (Lewisian) or even basal Moine of northwest Scotland. It is possible that the Johnston intrusives of South Pembrokeshire are of the same age (see Baker, 1971, Table 1). The Malvernian gneisses of the Welsh borderland may also be correlatives. Widespread orogenic events involving metamorphism, migmitization, and granitization are involved in these earlier phases of the history of the western approaches. Folding along north–south lines in Brittany could indicate a similar earlier orientation of depositional area from Brittany to Scotland, with perhaps no connection (regionally or time-wise) with the later Caledonian trough.

The widespread unconformity above these early basement rocks was followed by extensive vulcanicity in Pembrokeshire (Pebidian), Shropshire (Uriconian), and the Malverns. These mainly acid tuffs were largely water-deposited though there are some ignimbrites. No thick sediments follow the volcanics in southwest Wales, but in Shropshire there is the thick Longmyndian succession and in southeast Ireland there is a geosynclinal Cullenstown succession of greywackes, greenschists, etc. In Anglesey there is a 7000-m pile of eugeosynclinal spilites, cherts, turbidites, and acid extrusives (the great Monian complex). The similarity of the Monian with the Brioverian of Brittany is striking. A great marine area is indicated for later Precambrian times, involving an area extending even perhaps from Brittany northward to involve Wales, the Welsh borderland, and at least eastern Ireland. It must also have extended still further northward as the area of deposition of the Ingletonian of northern England and the thick pile of Moine sediments in northern Scotland. The "other side" of this great depositional area is probably indicated by the red Torridonian facies of northwest Scotland. With a great outpouring of acid extrusives at the start of this depositional chapter which later involved spilitic outpourings and development of pillow lavas, one is tempted to see signs of a Precambrian rival to the lower Paleozoic geosyncline of Britain. The trend of the Precambrian depositional area is not easy to reconstruct, and it is not certain that the Brittany region was continuous with the British depositional tract. In fact, Anglesey may have been over the site of an ultimate subduction zone (see Dewey, 1969) and the acid Dimetian intrusives of Pembrokeshire could be associated with movement (westward?) of the same (eastern?) plate. North–south trends of mid-Brioverian folds and late-Brioverian (Cadomian) structures could point to continuity of plate edges from Wales to Brittany.

One interesting correlation of the late Precambrian is of the Normandy and Brittany tillites (late Brioverian) with similar glacial deposits in the Dalradian of Scotland, Ireland, and Scandinavia. If the tillites are of low topographical origin, then difficult problems of global position arise. [On the other hand, mountain building could have uplifted Welsh and other "western approaches" areas sufficiently high to allow ice sheets to develop and move down to areas as far apart as Brittany and northern Britain (see Baker, 1971, p. 251).]

Dalradian deposition in northern Ireland and in Scotland transcends the Precambrian–Cambrian time boundary, so that our Precambrian depositional area (dare one say "ocean"?) persisted here. Cambrian deposits (largely rudites) are restricted in Brittany and are of shallow-shelf type (glauconitic sands plus shales) in the Welsh borderland. Land or shelf environments are therefore likely in the south of the western approaches and over southeast Wales and adjacent England. Westward and northwestward, however, marine deposition resumed after the late Precambrian earth movements, and there may even have been continuity of extrusive-sedimentary deposition in North Wales across the Precambrian–Cambrian junction. In Merionethshire and in County Wexford, thick Cambrian turbidites accumulated and the Lingula Flags of North Pembrokeshire indicate a similar environment. Between the Irish and Welsh trough, however, was the horstlike area of persistent uplift, noted by Crimes and Crossley (1968). Still further to the north was the Dalradian trough from Scotland to Ireland, with a northwestern border shelf on which the Durness Limestone of northwest Scotland was deposited. It is tempting to think of an expanding ocean during the Cambrian with Wales (on one side) and Scotland–northern Ireland (on the other) gradually moving further away from one another. Continued granitic injection at depth along an old subduction zone near Anglesey and southeast Ireland could account possibly for the horst uplifts in the southern Irish Sea area.

The more northeast–southwest trend of the Cambrian "ocean" over Britain, as compared with the more northerly orientation of the previous (Precambrian) depositional area, could be due to clockwise turning by the start of the Paleozoic Era.

Some contractions of this sea occurred before the start of Ordovician deposition in England, Wales, and southern Ireland. Moreover, quite widespread earth movements occurred in mid-Ordovician (pre-Caradocian) times in many areas and the Llandeilian stage is frequently unrepresented. These mid-Ordovician uplifts, with consequent erosion of deposits previously laid down, make it impossible to position the extent of earlier Ordovician deposition in areas like central England or southeast Wales.

The general picture of Ordovician times, however, is of a sea trough extending from Scandinavia across the British area and westward. Muds with cherts were deposited extensively and there was much vulcanicity, especially in southeast Ireland (Tramore), North Wales (Snowdonia and Cader Idris), Pembrokeshire, the Lake District, and the Southern Uplands of Scotland. Some of this vulcanicity was subaerial with many volcanic islands and archipelagoes, but there were also thick submarine volcanics, as at Tramore, Fishguard, and Ballantrae (Scotland). As Dewey (1969) has pointed out, there may have been contraction of the lower Paleozoic waterway with subduction zones in areas like the Southern Uplands and the Lake District. It is not as easy to reconstruct for the more southern areas. In Brittany, the Ordovician picture begins with the advancing beach deposition of the Grès Armoricain by a sea invading from the southeast. This advance could have spread into southwest England and the English Channel area. One must remember, of course, the intense shortening of areas like Brittany, the Channel, and southwest England by the Carboniferous orogeny. The point here is that these areas may well have been (depositionally) far south of their present position relative to areas like South Wales or southeast Ireland. An appreciable land area (dare one say a "plate"?) may have separated Wales from southwest England and Brittany. The violent volcanic outpourings in Pembrokeshire and southeast Ireland could be associated with subduction (perhaps as island arcs) somewhere south of Wales and Ireland. The continuing advance from the south of the sea into Brittany could then indicate a new sea-floor spreading somewhere south of that French area. This southern sea was to become the important area of deposition in the Devonian and Lower Carboniferous of central Europe.

In any attempt to correlate (through continental drift) from Europe across to North America, it is very important to remember the great crustal shortening, over the southernmost British Isles and northwest France, as a result of the Armorican (Hercynian) orogeny. This folding and thrusting may have virtually obliterated from present view an extensive area that was once somewhere between Wales and Brittany.

Over Brittany, Silurian deposition of fine muds indicate a fairly peaceful continuing of marine conditions with the southern ocean still expanding. Southwest England was again part of this southern (Brittany) picture, but important events were happening not far to the north. Widespread earth movements occurred at the start of Silurian times over South Wales, the Welsh borderland, and central England, with folding and uplift. Moreover, in Pembrokeshire there was appreciable vulcanicity. This vulcanicity extended into southwest Ireland (Dingle) and the Mendips. In Pembrokeshire, these Skomer Volcanics are situated very close to the zone of greatest uplift (though shortening by Armorican thrusting must again be remembered). This zone of uplift is bordered northward by thick developments of lower Paleozoics with, in the Lower Silurian, a great thickness of turbidites deposited in an elongate trough over Cardiganshire. As pointed out by Sanzen-Baker (1972, p. 2), there is a comparison here with the modern setting on the margin of a Japan Sea type basin. Some subduction over Pembrokeshire, and probably also over southwest Ireland, must be at least considered a possibility. Further north the main British waterway continued to contract with thick downwarped deposition over southern Scotland and the Lake District. Protected central belts of quiet mud deposition lay over Anglesey and East Ireland. By later Silurian times, the thick turbidites of West Wales had spread to northeast Wales (Denbigh-shire). The southeast edge of this trough ran along Shropshire with shallow shelf conditions over central England.

The end of the lower Paleozoic saw a great change in the geography of the British area, with the replacement of the important lower Paleozoic ocean by a rugged orogenic mountain belt. The greatest changes occurred over Scotland, northern England, North Wales, and Ireland. Granite intrusion took place into the roots of the orogen. South of the mountain belts, scattered rem-nants of the Late Silurian sea persisted at first over southeast Wales and central England, but these eventually became freshwater and merged into mudflats, estuaries, and river plains, now situated on the northern margin of the still growing "southern ocean." This southern sea spread marine muds, sands, cherts, and limestones into southwest England. Submarine vulcanicity also occurred. In North Devon one sees the periodic advances and retreats of the northern edge of this sea. Some orogenic unrest over this southern ocean is indicated by the thick flysch deposits of the Gramscatho unit in Cornwall, probably deposited off the flanks of a growing chain between Cornwall and Brittany, where there were thick sediment influxes from the north at times. The great influx of nonmarine sediments northward into a Kerry (southwest Ireland) trough is also a sign of some warping to the south of the British area.

This temporary contraction of the southern marine area was, however, followed by a renewed sea-floor spreading at the commencement of the Car-boniferous, with the southern waters spreading north over what had been fluviatile British basins. The marine invasion resulted in the deposition of the Lower Carboniferous carbonate sheets of England, Wales, and Ireland. Over southernmost Britain, however, growing unrest produced appreciable down-warping, especially over the "Carboniferous Slate" belt of southern Ireland. The radiometric evidence over Cornwall is indicative of early orogeny not too far south of present-day Cornubia. Serpentine intrusion might be associated with some developing subduction in the region of the Lizard belt. The trans-gression of the lowest Carboniferous (often deltaic) deposits across even Precambrian in Brittany, is also indicative of appreciable crustal unrest at this time. Later, Lower Carboniferous (Viséan) deposition was more extensive in Brittany and resembles that in southwest England (Culm), though it must be noted that some granitic intrusion occurred at depth in the French area.

At the beginning of the Upper Carboniferous (Namurian), uplifts over parts of Britain resulted in sand and mud spreading over the carbonate sheets and deltaic deposition predominated in South Wales and Ireland. Over south-west England, quiet mud deposition gave way to turbidites, spreading in mainly from uprising areas somewhere to the south. In Brittany there is no Namurian north of the Vendée, and limnic-type Stephanian Coal Measures were deposited in isolated basins eroded into a floor of much folded, granitized and migmatized Lower Carboniferous and earlier rocks. Much the same drastic change occurs in other French areas such as the Central massif and the Vosges. It is clear that the main Armorican orogeny had already reached its climax in these French areas, probably before the beginning of Westphalian times. It is significant that as Westphalian times progressed, molasse-type Pennant sandstones spread northward into South Wales and areas to the east. By this time the earth movements had probably spread to southwest England and the channel area.

The striking thing about the Carboniferous history of Western Europe is the rapid and drastic contrast between the widespread marine Lower Carboniferous and the post-orogenic higher Carboniferous, deposited in isolated areas on a deformed floor. Surely, rapid movements of crustal plates must have taken place during Middle Carboniferous times, the Lower Carboniferous marine deposits being caught up in folded rucks as a major southern (African?) plate moved north. Eventually, the rucking reached southwest Britain and some subduction is likely to have taken place somewhere near (or over) southwest England, with the emplacement of a granite magma.

A new continental phase ensued, embracing the Permian and Triassic periods, over northwest France and the British area. Some areas were under active erosion, others were low enough to receive screes, flood conglomerates, aeolian sands, and dusts. The environment was an arid or subarid one, winds were prevalently from the northeast, and daytime temperatures were in the region of 30–40°C. Important rivers developed by the early Triassic, draining the Brittany and channel uplands and carrying Ordovician quartzite pebbles northward to Devon and even to central England.

Late in the Triassic, the main (Rhaetian) marine advance into areas like Somerset and Devon and South Wales began. This transgression, heralding the more extensive Jurassic floodings of Britain and France, can probably be correlated with the opening of some part of the North Atlantic in the Late Triassic. The first water invasions may in fact have come from the southwest into the area that is now the Celtic Sea and the South Irish Sea and along the line of the present Bristol Channel into Gloucestershire and the Midlands (see Audley-Charles, 1970, p. 62).

It now seems likely (on the evidence of recent boreholes and recent sea-bottom studies around western Wales and in the Bristol Channel) that much more of Britain was drowned by Jurassic seas than was previously thought. Open-water conditions may well have existed over the Welsh, southwest England, and Celtic Sea areas. Important earth movements in Middle Cretaceous times, however, interrupted the continuity of Mesozoic deposition and important folding and fracturing occurred in areas such as the western English Channel, Bristol Channel, southwest England, and probably the South Irish Sea. Uplift and erosion removed the already deposited Mesozoic sheets over southwest England and at least some of them over Wales. Similar uplifts and erosion affected Brittany. These important Middle Cretaceous movements heralded a new and very major phase of Atlantic rifting and sea-floor spreading, followed by one of the most spectacular marine transgressions in British (and European) geological history—the marine advance that culminated in the deposition of the widely uniform Chalk.

Some cessation of Atlantic opening is indicated by the retreats of the end-Cretaceous waters, but renewed spreading took place during Paleogene times with extensive marine deposition over what are now the continental shelf areas of southwest Britain. Southeast England was also periodically drowned. It is tempting to postulate from (a) the Porcupine Sea Bight (off southwest Ireland) and (b) the north–south alignment of Eocene igneous centers down western Britain from western Scotland and northeast Ireland to Anglesey and Lundy, that Ireland and its western shelf was loath to drift eastward as much as eastern Britain and Europe, with crustal splitting along those north–south lines. When Spain had drifted south away from southwest Britain and northwest France is difficult to determine, but some separation seems to have taken place before the deposition of the Upper Cretaceous. Maybe it was this splitting that gave rise to the widespread crustal unrest in the Lower Cretaceous of southwest and southern Britain.

The Miocene earth movements folded the Mesozoic and Paleogene deposits of southern England and appreciably faulted them both on the present land and water areas of the western approaches. Over the shelf, however, Miocene sediments accumulated, though with slumping from fracturing, especially west and southwest of Brittany. Regional uplift of land areas in the Miocene was followed by some drowning of the land in the Pliocene. The extent and amount (in depth) of the drowning is a matter of controversy. The recovery of the land is marked by late Pliocene–Early Quaternary marine-cut platforms.

REFERENCES

Al-Shaikh, Z. D., 1969, Geophysical results from Caernarvon and Tremadoc Bays: *Nature*, v. 224: p. 897–899.

Al-Shaikh, Z. D., 1970, The geological structure of part of the central Irish Sea: *Geophys. J. Astr. Soc.*

Audley-Charles, M. G., 1970, Stratigraphical correlation of the Triassic rocks of the British Isles: *Quart. Journ. Geol. Soc. Lond.*, v. 126, p. 19–89.

Baker, J. W., 1969, Correlation problems of metamorphosed Precambrian rocks in Wales and southeast Ireland: *Geol. Mag.*, v. 106, p. 249–259.

Baker, J. W., 1971, The Proterozoic history of southern Britain: *Proc. Geol. Assoc.*, v. 82, p. 249–266.

Bamford, S. A. D., Blundell, D. J., and Davey, F. J., 1970, Geology of the South Irish Sea: *Proc. Geol. Soc. Lond.*, n. 1662, p. 79–80.

Barrois, C., 1900, *Int. Geol. Congr.* 8 Guide no. 7, Bretagne, p. 1–32.

Bishop, A. C., Bradshaw, J. D., Renouf, J. T., and Taylor, R. T., 1969, The stratigraphy and structure of part of west Finistère, France: *Quart. Journ. Geol. Soc. Lond.*, v. 124, p. 309–348.

Blundell, D. J., Davey, F. J., and Graves, L. J., 1970, Surveys over the South Irish Sea and Nymph bank: *Geol. Soc. Lond.*, Circular 161.

Bott, M. H. P., Day, A. A., and Masson-Smith, D., 1958, The geological interpretation of gravity magnetic surveys in Devon and Cornwall: *Phil. Trans. Roy. Soc. A*, v. 251, p. 161–191.

Brindley, J. C., and Gill, W. D., 1958, Summer field meeting in South Ireland: *Proc. Geol. Assoc.*, v. 69, p. 244–261.

Bullerwell, W., and McQuillin, R., 1968, Preliminary report on a seismic reflection survey in the southern Irish Sea, July 1968: Report No. 69/2, Inst. Geol. Sci., p. 1–7.

Capewell, J. G., 1957, The Stratigraphy and structure of the country around Sneem, Co. Kerry: *Proc. Roy. Irish Acad.*, v. 58B, p. 167–183.

Clarke, R. H., and Bailey, R. J., 1970, Seismic reflection survey of the Porcupine Seabight: *Proc. Geol. Soc. Lond.*, n. 1662, p. 85–87.

Coe, K., ed., 1962, Some aspects of the Variscan Fold Belt: Manchester University Press, p. 6–163.

Coe, K., 1969, The Geology of the minor instrusions of West Cork, Ireland: *Proc. Geol. Assoc.*, v. 80, p. 441–457.

Coe, K. and Selwood, E. B., 1963, The stratigraphy and structure of part of the Beara peninsula, Co. Cork: *Proc. Roy. Irish Acad.*, v. 63B, p. 33–59.

Coe, K., and Selwood, E. B., 1964, Some features of folding and faulting in the Sheep's Head peninsula, Co. Cork, Ireland: *Sc. Proc. Roy. Dublin Soc.*, v. A2, p. 29–41.

Coe, K. and Selwood, E. B., 1968, The upper Palaeozoic stratigraphy of West Cork and parts of South Kerry: *Proc. Roy. Irish Acad.*, v. 66B, p. 113–131.

Cole, G. A. J., 1922, Some features of the Armorican (Hercynian) folding in southern Ireland: Rept. Int. Geol. Congress, X111 (Belgium), p. 423–438.

Cornwell, J. D., 1971, Geophysics of the Bristol Channel area: *Proc. Geol. Soc. Lond.*, n. 1664, p. 286–289.

Crimes, T. P. and Crossley, J. D., 1968, The stratigraphy, sedimentology, ichnology and structure of the lower Palaeozoic rocks of part of northeastern County Wexford: *Proc. Roy. Irish Acad.*, v. 67B, p. 185–215.

Curry, D., Gray, F., Hamilton, D., and Smith, A. J., 1967, Upper Chalk from the seabed, south of Cork, Eire: *Proc. Geol. Soc. Lond.*, n. 1640, p. 134–136.

Curry, D., Hamilton, D., and Smith, A. J., 1970, Geological and shallow subsurface geophysical investigations in the western approaches to the English Channel: Report No. 70/3, Inst. Geol. Sci., p. 1–12.

Dawson-Grove, G. E., 1955, Analysis of minor structures near Ardmore, County Wexford, Ireland: *Quart. Journ. Geol. Soc. Lond.*, v. 111, p. 1–21.

Day, G. A., and Williams, G. A., 1970, Gravity surveys in the Celtic Sea: *Proc. Geol. Soc. Lond.*, n. 1662, p. 84–85.

Dearman, W. R., 1963, Wrench faulting in Cornwall and South Devon: *Proc. Geol. Assoc.*, v. 74, p. 265–287.

Dearman, W. R. 1969, An outline of the structural geology of Cornwall: *Proc. Geol. Soc. Lond.*, n. 1654, p. 33–39.

Dearman, W. R., 1970, Some aspects of the tectonic evolution of southwest England: *Proc. Geol. Assoc.*, v. 81, p. 483–491.

Dearman, W. R., Leveridge, B. E., and Taylor, R. G., 1970, Structural sequences and the ages of slates and phyllites from southwest England: *Proc. Geol. Soc. Lond.*, n. 1654, p. 41–45.

Dewey, J. F., 1969, Evolution of the Appalachian–Caledonian orogen: *Nature*, v. 222, p. 124–129.

Dixon, E. E. L., 1921, The Geology of the South Wales Coalfield, Part X111: The county around Pembroke and Tenby: Mem. Geol. Surv., U.K.

Dobson, M. R., 1971, A Review of the economic potential of part of the Welsh continental shelf: "Mineral exploitation and economic geology": Univ. of Wales Intercollegiate Colloquium, Univ. Coll. Cardiff, p. 49–56.

Dodson, M. H., 1961, Isotopic ages from the Lizard peninsula, South Cornwall: *Abstr. Proc. Geol. Soc. Lond.*, n. 1591, p. 133.

Dodson, M. H. and Rex, D. C., 1971, Potassium–argon ages of slates and phyllites from southwest England: *Quart. Journ. Geol. Soc. Lond.*, v. 126, p. 465–499.

Donovan, D. T., Lloyd, A. J., and Stride, A. H., 1971, Geology of the Bristol Channel: *Proc. Geol. Soc. Lond.*, n. 1664, p. 294–295.

Gardiner, P. R. R., 1970, Regional fold structures in the lower Palaeozoics of southeast Ireland: *Bull. Geol. Surv. Ireland*, n. 1, p. 47–51.

Gauss, G. A. and House, M. R., 1971, The Devonian successions in the Padstow area, North Cornwall: Geol. Soc. Lond. Circular 166, p. 4–5.

George, T. N., 1940, The Structure of Gower: *Quart. Journ. Geol. Soc. Lond.*, v. 96, p. 131–198.

Griffiths, D. H., King, R. F., and Wilson, G. D. V., 1961, Geophysical investigations in Tremadoc Bay, North Wales: *Quart. Journ. Geol. Soc. Lond.*, v. 117, p. 171–191.

Hamilton, D. and Blundell, D. J., 1971, Submarine geology of the approaches to the Bristol Channel: *Proc. Geol. Soc. Lond.*, n. 1664, p. 297–300.

Hendriks, E. M. L., 1959, Summary of present views on the structure of Devon and Cornwall: *Geol. Mag.*, v. 96, p. 253–257.

Hill, M. N. and King, W. B. R., 1953, Seismic prospecting in the English Channel and its geological interpretation: *Quart. Journ. Geol. Soc. Lond.*, v. 109, p. 1–19.

Hill, M. N. and Vine, F. J., 1965, A Preliminary magnetic survey of the western approaches to the English Channel: *Quart. Journ. Geol. Soc. Lond.*, v. 121, p. 463–475.

Hodson, F. and Lewarne, G. C., 1961, A mid-carboniferous (Naumurian) basin in parts of the counties of Limerick and Clare, Ireland: *Quart. Journ. Geol. Soc. Lond.*, v. 117, p. 307–333.

Holland, C. H., 1969, The Irish counterpart of the Silurian of Newfoundland: *Amer. Assoc. Petrol. Geol. Memoir*, v. 12, p. 298–308.

Horne, R. R., 1970, A Preliminary re-interpretation of the Devonian palaeogeography of western County Kerry: *Bull. Geol. Surv. Ireland*, n. 1, p. 53–60.

Jones, O. T., 1938, Evolution of a Geosyncline: *Quart. Journ. Geol. Soc. Lond.*, v. 94, p. lx–cx.

Jones, O. T., 1951, The drainage system of Wales and the adjacent regions: *Quart. Journ. Geol. Soc. Lond.*, v. 107, p. 201–225.

Jones, O. T., 1956, The Geological evolution of Wales and the adjacent regions: *Quart. Journ. Geol. Soc. Lond.*, v. 111, p. 323–351.

King, W. B. R., 1954, The geological history of the English Channel: *Quart. Journ. Geol. Soc. Lond.*, v. 110, p. 77–101.

Lambert, J. L. M., 1962, A reinterpretation of part of the Meneage crush zone: *Proc. Ussher Soc.*, v. 1, p. 24–25.

Lambert, J. L. M., 1964, The unstratified sedimentary rocks of the Meneage area, Cornwall: Circular 118, Proc. Geol. Soc. Lond., p. 3.

McKerrow, W. S., 1962, The chronology of Caledonian folding in the British Isles: *Proc. Nat. Acad. Sc.*, v. 68, p. 1905–1913.

Miller, J. A. and Green, D. H., 1961, Age determinations of rocks in the Lizard (Cornwall) area: *Nature*, v. 192, p. 1175–1176.

Owen, T. R., 1971a, The relationship of Carboniferous sedimentation to structure in South Wales: *Compte Rendu de Congres Intern. Strat. Geol. Carbonif. Sheffield 1967*, v. 111, p. 1305–1316.

Owen, T. R., 1971b, The structural evolution of the Bristol Channel: *Proc. Geol. Soc. Lond.*, n. 1664, p. 289–294.

Owen, T. R., Bloxam, T. W., Jones, D. G., Walmsley, V. G., and Williams, B. P. J., 1971, Summer (1968) field meeting in Pembrokeshire, South Wales.: *Proc. Geol. Assoc.*, v. 82, p. 17–60.

Philcox, M. E., 1963, Compartment deformation near Buttevant, Co. Cork, Ireland and its relation to the Variscan thrust front: *Sc. Proc. Royal Dublin Soc.*, n. A2, p. 1–11.

Pruvost, P., 1949, Les mers et les terres de Bretagne aux temps palaeozoiques: *Annls. Hebert Hang.*, v. 7, p. 345–362.

Ramsbottom, W. H. C., 1970, Carboniferous faunas and palaeogeography of the southwest England region: *Proc. Ussher Soc.*, v. 2, Pt. 3, p. 144–157.

Reading, H. G., 1965, Recent finds in the Upper Carboniferous of southwest England and their significance: *Nature*, v. 208, p. 745–747.

Sanzen-Baker, I., 1972, Stratigraphical relationships and sedimentary environments of the Silurian—Early Old Red Sandstone of Pembrokeshire: Geologists Association Circular No. 741.

Stewart, F. H. and Johnston, M. R. W., eds., 1963, The British Caledonides: Oliver and Boyd, Edinburgh, 250 p.

Stride, A. H., Curray, J. R., Moore, D. G., and Belderson, R. H., 1969, Marine geology of the Atlantic continental margin of Europe: *Phil. Trans. Royal Soc.*, Series A, v. 264, p. 31–75.

Van Straaten, L. M. J. U., ed., 1964, Developments in Sedimentology: Elsevier, Amsterdam, 464 p.

Walsh, P. T., 1966, Cretaceous outliers in southwest Ireland and their implications for Cretaceous palaeogeography: *Quart. Journ. Geol. Soc. Lond.*, v. 122, p. 63–84.

Walsh, P. T., 1968, The Old Red Sandstone west of Killarney, Co. Kerry, Ireland: *Proc. Roy. Irish Acad.*, v. 66B, p. 9–26.

Wilson, H. E., 1968, Geology of the Irish Sea area: *Irish Nat. Journ.*, v. 16, p. 102–105.

Wood, A. and Woodland, A. W., 1968, Borehole at Mochras, west of Llanbedr, Merioneth-shire: *Nature*, v. 219, n. 5161, p. 1352–1354.

Ziegler, A. M., McKerrow, W. S., Burne, R. V., and Baker, P. E., 1969, Correlation and environmental setting of the Skomer Volcanic Group, Pembrokeshire: *Proc. Geol. Assoc.*, v. 80, p. 409–439.

Chapter 9

THE GEOLOGY AND SEDIMENTATION HISTORY
OF THE BAY OF BISCAY

M. Vigneaux

Institut de Géologie du Bassin d'Aquitaine
Université de Bordeaux
Talence, France

I. INTRODUCTION

The Bay of Biscay (or Gulf of Gascony) is a large wedge-shaped re-entrant of
the Atlantic Ocean, bordered by dissimilar continental areas. It is floored by
an abyssal plain interrupted by numerous abyssal hills with a broad continental
shelf to the east and a much narrower shelf to the south. There is a long record
of varied geological investigation which serves to demonstrate the complex
pattern of hydrology, geomorphology, and sedimentation within the bay.
It is only in recent years, however, that the region has been intensively studied
using a variety of different techniques, and while extensive data are available
it would be premature to attempt a comprehensive survey at this time. Instead,
we plan to review the geological framework of the bay, to examine the available geophysical data, and to accent certain aspects of the geomorphology
and sedimentation pattern of the littoral, continental shelf and bathyal regions.

II. THE CONTINENTAL MARGINS AND GEOLOGICAL FRAMEWORK OF THE BAY OF BISCAY

To discuss the geological framework of the Bay of Biscay in detail would require a complete review of the regional history from the Precambrian to the present day. Treatment on such a scale is clearly beyond the scope of this review. It is possible, however, to define a number of major geological units and investigate their role in the pattern of sedimentation which unfolds during the Mesozoic and Cenozoic. This should provide the limits within which the development of this part of the ancient Atlantic Ocean may be studied.

Examination of the relevant geological maps of France and Spain allows us to distinguish four geological units on the basis of their geological history. These units are (Fig. 1):

1. The Armorican massif
2. The northwestern part of Spain (Galicia and Asturia)
3. The Pyrenean–Cantabrian region
4. The Aquitaine basin

The first two units are symmetrically disposed about the axis of the bay forming the sides of a wedge. They have many similarities and may share a common geological history. The latter two, which border one another, comprise the wedge and have quite different geological histories.

For many years, writers have pointed out the similarities in deep structure of the Armorican and Galician regions. In light of current views concerning the opening of the Bay of Biscay, it appears possible that the two areas were continuous at the end of the Hercynian orogeny (Bard *et al.*, 1970) forming a single Hercynian Ibero-Armorican arc (Cogné, 1970). Tectonic geologists have speculated that subsequent to the Hercynian orogeny, the Iberian peninsula was considerably displaced with respect to the European landmass. The mechanics of such a displacement are far from simple and several hypotheses have been proposed involving rotational or translational movements (Mattauer and Seguret, 1970; Le Pichon *et al.*, 1970). Combining rotational and translational motion, Bard *et al.* (1970) envisioned an opening of the Bay of Biscay compatible with what is known of the Hercynian and Pyrenean orogenies. According to this hypothesis, the geological history of these marginal blocks was essentially determined prior to the formation of the embryonic Bay of Biscay. It would thus seem that from the Jurassic to the present time only relatively minor land–sea interactions occurred. The relatively minor Eocene transgression in southern Brittany (Durand, 1962), for example, did little to modify the Armorican massif which acquired its form at the end of the Paleozoic.

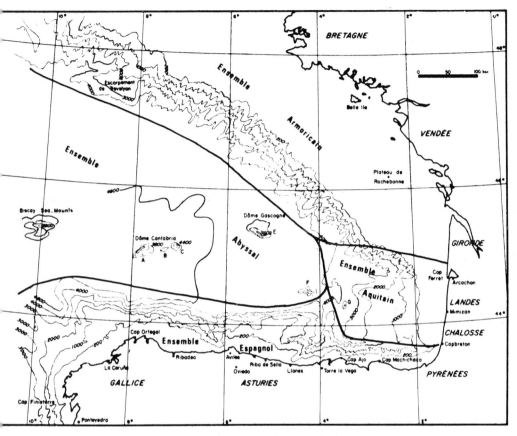

Fig. 1. Bathymetric map of the Bay of Biscay.

It would be tempting to consider the Armorican massif of Brittany and the massif of northern Spain as stable blocks which funneled marine transgressions eastward into the subsiding, more flexible zones which were the scene of frequent transgressions and regressions. In fact, the structure of the first of these "flexible" zones, the Pyrenean–Cantabrian region, is complex and is the result of a series of events beginning in the Late Jurassic. In particular, as Feuillée and Rat (1970) have shown, the role of the Hercynian basement is important, as basement mobility controlled both the paleogeography and the major structural trend in the Pyrenees from Upper Jurassic to Miocene times.

Within the Pyrenean–Cantabrian region, Feuillée and Rat (1970) established three subdivisions:

(a) The Basque Arc, a region composed of Paleozoic rocks extended westward by folds as far as Bilbao.

(b) The Navarre–Cantabrian zone, a system of synclines with cores of Cretaceous and Tertiary rocks whose center lies in Vitoria province.

(c) The peri-Asturian zone, lying in the provinces of Santander and northern Burgos, where the effects of the Asturian Paleozoic basement are clearly evident despite cover of Jurassic and Lower Cretaceous deposits.

Until the Upper Jurassic, the paleogeography of the Pyrenean–Cantabrian region was closely related to the Mediterranean. Only after general uplift in Upper Jurassic times did a new sedimentary regime become established revealing for the first time evidence of the existence of the Bay of Biscay. The maps of Feuillée and Rat (1970) clearly illustrate the limited extent of these basins during the course of the Lower Cretaceous. In contrast, transgression during the Upper Cretaceous was more extensive, and a marine link across the Ebro ridge established a connection between the bay and the Mediterranean. With the regression at the end of the Cretaceous this link was broken, and only the present coastal regions were submerged. Thereafter, Tertiary transgressions covered only restricted areas, and the last marine traces are of Oligocene age with deposits comparable to those in the Biarritz section.

The remaining region, the Aquitaine basin, presents an entirely Atlantic association, which, as its geological history shows, was acquired from Upper Cretaceous times. As the Aquitaine basin has been described in numerous publications (Vigneaux, 1962; Winnock, 1970), only the essential elements which characterize and relate it to the Bay of Biscay will be given here.

The generally triangular Aquitaine basin has a marked asymmetry with a broad eastern margin on which are preserved Tertiary deposits, and a narrow steep southern margin in the coastal region of Spain. Compared with the other parts of the Bay of Biscay, the Aquitaine basin gives an impression of relative tectonic simplicity. The impression is false since the majority of tectonic features are masked by the Tertiary cover which is only rarely affected by anticlinal folds. It is this Tertiary sequence, several hundred meters thick in the north and thickening to over 1000 m in the south, which gives the basin one of its distinctive features.

After the Upper Cretaceous transgression (Bonnard et al., 1958) when the waters of the Atlantic washed against the Central massif, the Aquitaine basin was the site of transgressions and regressions of decreasing amplitude during the Cenozoic, the Oligocene is regressive with respect to the Eocene, and the Miocene with respect to the Oligocene. The last Pliocene transgression closely approached the present coastline (Moyes, 1965). During the Quaternary the present aspect of the coast was established. The history of the Aquitaine basin is thus one of shrinking toward the southwest, and it would then appear that the Aquitaine basin may be regarded as an emergent part of the Bay of Biscay and thus of considerable importance in providing historical evidence of the behavior of the latter. The sedimentary zones in the northern part of the basin can be traced out on to the Armorican–Aquitaine continental shelf without break or sudden change in sedimentary type.

III. GEOMORPHOLOGY OF THE BAY OF BISCAY

Using the bathymetric maps of the Institut Scientifique et Technique des Peches Maritimes, Berthois, Brenot, and Ailloud (1965) attempted a preliminary interpretation of the morphology of the Bay of Biscay. They recognized three units (Fig. 1), a northern or Armorican unit lying to the north of the Arcachon basin, a central unit extending from the Arcachon basin to the Gouf de Cap Breton, the Aquitaine unit, and a third Asturian–Cantabrian unit comprising the Spanish coastal region. The scheme was completed by the subsequent addition of a fourth unit comprising the abyssal plains (Berthois and Brenot, 1966) and the whole reinvestigated by Prud'homme and Vigneaux (1970). These studies did not include the continental shelf in the absence of satisfactory data. This gap has been filled through the work of Vanney (1969) and of the Geological Institute of the Basin of Aquitaine (Vigneaux, 1971) insofar as the first two units are concerned.

A. The Armorican Unit

The continental platform is broad, from 65 to 85 miles wide dipping regularly seaward with but a single important morphological anomaly (the Rochebonne seamount lat. 46°10′ N, long. 2°15′ W). The general lack of submarine relief (Vanney, 1969) has forced geomorphologists to use high-resolution bathymetry with 5-m contour intervals and frequent soundings (Vanney, 1969; Caralp et al., 1971). This detail has allowed students of the area to locate the principal sea level stands of the Quaternary. Over the whole shelf Pinot (1968) and Naudin (1971) recognize two shorelines, one at −100 to −85 m and the other at −50 to −40 m, which have been confirmed by sedimentological and paleontological work (Caralp et al., 1971; Lapierre, 1969).

Naudin (1971) carried out a morphostructural study based on the network of grooves which seam the surface of the continental shelf. From these grooves he presented a hypothetical representation of deep structure on the supposition (Naudin and Prud'homme, 1971) that all structural and lithological anomalies (synclines, anticlines, faults, joints, indurated beds) have a direct and visible effect on the pattern and distribution of grooves. He concluded that the whole

region up to lat. 45°10′ N had a marked similarity to the structural disposition
of the Armorican unit as a whole in the sense that a relatively thin sedimentary
sequence with more or less monoclinal form rested on basement at shallow
depths and was characterized by numerous discontinuities. These conclusions
are currently being tested by other techniques (Caralp *et al.*, 1971; Vanney
et al., 1970; Montadert *et al.*, 1970).

The continental slope is appreciable, essentially continuous from summit
(−200 m) to base (−4600 m) and cut by numerous canyons. Prud'homme and
Vigneaux (1970) carried out an analysis of the pattern of surface grooves to
demonstrate a vertical evolution of the continental slope and to divide the
Armorican unit into three subdivisions. The structural map (Fig. 2) they
propose can be used as a basis of comparison against the geophysical results
(Fig. 3) (Vigneaux, Naudin, and Prud'homme, 1971). Many similarities are
apparent, in particular the alignment of paleorelief at the base of the continental slope (Montadert *et al.*, 1970; Le Mouel and Le Borgne, 1970).

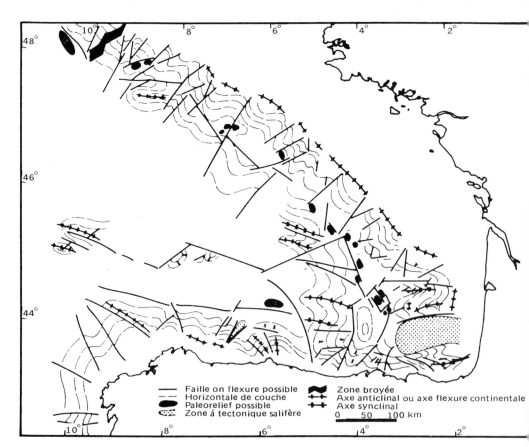

Fig. 2. Structural map of the Bay of Biscay.

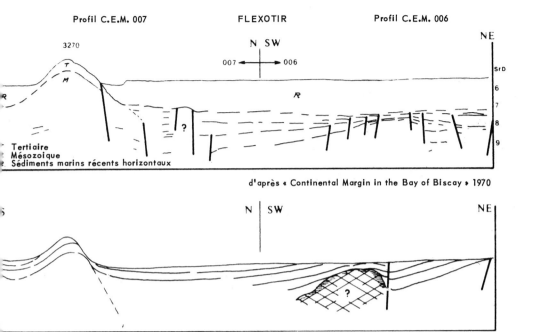

Fig. 3. Flexotir profile and interpretation (CEM 007 and 006) (bathymetry from Berthois and Brenot).

B. The Aquitaine Unit

The continental shelf is considerably reduced in width south of lat. 45°10′ N. At the same time its lower limit is defined by the 150-m isobath. It is a region of low relief which makes morphological studies difficult. Naudin (1971) has shown that the former shorelines established in the Armorican unit undergo important changes here. It appears the farther south one goes, the less clearly the two marine phases can be distinguished and eventually at about lat. 44°10′ N they appear to merge into a single line at a depth of about 80 m. The deep structure of the shelf is masked by a clearly defined succession of anticlines and synclines, which can be used to divide the unit. The morphological interpretation of the Aquitaine continental shelf is confirmed by the results of oil exploration (Dardel and Rosset, 1970; Winnock, 1970; Valery et al., 1970) and by isopach maps of the base of Tertiary (Pratviel, 1967).

The Aquitaine continental slope has morphological complexities which distinguish it from the classic pattern. It is divided into three zones by Berthois et al. (1965) as follows:

1. A restricted but sharp upper slope, oriented north–south, from depths of −180 to −1000 m.

2. A large platform between −1000 and −2000 m.

3. A lower slope extending from −2000 m to −4000 m.

There are two important submarine canyons, the Cap-Ferret canyon to the north and the Gouf de Cap Breton in the south. The latter is deeply incised into the continental shelf and has an essentially east–west trend, and beyond about long. 3°30′ W it is continued by the north–south-trending Santander canyon. The canyons, the occurrence of thrust faults, and halokinetics are characteristics of the Aquitaine continental slope.

On the basis of an analysis of the morphology, Berthois et al. (1965) hypothesized that the whole of this part of the Bay of Biscay has sunk or has been tilted between two major discontinuities toward the abyssal plains by means of a series of north–south fractures, the faults probably being reactivated in recent times. Prud'homme and Vigneaux (1970) described fractures and folds which suggest possible torsional and hinge-type movements and lead the authors to suggest a relative rotation and easterly translation of the Iberian peninsula with respect to the rest of Europe. It is interesting that these hypotheses based upon the analysis of bathymetric data accord with the general interpretation of the structural history of the Bay of Biscay (Muraour, 1970; Mattauer and Seguret, 1970; Valery et al., 1970).

C. The Asturian–Cantabrian Unit

The general character of the Spanish coast is the narrow, continental shelf, which passes abruptly into a regular continental slope providing a rapid transition into the abyssal plain (−4600 m). The morphology is complex with numerous shoal areas and canyons. According to Berthois et al. (1965), three subdivisions can be made as follows: an eastern Cantabrian zone, a central Asturian zone, and a western Galician zone.

The eastern zone appears to have its morphology controlled by structures representing the continuation of the axial zone of the Pyrenees, while in the Asturian zone there is a passage from crustal mobility into one of basement stability. In the Galician zone the shelf is broader and deeper, and it is possible to trace out onto it structures found in the continent (Boillot et al. 1970; Berthois et al., 1965), e.g., the position of the canyons seem controlled by fractures seen on the continent (Prud'homme, 1970).

D. The Abyssal Plains Unit

The maps of Berthois and Brenot (1966) illustrate that despite the general morphological monotony of the abyssal plains, there are some morphological features of note such as the Biscay seamounts, the Gascony and Cantabria domes, and a number of protuberances lettered B, C, D, F, and G. In the deepest part (−4800 m) horizontal bedded sediments obscure most topographic relief, whereas in the peripheral zone at depths −4800 to −4300 m the reduced sediment thickness exposes an undulating surface with minor relief. A piedmont zone built up of sediments occurs where the canyons open on to the abyssal plains.

Despite the problems of interpreting the relief, Prud'homme and Vigneaux (1970–1971) point out that the majority of seamounts correspond to horst blocks defined by faults trending at 110° and at 40°. These fracture systems appear to be responsible for the broad structural pattern of this part of the bay, and faults with this trend can be traced from the Cap Ferret canyon to the Biscay seamounts. The line of fracture is well defined in the peripheral zone (Berthois and Brenot, 1966) and is found in geophysical exploration (Frappa et al., 1970). This mid-Biscayan fault alignment forms a conspicuous separation between the Armorican and Spanish areas.

E. Conclusions

In this rapid review of the results of geomorphologic studies in the Bay of Biscay it is clear that even if the same pattern constantly reappears, there are significant variations. The significance of the variations raises important problems. Although there may be a greater or lesser displacement between the geomorphological feature and its underlying cause, geophysical exploration shows that the topographic anomalies remain sensibly parallel to deep structural trends. There is, however, seldom any relationship between the magnitude of cause and its effect, that is, there is seldom any morphological argument for distinguishing between a fault and a diaclasis.

Finally, there seems to be a certain uniqueness in the data analyzed and consequently in the phenomena disclosed by analysis. Thus the study of relief furnishes much information about the superficial part of the ocean floor, while the analysis of the network of grooves provides information on the deeper, older regions. Thus the geomorphological study, while it permits a rapid reconstruction of the history of a basin in the horizontal sense has severe limitations on what may be interpreted in a vertical sense insofar as interpretation of the anomalies is concerned. The verification of certain of the conclusions drawn from the morphological work, however, lends some credence to this method of investigation.

One of the most important features in the structural pattern of the Bay of Biscay is the existence of the Cap Ferret fault in the general region of the Arcachon basin. Its effects are apparent in most methods of investigation for it marks a discontinuity in the continental platform from all points of view. It occurs at the point where the shelf is narrowest and can be recognized in surface geology not only by structural features but also because of its effects on hydrology. It forms the line of separation between the homogeneous Hercynian or neo-Hercynian region to the north and the zone to the south whose structures are controlled by the Pyrenean chain, and is apparent both onshore and in the continental platform and slope (Prud'homme, 1972).

The fault is thus a regional feature separating the earlier Hercynian ensemble from the Pyrenean–Cantabrian region dominated by recent, tectonic events.

IV. THE CONTRIBUTION OF GEOPHYSICS TO THE STUDY OF THE BAY OF BISCAY

A variety of geophysical techniques have been used in the Bay of Biscay. Not all bathymetric zones of the bay have been studied to the same extent; in some there have been detailed reconnaissance surveys; others have only been cursorily examined. Chronologically the first to be used, seismic techniques are certainly the most widely used, and for that reason will be discussed first.

A. Seismic Reflection Results

The continental shelf north of La Rochelle is more than 200 km wide and dips shallowly toward the continental slope. It has been studied in detail here by Bouysse and Horn (1968) and Vanney *et al.* (1970). They show a relatively thin cover of post-Paleozoic sedimentary rocks thickening southward. The basement dips seaward and at the margin of the continental shelf is affected by important faults with a south Armorican trend. Toward the center of the shelf one of the NNW–SSE faults is itself affected by transverse displacement.

The sedimentary prism has an isoclinal form, with progressively younger beds outcropping in a traverse seaward. The sedimentary sequence begins with Paleogene formations to the north of Noirmoutier, whereas to the south Mesozoic beds have been recognized in continuous profiling. This suggests a tilting seaward of the whole shelf, which becomes more pronounced toward the continental slope.

Folds with a Variscan trend which occur on land have not as yet been found in seismic profiles of the shelf region.

To the south of La Rochelle the same fundamental pattern is continued although there are some distinctive features. The presence of the shoal area of Rochebonne results from a basement uplift along faults with a south Armorican trend. In the region of Pertuis Vendéens, Mesozoic prolongation of the northern margin of the Bay of Biscay can be recognized (Martin *et al.*, 1968, 1971). South of the Gironde estuary only Tertiary beds are found and at the margin of the continental slope, there is a zone where the beds are strongly affected by deformation following a northwest–southeast trend. Seaward of the Landes area the Tertiary deposits reach an appreciable thickness.

On the continental shelf of the Spanish coast which is here poorly developed, continuous sparker profiles have been recorded (Boillot *et al.*, 1970). The prolongation of the pyrenean chain as far as Cape Ortegal (long. 9° W) was thus demonstrated. Furthermore, the facies and orientation of the folds of the Lower Cretaceous on the shelf are the same as those in the Pyrenees. The Tertiary beds, however, trend obliquely with respect to the Lower Cretaceous, favoring the existence of a Cantabrian fault. West of Cape Ortegal the continental shelf is cut out of the crystalline basement rocks of Galicia with only a thin sedimentary veneer.

There have been numerous reconnaissance profiles of the abyssal plains of the Bay of Biscay using a variety of methods by several organizations. The first studies using an air gun or sparker date from about a decade ago and focused attention on the complexity of the structures in the abyssal plains and continental slope. Subsequently, the development of a more powerful seismic source, the "flexotir" by the Institut Français du Petrole, marked a distinct advance in the knowledge of the geological structure in these areas (cf. Montadert *et al.*, 1970; Le Pichon *et al.*, 1970; Malzac, 1970; Muraour *et al.*, 1970). These results, presented at a conference on the structural history of the Bay of Biscay, established a number of units and lead to a provisional structural map.

Four distinct formations were recognized as follows:

Formation 1: Post-Eocene deposits, primarily turbidites which cover the abyssal plain and pass laterally into formations of pelagic origin.

Formation 2: Lower Tertiary–Upper Cretaceous rocks overlain with slight disconformity by formation 1, and often affected by faults restricted to the formation.

Formation 3: Top apparently Lower Cretaceous, bounded by discordances above and below. It fills depressions in formation 4.

Formation 4: The seismic basement. It is probably composed of rocks of variable lithology. In some places a farther horizon intervenes between it and formation 3. This additional horizon is considered to consist of evaporites.

Structurally, the seismic profiles indicate a central region where the basaltic basement is at shallow depths. It is bordered to the north and south

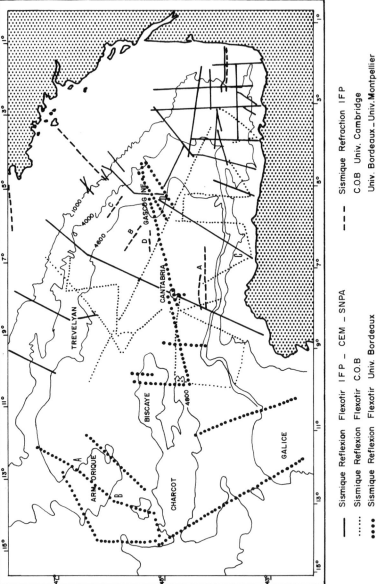

Fig. 4. Location map of the seismic profiles.

Sismique Reflexion Flexotir IFP – CEM – SNPA   Sismique Refraction IFP

Sismique Reflexion Flexotir C.O.B   C.O.B Univ. Cambridge

Sismique Reflexion Flexotir Univ. Bordeaux   Univ. Bordeaux – Univ. Montpellier

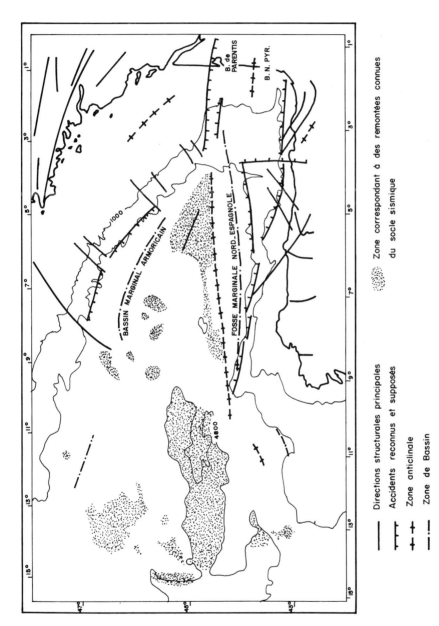

Fig. 5. Structural section of the Bay of Biscay.

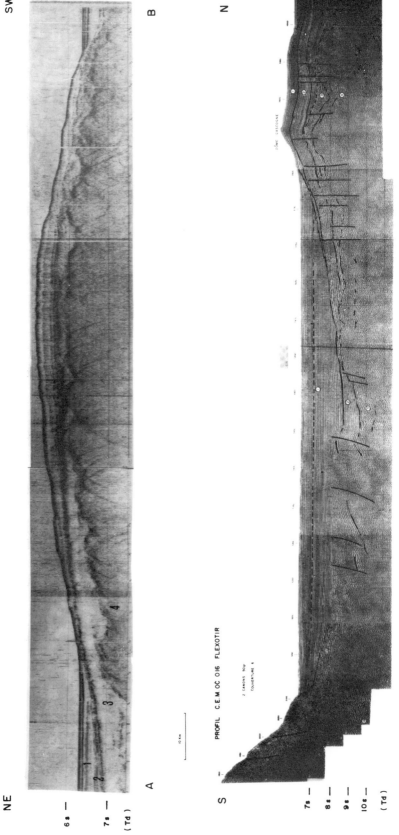

Fig. 6a. (AB) Seismic profile of the Amorican mountains; (CD) seismic profile from the Spanish continental slope to the Gascony dome.

by troughs, the Armorican and north Spanish troughs (Fig. 4 and 5). The basement of the troughs appears to be formed by older sedimentary horizons and not by basalt. Within the troughs diapiric structures are observed.

The transition from the central high into the north Spanish trough is complex. Traversing from north to south there is a zone of high relief formed by the Biscay seamounts, the Cantabria and Gascony domes, and the South Gascony ridge whose northern continuation reaches to the sill which limits the Parentis basin. South of the ridge the beds thicken and are disturbed. The cause of the disturbance is not known. It has recently been shown that the seamounts at the western limits of the Bay of Biscay (Armorican mount) which rest upon an unknown but presumably basaltic basement, contain all the horizons recognized elsewhere (Fig. 6a).

B. Seismic Refraction Results

This technique has been used relatively infrequently in the Bay of Biscay, but the few studies which have been carried out are extremely important. There are results both from the abyssal plains (Ewing and Zauere, 1964; Bacon et al., 1969–1970; Damotte et al., 1970; Sichler et al., 1970) and from the continental shelf region (Frappa and Horn, 1971; Martin et al., 1968; Muraour et al., 1971). With the use of seismic velocity data, depths to the various horizons can be calculated so that the refraction profiles provide a depth scale for the reflection data. In the center of the bay the Mohorovičic discontinuity lies at a depth of 12 km and thus oceanic crust exists in this region. Repeated profiles in the same area (lat. 45° N, long. 6° W) by the I.F.P. and the University of Cambridge lead to slightly different results with the M discontinuity at 10 km overlain by 5.5 km of crust and a sedimentary cover of the order of 2.2 km thick.

C. Paleomagnetic Results

At the present time paleomagnetic results are relatively few in number (Girdler, 1965; Van Dongen, 1967; Van der Voo, 1967, 1969). They are derived from rocks ranging from Lower Silurian to Eocene in age collected from a number of sites on the Iberian peninsula. Comparable results are available from the stable parts of Europe (Zijderveld, 1967; Zijderveld and Van der Voo, 1970) from rocks of Upper Carboniferous–Lower Permian and of Cretaceous–Eocene age.

The results from the rocks of Upper Carboniferous–Lower Permian age show differences of the order of 30° in magnetic declination, whereas the Upper Cretaceous and Eocene results are essentially the same. The standard interpretation proposes a displacement of the Iberian peninsula with respect to Europe, presumably by a combination of translational and rotational motion some time during the time interval Triassic to Eocene.

D. Gravimetric Results

Several different teams have carried out gravimetric work in the Bay of Biscay (Bacon and Gray, 1970; Day and Williams, 1970; Sibuet and Le Pichon, 1970) and a free-air anomaly map has been produced for the region between long. 3–12° W. The quantitative examination of this map permits four distinct anomaly zones to be recognized.

In the north there is a 50-mgal minimum which trends northwest–southeast along the foot of the continental slope then extends essentially east–west between long. 6°30′ and 8°30′ W. The minimum overlies the South Armorican trough, the importance of which has been demonstrated seismically (Bacon *et al.*, 1969; Montadert *et al.*, 1971). A similar, but greater negative anomaly (greater than 100 mgals, Boillet and D'Ozouville, 1970) occurs in the south stretching east–west along the foot of the north Spanish slope between long. 4° and 9° W. Again the seismic results show that the gravity minimum coincides with a deep sedimentary trough. Several authors have pointed out that the gravity anomaly computed for a fixed density–thickness model of the sedimentary trough is greater than the observed gravity anomaly (Bacon *et al.*, 1969; Sibuet and Le Pichon, 1970; Sichler *et al.*, 1970). This difference is further increased if it is assumed that the base of the sedimentary column is formed of evaporites (2.15 g/cm^3, Pautot *et al.*, 1970; Sichler *et al.*, 1970) instead of normal sediments (density 2.4 g/cm^3). The low anomaly may be explained by a thinning of the oceanic crust (2.84 g/cm^3) or by an increased mantle density immediately below the trough (Bacon *et al.*, 1969; Sibuet and Le Pichon, 1970). The latter suggests that the north Spanish basin is an ancient oceanic trough presently filled with sediments, the trough itself being the consequence of subduction resulting from the relative approach of the oceanic part of the bay to the north and the Iberian continental part to the south during the compressive phase of Pyrenean tectonism.

In the deeper parts of the bay a series of positive anomalies are found linked for the most part with areas of shoaling, such as the Gascony and Cantabria domes, etc. This is interpreted by Sibuet and Le Pichon (1970) as due to an up-arching of the oceanic crust along the margin of the Spanish trough. According to Walcott (1970) such an up-arching in front of an oceanic trough should be marked by a topographic up-arching. Similar interpretations have been made for active oceanic troughs such as Tonga, Kermadec (Talwani *et al.*, 1961).

A zone of positive anomalies in the east corresponds to the submerged Landes platform. This maximum abruptly interrupts the anomaly pattern characteristic of the north Spanish trough at long. 3°30′ W. East of long. 3°30′ W a gravity minimum comparable in amplitude with that of the north Spanish trough is found over the continuation of the Cap Breton canyon. Sibuet and Le Pichon (1970) actually equate the two and assume an offset of 100 km along a north–south transform fault lying in the longitude of the Santander canyon.

E. Magnetic Measurements and Chronology

The magnetic surveys of Matthews and Williams (1968) show the existence of a series of positive and negative magnetic lineations. These have a fanlike pattern and are found in the central part of the bay (Fig. 6b). The authors propose that they are the results of the formation of new oceanic crust during the rotation of Spain. The distribution of the axes of the magnetic anomalies is consistent with the idea of the progressive injection of basaltic material

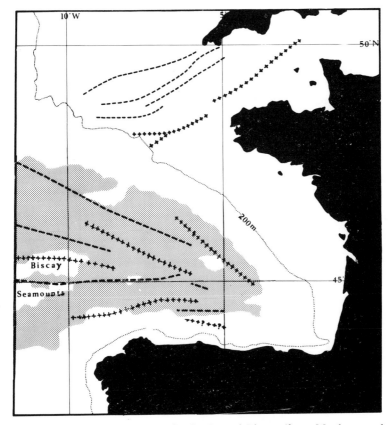

Fig. 6b. Magnetic anomaly axes in the Bay of Biscay (from Matthews and Williams).

along a line bisecting the bay. The injection mechanism would have been discontinuous. This interpretation explains the presence of magnetic anomalies in the bay by a process of expansion similar to that occurring along mid-oceanic ridges.

The anomalies can be located with more precision using the aeromagnetic map of the Bay of Biscay (Le Borgne and Le Mouel, 1970; Le Mouel and Le Borgne, 1970) than their general disposition as given by Matthews and Williams (1968). There is no clear relationship between the topography of the abyssal region and the strong magnetic relief in the central part of the bay. Although the magnetic relief is usually attributed to the remanent magnetization of the injected basaltic material under conditions of alternating field polarity, it is worth remarking that they might be due to susceptibility variations or to a mechanism of alteration such as that proposed by Pecherskiy and Mirlin (1971).

Williams (1970) considered the chronological relationship of the bundle of anomalies in the bay and the north–south anomalies in the Atlantic beyond long. 15° W and concluded that the Bay of Biscay opened prior to the opening of the North Atlantic. Applying the Heirtzler *et al.* (1968) time scale of reversals and the sequence of Helsley and Steiner (1969), the opening took place prior to, or during, the Upper Cretaceous (Williams, 1970). This, however, conflicts with the current view of the opening of the North Atlantic about 180 m.y. ago (Le Pichon and Fox, 1971) and with the reconstructions of Dietz and Holden (1970).

V. SEDIMENTATION HISTORY

The Bay of Biscay, whose morphostructural characteristics indicate instability, is thus a key zone, for surrounded by unstable areas of folding and faulting, it forms a natural sink for sediments transported from the adjoining continental regions. Within it, the sediments are protected by those same adjoining land areas which act as a screen. Thus post-Jurassic sediments, and in particular those of Cenozoic age, should have preserved their original characteristics. For this reason, the history of sedimentation in the Bay of Biscay has a particular interest.

To facilitate the description of the results which have been obtained, the pre-Quaternary stratigraphy will first be described followed by a discussion of the more recent deposits.

A. Pre-Quaternary Stratigraphy

There is a considerable body of data concerning the shelf areas of Aquitaine, Brittany, and northern Spain. The data for the continental slope are much more sporadic, while only the more elevated parts of the abyssal plains appear to have been explored.

In a general sense, the various stratigraphic horizons recognized on the continental shelf have facies comparable in age and type with those found in the adjoining continental areas. In the deeper parts of the Bay of Biscay, however, the outcrops of older rocks commonly reflect conditions of deep-water sedimentation, facies scarcely known in basins exposed at the present time. In the Miocene, for example, this distinction between the facies of the continental shelf and the deeper parts of the continental slope led Klingebiel, Pujol, and Vigneaux (1970) to conclude that the sea-floor topography along the coasts of Brittany and Aquitaine was quite comparable with that existing at the present day.

The outcrop map of the Bay of Biscay (Fig. 7) is based upon a synthesis of the results of Berthois (1955), Guilcher (1967), Jones *et al.* (1968), Andreieff *et al.* (1968), Klingebiel *et al.* (1968), Dupeuble *et al.* (1969), Stride *et al.* (1969), Boillot *et al.* (1970), Barusseau *et al.* (1971), Boillot *et al.* (1971), Caralp *et al.* (1971), and Dupeuble *et al.* (1971). The map shows the widespread occurrence of Miocene outcrops both on the shelf and continental slope. Further, the hypothetical limit of the Miocene deposits (Caralp *et al.*, 1969) appears to be a continuation to the northwest of those known in detail in the northern part of Aquitaine (Alvinerie, 1969).

There is a great difference between the Breton–Aquitaine coastal margin and that of northern Spain. As Boillot (in Boillot *et al.*, 1971) indicated the Spanish shelf is formed of strongly folded Cretaceous flysch discordantly overlain by Tertiary rocks, whereas in the north the shelf is less affected by tectonic movements but forms a major Cretaceous/Tertiary monocline.

Interesting as these results are, they are nonetheless fragmentary and in the majority of cases it would be foolhardy to base a comparison of the marine and continental regions on the few known and dated submarine crops. It then seems preferable to restrict consideration to a form less interpretative than a geological or structural map. A number of maps have been produced for restricted areas, but these rely for a great part upon geophysical interpretation (Vanney *et al.*, 1970).

The sampling of deeper horizons by coring is unfortunately in its infancy in the Bay of Biscay. Two examples have been chosen—one on the continental shelf, the other in the abyssal plain—to illustrate the opportunities such a program would open out.

A relatively modest program of drilling off the Landes coast was carried out in 1968. Holes were drilled in water depths of 37–56 m with cores of 30–50-m length recovered. These were exploratory borings designed to provide information needed for the construction of drilling platforms for oil extraction. The cores revealed a 20–30-m thickness of essentially sandy Quaternary beds

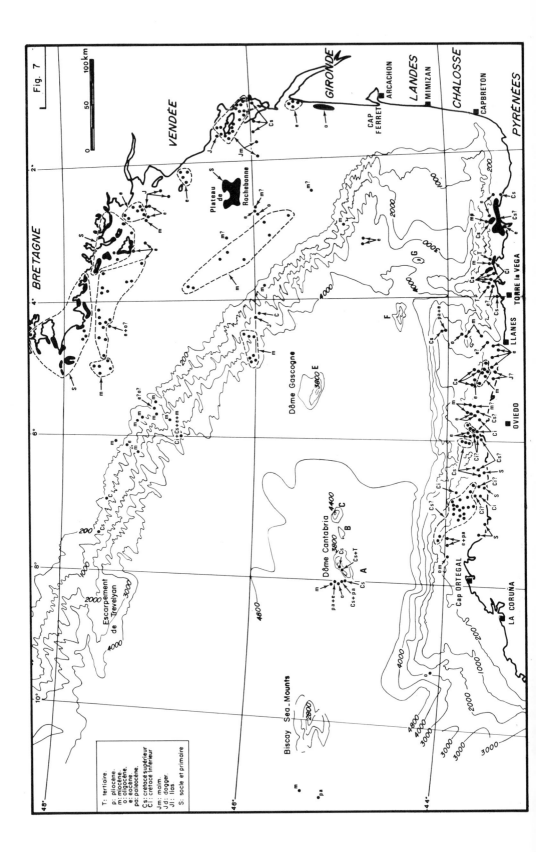

Fig. 7

containing an infralittoral microfauna resting upon Pliocene (s.l.) containing circalittoral and infralittoral faunas and an infralittoral upper Miocene which from a mineralogical and micropaleontological point of view may be compared with onshore outcrops at Salles (Alvinerie *et al.*, 1967). The same horizons have been found in shallow seismic exploration. The Pliocene–Pleistocene contact is an erosional surface with ancient valleys now filled with sediment. One of these valleys probably corresponds to the former course of the Eyre, a small coastal stream which now has its estuary in the Arcachon basin.

In the same region there are records of 11 holes drilled for oil representing a total of 30,909 m of drilling. From these records it is possible to infer a prolongation in the north of the Aquitaine Jurassic shelf where Upper Creta-ceous rests directly upon an eroded Jurassic surface. In the central region an extension of the Parentis basin is recognizable. Further to the south drillings disclose a highly folded zone with the development of diapirs.

According to Dardel and Rosset (1970) the Parentis basin marks the eastern end of a subsiding trough formed by crustal extension during the course of the Lower Cretaceous. It therefore marks the eastern limit of the deep fractures which resulted in the formation of the Bay of Biscay in a region where up to that time the continental coast had remained essentially un-deformed since the beginning of the Mesozoic.

There are results from the two holes drilled in the abyssal plains during the course of leg 12 of the *Glomar Challenger* cruise. The selection of the sites, one on the summit of the Cantabria seamount and the other a degree further west, was guided by the results of earlier seismic surveys (Montadert *et al.*, 1970). A summary of the provisional results was given by Laughton and Berggren (1970); they claim that Cretaceous was not reached (the deepest penetration, 761 m, ended in Paleocene), and assume that the Cretaceous material previously recognized by Jones and Funnell (1968) must have been transported. The holes thus provide no information on the pre-Cenozoic history of the Bay of Biscay.

In summary, an appreciable amount of data is available on surface out-crops such that a surface distribution map may soon be possible. Knowledge of the vertical succession remains extremely poor, largely because of the tech-nical problems and high costs of drilling in deep water.

B. Recent Sedimentation

1. *Recent Sedimentation on the Continental Shelf*

In a preliminary synthesis of the geology of the continental shelf, Berthois (1955) was able to define the principal sedimentological characteristics. He showed the existence of sand bodies each with a specific sedimentological

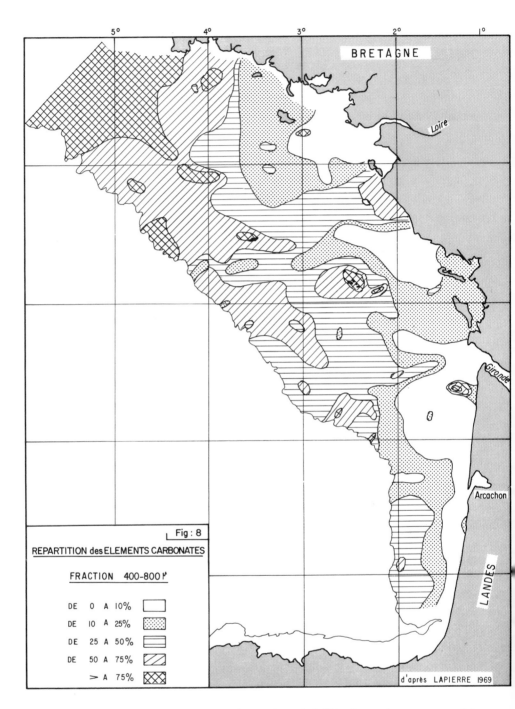

Fig. 8. Distribution of carbonates on the continental shelf in the northeastern part of the
Bay of Biscay: fraction from between 400 and 800 m.

significance and that the older sand deposits between Brest and the shoals of La Chapelle indicated a sedimentary regime different from that now existing. He also showed that an important difference can be recognized in the morphology at the quartz grains derived from the Brittany and Vendée coasts on the one hand and the region south of Oléron on the other. Subsequently, the existence of two zones of uniform pebble distribution along the edge of the continental slope and along the −50 to −60 m as well as restricted areas of coarse-grained sediments was demonstrated (Berthois and Le Calvez, 1959). These regions of arenaceous sediments are interpreted as littoral beach zones bordered by dunes behind which developed broad alluvial plains. In the latter sheltered zone, fine-grained sediments mixed with reworked sands formed mud flats.

The preservation of the beach and dune facies with some reworking implies that since the drowning of this littoral margin, the input of sediment on to the shelf has been small or nonexistent, although the introduction of fine-grained sediments, in particular close to the Brittany shore and at the mouth of the Gironde, cannot be excluded.

Detailed sedimentological studies of grain size, heavy minerals, and the carbonate fraction (Fig. 8) have since been made by Lapierre (1969). All three techniques lead him to conclude that there was a southward movement of material, the heavy minerals even providing evidence of the dispersion of the sediments introduced by the Garonne River.

The clay minerals have been studied by Klingebiel, Lapierre, and Latouche (1966, 1967) who show the homogeneity of composition despite the variation in facies. Two forms predominate, illite and chlorite, the latter often partly altered and interlayered with a chlorite–vermiculite. Montmorillonite and kaolinite are poorly represented [M/I ratio 0.2(average)]. In this the clays of the Bay of Biscay show a parallel development to that found in the Atlantic, although Latouche (1971) pointed to some features in the coastal region which reflect the proximity of the continental area. The study of a restricted region off the mouth of the Loire led Rumeau and Vanney (1969) to conclude that the mud flats are in equilibrium under present-day conditions.

Up to the present time few studies of the microfauna treat the shelf as a whole. The only general study of the ostracods (Moyes and Pachier, 1968) shows that a greater proliferation of species exists in the littoral domain and little diversified but widespread infra- and circalittoral domains. Large regions, which correspond roughly to the outcrop of coarse sand, are devoid of ostracods.

The benthonic foraminifera of restricted regions have been studied by Schnitker (1969), Rouvillois (1970), and Pujos (1971), from which certain associations with bathymetry (infra- and circalittoral zones) and with the nature of the substratum have been elicited. The planktonic foraminifera of the bay have been studied as a unit (Caralp, 1971). On the continental shelf,

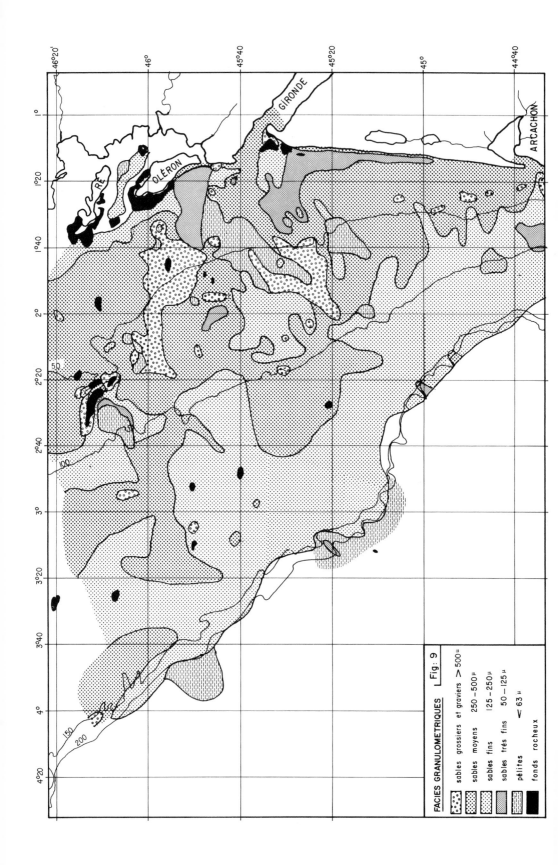

FACIES GRANULOMETRIQUES | Fig: 9

sables grossiers et graviers > 500 μ

sables moyens 250 - 500 μ

sables fins 125 - 250 μ

sables très fins 50 - 125 μ

pélites < 63 μ

fonds rocheux

GIRONDE

RE   OLÉRON

ARCACHON

the eastward extension of characteristic assemblages is limited by the -100-m isobath. The rich circalittoral assemblage varies greatly along the margin of the shelf and seems to reflect the existence of a current controlling the transportation and deposition of plankton in the circalittoral zone (Caralp, 1971; Andreieff et al., 1971).

A detailed sedimentological and microfaunal study of the continental shelf of the northern part of the Bay of Biscay by Andreieff et al. (1971) provides information on the evolution of the shelf during the latter part of the Quaternary. They report practically zero deposition so that the sedimentary zones recognized are relict. Based upon grain size analysis they recognize three sand horizons: coarse, mixed, and fine-grained overlain by a silty mud. Only the fauna of the mud is appropriate to the recorded bathymetry; the sand horizons now at depths of 130–150 m are regarded as marking the first stages of the Flandrian transgression.

In a detailed sedimentological study of the region west of the Gironde (Vigneaux et al., 1970; Vigneaux, 1971; Caralp et al., 1971) it was also shown that only along the littoral fringe is there any correspondence between the nature of the sediments and the hydrodynamic conditions now existing. At depth of about -90 m, the pattern of sediment distribution (Fig. 9) suggests the location of an ancient shoreline. The coarse sands and shelly gravels aligned along the -50-m isobath seaward of the Ile d'Oléron and west of Medoc represent a further ancient littoral zone. The changing proportion of carbonate material (Lapierre, 1969) is consistent with this. The clay minerals (Latouche, 1971) are consistent with those found in other parts of the bay, but reflect variations caused by the introduction of montmorillonite by the Gironde.

The distribution of geochemical facies (Fig. 10) provides further information on the paleogeographic evolution of the shelf during the Quaternary. The older Quaternary shelf had a largely arenaceous cover, and the occurrence of more silicate-rich facies reflect the latter stages of sea-level change. The presence of fossil alluvial-filled channels mark former river courses during regressive phases.

Thus to summarize the results of these two detailed studies, it is clear that recent sediments are sparsely represented, and the introduction of fine-grained alluvial material from the rivers is a relatively recent phenomenon affecting only the upper part of the Holocene. The distribution of the surface sediments is directly related to the Quaternary history of the continental shelf, that is to say, fashioned by transgression and regression linked to glaciation.

The study of older sediments on the continental shelf is handicapped not only by short cores (2–3 m in length) but by sampling sites so scattered that they are ill suited to provide a general idea of the stratigraphy. There are only

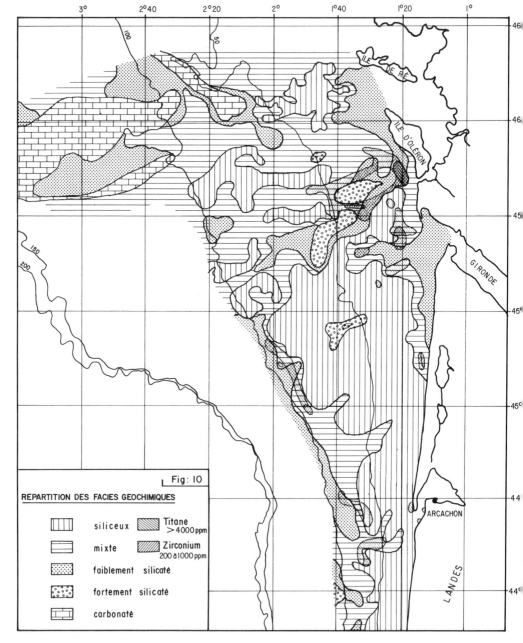

Fig. 10. Distribution of geochemical facies on the Aquitaine continental shelf.

two regions on the shelf where there is a sufficient core density to make study feasible; those are the regions between Penmarc'h and L'Ile de Croix and west of the Gironde estuary. At the first of these two localities the cores present evidence in terms of progressively finer grain size of the Flandrian transgression. The facies variation west of the Gironde from aeolian beach sands through sandy mud to fine mud is likewise evidence of the last transgression since 10,000 yr B.P. as Bouysse and Le Calvez (1967) have recognized in the Armorican region. The stages in transgression can also be documented from the ostracod microfauna (Carbonel, Moyes, and Peypouquet, 1972; Vigneaux, 1971).

2. Recent Sedimentation on the Continental Slope and Abyssal Plains

The farther from the continental shelf, the less information there is on the nature of the sediments and the more clearly impossible the construction of any distribution map. On the whole, more information is available from the disturbed regions such as the abyssal hills or canyons than there is from the monotonous abyssal plains. We are therefore constrained to a discussion of the details available for a single region between lat. 45° and 47° N (Vigneaux, 1971).

As might be expected, the proportion of sand in the surficial sediments falls progressively with increasing distance from the continental shelf from 70 to 95% on the shelf to <10% (often less than 2%) on the abyssal plain. The exceptions reflect the local transport of turbidites down submarine canyons and thus have no real geographic significance. The distribution of carbonate matter shows the converse relationship passing from 2 to 20% at the head of the continental slope to >30% below 3000 m, although these may be local variations.

The examination of the benthonic foraminifera permitted Caralp, Lamy, and Pujos (1970) to recognize a series of bathymetric associations. This zonation is useful (Vigneaux, 1971) in establishing whether a given microfauna is in accord with the observed bathymetry. The mixture of benthonic fauna along the axes of canyons can be resolved by this means (Caralp, Moyes, and Vigneaux, 1970). In contrast, the planktonic foraminifera are homogeneously distributed and the proportion of species is in general agreement with that found in other temperate regions of the world. Caralp (1971) distinguished two associations, one oceanic deposited in bathyal and mesobathyal zones and the other neritic found on the shelf and upper part of the continental slope. The differences lie in the proportion of certain species, and the recognition of these differences helps to distinguish recent from nonrecent assemblages from outcrops on the slope, or to separate recent associations from different climatic regimes.

If only the absolute abundances of the elements, trace elements in the bathyal and abyssal zones are considered, then it is possible to demonstrate low-amplitude, apparently continuous variations (Vigneaux, 1971; Caralp *et al.*, 1971) between the northern and southern parts of the bay (Table I). If, on the other hand, the same details are examined with respect to possible mineral provenance, then it would appear that the relationship of the northern and southern parts of the bay are quite different. Latouche and Parra (1970) show that in the north, elements such as Pb, Rb, Ni, Ba, and especially Zn are exclusively associated with the clay fraction and generally diminish in quantity seaward. In the south the relationship of Zn to the fine grained fraction is not good and implies other hosts than the clay fraction. The distinction resides in the differing nature of the adjoining continental region—Hercynian massif in the north, Pyrenees in the south. The authors leave open the question of introduction of material from the Atlantic.

In addition to the JOIDES drilling, there are a number of short cores (10–12 m) from the region of the Cantabrian dome or the Biscay seamounts. They provide important details of Quaternary stratigraphy, and since the water depth has always been appreciable and sedimentation low there is the prospect of finding well-preserved upper Quaternary marine horizons. In the continental shelf, however, repeated transgressions and regressions, the reworking, erosion, and redeposition of material, makes it impossible to find a complete Quaternary section, particularly if one follows Berthois and Le Calvez (1959) and assumes that the maximum regression reached the border of the shelf (-200 m).

Most of the information presently available is from cores from the Gascony dome at a depth of 3920 m. The stratigraphic control, is by means of planktonic foraminifera, and as these are directly influenced by temperature (Bé and Hamlin, 1967) they provide a record of climatic variation (Ericson, Ewing, and Wollin, 1964), the result of Quaternary glaciations well known in the adjoining continental areas. In the Bay of Biscay the stratigraphy has been established by Caralp (1971). She records a continued alternation of climatic conditions varying between temperate (similar to the present) and arctic, back to the Riss–Wurm interglacial (Fig. 11). A radiocarbon date from a depth of 4 m on a core from the Gascony dome gave an age of $37,500 \{ {}^{+3900}_{-2500}$ yr B.P. which corresponds to the base of the Wurm II–Wurm III Interstadial (Bordes, 1968; Lumley, 1969). Other groups have been examined, and the benthonic foraminifera were found to display quantitative variations which could also be directly linked to temperature (Pujos-Lamy, 1971).

There are no marked sedimentological changes in the cores. The amount of carbonate present reflects the abundance of the foraminiferal fauna which is at a maximum during warm intervals. In the cold or very cold horizons, although sand is more abundant, it never exceeds 12–15%. During the warm

TABLE I

Average Geochemical Characteristics of the Regions Studied

Domains	%		ppm										%		%			
	pH	C. org.	Pb	Rb	Ti	Mn	Sr	Ba	Zn	Ni	Cu	Zr	P$_2$O$_5$	S	CO$_3$Ca	<2 (μ)	2÷50 (μ)	>50 (μ)
Canyon Gascogne	8.57	0.998	32	75	2398	290	365	330	69	25	23	103	0.120	0.155	24	32	36	32
Canyons de Noirmoutier et des Sables d'Olonne	7.81	0.785	41	80	2098	290	370	330	72	20	25	63	0.170	0.105	24	23	37	40
Pente continentale armoricaine	8.68	0.627	32	87	2093	315	378	289	33	20	16	56	0.134	0.083	—	42	43	15
Canyon du Cap Ferret	8.56	0.799	47	95	2225	418	429	267	65	26	31	66	0.163	0.113	21.5	44	44	12
Couf de Cap Creton	8.04	2.55	89	66	3297	205	360	305	180	12	22	108	0.150	0.240	16	39	50	11
Dôme Gascogne	8.42	0.665	28	72	2211	688	458	262	48	21	34	58	0.108	0.040	—	48	41	11
Moyenne du golfe	—	1.64	60	95	3300	355	330	360	119	28	24	85	0.130	0.195	20	33	37	30

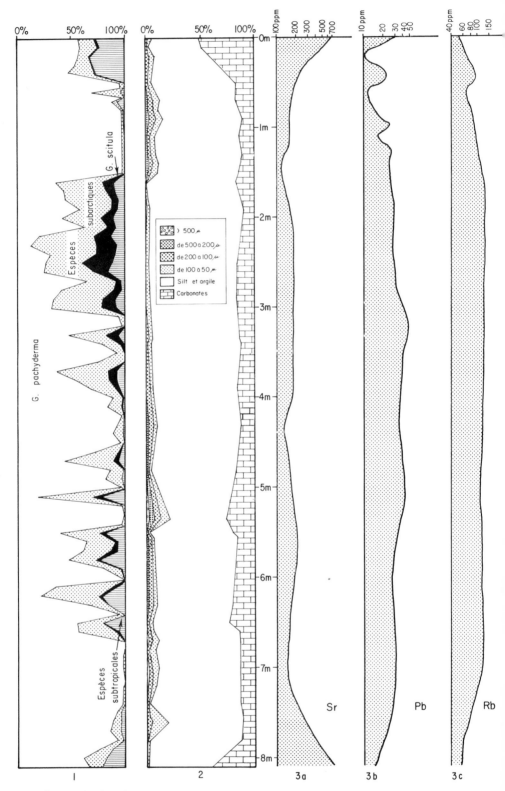

Fig. 11. Study of core CH. C. 6928 (Gascony dome): (1) distribution of planktonic
foraminifera; (2) lithology; (3) geochemical character.

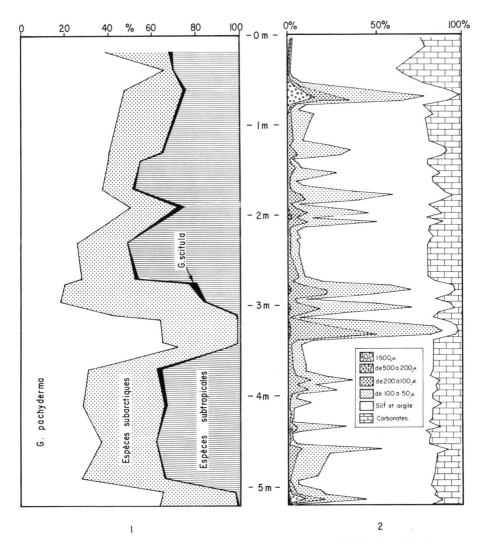

Fig. 12. Study of core Ch. C. 6714 (Sables d'Olonne canyon): (1) distribution of planktonic foraminifera; (2) lithology.

periods there is an enhanced abundance of strontium, presumably related to the fauna, and an impoverishment of lead, titanium, and iron (Vigneaux, 1971). Chenouard, Berthois, and Lalou (1969) show that the abundance of potassium and thorium is also related to the climatic conditions of sedimentation, and this abundance variation provides a rapid means of determining large changes in the mode of sedimentation.

Local variations in sedimentation show either in the thickness of stadial or interstadial deposits or in the type of sedimentation. One of the most obvious is the increased thickness found in the lower parts of canyons where they debouch on to the abyssal plain (see Figs. 11 and 12) almost certainly due to the sediment transported down the canyon and spread out by turbidity currents.

VI. CONCLUDING REMARKS

In the form of a conclusion to this rapid and incomplete survey of the state of geological knowledge concerning the Bay of Biscay, it is convenient to underline the two main features of the region, the one its history with respect to the North Atlantic basin as a whole, the other the Bay as a particular sedimentary environment.

The origin of the Bay of Biscay, the time and its mode of origin, have been much debated recently (see Symposium I.F.P. and C.N.E.X.O. Rueil-Malmaison, 1970). In a general way, two hypotheses have been proposed, which perhaps ought to be combined (Bard et al., 1970). One is based on the remarkable similarities which exist between the Hercynian of the northwest extremity of the Iberian peninsula and the Armorican massif of France. It proposes that at some time subsequent to the Hercynian orogeny, drift of the Iberian peninsula with respect to continental Europe occurred with rotation about a point lying in the apex of the bay (Choubert, 1935; Cogné, 1970). The second hypothesis, which takes into consideration the Pyrenees, combines a partial Mesozoic rotation with a late Hercynian eastward translation of the Iberian peninsula (Muraour, 1970; Mattauer and Seguret, 1970; Le Pichon et al., 1970). Recent geomorphological studies (Prud'homme, 1972) indicate that the line of demarcation between the Hercynian and Pyrenean structures follows the line of the Cap Ferret fault.

As far as the timing of the opening of the Bay of Biscay is concerned, the information is both abundant and diverse. According to paleomagnetic results the drift process must have occurred between Triassic and Eocene times. Williams (1970), after a consideration of the magnetic reversal chronology, regarded the widening of the Bay of Biscay as beginning prior to or during the Upper Cretaceous, the time of the maximum marine incursion in the Aquitaine basin.

Recently, Latouche (1971) demonstrated the existence of an attapulgite facies in the Paleocene deposits on the Cantabrian dome. This facies implies a specific (hypersaline) environment, i.e., deposition in water isolated from open-sea conditions. It would appear from this that during the Paleocene the opening of the bay had not affected the extreme south where epicontinental conditions still held sway.

Thus one is led to the conclusion that the opening of the Bay of Biscay may have occupied a relatively long time, from Lower Cretaceous to Eocene, and not necessarily have occurred at a uniform rate.

As a sedimentary environment in a temperate region the Bay of Biscay is both complex and intrinsically of interest. Clearly there are many features about the Bay of Biscay which make it distinctive with respect to the North Atlantic, although it is only a somewhat isolated part of that ocean. However, the only evidence of true "oceanic" features is found seaward of the Biscay seamounts. The wide continental shelf displaces the deep-water zone far to the west, although the deep incision of the Cap Breton canyon leads to the presence of deep-water conditions at a relatively short distance from the actual coastline. Apart from that, however, studies indicate that the region was protected to some extent from factors which could appreciably modify the conditions of sedimentation. The Bay of Biscay is thus a theater of sedimentation wherein developed a great variety of environments whose interpretation provides valuable clues toward the general understanding of ancient environments.

ACKNOWLEDGMENTS

I wish to express gratitude to all the members of the Institut de Geologie du Bassin d'Aquitaine for their help and cooperation in the preparation of this chapter. Particular thanks are due to Drs. Naudin and Prud'homme for their help with the geomorphology, and to Drs. Frappa, Martin, and Muraour for their assistance with the section on the geophysics of the Gulf of Aquitaine. Thanks are also due to Mlle. Caralp for her help in the general editing of the text.

REFERENCES

Alvinerie, J., 1969, Contribution sédimentologique à la connaissance du Miocène aquitain; interprétation stratigraphique et paléogéographique: *Thèse Doctorat d'Etat*, Fac. Sc. Bordeaux, n. 218, 2 tomes, 213 p.

Alvinerie, J., Barrier, J., Caralp, M., Ittel, D., Klingebiel, A., Magné, J., and Moyes, J., 1967, Reconnaissance des fonds marins et des séries superficielles de la plateforme continentale au large de la côte landaise (Golfe de Gascogne): *C.R. 86e Congrès, Ass. Fr. Av. Sc., Bordeaux—Actes Soc. Linn., Bordeaux*, vol. sp., p. 121–136.

Andreieff, P., Boillot, G., Buge, E., and Gennesseaux, M., 1969, La couverture sédimentaire tertiaire à l'Ouest et au Sud-Ouest du massif armoricain: *Bull. B.R.G.M.* (2e sér.), section IV, n. 4, p. 23–37.

Andreieff, P., Bouysse, P., Chateauneuf, J., Lhomer, A., and Scolari, G., 1971, La couverture sédimentaire meuble du plateau continental externe de la Bretagne méridionale: *Cahiers Océanogr.*, XXIII, 4, p. 343–381.

Andreieff, P., Bouysse, P., Horn, R., and Monciardini, C., 1970, Géologie des approches occidentales de la Manche: *C.R. Acad. SC.*, Paris, t. 270, p. 2756–2759.

Andreieff, P., Bouysse, P., Horn, R., and Lhomer, A., 1968, Données récentes sur l'Eocène au large de la Bretagne méridionale: *C.R. Som. Soc. Géol. Fr.*, 5, p. 161–162.

Bacon, M. and Gray, F., 1970, A gravity survey in the eastern part of the Bay of Biscay: *Earth and Planet Sc. Letters*, v. 10, n. 1, p. 101–105.

Bacon, M., Gray, F., and Matthews, D. H., 1969, Crustal structure studies in the Bay of Biscay: *Earth and Planet Sc. Letters*, v. 6, n. 5, p. 377–385.

Bard, J. P., Capdevila, R., and Matte, P., 1970, La structure de la chaine hercynienne de la meseta ibérique: comparaison avec les segments voisins: *Symposium sur l'histoire structurale du Golfe de Gascogne*, Rueil Malmaison, 14-16 déc. 1970, Edit. Technip, n. 22, t. I, 1971, p. 1-4-1-1-4-68.

Barusseau, J. P. and Martin, G., 1971, Esquisse géologique et structurale des Pertuis Charentais et de leurs abords: *Rev. Geogr. phys. et Geol. dyn.*, Paris(2), XIII, f. 4, p. 403–412.

Be, A. W. and Hamlin, W. H., 1967, Ecology of recent planktonic foraminifera: part. 3. Distribution in the North Atlantic during the summer of 1962: *Micropal. U.S.A.*, v. 13, n. 1, p. 87–106.

Berthois, L., 1955, Contribution à l'étude de la sédimentation et de la géologie sous-marines dans le Golfe de Gascogne: *Bull. Ass. Fr. Etude Grande Prof. Océan*, n. 5, p. 11–20.

Berthois, L. and Brenot, R., 1966, Morphologie de la partie Est de la plaine abyssale du Golfe de Gascogne: *Bull. Ass. Fr. Etude Grande Prof. Océan.*, n. 5, p. 11–20.

Berthois, L., Brenot, R., and Ailloud, P., 1965, Essai d'interprétation morphologique et tectonique des levées bathymétriques exécutées dans la partie Sud-Est du Golfe de Gascogne: *Rev. Trav. Inst. Pêches Marit.*, v. 29, n. 3, p. 321–342.

Berthois, L., Brenot, R., and Debyser, J., 1968, Remarques sur la morphologie de la marge continentale entre l'Irlande et le Cap Finisterre: *Rev. Inst. Petr. Ann. Comb. Liquides*, v. 23, p. 1046–1049.

Berthois, L., Brenot, R., Debyser, J., 1969, Carte bathymétrique de la marge continentale du Golfe de Gascogne et de la mer Celtique (à l'échelle du 1.000.000e): *Institut Français du Pétrole*, rap. n° 17419, 5 p.

Berthois, L. and Le Calvez, Y., 1959, Deuxième contribution à l'étude de la sédimentation dans le Golfe de Gascogne: *Rev. Trav. Inst. Pêches Marit.*, v. 23, n. 2, p. 323–377.

Beuther, A., 1966, Geologische Untersuchungen in Wealden und Utrillas Schichten im Westteil der Sierra de los Cameros (Nordwestliche Iberische Ketten): *Beih. geol. Jahrb.*, v. 44, p. 103–121.

Boillot, G. and D'Ozouville, L., 1970, Etude structurale du plateau continental nord-espagnol entre Aviles et Llanes: *C.R. Acad. Sci. Paris*, t. 270, sér. D, p. 1865–1868.

Boillot, G., Dupeuble, P. A., Lamboy, M., D'Ozouville, L., and Sibuet, J. C., 1971, Les prolongements occidentaux de la chaine pyrénéenne sur la marge continentale nord espagnole entre 4° et 9° de longitude Ouest: *C.R. som. Géol. Fr.*, Paris, I, p. 11.

Boillot, G., Dupeuble, P. A., Le Lann, F., D'Ozouville, L., 1970, Etude stratigraphique des terrains affleurant sur le plateau continental Nord-espagnol entre Aviles et Llanes: *C.R. Som. Soc. Géol. Fr.*, Paris, v. 3, p. 78–80.

Boillot, G. and Rousseau, A., 1971, Etude structurale du plateau continental nord-espagnol entre 2°20 et 3°30 de longitude Ouest: *C.R. Acad. Sci. Paris*, t. 272, sér. D, p. 2056–2059.

Bonnard, E., Debourle, A., Hlauschek, H., Michel, P., Perebaskine, V., Schoeffler, J., Seronie-Vivien, R., and Vigneaux, M., 1958, The aquitanian basin, southwest France: *Habitat of Oil. A.A.P.G.*, p. 1091–1122.

Bordes, F., 1968, Le Paléolithique dans le monde: l'*Univers des connaissances*: Hachette Edit., Paris, 256 p.

Bott, M. P., 1967, Solution of the linear inverse problem in magnetic interpretation with application to oceanic magnetic anomalies: *Geophys. J. R. Astr. Soc.*, v. 13, p. 313–323.

Boulanger, D., 1968, Révision du Nummulitique de la Chalosse, du Béarn et du Bas-Adour

(Landes et Basses Pyrénées): *Thèse Doct. Sci. Nat.*, Paris, ronéotypée, Archives Centre Document. C.N.R.S. n. 2838, 2 vol., 369 p.

Bouysse, P. and Horn, R., 1968, Etude structurale du plateau continental au large des côtes méridionales de la Bretagne, France: *Colloque du C.N.R.S.* à Villefranche, Monaco, 21 p.

Bouysse, P. and Le Calvez, Y., 1967, Etudes des fonds marins compris entre Penmarc'h et Groix (Sud Finistère): *Bull. B.R.G.M.*, 1ère sér., 2, p. 39–73.

Brenot, R. and Berthois, L., 1962, Bathymétrie du secteur atlantique du banc Porcupine (ouest de l'Irlande) au cap Finisterre (Espagne): *Rev. Trav. Inst. Pêches Marit.*, v. 26, n. 2, p. 219–246.

Bullard, É. C., Everett, J., and Gilbert-Smith, A., 1965, A symposium on continental drift: *Phil. Trans. Roy. Soc. London*, série A., v. 258, p. 41–51.

Caralp, M., 1971, Les Foraminifères planctoniques du Pléistocène terminal dans le Golfe de Gascogne. Interprétation biostratigraphique et paléoclimatique: *Thèse Doctorat ès-Sci. Nat., Univ. Bordeaux I*, n. 333 et *Bull. Inst. Géol. Bassin Aquitaine*, n. 11-1, 187 p.

Caralp, M., Dumon, J. C., Frappa, M., Klingebiel, A., Latouche, C., Martin, G., Moyes, J., Muraour, P., Prud'homme, R., and Vigneaux, M., 1971, Contribution à la connaissance géophysique et géologique du Golfe de Gascogne. Bilan des travaux effectués du 1er octobre 1969 au 30 juin 1971: *Bull. Inst. Géol. Bassin Aquitaine*, Bordeaux, n° spécial, 142 p.

Caralp, M., Lamy, A., and Pujos, M., 1970, Contribution à la connaissance de la distribution bathymétrique des Foraminifères dans le Golfe de Gascogne: *Rev. Esp. Micropaleontologia, Madrid*, v. II, p. 55–84.

Caralp, M., Moyes, J., Pujol, C. and Pujos, M., 1969, Sur la continuité du Miocène inférieur au Nord Ouest du bassin d'Aquitaine (Golfe de Gascogne): *C.R. Som. Soc. Géol. Fr.*, Paris, p. 244–245.

Caralp, M., Moyes, J., and Vigneaux, M., 1970, Essai d'utilisation des mélanges de micro-organismes benthiques dans la reconstitution des environnements. Application à un canyon du Golfe de Gascogne: *Deep Sea Research*, v. 17, p. 661–670.

Caralp, M., Vigneaux, M., 1960, Aspect structural du Médoc atlantique: *Bull. Soc. Géol. Fr.*, 7e série, 2 (6), p. 796–800.

Carbonel, P., Moyes, J., and Peypouquet, J. P., 1972, The use of ostracodes in the discovery of a Holocene Lagoon west of the Gironde estuary in the Bay of Biscay: *Symposium sur la biologie et la paléobiologie des Ostracodes*, 14–17 Août 1972, 1972, Université de Delaware, U.S.A. (à paraître).

Carey, S. W., 1958, A tectonic approach to continental drift: *Symposium Continental Drift*, Univ. Tasmania, Hobart, p. 177–355.

Chenouard, L., Berthois, L., and Lalou, C., 1969, Etude d'une carotte sédimentaire de la pente continentale atlantique: *C.R. Acad. Sc. Paris*, t. 268, p. 1470–1473.

Choubert, B., 1935, Recherches sur la génèse des chaines paléozoiques et antécambriennes: *Rev. Géogr. Phys. et Géol. dynam.*, VIII, 1, p. 5–50.

Ciry, R., Rat, P., Mangin, P., Feuillee, P., Amiot, M. and Delance, J. H., 1967, Evolution paléogéographique et structurale de la région basco-cantabrique: *C.R. som. Soc. Géol. Fr.*, p. 391–394.

Cogné, J., 1957, Schistes cristallins et granites en Bretagne méridionale: le domaine de l'anticlinal de Cornouaille: *Mém. explic. Carte Géol. Fr.* (1960), 382 p.

Cogné, J., 1970, Le massif Armoricain et sa place dans la structure des socles Ouest-européens: l'arc hercynien ibéro-armoricain: *Symposium sur l'histoire structurale du Golfe de Gascogne*, Rueil Malmaison, 14-16 déc. 1970, Edit. Technip, n. 22, t. I, 1971, p. 1.1.1–1.1.24.

Collette, B. J. and Rutten, K. W., 1970, Differential compaction vs. diapirism in abyssal plains: *Marine Geoph. Res.*, v. I, p. 104–107.

Collette, B. J., Shouten, J. A., and Rutten, K. W., 1971, Sediment distribution in the West European Basin Atlantic Ocean: *Symposium sur l'Histoire structurale du Golfe de Gascogne*, Rueil Malmaison, 14-16 déc. 1970, Edit. Technip, n. 22, t. II, p. VI-7-1–VI-7-10.

Collette, B. J., Shouten, J. A., Rutten, K. W., Doornbos, D. J., and Staverman, W. H., 1971, Geophysical investigations off the Surinam coast: *Hydrographic Newsletter*.

Coron, S., Guillaume, A. and Bouvet, J., 1970, Contribution gravimétrique à l'étude du Golfe de Gascogne et des Pyrénées: *C.R. Acad. Sc. Paris*, t. 271, série D, p. 756–759.

Curray, J. R., Moore, D. G., Belderson, R. H., and Stride, A. H., 1966, Continental margin of western Europe: slope progradation and erosion: *Science*, v. 154, p. 265–266.

Curry, D., Hamilton, D., and Smith, J., 1970, Geological and shallow subfarce geophysical investigations in the western approaches to the English Channel, Natural environment research council: *Institute of Geological Sciences*, rep. n. 70/3, 12 p.

Cuvillier, J., 1956, Stratigraphic correlations by microfacies in western Aquitaine: *E. J. Brill*, édit. Leiden, 33 p.

Damotte, B., Debyser, J., Montadert, L., and Delteil, J. R., 1969, Nouvelles données structurales sur le Golfe de Gascogne obtenues par sismique réflexion "Flexotir": *Rev. I.F.P. et Ann. Combust. Liquides*, v. 24, p. 1061–1072.

Damotte, B., Grau, G., Gray, F., Limond, W. Q., and Patriat, P., 1970, Utilisation conjointe des méthodes sismiques réflexion et réfraction pour la détermination des vitesses: *Symposium sur l'Histoire structurale du Golfe de Gascogne*, Rueil Malmaison 14-16 dec. 1970, Edition Technip, n. 22, 1971, t. 2, p. VI-15-1–VI-15-26.

Dardel, R. A. and Rosset, R., 1970, Histoire géologique et structurale du Bassin de Parentis et de son prolongement en mer: *Symposium sur l'histoire structurale du Golfe de Gascogne*, Rueil Malmaison, 14-16 déc. 1970, Edit. Technip, n. 22, t. I, 1971, p. IV-2-1–IV-2-28.

Day, G. A. and Williams, C. A., 1970, Gravity compilation in the Nord-Est Atlantic and interpretation of gravity in the Celtic sea: *Earth and Planet, Sc. Letters*, v. 8, p. 205–213.

Deloffre, R., 1965, Etude géologique du flysh crétacé supérieur entre les vallées de l'Ouzom et du gave de Mauléon (Basses Pyrénées): *Imp. Bierre*, Bordeaux.

Dietz, R. S. and Holden, J. C., 1970, The breakup of Pangaea: *Scientific American*, v. 223, n. 4, p. 30–41.

Dupeuble, P. and Lamboy, M., 1969, Le plateau continental au nord de la Galice et des Asturies: premières données sur la constitution géologique: *C.R. Acad. Sc. Paris*, t. 269, D, p. 548–551.

Dupeuble, P. and Rousseau, A., 1971, Stratigraphie des terrains affleurant sur le plateau continental Nord espagnol entre Santander et Guernica: *C.R. Acad. Sc. Paris*, t. 272, D, p. 1952–1955.

Durand, S., 1962, Le Paléogène du Nord Ouest de la France: *Colloque sur le Paléogène*, Bordeaux, sept. 1962. *Mém. B.R.G.M.*, n. 28, t. I, p. 517–529.

Du Toit, A. L., 1937, *Our Wandering Continents*: Oliver and Boyd, Edinburg, 366 p.

Ericson, D. B., Ewing, M., and Wollin, G., 1964, The Pliocene in deep-sea sediments: *Science*, v. 146, n. 3645, p. 723–732.

Ewing, J. I. and Jaunere, R., 1964, Seismic profiling with a pneumatic sound source: *J. Geophys. Res.*, v. 69, p. 4913–4915.

Feuillée, P., 1967, Le cénomanien des Pyrénées basques aux Asturies. Essai d'analyse stratigraphique: *Mém. Soc. Géol. Fr.*, N.S., t. XLVI, fasc. 3, n. 108.

Feuillée, P. et Rat, P., 1970, Structures et paléogéographies pyrénéo-cantabriques: *Symposium sur l'histoire structurale du golfe de Gascogne*, Rueil Malmaison, 14-16 déc. 1970, *Edit. Technip*, n. 22, t. II, 1971, p. V-1-1–V-1-48.

Frappa, M. and Horn, R., 1971, Etude par sismique réfraction du plateau continental au large d'Ouessant: *Bull. Inst. Géol. Bassin Aquitaine*, Bordeaux, n. 11/2, p. 401–410.

Frappa, M., Klingebiel, A., Malzac, J., Martin, G., Muraour, P., and Vigneaux, M., 1970, Remarques sur la structure des montagnes sous-marines de Biscaye (Golfe de Gascogne), à la suite d'une étude sismique par réflexion: *C.R. Soc. Geol. Fr.*, fasc. 5, p. 149–150.

Girdler, R. W., 1965, Continental drift and the rotation of Spain: *Nature*, v. 207, p. 395–397.

Girdler, R. W., 1968, A paleomagnetic investigation of some Late Triassic and Early Jurassic volcanic rocks from the northern Pyrenees: *Ann. de Geophys.*, v. 24, n. 2, p. 1–14.

Gray, F., and Stacey, A. P., 1970, Gravity and magnetic interpretation of Porcupine bank and Porcupine bight: *Deep Sea Res.*, v. 17, p. 467–475.

Guilcher, A., 1967, Morphosédimentologie de la proche plateforme continentale atlantique entre Ouessant et les Sables d'Olonne: *Rev. Géogr. Phys. et Géol. Dyn.* (2), 9, 3, p. 181–190.

Guilcher, A., 1969, Progrès de la connaissance de la plateforme continentale atlantique ouest et sud-armoricaine: *Ac. Géogr. Lov.* à paraître.

Gygi, A. and McDowell, F. W., 1970, Potassium-argon ages of glauconites from a bio-chronologically dated Upper Jurassic sequence of northern Switzerland: *Eclogae geol. Helvetiae*, v. 63, n. 1, p. 111–118.

Heirtzler, J. R., Dickson, G. D., Herron, E. M., Pitman, W. C., and LePichon, X., 1968, Marine magnetic anomalies, geomagnetic field reversals and motions of the ocean floor and continents: *J. Geophys. Res.*, v. 73, n. 6, p. 2119–2136.

Helsley, C. E. and Steiner, M. B., 1969, Evidence for long intervals of normal polarity during the Cretaceous Period: *Earth Planet. Sci. Lett.*, v. 5, p. 325–332.

Hempel, P. M., 1967, Der Diapir von Poza de la Sal (Nordspanien): *Beith. geol. Jahrb.*, v. 66, p. 95–126.

Iaga, 1969, International geomagnetic reference field 1965: *O. J. Geophys. Res.*, v. 74, p. 4407–4408.

Jones, E. J. W., 1966, Acoustic subbottom profiling in the northeastern Atlantic: *Proceed. I.R.E. conf. n. 8, on Electronic Engineering in Oceanography* (12–15 sept. 1966) University Southampton (supplement vol.), paper n. 23, p. 1–7.

Jones, E. J. W., 1968, Continuous reflection profiles from the european continental margin in the Bay of Biscay: *Earth Planet. Sc. Letters*, v. 5, n. 2, p. 127–134.

Jones, E. J. W. and Ewing, J. I., 1969, Age of the Bay of Biscay: evidence from seismic profiles and bottom samples: *Science*, v. 166, p. 102–105.

Jones, E. J. W. and Funnel, B., 1968, Association of a seismic reflector and upper Cretaceous sediment in the Bay of Biscay: *Deep Sea Res.*, v. 15, p. 701–709.

Klingebiel, A., Lapierre, F., Larroudé, J., and Vigneaux, M., 1968, Présence d'affleurements de roches d'âge miocène sur le plateau continental du Golfe de Gascogne: *C.R. Acad. Sc. Paris*, t. 266, p. 1102–1104.

Klingebiel, A., Lapierre, F., and Latouche, C., 1966, Sur la nature et la répartition des minéraux argileux dans les sédiments récents du Golfe de Gascogne: *C.R. Acad. Sc. Paris*, t. 263, p. 1293–1294.

Klingebiel, A., Lapierre, F., and Latouche, C., 1967, Les minéraux argileux des sédiments récents du Golfe de Gascogne: *Bull. Inst. Géol. Bassin Aquitaine*, Bordeaux, n. 3, p. 127–135.

Klingebiel, A., Pujol, C., and Vigneaux, M., 1970, Sur la stabilité des positions relatives du plateau et du talus continental dans le Golfe de Gascogne depuis le Miocène moyen: *C.R. Acad. Sc. Paris*, t. 270, 26, p. 3179.

Lapierre, F., 1969, Répartition des sédiments sur le plateau continental du Golfe de Gascogne:

intérêt des minéraux lourds: *Thèse Fac. Sc.* Univ. Bordeaux, n. 256, 2 tomes, t. 1; 182 p., t. 2, 90 p.

Lapierre, F., 1970, Fleuves et rivages préflandriens sur le plateau continental du Golfe de Gascogne: *Quaternaria*, XII Roma, 1970, p. 207–217.

Lapierre, F., Robert, J. P., and Ville, P., 1970, Esquisse géologique de la Manche orientale: *C.R. Acad. Sci., Paris*, t. 271 (série D), p. 381–384.

Latouche, C., 1971, Découverte d'attapulgite dans des sédiments carottés sur le Dôme Cantabria (Golfe de Gascogne). Conséquences paléogéographiques: *C.R. Acad. Sci., Paris*, t. 272, p. 2064–2067.

Latouche, C., 1971, Les argiles des bassins alluvionnaires aquitains et des dépendances océaniques. Contribution à l'étude d'un environnement: *Thèse Doctorat d'Etat, Univ. Bordeaux I*, n. 344, 2 tomes, 415 p.

Latouche, C. and Parra, M., 1970, Contribution à la connaissance des caractéristiques géochimiques des vases récentes et subactuelles du Golfe de Gascogne: *Bull. Inst. Géol. Bassin Aquitaine*, Bordeaux, n. 9, p. 167–185.

Laughton, A. S. and Berggren, W. A., 1970, Deep sea drilling in the Bay of Biscay (DSDP Leg XII): *Symposium sur l'histoire structurale du Golfe de Gascogne*, Rueil Malmaison, 14-16 déc. 1970, Edit. Technip, n. 22, t. II, 1971, p. VI-I-1–VI-I-4.

Laughton, A. D., Berggren, W. A., Bensen, R., Davis, T. A., Frantz, U., Musich, L., Perch-Nielsen, K., Ruffman, A., Van Hinte, J. E., and Whitmarch, R. B., 1970, Deep Sea Drilling project: leg 12: *Geotimes*, v. 15, n. 9, p. 10–14.

Le Borgne, E. and Le Mouel, J., 1970, Cartographie aéromagnétique du Golfe de Gascogne: *C.R. Acad. Sci. Paris*, t. 271, sér. D, n. 14, p. 1167–1170.

Le Mouel, J. L. and Le Borgne, E., 1970, Carte aéromagnétique du Golfe de Gascogne. Présentation et traitements: *Symposium sur l'histoire structurale du golfe de Gascogne*, 14–16 déc. 1970, Rueil-Malmaison. Technip Edit., n. 22, t. 2, p. VI-3-1–VI-3-12.

Le Pichon, X., Bonnin, J., and Sibuet, J. C., 1970, La faille nord-pyrénéenne: faille transformante liée à l'ouverture du golfe de Gascogne: *C.R. Acad. Sci. Paris*, D. 271, p. 1941–1944.

Le Pichon, X., Bonnin, J., Francheteau, J., and Sibuet, J. C., 1970, Une hypothèse d'évolution tectonique du golfe de Gascogne: *Symposium sur l'histoire structurale du golfe de Gascogne*, 14–16 déc. 1970, Rueil-Malmaison. Technip Edit., n. 22, t. 2, p. VI-11-1–VI-11-44.

Le Pichon, X. and Fox, P. J., 1971. Marginal offsets, fracture zones, and the early opening of the North Atlantic: *Journ. Geophysical Research*, v. 76, n. 26, p. 6294–6308.

Ludwig, W. J., Ewing, J. I., Ewing, M., Murauchi, S., Den, N., Asano, S., Hotta, H., Hayakawa, M., Asanuma, T., Tchikawa, K., and Noguichi, I., 1966, Sédiments and structure of the Japan trench: *J. Geophys. Res.*, v. 71, n. 8, p. 2121–2137.

Lumley, H. de, 1969, Les civilisations préhistoriques en France. Corrélations avec la chronologie Quaternaire: *Ed. C.N.R.S., Etudes Françaises sur le Quaternaire, VIIIe Congrès Intern.* INQUA, n. 13, p. 151–169.

Malzac, J., 1970, Apport d'une campagne de sismique réflexion Flexotir à l'étude géophysique du golfe de Gascogne: *Bull. Inst. Géol. Bassin d'Aquitaine*, Bordeaux, n. 9, p. 265–267.

Mangin, J. Ph., 1958, Le nummulitique sud-pyrénéen à l'ouest de l'Aragon. *Edit. Libreria general*, San Miguel, 4, Zaragoza.

Martin, G., Muraour, P., and Riccolvi, M., 1968, Etude par sismique réfraction du plateau continental au large de Belle Ile: *Trav. Lab. Géophysique.* Montpellier, fasc. 2.

Martin, G. and Vanney, J. R., 1971, Interprétation structurale du plateau de Rochebonne (golfe de Gascogne): *C.R. Acad. Sci.*, Paris, t. 273, p. 8, 11.

Martinez-Alvarez, J. A., 1968, Consideraciones respecto a la zona de fractura ("Falla cantabrica") que se desarolla desde Avilés (Asturias) hasta Cervera del Pisuerga (Palencia): *Act. geol. hisp.*, t. III, 5, p. 142–144.

Mattauer, M., 1968, Les traits structuraux essentiels de la chaîne pyrénéenne. *Rev. géogr. phys. et géol. dyn.*, X, 1, p. 3–12.

Mattauer, M. and Seguret, M., 1970, Les relations entre la chaîne des Pyrénées et le Golfe de Gascogne: *Symposium sur l'histoire structurale du Golfe de Gascogne*, Rueil Malmaison, 14–16 déc. 1970, Edit. Technip, n. 22, t. I, 1971, p. IV-4-1–IV-4-24.

Matthews, D. H. and Williams, C. A., 1968, Linear magnetic anomalies in the Bay of Biscay; a qualitative interpretation: *Earth and Planetary Sc. Letters*, v. 4, n. 4, p. 315–320.

Mensink, H., 1966, Stratigraphie and Palägeographie des marinen Jura in den nordwestlichen Iberischen Ketten (Spanien): *Beih. geol. Jahrb.*, v. 44, p. 55–102.

Montadert, L., Damotte, B., Debyser, J., Fail, J. P., Delteil, J. R., and Valery, P., 1970, Continental margin in the Bay of Biscay: S.C.O.R. *Symposium on the Geology of the East Atlantic Continental Margin*, Cambridge Delany Edit., *Institute of Geological Sciences*, Report n. 70/15, p. 43–74.

Montadert, L., Damotte, B., Delteil, J. R., Valery, P., and Winnock, E., 1970, Structure de la marge continentale septentrionale du Golfe de Gascogne (Bretagne et entrées de la Manche): *Symposium sur l'histoire structurale du Golfe de Gascogne*, Rueil Malmaison, 14–16 déc. 1970, Edit. Technip, n. 22, t. I, 1971, p. III-2-1–III-2-22.

Moyes, J., 1965, Les Ostracodes du Miocène aquitain. Essai de paléoécologie stratigraphique et paléogéographie: *Thèse Doctorat Etat, Fac. Sciences Bordeaux*, Impr. Drouillard, Bordeaux, 312 p.

Moyes, J. and Pachier, E., 1968, Répartition de quelques associations fauniques d'Ostracodes du Golfe de Gascogne: *Bull. Inst. Géol. Bassin Aquitaine*, Bordeaux, n. 5, p. 83–86.

Muraour, P., 1970, Considérations sur la génèse de la Méditerranée occidentale et du Golfe de Gascogne (Atlantique): *Tectonophysics*, v. 10, p. 663–677.

Muraour, P., Malzac, J., Frappa, M., and Martin, G., 1970, Contribution à l'étude géophysique du Golfe de Gascogne: *C.R. Acad. Sci. Paris*, t. 270, série D, n. 12, p. 1552–1554.

Muraour, P., Martin, G., Riccolvi, M., and Frappa, M., 1971, Etude par sismique réfraction de la marge continentale des Landes: *Bull. Inst. Géol. Bassin Aquitaine*, Bordeaux, n. 10, p. 101–109.

Nafe, J. E. and Drake, C. L., 1957, Physical properties of crystal materials as related to compressional wave velocities: *Annual Meeting of Soc. Expl. Geophys*, Dallas, Texas (unpublished).

Naudin, J. J., 1971, Etude morphostructurale du plateau continental aquitain: *Thèse Doctorat Spécialité*, Université Bordeaux I, n. 936, 114 p.

Naudin, J. J. and Prud'homme, R., 1971, Méthodes d'analyses morphologiques et morphostructurales d'interprétation des topographies et des bathymétries dans les domaines continentaux et marins: *Bull. Inst. Géol. Bassin Aquitaine*, Bordeaux, n. 10, p. 111–144.

Nesteroff, W., Duplaix, S., Sauvage, J., Lancelot, Y., Melieres, F., and Vincent, E., 1968, Les dépôts récents du canyon de Cap Breton: *Bull. Soc. Géol. Fr.*, v. X, n. 2, p. 218–252.

Netherlands Hydrographer, 1967, Navado III bathymetric, magnetic and gravity investigations. H. Neth M. S. *Snellius* 1964–1965.

Parga-Pondal, I., 1967, Carte géologique du nord-ouest de la péninsule ibérique: *Serv. géol. Portugal*.

Pautot, G., Auzende, J. M., and Le Pichon, X., 1970, Continuous Deep sea layer along North Atlantic margins related to early phase of rifting: *Nature*, v. 227, p. 361–354.

Pecherskiy, D. and Merlin, Y. E., 1971, Magnetization of rocks and nature of magnetic anomalies in the rift zone of the Atlantic Ocean: *Izvestiya*, n. 5, p. 326.

Peypouquet, J. P., 1970, Les Ostracodes de la région de Cap-Breton. Intérêt écologique et paléoécologique: *Thèse Doctorat Spécialité. Fac. Sc. Bordeaux*, n. 805, 267 p.

Pinot, J. P., 1968, Littoraux würmiens submergés à l'Ouest de Belle Ile: *Bull. Assoc. Fr. pour l'étude du Quaternaire*, v. 3, p. 197–216.

Poignant, A., 1965, Révision du Crétacé inférieur en Aquitaine occidentale et méridionale: *Impr. de Lagny*, Seine et Marne.

Pratviel, L., 1967, Topographies souterraines de l'Aquitaine occidentale: *Actes Soc. Lin. Bordeaux*, t. 104, sér. B, n. 25, 1 p.

Prud'homme, R., 1970, Les lignes de contours de thalwegs. Une méthode d'étude morphologique à partir des cartes topographiques: *Bull. Soc. de Borda*, Dax, n. 337, p. 77–80.

Prud'homme, R., 1972, Analyse morphostructurale appliquée à l'Aquitaine occidentale et au Golfe de Gascogne. Définition d'une méthodologie cartographique interprétative: *Thèse Doctorat ès-Sci. Nat., Univ. Bordeaux I*, n. 353, 363 p.

Prud'homme, R. and Vigneaux, M., 1970, Etudes géomorphologiques et morphométriques du substratum sous-marin profond du Golfe de Gascogne: *Actes Acad. Bordeaux*, 4e sér., t. XXV, p. 1–19.

Prud'homme, R. and Vigneaux, M., 1971, Hypothèse sur l'organisation structurale du Golfe de Gascogne en fonction de l'analyse morphologique: *C.R. Acad. Sci. Paris*, t. 272, p. 527–530.

Pujos, M., 1971, Quelques exemples de distributions des Foraminifères benthiques sur le plateau continental du Golfe de Gascogne. Relation entre microfaune et environnement: *Cahiers Océanogr.*, XXIII, 5, p. 445–453.

Pujos-Lamy, A., 1971, Les Foraminifères benthiques abyssaux: leur utilisation pour la mise en évidence des variations climatiques dans une carotte du Quaternaire récent: *C.R. Acad. Sc. Paris*, t. 272, p. 215–218.

Raitt, R. W., Fisher, R. L., and Mason, R. G., 1955, Tonga Trench: *Geol. Soc. Am. Spec. Pap.*, 62, p. 237–254.

Robert, J. P., 1969, Géologie du plateau continental français: *Revue I.F.P.*, v. 24, n. 4, p. 383–440.

Rouvillois, A., 1970, Biocénose et taphacoenose de Foraminifères sur le plateau continental atlantique au large de l'Ile d'Yeu: *Rev. Microp.*, Paris, v. 13, n. 3, p. 188–204.

Rumeau, J. L. and Vanney, J. R., 1969, Caractères géochimiques et origine des sédiments récents du plateau continental atlantique dans le nord du Golfe de Gascogne: *Bull. Centre Rech. Pau*, S.N.P.A., v. 3, n. 1, p. 125–146.

Saint-Requier, A., 1970, La baie d'Audierne, étude de morphologie et de sédimentologie sous-marines: *Thèse 3e cycle*, Paris, n. 223.

Schnitker, D., 1969, Distribution of Foraminifera in a portion of the continental shelf of the Golfe (Gulf of Biscay): *Bull. Centre Rech. Pau*, S.N.P.A., v. 3, n. 1, p. 33–64.

Schoeffler, J., 1965, Une hypothèse sur la tectogénèse de la chaine pyrénéenne et de ses abords: *Bull. Soc. Géol. Fr.*, p. 917–920.

Schoeffler, J., 1969, Le Gouf de Cap Breton de l'Eocène inférieur à nos jours: *Submarine Geology and Geophysics*, Colston papers, vol. XVII, p. 265–268.

Schouten, J. A., 1971, A Fundamental analysis of magnetic anomalies over oceanic ridges: *Marine Geophys. Res.*, v. 1, p. 111–144.

Schouten, J. A., Rutten, K. W., and Collette, B. J., 1970, Magnetic anomaly symmetry in the Bay of Biscay: *Symposium sur l'histoire structurale du golfe de Gascogne*. Rueil Malmaison. Technip Edit., n. 22, t. 2, p. VI-13-1–VI-13-10.

Scrutton, R. A., Stacey, A. P., and Gray, F. (in preparation): Evidence for the mode of formation of Porcupine Seabight.

Sibuet, J. C. and Le Pichon, X., 1970, Structure gravimétrique du golfe de Gascogne et le fossé marginal nord-espagnol: *Symposium sur l'histoire structurale du golfe de Gascogne*, 14–16 déc. 1970, Rueil-Malmaison, Technip Edit., n. 22, t. 2, p. VI-9-1–VI-9-18.

Sibuet, J. C., Pautot, G., and Le Pichon, X., 1970, Interprétation structurale du golfe de Gascogne à partir des profils sismiques: *Symposium sur l'histoire structurale du golfe de Gascogne*, 14–16 déc. 1970, Rueil-Malmaison, Technip Edit., n. 22, t. 2, p. VI-10-1–VI-10-32.

Sichler, B., Martinais, J., Sibuet, J. C., and Le Pichon, X., 1970, Vitesses des ondes sismiques dans les couches sédimentaires profondes du golfe de Gascogne: *Symposium sur l'histoire structurale du golfe de Gascogne*, 14–16 déc. 1970, Rueil-Malmaison, Technip Edit., n. 22, t. 2, p. VI-8-1–VI-8-20.

Stride, A. H., Curray, J. R., Moore, D. G., and Belderson, R. H., 1969, Marine geology of the atlantic continental margin of Europe: *Phil. Trans. Roy. Soc. London*, A, v. 264, p. 31–75.

Talwani, M., Sutton, G. H., and Worzel, J. L., 1959, A crustal section across the Puerto Rico trench: *J. Geophys. Res.*, v. 64, p. 1545–1555.

Talwani, M., Worzel, J. L., and Ewing, M., 1961, Gravity anomalies and crustal section across the Tonga trench: *J. Geophys. Res.*, v. 66.

Talwani, M., Worzel, J. L., and Landisman, M., 1959, Rapid gravity computations for two-dimensional bodies with application to the Mendocine submarine fracture zone: *J. Geophys. Res.*, v. 64, p. 49–59.

Tarling, D. H. and Symons, D. T. A., 1967, A stability index of remanence in paleomagnetism. *Geophys. Jour. Roy. Astr. Soc.*, v. 12, p. 443–448.

Tischer, G., 1966, Über die Wealden-Ablagerung und die Tecktonik der östlichen Sierra de los Cameros in den nordwestlichen Iberischen Ketten (Spanien): *Beih. geol. Jahrb.*, v. 44, p. 123–164.

Valery, P., Delteil, J. P., Cottencon, A., Montadert, L., Damotte, B., and Fail, J. P., 1970, La marge continentale d'Aquitaine: *Symposium sur l'histoire structurale du Golfe de Gascogne*, Rueil-Malmaison, 14–16 déc. 1970, Edit. Technip., n. 22, t. I, 1971, p. IV-8-1–IV-8-24.

Van der Voo, R., 1967, The rotation of Spain: paleomagnetic evidence from the Spanish Meseta: *Paleogeogr., Paleoclimatol., Paleoecol.*, v. 3, n. 4, p. 393–416.

Van der Voo, R., 1968, Geology and paleomagnetism of an anticlinal structure in Lower Triassic sediments near Atienza (Guadalajara, Spain): *Geol. Mijnbouw*, v. 47, n. 3, p. 186–190.

Van der Voo, R., 1968, Comments on a paper by N. D. Watkins and A. Richardson on "Paleomagnetism of the Lisbon volcanics": *Geophys. J.*, v. 16, p. 543–547.

Van der Voo, R., 1969, Paleomagnetic evidence for the rotation of the Iberian peninsula: *Tectonophysics*, v. 7, p. 5–56.

Van der Voo, R. and Zijderveld, J. D. A., 1971, A renewed paleomagnetic study of the Lisbon volcanics, and its implications for the rotation of the Iberian Peninsula: *J. Geophys. Res.*, sous presse.

Van Dongen, P. G., 1967, The rotation of Spain: paleomagnetic evidence from the eastern Pyrenees: *Paleogeogr., Paleoclimatol., Paleoecol.*, v. 3, n. 4, p. 417–432.

Vanney, J. R., 1967, La montagne sous-marine "Cantabria" (Golfe de Gascogne): *Trav. C.R.E.O.*, v. 7, n. 2, p. 19–23.

Vanney, J. R., 1969, Le précontinent du centre du Golfe de Gascogne. Recherches géomorphologiques: *Ecole Pratique Hautes Etudes*, Lab. Géomorph. Dinard, Mém. n. 16, 365 p.

Vanney, J. R., Scolari, G., Lapierre, F., Martin, G., and Dieucho, A., 1970, Provisional geological map of the Armoricain shelf: *Colloque I.F.P., CNEXO*, Rueil-Malmaison, 14–16 déc. *Symposium sur l'histoire structurale du Golfe de Gascogne*, Edit. Technip, n. 22, t. 1, 1971, p. III-1-1–III-1-20.

Vekua, L. V., 1961, Certain results of paleomagnetic studies of effusive rocks of Georgia: *Akad. Nauk. SSSR Izv., Geophys., Ser.*, p. 1668–1673.

Vigneaux, M., 1962, Le Bassin d'Aquitaine: *Colloque sur le Paléogène*, Bordeaux, sept. 62, *Mém. B.R.G.M.* n. 28, t. I, p. 177–226.

Vigneaux, M., 1971, Bilan d'études d'environnement marin et application dans le Golfe de Gascogne: *Colloque International C.N.E.X.O.*, Bordeaux, thème III, t. 2, G1-05, 66 p.

Vigneaux, M., Caralp, M., Dumon, J. C., Klingebiel, A., Latouche, C., Moyes, J., and Prud'homme, R., 1970, Bilan des travaux de géologie marine effectués en 1969 dans le Golfe de Gascogne par l'Institut de Géologie du Bassin d'Aquitaine: *Bull. Inst. Géol. Bassin Aquitaine*, n. sp., 107 p.

Vigneaux, M., Naudin, J. J., and Prud'homme, R., 1971, Interprétation géologique des données bathymétriques comme accès à la connaissance architecturale des bassins océaniques. Application au Golfe de Gascogne: *Colloque International C.N.E.X.O.*, Bordeaux, thème III, t. 2, G1-01 bis, 33 p.

Vollstadt, H., Rother, K., and Nozharov, P., 1967, The paleomagnetic stability and the petrology of some Coenozoic and Cretaceous andesites of Bulgaria: *Earth Plan. Sci. Lett.*, v. 3, p. 399–408.

Walcott, R. O., 1970, Flexural rigidity, thickness, and viscosity of the lithosphere: *J. Geophys. Res.*, v. 75, p. 3941–3954.

Watkins, N. D. and Richardson, A., 1967, Paleomagnetism of the Lisbon volcanics: *Geophys. Jour.*, n. 15, p. 287–304.

Williams, C. A., 1970, Magnetic survey west of Biscay: *Symposium sur l'histoire structurale du Golfe de Gascogne*, Rueil-Malmaison, Technip Edit., n. 22, t. 2, p. VI-4-1–VI-4-6.

Winnock, E., 1970, Géologie succincte du bassin d'Aquitaine (contribution à l'histoire du Golfe de Gascogne): *Symposium sur l'histoire structurale du Golfe de Gascogne.* Rueil-Malmaison, 14–16 déc. 1970, Edit. Technip, n. 22, t. I, 1971, p. IV-1-1–IV-1-30.

Worzel, J. L., 1965, Deep structure of coastal margins and mid-oceanic ridges, in: *Submarine Geology and Geophysics, Colston Paper 17*: Butterworths Scientific Publications, London, XVII, p. 335–361.

Zijderveld, J. D. A., 1967, The natural remanent magnetizations of the Exeter volcanic traps (Permian, Europe): *Tectonophysics*, v. 4, n. 2, p. 121–153.

Zijderveld, J. D. A. and Van der Voo, R., 1970, Golfe de Gascogne. Fiches synthétiques des forages marins in *Rapport annuel* 1970. Recherches et production d'hydrocarbures en France. Service de conservation des gisements d'hydrocarbures, Rueil-Malmaison.

Chapter 10

GEOLOGY OF WEST AFRICA AND CANARY AND CAPE VERDE ISLANDS

William P. Dillon*

Department of Geology
California State University, San Jose
San Jose, California

and

Jean M. A. Sougy

Laboratoire de Géologie Structurale
Université de Provence, Centre de St. Jérôme
Marseille, France

I. OVERVIEW OF WEST AFRICAN GEOLOGY

West Africa consists essentially of a Precambrian granitized craton (radiometric ages of 2700 to 1600 m.y.) covered by a thin sedimentary blanket, the oldest sedimentary rocks being as much as 1000 m.y. old. The Precambrian basement has been warped into three major uplifts trending approximately east-northeast (Fig. 1): the Léo uplift (Dorsale de Léo, also known as the Liberia–Upper Volta uplift, and Bouclier éburnéen); the Reguibat uplift; and the Anti-Atlas uplift. Deposition during most of Paleozoic time apparently occurred in epicontinental seas. The principal preserved depositional areas are the Taoudeni basin in Mauritania and Mali (Fig. 2); the Tindouf basin in

* Present address: Office of Marine Geology, U.S. Geological Survey, Woods Hole, Massachusetts 02543

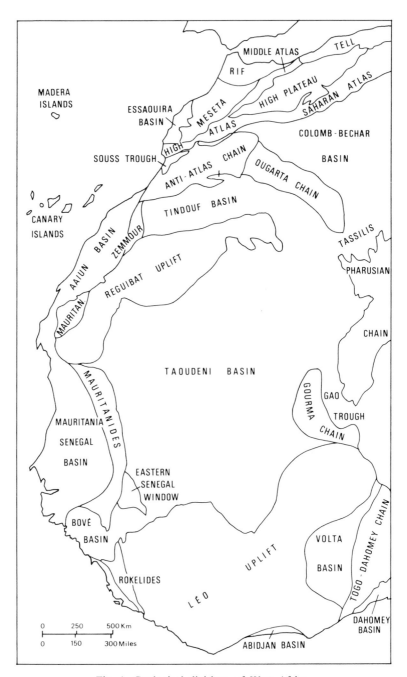

Fig. 1. Geological divisions of West Africa.

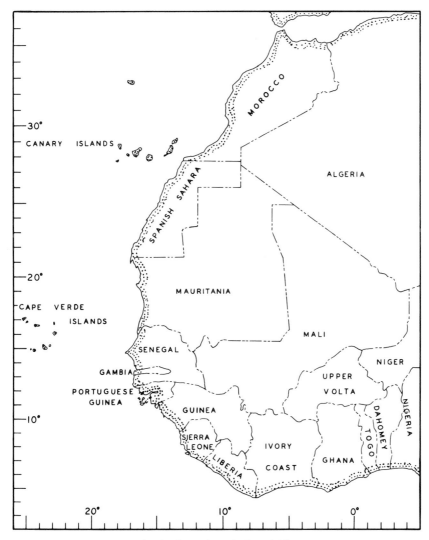

Fig. 2. Countries of West Africa.

Morocco, Algeria, Mauritania, and Spanish Sahara; the Volta basin in Ghana and Upper Volta; and the Bové basin in Guinea and Portuguese Guinea.

In late Paleozoic time orogenies folded and thrust eastward the rocks of the western parts of the Taoudeni and Tindouf basins forming the Mauritanides and the folded zone of the Morroccan Meseta.

Along the coast of west Africa, these structures are covered by rocks of a series of continental margin depositional areas, the Senegal basin (also known as the Mauritania-Senegal basin), the Aaiun basin (also known as the Rio basin or, in its northern part, as the Tarfaya basin), and the Essaouira basin.

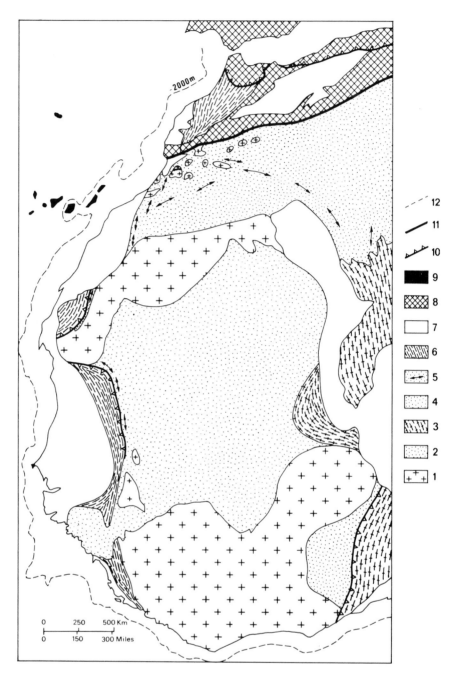

Fig. 3. Summary of West African structural geology. Legend: (1) Granitized basement > 1000 m.y.; (2) unfolded upper Precambrian rocks < 1000 m.y.; (3) Pan-African fold belts (650–550 m.y.) (reactivated basement and folded upper Precambrian cover); (4) (same symbol as 2) unfolded Paleozoic cover 570–280 m.y.; (5) folded upper Precambrian and Paleozoic cover of Hercynian foreland; (6) Hercynian orogenic belts (folded Paleozoic cover and older rocks; metamorphic and allochthonous zones); (7) Mesozoic–Cenozoic cover; (8) Alpine orogenic belts (folded Mesozoic and lower Cenozoic cover and reactivated Hercynian basement); (9) Mesozoic to Holocene volcanic rocks; (10) overthrust; (11) major fault; (12) −2000 m isobath.

These are partially separated by the basement uplifts, but actually these basins and the continental shelf formed a connected zone of sedimentation along the western continental margin of Africa. The present area of the coastal basins apparently received sediment through most of Paleozoic time as did the interior basins, and the two sets of basins probably formed continuous depositional zones at that time. That is, the Senegal and Taoudeni basins were probably connected as were the Aaiun and Tindouf basins.

During most of the Mesozoic era the coastal basins subsided and received sediment, while the interior basins generally did not. Little deposition has occurred on the continental block since Eocene. In the northern part of the area Tertiary tectonism in the Atlas and Rif mountains has complicated this simple picture. A summary of West African geology is presented in Fig. 3.

II. THE PRECAMBRIAN FRAMEWORK

Northwest Africa, as far north as the northern limit of the Anti-Atlas Mountains in Morocco, consists of a Precambrian granitized basement which

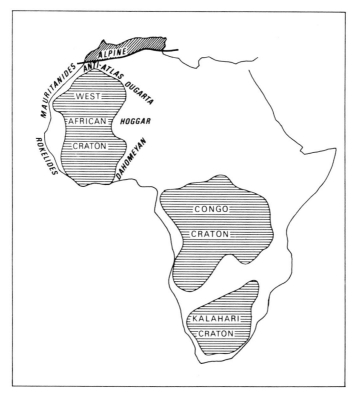

Fig. 4. Cratons of Africa, and approximate locations of orogenic zones of West Africa. Orogenic zones indicated in italics (after Clifford, 1966, 1970).

forms part of the largest shield area in the world, the African shield (Fig. 4). Precambrian rocks are rare to the north of the South Atlas fault which forms the boundary between the Anti-Atlas and Atlas mountains. While the granitized shield is well exposed in the region south of a line from Suez, Egypt to Accra, Ghana, toward the northwest it is increasingly masked by a blanket of unfolded or little folded sedimentary rocks. The three zones of basement exposed in the uplifts of West Africa follow a pattern of decreasing exposure toward the north (Fig. 5). The northwestern part of the shield is the West African craton (Kennedy, 1965) which was consolidated during the final Birrimian granitizations at about 1800 m.y. but contains much older rocks which were metamorphosed and migmatized about 2700 m.y. ago. The cores of cratons in central and southern Africa have ages of about 3000 m.y., but this age has only been reported at scattered locations in Liberia and Sierra Leone in West Africa (Leo and White, 1968; Choubert and Faure-Muret, 1971d). The tectonic trends are generally north–south but tend to change to northwest in the northwestern part of the craton (Fig. 5).

A. Zone of Outcrop

1. The Léo Uplift

The Léo uplift is separated into two zones by the Sassandra River fault, a major mylonitized fracture (Fig. 5). To the west, rocks of Archean age are dominant. These are catametamorphic quartzites, ferruginous quartzites, gneisses, charnockites, granulites and migmatites with a minimum age, for the most part, of 2600–2750 m.y., with dates of 2850 m.y. also reported (Hurley et al., 1966, Allen Snelling and Rex, 1967; Papon, Roques and Vachette, 1968; Andrews-Jones, 1971). This ancient basement occupies the western part of the Ivory Coast, eastern Guinea, Liberia, and Sierra Leone. It is referred to as the Guinéo-Libérian craton, or in part, as the Man massif (Choubert and Faure-Muret, 1971c) or the Liberianides (Tagini, 1971). In Liberia the older (2700 m.y.) rocks have been distinguished from younger (2000 m.y. and 550 m.y.) rocks by mapping of aeromagnetic trends (Behrendt and Wotorson, 1971).

East of the Sassandra River fault lie "Birrimian" flysch, schist, and epimetamorphic graywackes and greenstones showing a north-northeast tectonic trend. These were invaded by synkinematic granites dating from 2000 to 1900 m.y. (Eburnean cycle) (Arnould, 1961; Bonhomme, 1962), with the latest intrusion at 1650 m.y. (Vachette, 1973, oral comm.).

2. The Minor Massifs of Eastern Senegal and Kayes

The granitic basement also reaches the surface in the eastern Senegal and Kayes massifs (Fig. 5). Cross-cutting granites give a minimum age of 2045 m.y.

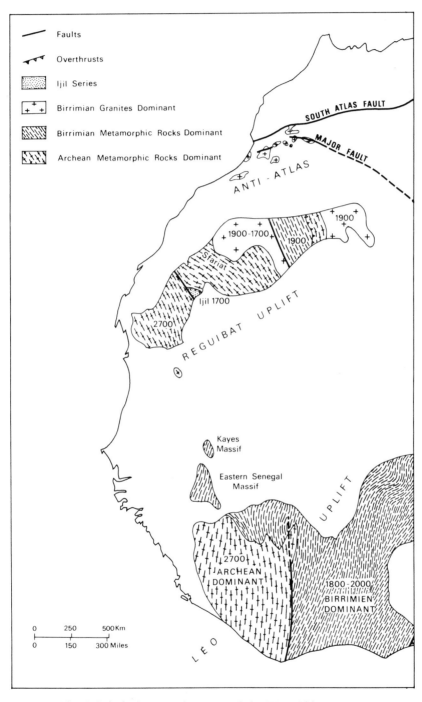

Fig. 5. Principal zones of outcrop of the West African craton.

(Bassot, Bonhomme, Rogues, and Vachette, 1963). The tectonic trends are usually north-northeast and the rocks are generally assigned to the Birrimian, although they crop out west of the trend of the Sassandra River fault. Other outcrops of basement occur in the Mauritanides, which were exposed due to later overthrusting or erosion.

3. *The Reguibat Uplift*

The Reguibat uplift (Blanchot, 1955, Rocci, 1957; Sougy, 1960; Barrere, 1967; Clifford, 1970; Bronner, 1972) has a somewhat similar outcrop pattern to the Léo uplift. A major fault separates an eastern zone with Birrimian characteristics (black graywackes of the Aguelt Nebkha Series and Yetti granites) from a western zone comprising several groups with minimum age of 2600 m.y. (Rhallamane, Tiris, Tasiast, and Amsaga series) (Fig. 5). The western zone was invaded heavily by post-tectonic granite of ages in the range 2000-1700 m.y., however (Vachette, *et al.*, 1973). The old rocks of the western zone belong to the cata- and mesozones of metamorphism just as do those found in the western section of the Léo uplift, with charnockites, gneisses, quartzites, and magnetitic quartzites present. Structural trends at the southwest end of the uplift are north-northeast as in the southwestern part of the Léo uplift, but swing to the northwest on approaching the central part of the outcrop (Fig. 5). The linear Sfariat chain, characterized by vertical bands of magnetitic quartzite, takes this new direction.

Rocks of the Reguibat uplift are invaded by nonoriented, sheared granites (Rocci, 1957; Arribas, 1958; Sougy, 1960). During a recent study, ages of 2000–1800 m.y. were determined for these granites (Vachette, *et al.*, 1973). In the eastern part of the uplift are found nuclei of the older series (Chegga Series) which have been thrust over the beds of Birrimian age (Marchand *et al.*, 1971; p. 65–65 and pl. 15) and which also have been intruded by the 2000–1700 m.y. granites. East of long. 9° W a group of acidic volcanic–detrital rocks which have also been granitized (Imourene, Aioun, and Abd el Malek series) were involved in the latest Birrimian folding and may represent a molasse deposit.

In the west-central part of the Reguibat uplift, within the zone of old rocks, lies a small area of hematitic quartzites (itabirites), mica shists, and tectonic breccias which have been provisionally radiometrically dated at 1200–1700 m.y. (Fig. 5) (Bronner and Vachette, oral commun. 1973). These rocks appear to represent an ungranitized, allochthonous, synclinal klippe, preserved along the northern margin of a large west-northwest-trending fault. This fault separates different tectonic zones within the magnetitic quartzites of the old rocks (Tiris Series). Thus a unique accident seems to have preserved evidence of another orogenic cycle.

4. *The Anti-Atlas Uplift*

The Anti-Atlas Mountains contain a series of massifs of basement, along another east-northeast-trending uplift (Fig. 5). The style of occurrence apparently results from the intersection of several tectonic trends, perhaps even including Hercynian and Alpine ones. The Anti-Atlas zone seems to be a very complex one, perhaps because it existed at the northern margin of the ancient continent.

Again, the most ancient rocks are found predominantly in the western part of the chain. These are mica schists (Kerdous System), trending east–west with ages of 2700–2500 m.y. and degree of metamorphism increasing markedly from north to south (Choubert, 1952; Choubert, 1963; Charlot *et al.*, 1970; Choubert and Faure-Muret, 1971a). Locally intense migmatization has occurred.

Rocks, mainly schists, trending north–south of age 1950–1750 m.y. (Zenaga System) crop out in the central part of the Anti-Atlas to the south of the major fault (Fig. 5). These are, of course, comparable in age to the rocks of the eastern parts of the Reguibat and Léo uplifts. The structural relationships of these rocks to the older ones is unknown, as their contacts are masked by more recent formations.

Perhaps only the series discussed above should be considered as part of the West African craton (Caby, 1970; Sougy, 1972, 1970). However, a number of more recent tectonic intervals may have occurred in the Anti-Atlas, and their overprints create considerable problems in dating rocks (Charlot *et al.*, 1970). Rocks affected by these younger events possibly are represented by: (1) schists, marbles, and metavolcanic rocks of 1670–1600 m.y. age; (2) epicontinental (miogeosynclinal) rocks in the western Anti-Atlas which underwent orogeny at about 1500–1440 m.y. age; and (3) rocks of a eugeosynclinal–miogeosynclinal pair in the eastern Anti-Atlas, where dates on the associated granites produce ages of about 1050–950 m.y. Perhaps other periods of emplacement of granites and pegmatites have taken place in the area also, but this subdivision into many orogenic episodes is quite speculative.

B. Evolution of the West African Craton

The development of West Africa may be outlined, but conclusions must be accepted with care, particularly for pre-Silurian times. Except for Morocco, there are very little paleontological data and many of the problems are more outlined than resolved.

The western parts of the present uplifts were probably sections of a continuous cratonized, metamorphosed, granite-intruded Precambrian nucleus, which was stabilized about 2700 m.y. ago. This nucleus is characterized by

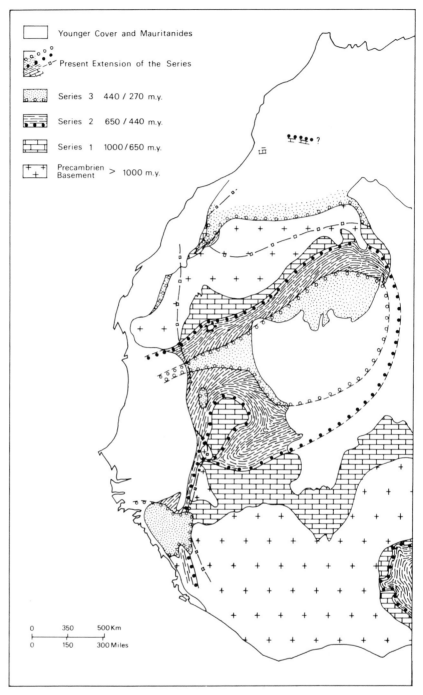

Fig. 6. Zones of outcrop and probable extensions of upper Precambrian and Paleozoic sedimentary series. Series 1 is about 1000–650 m.y. old and includes Stromatolitic limestones. Series 2 has a basal tillite of about 650–620 m.y. Series 3 has a basal tillite of about 440 m.y. and includes Upper Ordovician to Carboniferous beds.

complex folds and magnetitic quartzites. Subsequently, a second phase of cratonization at 2000–1700 m.y. ago affected the rocks to the east of the older section. Tectonic trends of both orogenic zones were approximately north–south. The West African craton has probably reacted as a stable block since the granitic intrusions of 1700 m.y. ago.

Complex ideas have been presented regarding Birrimian geosynclines in the Léo uplift region (Choubert and Faure-Muret, 1971c). According to Tagini (1971) a continental margin probably existed to the east of the most ancient rocks. The first fairly clear evidence for continental margin depositional conditions is in the sediments of the northern Anti-Atlas, which were deposited between the orogeny of 1500–1400 m.y. ago and that of about 1000 m.y. ago (Choubert and Faure-Muret, 1971a; Schenk, 1971). In this northern zone of the Anti-Atlas, sediments typical of the continental rise (black and gray shale, flysch, and basaltic volcanic rocks) occur to the north and shelf-type sediments (marine carbonates, calcareous shales and sandstones, with continental red beds and rhyolitic volcanics) crop out to the south (Schenk, 1971). The area north of the Anti-Atlas has been accreted to the West African craton since 1000 m.y. ago.

Geochronological studies seem to indicate that the two main orogenic zones of the West African craton are continued on the northeast coast of Brazil and in the Guiana shield (Hurley et al., 1967). Certain other features of the West African basement may become useful in intercontinental comparison, such as the northwest-trending Sfariat chain in the Reguibat uplift, which trends approximately normal to the present continental margin. The Archean–Birrimian boundary is also a potentially valuable marker. Allochthonous material found in nappes of the Mauritanide zone (sericite schists, mica schists, itabirites with hematite of age 1700 (?) m.y.) (Bronner, 1972; Vachette, pers. comm.) and not known in situ, apparently had its origin to the west and also presents the possibility of intercontinental correlations.

III. LATE PRECAMBRIAN–EARLY PALEOZOIC DEVELOPMENT OF WEST AFRICA

A. The Upper Precambrian Cover (1000–650 m.y.)

The cratonized basement was peneplaned during the course of a long continental evolution. Upon this continental paleosurface, which is characterized by scattered aeolian pebbles (Bongrand et al., 1961), a sedimentary blanket was deposited (Fig. 6). Deposition occurred in a series of basins separated by gradual warpings of basement. In addition to the Taoudeni and Tindouf basins these include the smaller Volta and Bové (or Bafata, or Bowé) basins, and the continental shelf area of the Anti-Atlas (Fig. 1). The sediments in these basins are upper Precambrian (about 1000 m.y. old) to Upper Carboniferous. They are generally epicontinental marine and continental types with unconformities and some discordances resulting from minor adjustments of the craton, but are unfolded or only slightly folded in the basins. The deposits also extend beyond the borders of the craton into more unstable zones, where their structure reveals evidence of successive orogenic episodes (Pan-African, Caledonian–Taconic, Hercynian–Acadian, and Alpine).

Here we will consider the deposits formed prior to the Pan-African episode and the Eocambrian glacial epoch of 620–650 m.y. Despite their variations, the principal stratigraphic outline established along the northwest margin of the Taoudeni basin seems generally valid (Trompette, 1969a, 1969b, 1970).

Sediments deposited from 100 to 650 m.y. ago form a series of sandstone–shale–limestone beds with very thick development of stromatolitic limestones in some beds (series 1, Fig. 6). At least three groups, separated by discontinuities, can be recognized. The lowermost group generally rests on peneplaned basement of the West African craton and is characterized by sandstones and a lagoonal facies. Above an unconformity, the lower part of the second group consists of sandstones with some evidence of glaciation. Immediately above these are transgressive shales and then stromatolitic limestones. Above another unconformity, the uppermost group is of shale–sandstone composition. All the rocks of these lower groups show indications of deposition in either continental or shallow marine conditions, except perhaps in the Hoggar (Fig. 4) (Caby, 1970). Conditions for development of the stromatolite reefs may have been warm, calm, high-salinity, shallow water with low sediment supply (Trompette, 1969b). Each of the groups has a slightly different depocenter (Marchand et al., 1971) indicating some warping of the craton. In the west the groups of this upper Precambrian series were eroded prior to the deposition of the second cycle, and in some cases the latter may rest directly on basement. The series is not known along most of the south side of the Tindouf basin, but an ancient shoreline of this age is reported at the basin's southwest corner (Sougy, 1964).

In the Anti-Atlas region Choubert (1952, 1955) and Choubert et al. (1966, 1968) have correlated the carbonates of the Adoudounian (about 800–570 m.y., Choubert and Faure-Muret, 1971a) to this first series of sediments deposited upon the West African craton. In contrast, Sougy (1970, 1972) and Caby (1970) have suggested that the series might be equivalent to the older (Precambrian II$_2$) rocks which lie folded against the southwest side of the major fault in the massifs of the central Anti-Atlas.

The old deposits are predominantly sandy in the Volta basin and separated by an unconformity from the Eocambrian tillites above (Leprun and Trompette, 1969; Sougy, 1971).

B. The Upper Precambrian and Lower Paleozoic Cover (650–440 m.y.)

The late Precambrian deposition (sometimes called "Infracambrian" which includes late Precambrian and Early Cambrian) was interrupted by a glacial episode called "Eocambrian" at about 620–650 m.y. A second major glaciation by continental ice sheets took place at the end of the Ordovician (440 m.y.). These glacial episodes provide useful markers in West African deposition.

The Eocambrian event is known at many places in the world and, in Africa, evidence is abundant around the Taoudeni basin (Zimmerman, 1960) and along the western flank of the Volta basin (Leprun and Trompette, 1970; Sougy, 1971). Ice movement was apparently from north to south in West Africa, based on evidence from the glacial erosion surface on the south flank of the central Reguibat uplift (Marchand *et al.*, 1971).

Beds in the Taoudeni basin directly above the surface of glacial erosion are glacial conglomerates containing striated pebbles. Above these are rocks deposited during transgression, including limestone, then shale with flint, and a few rare stromatolites (series 2, Fig. 6). In the western part of the basin this series 2 transgresses directly onto basement. A widespread red-bed section occurs above the shale, which may be intercalated with white sandstone to the west. This is succeeded by sandstone which is assigned to the Tremadocian (base of the Ordovician) (Late Cambrian) (Trompette, 1970). The shales may reflect transgression after the end of the glaciation and the coarser sediments may reflect subsequent uplift of new sediment sources. The red beds might be considered a molasse (Trompette, 1972), as similar rocks in the Dahomey–Togo–Ghana area and in the Hoggar seem to represent such deposits. The Cambrian, if it exists, is not paleontologically distinguishable from older rocks in the Taoudeni basin except by inarticulate brachiopods (*Westonia*) of doubtful age, but assigned to the upper Cambrian by Legrand (1969).

In the Tindouf basin, only the upper part of the series 2 sequence is represented, the basal tillite being unknown. The area may have been an upland area, undergoing glacial erosion and supplying sediment to the southern regions. Certainly the Reguibat uplift was being denuded at this time, as blocks of Rapakivi-type granite (El Archeouat Granite) from the uplift are found in the basal tillite on the northern flank of the Taoudeni basin (Marchand *et al.*, 1970).

In the Anti-Atlas of Morocco, rhyolites of the Precambrian III (1050–900 m.y. to 550 m.y.) (Choubert and Faure-Muret, 1971*a*) apparently subsided strongly, and carbonate accumulation took place (Adoudounian) including a red-bed sequence ("Lie de Vin" Series). Sougy (1972) correlates these to the second series (Fig. 6) by comparing them to the limestones (Monod, 1952; Trompette, 1970) and red beds (Villemur, 1967; Marchand *et al.*, 1972, p. 54) of the Taoudeni basin. Speculatively, the Lie de Vin Series might also represent molasse associated with the Pan-African event.

Above these upper Precambrian–Infracambrian rocks lies a Cambrian–Ordovician series beginning with *Archaeocyathus* limestones which are well developed in the Anti-Atlas (Choubert and Debrenne, 1964) and present in the Tindouf basin and (Querol, 1966), and also part of the series 2 (Fig. 6). These were apparently epicontinental (shelf) deposits, although they are very thick and contain andesitic volcanic sediments. In the High Atlas Mountains, a monotonous shale facies is present (Destombes and Jeannette, 1955; Choubert and Marcais, 1952). Flysch, graywackes, volcanic sediments, and turbidites again suggest a eugeosynclinal, slope–island arc environment in this area.

Ordovician sediments of series 2 which can be seen in Africa generally represent deposition on a shallow shelf, and are dominantly clastic, much of them fairly coarse (sand). In the Anti-Atlas, Ordovician sediments are represented by clean sandstones interbedded with some shale. Four periods of transgression may have occurred, in which shales were laid down (Destombes, 1971) and separated by periods of regression, in which sands were deposited. In the Tindouf basin the Ordovician sediments are quartzitic sandstone at bottom and top with marine green shales between them (Sougy, 1955, 1956, 1964; Querol, 1966). To the south, the Ordovician (?) rocks are mainly continental sandstone (Carrington da Costa, 1951; Destombes, Sougy and Willefert, 1969), with 3000 ft of white quartzose sandstone in the Bafata basin, which is siliceous, calcareous, or sometimes cemented by salt. The literature implies a general decrease in marine affinities of the sediments on moving southward, but this may simply be a result of the restricted zone of Ordovician rocks normally observed by geologists.

C. Pan-African Tectonism

Kennedy (1964, 1965) has shown that the African cratons are surrounded by belts of rocks with radiometric ages which are generally in the range 500–650 m.y. (550 ± 100 m.y., Clifford, 1970). This period is referred to as the Pan-African (or Damaran) thermotectonic event. The Pan-African chains resulted from remobilizing of the craton's margin and folding of its sedimentary cover, without, however, any significant granitization. The more important effects seem to have occurred at about 600 m.y. after the formation of the rocks of the series 1 (Fig. 6).

Pan-African chains are found everywhere around the West African craton, except along its southern margin. East of the craton the Dahomeyan chain is interrupted on the south at the continental margin and has thrusts directed to the west toward the Volta basin. In northwestern Nigeria significant orogenic activity was recorded (Grant, 1967, 1969a, 1969b; McCurry, 1971). The Pharusian chain (Fig. 1) of the Hoggar (Fig. 4) is a metamorphic series on the east side of the craton which is also related to the Pan-African event (Caby, 1970). North of the craton, in the Anti-Atlas, a thick series of quartzites and stromatolitic limestones and shales, conglomerates, and volcano-detrital sediments were affected by the tectonism. These are much less granitized than the shield, but ages of rhyolites and granites which exist are about 500 m.y. (Charlot et al., 1970; Choubert et al., 1965).

To the west of the craton the Pan-African event was very important in Sierra Leone (Allen, 1965, 1968, 1969; Andrews-Jones, 1971). There, in the Rokelides (Figs. 1 and 4) the Precambrian deposits are progressively more metamorphosed and folded toward the west, and volcanic rocks become dominant (Rokelian assemblage: Rokel River Series, Marampa Schists, and Kasila Series from east to west). The higher metamorphic grade rocks to the west were thrust eastward over each other and over the old craton. Similar K/Ar ages of about 550 m.y. have been reported from the Marampa Schists and Kasila Series as well as from the reactivated old cratonic rocks involved in the orogeny (Allen, Snelling and Rex, 1967; Allen, 1969; Andrews-Jones, 1971).

The trend of Pan-African tectonism apparently continues northward in the rocks of the Saionya Scarp Series in the Mauritanides zone of Senegal and Mauritania. These outcrops are separated from the Rokelian assemblage by the Paleozoic deposits of the Bové basin, and interpretation of ancient tectonism is complicated by overprints of two later orogenic episodes (Sougy, 1962a, 1969; Lécorché and Sougy, 1969). However, dates of 645 m.y. and 560 m.y. (Rb/Sr) have been reported in the Mauritanides (Bassot et al., 1963). As in the Rokelides, the Falémé Series to the east is unmetamorphosed and rests directly on basement; to the westward it is found to be overthrust by metamorphic units such as the Bakel Series (LePage, 1972). Querol (1966) cites an age of 650 m.y. for basement rocks from an oil well drilled in Spanish Sahara, suggesting the possibility of a Pan-African effect there.

D. Development of West Africa During the Pan-African Period

Obviously the Pan-African thermotectonic event was important in the border regions of the Precambrian craton. Although often considered essentially a thermal event, it certainly included considerable thrusting in Sierra Leone, Dahomey, and Nigeria. Its effects cannot be traced offshore from the Anti-Atlas, where the old craton is apparently terminated by faults paralleling the coast (Dillon, 1969). The event appears to be of the same age as the Avalonian event of North America, dated as 550–650 m.y. before present. This influenced basement rocks of central Florida, the Cape Hatteras area, eastern New

England, Nova Scotia, and eastern Newfoundland (King, 1969, Fig. 10; Denison, Raveling, and Rouse, 1967). In Brazil an orogeny (Caririan) of similar age has been recorded (Hurley *et al.*, 1967). These various events are probably synchronous (Schenk, 1971, p. 1238) and movements between the North American, South American–West African plates might be suggested to account for the metamorphic and tectonic effects. McCurry (1971) concludes that another distinct plate, the Gabon plate, was being thrust against the southeastern border of the West African plate. The discordance found in the sedimentary cover of the basins on the West African craton may record various orogenic pulses produced by movements between these and perhaps other plates. The pillow lavas, andesites, and spilitelike diabases of western Sierra Leone (Allen, 1969) may be related to zones of ocean-floor spreading and subduction formed during these plate movements. For Pan-African time, we have the first evidence of a West African continental margin which seems to be related in its location and trend to the subsequent continental margins.

Whatever the nature of the plate movements, the Pan-African events clearly caused accretion of subsequently stable crust to the older cratonic nucleus of West Africa (Black and Girod, 1970; Clifford, 1970). Since that time much of Africa has reacted as a stable piece of the earth's crust.

IV. PALEOZOIC DEVELOPMENT OF WEST AFRICA

A. Upper Ordovician to Carboniferous Rocks

1. *Upper Ordovician*

Glacial sediments were deposited at the end of Ordovician time everywhere in the Anti-Atlas, Tindouf, Taoudeni, and Bové basins, forming the basal deposits of series 3 (Fig. 6) (Sougy, 1955; Sougy and Lécorché, 1963; Destombes, 1968a, 1968b; Dia *et al.*, 1969; Beuf *et al.*, 1971; Deynoux *et al.*, 1972). Directions of ice movement in this glaciation were from the southeast to the northwest in the Reguibat area. A similar tillite seems to exist in Sierra Leone in the Saionya Scarp Series of sandstones which rest discordantly upon the Rokelides (Pollett, 1956, p. 9; Reid and Tucker, 1972). Stratigraphic relationships suggest that the Bové basin did not begin to form until the Late Ordovician (Sougy, 1969, Fig. 5a).

In Morocco, sandstones of this age pass northeastward into a marine conglomeratic facies, indicating the flow of the ice sheet into the ocean. No such marine facies development is observed in the glacial sandstones of Spanish Sahara for this time (Bronner and Sougy, 1969b).

2. *Silurian*

During the Silurian a graptolitic, black shale with a high organic content was deposited in the basins of West Africa (Kilian, 1926; Menchikoff, 1930; Remack-Petitot, 1960) and Spain. This is found in the Atlas, Anti-Atlas (Hollard and Jacquemont, 1958), Tindouf (Gevin, 1960), and Zemmour (Sougy *et al.*, 1964) regions (Fig. 1) on the southern flank of the Reguibat uplift (Marchand *et al.*, 1972), in southern Senegal, and in Guinea (Sinclair, 1920; de Chételat, 1938). Some bituminous limestone is present in the Spanish Sahara as well as the shales (Querol, 1966). The rocks occur as black pyritiferous schists in eastern Senegal (Maugis, 1954), apparently having been metamorphosed during the Hercynian orogeny in Devonian time. A white sandstone in the western Taoudeni basin is also considered to be of Silurian age. The slaty, pyritiferous, graptolitic black shale in the Bafata basin generally amounts to 45–100 m in Guinea and Portuguese Guinea, but over 150 m has been cored in the Casamance area of southern Senegal (oil company reports).

The Silurian apparently was a time of transgression and low rate of sediment supply over most of West Africa. The transgression may be related to melting of the Ashgillian Ice Sheets (Lécorché and Sougy, 1969).

3. *Devonian*

During the Devonian West Africa was subject to a series of marine transgressions which extended as far to the southeast as Nigeria during Middle Devonian. The greatest subsidence and most complete sequences accumulated to the northwest in the area of the western Anti-Atlas (Hollard, 1967), although in northern Morocco there were apparently emergent areas after deposition of Lower Devonian flysch (De Koning, 1957). Toward the southeast, disconformities and continental facies become progressively more important (Marchand *et al.*, 1972).

An unusually thick Devonian section is present in southernmost Morocco. There, 4000 m of neritic shale, ferruginous sandstone, and limestone are present at the northern edge of the Tindouf basin. Lower Devonian rocks are typically shale, siltstone, and shaly to calcareous sandstone in the Tindouf basin. Middle Devonian deposits are shale, siltstone, and limestone, formed as well-developed bioherms and biostromes (Hazzard, 1961; Sougy, 1964; Querol, 1966, Hollard, 1967). In Late Devonian, some shale, siltstone, and very impure sandstone was deposited in the Tindouf basin, but this was a time of regression. The barrier of the Mauritanides and restrictions due to closing of the ocean may have been responsible for evaporites deposited with the sediments in the Tindouf basin. Where they are sampled in the Senegal basin, rocks appear to be similar, although metamorphosed (Maugis, 1954). In the Taoudeni basin, Lower Devonian ferruginous sandstone is about 100 m thick.

The Bafata basin, unlike most of the areas to the north, apparently received an influx of coarser clastics (about 100 m of sandstone) in Early and Middle Devonian with deposition of finer clastics (200–300 m of shale and sandy shale) in Late Devonian. Apparently a new source for sediment became available in Early Devonian for the Bafata basin. At the end of Devonian time the basin was also uplifted by Hercynian movements (Carrington da Costa, 1951).

West African fauna show affinities with American and European fauna after Lower Devonian time (LeMaitre, 1950, 1961; Villemur and Drot, 1957; Drot, 1966; Hollard, 1967; Boucot et al., 1969).

4. Carboniferous

Where Carboniferous rocks are present in the series 3, the Lower Carboniferous deposits are a marine transgressive facies (Productid limestone of Visean age–late Early Carboniferous) and the Middle and Upper Carboniferous rocks are continental; mainly sandstone and siltstone (Alía Medina, 1949; Maugis, 1954; Fabre and Villemur, 1959; Querol, 1966; Villemur, 1967). Some reef carbonates have been drilled near the center of the Tindouf basin.

Hercynian uplift is demonstrated by the thickening of Carboniferous beds toward the east in the Tindouf basin, where earlier beds had thickened toward the northwest, away from the Reguibat uplift. There is a notable absence of Carboniferous deposits along the western part of the Paleozoic basins (Fig. 7). The progressive decrease of grain size as well as increase in thickness toward the east suggests that the Mauritanide zone was emergent at this time (Gevin, 1960). This is probable, as the last movements of the Hercynian orogeny are post-Frasnian (early Late Devonian), although a later upwarping might have resulted in removal of Carboniferous deposits in the western parts of the Paleozoic basins.

The Paleozoic succession series 3 (Fig. 6), apparently ends with Devonian deposits in the Bové basin and the western extensions of the Tindouf and Taoudeni basins. Although Carrington da Costa (1951) reported Carboniferous with questionable Calamites in the Bové basin, this apparently has not been confirmed by more recent petroleum exploration work.

The Middle Carboniferous deposits signal the beginning of a major regression affecting all of the West African craton. The sea retreated from the Anti-Atlas and Ougarta chains as they were uplifted and they separated the Tindouf basin from the Colomb–Bechar basin to the northeast and the old continental margin to the north. These chains have remained as cratonized positive regions since then. Sediments have been continental since late Visean in the Tindouf basin and since mid-Westphalian (late Middle Carboniferous) in the Colomb–Bechar basin (Pareyn, 1963). Taoudeni basin sedimentation also became continental after the transgression during which Visean limestone

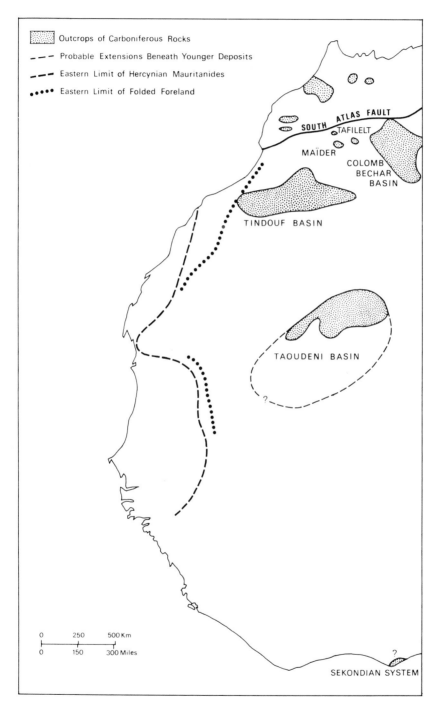

Outcrops of Carboniferous Rocks

Probable Extensions Beneath Younger Deposits

Eastern Limit of Hercynian Mauritanides

Eastern Limit of Folded Foreland

SOUTH ATLAS FAULT

TAFILELT

MAÏDER

COLOMB
BECHAR
BASIN

TINDOUF BASIN

TAOUDENI BASIN

?

0   250   500 Km
0   150   300 Miles

?

SEKONDIAN SYSTEM

Fig. 7. Zones of Carboniferous outcrops.

were deposited (Fabre and Villemur, 1959). In the High Atlas, which was emergent earlier, Carboniferous molasse and lacustrine deposits are present. Marine Devonian deposits and lacustrine beds (possibly Carboniferous) are present along the coast of Ghana (Crow, 1952; Cox, 1964, 1968).

B. The Mauritanide Chain and the Moroccan Hercynian Chain

The Taoudeni and Tindouf basins and the Reguibat uplift are sharply transected on their west by a complex fold belt, the Mauritanide chain (Figs. 1 and 8) defined by Sougy (1962a). The folds in the main, southern section of the Mauritanides (Fig. 8) and the nappes in the northern section were shoved eastward over the West African craton and its cover (Tessier et al., 1961; Giraudon and Sougy, 1963; Sougy, 1962b; Sougy and Bronner, 1969) and produced folds in the autochthon (Lécorché and Sougy, 1969; Rippert, 1973). The allochthonous material includes Precambrian basement and Devonian rocks. These folds of the foreland may be used to trace the extension of the Mauritanide chain in areas where it has been covered by younger sediments.

1. The Main Parts of the Mauritanide Chain

The Mauritanides appear in two sections (Fig. 8). The smaller northern section is separated from the southern part at the Requibat uplift, where the folding in both sections swings westward. Precambrian, Pan-African (Rokelide) and Caledonian (Taconic) orogenies probably affected the Mauritanides area, but an early phase of the Hercynian orogeny was the most significant one. Basement and cover as young as Upper Devonian is folded in most of the chain, while in the southernmost part, the sedimentary cover, beginning with the Upper Ordovician glacial sandstones, is undisturbed and overlies the Mauritanide folds above an angular unconformity.

In the north, large nappes of undated, but probably Precambrian metamorphic rocks cover the Reguibat uplift. These include remobilized basement material (Giraudon and Sougy, 1963) as well as epimetamorphic hematitic quartzites, sericite schists, and greenstones. These rocks were thrust along low-angle faults, and now cover beds as young as Upper Devonian which they fold at the contact (Sougy, 1962b; Lécorché and Sougy, 1969; Bronner and Sougy, 1969a; Sougy and Bronner, 1969).

The nappes of the northern part of the chain in Spanish Sahara and Mauritania overlie a shelf facies cover of the craton. Stromatolite limestones were eroded prior to the deposition of the tillite at the base of the second series (Trompette, 1970), and this erosion surface was transgressed westward and covered by younger sediments. In contrast, to the south, Bassot (1966, 1969) reports formation of shales and greenstone which are time equivalent

Fig. 8. Zones of exposure of Mauritanides and Hercynian chain of Morocco. Location of refraction profile reported by Sheridan *et al.* (1969).

to the second series. These probably represent deposits of the continental slope and rise.

2. *Extension of the Mauritanide Chain to the South*

In southern Senegal the Mauritanide chain seems to bifurcate (Sougy, 1962*a*, 1969; Bassot, 1969). The eastern branch was defined for the most part, prior to Late Ordovician, as noted above. It passes beneath beds of that age in the Bové basin (Reid and Tucker, 1972) (Fig. 1) and probably reappears as the Rokelides (Allen, 1969; Templeton, 1971). The Upper Ordovician, Silurian, and Devonian deposits of the Bové basin were almost unaffected by post-Ordovician activity (Sougy, 1969; Templeton, 1971), although the synclinal shape of the basin and its uplift have been laid to weak Hercynian forces (Carrington da Costa, 1951).

The western branch of the Mauritanides includes rocks affected by the Caledonian–Acadian and probably early-stage Hercynian orogenies and trends into the northern side of the Bové basin and southern Senegal basin. The chain might extend westward beneath sediments of the Southern Senegal basin parallel to the magnetic anomalies of the area (petroleum company reports; Templeton, 1971).

An extension of the western branch of the Mauritanides beneath the continental margin has been postulated by Sheridan, Houtz, Drake, and Ewing (1969). Their conclusion is based on data from a refraction line (location shown in Fig. 8), which they believe shows a basement high associated with the seaward bulge of the 2000-m contour. Actually, their profile is not situated on an extension of the Mauritanide trend, but rather off the almost unfolded Bové basin. Furthermore, interpretation of a magnetic intensity survey (McMaster, De Boer, and Ashraf, 1970) indicates that basement is deeper in the area of the bulge than it is to the south. Unmetamorphosed basement rocks of Ordovician to Devonian age have been drilled in the southern Senegal basin. Thus the proposed extension of the Hercynian Mauritanides to the continental margin (Rodgers, 1970, p. 201) although logical, does not seem to be confirmed by evidence now available.

3. *Extension of the Mauritanide Chain to the North*

North of the northern portion of the Mauritanides, the chain disappears beneath the sediments of the Aaiun basin. However, it can be traced farther north through Spanish Sahara by following the north-northeast trending folds of the foreland (Fig. 8), and also by the magnetic expression thought to result from a belt of shallow intrusions which apparently follow the Mauritanide trend (oil company reports).

Overthrusting of Hercynian age has been interpreted in the Anti-Atlas of southwestern Morocco (Dra massif area) by Mazéas and Pouit (1968), but this view is contested by Choubert and Faure-Muret (1969). This could be related to a general folding of sedimentary cover on detachment planes over the basement (Choubert and Marcais, 1952; Choubert, 1963; Ambroggi, 1963). This folding took place in several phases from the end of Devonian to the end of Carboniferous and produced only an incipient schistosity aligned N 30° E (Massacrier, 1972).

4. *The Hercynian Chain of Morocco*

Younger Hercynian tectonic effects are pronounced north of the South Atlas fault in the High Atlas Mountains (Fig. 1). The basement which crops out in the central part of the range consists of Hercynian folded Paleozoic rocks (Schaer, 1964; Leblanc, 1969), with folds trending approximately north–south, as in the Anti-Atlas area (DeSitter, 1956; Choubert, 1963). Thus the central High Atlas Mountains are mainly Hercynian structures upwarped by Tertiary orogenies. Uplift to the west may have begun in the Devonian. Nappes within the Visean rocks (upper Lower Carboniferous) of the northern High Atlas were displaced to the west, however, as in the Appalachians (Huvelin, 1967, 1970a, 1970b), indicating that Carboniferous uplift was centered to the east of prior upwarping.

North of the High Atlas, Hercynian folding is important in the Moroccan Meseta (Allary *et al.*, 1973) (Fig. 1). The folds, referred to as "Mesetan," are aligned north-northeast along the coast, but swing toward the northeast in the eastern part of the Meseta. Visean rocks of the Meseta region generally were deposited over folded beds which might be considered to represent a final orogenic phase contemporary with the final Mauritanide folding. The structures in the Visean rocks probably represent gravity gliding features.

The rocks involved in the folding of the High Atlas and Moroccan Meseta are quite different, of course, from those south of the South Atlas fault. They are the thick eugeosynclinal (slope–rise) equivalents of the miogeosynclinal (shelf) rocks to the south. In the High Atlas they include Lower Cambrian (Georgian) limestones and thick Cambro-Ordovician shales and volcanic sediments which have been slightly metamorphosed and cut by granites.

C. Development of West Africa in the Paleozoic

During the erosion of the Pan-African chain, Africa, North America, and South America may have formed an immense continent. The early Paleozoic deposits were laid down on the West African craton in shallow intracratonic basins with periods of deposition alternating with periods of erosion.

Except in Morocco, fossil fauna older than Silurian are extremely rare, and the Cambrian has been defined only in Morocco. Thus correlations are difficult, but without doubt, the cratonic area was enlarged by the Pan-African tectonism and Ordovician beds subsequently covered the eroded Pan-African chain. These deposits crop out as the Ordovician sandstones of the northern Hoggar and the Saionya Scarp Series of Sierra Leone.

The clearest and most widespread unconformity in the Paleozoic rocks is the Upper Ordovician glacial surface. Glacial features of this age (moraines, eskers, etc.) are found everywhere on the craton. They were derived from a large continental ice sheet which resulted in transport of glacial detritus from far to the south consistently toward the north or north-northwest in Mauritania and Algeria (Beuf et al., 1971; Trompette, 1970). During the Ordovician glaciation, West Africa and Brazil may have formed a single surface tilted down to the north. Schenk (1972) suggests that the same Ordovician marker is found in Nova Scotia (White Rock Formation) and concludes that Nova Scotia was formerly a microcontinent which has been included in the Canadian Appalachians (Schenk, 1971).

The north side of the Anti-Atlas region (present position of the South Atlas fault, Fig. 8) may mark the location of the shelf edge. The Anti-Atlas Mountains, the northern Tindouf basin, and, probably, areas to the west in Spanish Sahara underwent a strong subsidence with deposition of shelf (mio-geosynclinal) facies (Choubert, 1952; Sougy, 1956). To the north of this area the volcano-sedimentary (eugeosynclinal) sequence appears in the early Paleozoic rocks of the High Atlas. The area probably was a continental margin during early and middle Paleozoic time, perhaps until Early Carbon-iferous.

The very abrupt change in lithology across the South Atlas fault might also indicate that the eugeosynclinal region to the north of the Anti-Atlas is in fault contact with the miogeosynclinal area to the south. The rocks could have been derived from an oceanic zone to the west, and have reached their present position by sliding along the South Atlas fault (Schenk, 1971), prob-ably after the Hercynian orogeny but before Triassic (Proust and Tapponier, 1973).

On moving toward the south and southeast from the former shelf area of the Anti-Atlas and Tindouf basin, the transgressive periods apparently became shorter, until only the major transgressions are recorded. These major ones are the early Early Ordovician (Tremadocian–Arenigian) transgression, the Silurian transgression which may have been caused by a rising sea level pro-duced by melting of the Late Ordovician ice sheet (Sougy and Lé Corché, 1963; Beuf et al., 1966; Berry and Boucot, 1973) and which resulted in major black graptolitic shale deposition, and the transgression of the Middle Devo-nian.

South of the Tindouf basin no evidence exists for continental slope–rise deposits west of the presently exposed cratonic area during Ordovician to Devonian time. Rather, in central Mauritania, the sandy Silurian deposits contain many hiatuses, and are apparently shallow water deposits. However, a paleo-North Atlantic ocean probably did exist to the west in early Paleozoic (Schenk, 1971).

Sediments in the Mauritanides fold belt contain ophiolites and volcanic-derived sediments (Sougy, 1969), and a volcanic trend is reported in the central part of the Mauritanide fold belt outcrop (Chiron, 1969). Thus, presence of a continental margin subduction zone off the African coast seems possible during Silurian and parts of Ordovician and Devonian. However, it is possible that the greenstones are remobilized older rocks.

A belt of positive gravity anomalies parallels the Mauritanides, slightly west of the axis of the outcrop in Senegal, Mauritania, and Spanish Sahara (Blot *et al.*, 1962; Sougy, 1966). Blot *et al.* (1962) suggested that the gravity high was due to basic intrusion or a refold of the basaltic basement in Senegal and Mauritania. Bird and Dewey (1970, Figs. 2 and 3) show such bands of positive gravity anomalies just east (seaward) of their ancient (pre-Taconian) shelf margin. It is possible that such gravity highs result from remnants of oceanic lithosphere which might be expected to exist seaward of the ancient continental shelves. Therefore, the early Paleozoic margin of the African continent might be delineated approximately by this anomaly. To carry this speculation one step further, the band of negative gravity anomalies which runs along the coast may represent relatively low-density (compared to oceanic crust) volcanic rocks, and therefore indicate the trends of an ancient island arc. Then the suggested zones of early Paleozoic tectonism (Chiron, 1969; Bassot, 1969; Sougy, 1969) for southern Mauritania and the eastern branch of the southern Mauritanides might be related to subduction and may really be quite comparable to the Taconian activity of North America as viewed by Bird and Dewey (1970).

South of the Mauritanides, epicontinental deposits accumulated in the Bové basin of Guinea and Portuguese Guinea from Late Ordovician to Devonian time. Similar unfolded deposits (sandstones, orthoquartzites, and black and red shales of Ordovician to Silurian age) are present beneath a Mesozoic cover in northern Florida and southernmost Georgia (Applin, 1951; Bridge and Berdan, 1952; Puri and Vernon, 1964; Rodgers, 1970, p. 198) and probably rest on a Precambrian or lowest Paleozoic basement (Bass, 1969). Furthermore, Ordovician trilobites of an African type (Selenopeltis fauna) are found in North America only in these Florida rocks (Whittington, 1966). Thus the Florida block may have been part of Africa during the first opening of the Atlantic. Eugeosynclinal rocks deposited on 500–600 m.y.-old basement and on oceanic crust are present in the northern Appalachians (Naylor, 1969) and may signify the location of the plate suture produced from the paleo-Atlantic ocean basin. It would appear that a paleo-Atlantic existed during the early Paleozoic, and that the location of the suture resulting from its closure may be located west of the present African coast in the Bové basin area and perhaps west or northwest of the Florida block which may have formed part of the African plate at that time.

The matching of the Bové basin to north Florida, which is advanced tentatively above, would require a northward adjustment of Africa relative to North America from the generally accepted prerift fit of Bullard et al. (1965). However, other evidence for such an adjustment in the prerift positions of the continents has been cited by Le Pichon and Fox (1971) based on matching of fracture zones and by Cramer (1971) based on Silurian phytoplankton assemblages. Furthermore, such an adjustment would cause the general folding trend of the Mauritanides to appear parallel to that of the Appalachians, a parallelism which would be expected if folding resulted from collision of continental plates (Dewey and Bird, 1970).

The Mauritanides underwent a Caledonian–Taconic orogenic phase before the end of the Ordovician in the central and southern Mauritanides (Dia et al., 1969; Dia, 1972; Bassot, 1969) and an early Hercynian phase which occurred at the end of the Devonian (post Frasnian) in the entire belt from Senegal to Spanish Sahara. The Caledonian orogeny is dated at about 435 m.y. in the southern Mauritanides (Bassot, 1969), and the Hercynian phases are dated at around 355 m.y. (end of Devonian) to 205 m.y. (Permian) in the entire chain (Bonhomme, 1962; Vachette, 1964; Bassot, 1969).

It seems likely that the orogenies of the Paleozoic resulted from the closing of an ocean basin and opening of a second one at a rift zone located somewhat east of the first (relative to the continental plates involved) (Wilson, 1966). Although the orogenies may not be strictly time correlative to those of the Appalachians, the general scheme of Bird and Dewey (1970) would appear possible. The first orogeny (Caledonian) may result from subduction of the oceanic plate, the early Hercynian phase from continental plate collision, and the late Hercynian phases from events associated with early stages of continental rifting to form the present Atlantic Ocean basin.

In the Atlas, as in the Appalachians, uplift began in late Early Devonian, but the sliding of nappes toward the west did not occur until late Early Carboniferous. In the Anti-Atlas and along the northern flank of the Tindouf basin uplift began at the end of Devonian time and was active in the middle Carboniferous. The youngest radiometric ages seem to be concentrated in the northern part of the Atlas Range (Choubert et al., 1965). In a very general way, age of the most recent orogenic activity decreases on moving northward along the Paleozoic ranges which parallel the coast of West Africa.

After the late Early Carboniferous (Visean) transgression, further deposition over most of West Africa was continental until the end of Paleozoic time.

Ratschiller (1966–67) suggests that the trend of Paleozoic folding of the Mauritanides extends in a curve through the Anti-Atlas chain of Morocco and the Ougarta chain of Algeria (Fig. 1), rather than extending northward into the Moroccan Meseta. It is possible that early Paleozoic tectonic trends did follow this pattern, as post-Lower Cambrian thrusting in the central Anti-Atlas is directed toward the southeast (Leblanc, 1968). This trend, which follows the boundary of the craton, as did the Pan-African orogenic trends, may have been a branch of the early Paleozoic tectonism or may even represent the main early Paleozoic tectonic zone. The intersection of north–south (Mauritanide) and east northeast–west southwest tectonic trends has given rise to the brachyanticlinal (boutonnière) structures of the Anti-Atlas.

The Hercynian orogeny of the Mauritanides and the Moroccan Meseta, probably resulting from collision of the North American and African plates and early developmental stages of the new ocean basin, produced an elevation of the western part of the present West African area. This initiated a period of intense erosion. Terrigenous clastic sediments accumulated in the Tindouf, Taoudeni, and Colomb–Bechar basins, which were first marine, but soon became continental. The uplift resulted in the erosion of parts of the basins during Middle and Upper Carboniferous time, so that it is not possible to define their original extent. Permian deposits are not found south of the Atlas chain, and this Carboniferous–Permian erosion surface is a very important marker in West Africa. At this time, the end of the Paleozoic, Africa, North America, and probably Europe constituted a single continent.

V. LATE PALEOZOIC TO HOLOCENE DEVELOPMENT OF WEST AFRICA

A. Sediment Deposition and Tectonism

Lithology and time relationships of sedimentary formations of West Africa have been graphically summarized by Reyre (1966) and Ratschiller (1966–67).

1. *Permian and Triassic*

Permian sediments are quite scarce in Africa. They generally take the form of red sandstones and conglomerates (King, 1967; Ambroggi, 1963; Société Chérifienne des Pétroles, 1966) and are referred to as nonmarine or lacustrine

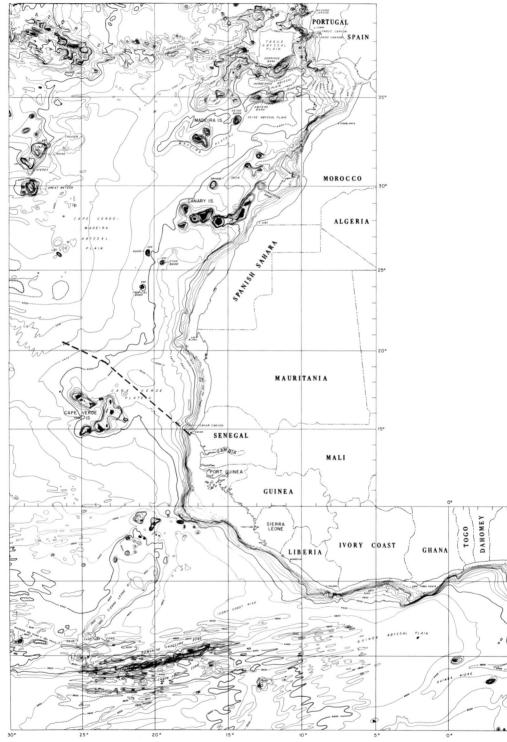

Fig. 9. Bathymetric contours of sea floor off West Africa. Figure was adapted from Uchupi (1971) with permission of the author. Location of seismic reflection profile in Fig. 10 is indicated by the dashed line.

deposits (Ambroggi, 1963). Such rocks are preserved in thick sections in grabens in the Atlas Mountains (Société Chérifienne des Pétroles, 1966; Schenk, 1971). These deposits probably are equivalent to the widespread but discontinuous "Série inférieure" of the "Trias détritique" of Algeria, which is thought actually to be of Upper Permian age by Demaison (1965) and which includes volcanic rocks.

Very generally, Moroccan Triassic deposits consist of Early Triassic coarse clastics (conglomerates and sandstones) followed by deposition of red muds and then salt in Late Triassic with vast basalt flows at the end of the period (Ambroggi, 1963; Société Chérifienne des Pétroles, 1966; deLoczy, 1951). The evaporitic sequence continues into Liassic (Early Jurassic) above the basalt. In detail the sequence is not so simple, with shale beds throughout. The upper part of the interval also includes anhydrite and dolomite (Salvan, 1968). Similar rocks are present in the Triassic of Algeria (Demaison, 1965; Ali, 1973). DeLoczy (1951) contends that *all* Moroccan salt is of Triassic age and that any which is thought to be younger is simply Triassic material which has been intruded.

Triassic rocks crop out only in Morocco, but gypsiferous marls, anhydrite, and gypsum of this age are found in drilling in the Spanish Sahara, and salt structures found there are also considered to be Triassic (Reyre, 1966). Salt domes have also been interpreted for regions off the Atlas Mountains (Robb, 1971) and off the Souss trough (Dillon, 1969). A group of salt domes is present beneath the continental shelf of southern Senegal, off the Casamance River (Aymé, 1965), in a zone where sediments are extremely thick, probably over 10,000 m (Association des Services Géologiques Africains (A.S.G.A.)–U.N.E.S.C.O., 1968). This salt has also been dated as Triassic based on palynology (Templeton, 1971).

Deep-ocean diapiric salt features have been reported for areas west of the Canary Islands at about long. 25° W, lat. 25° N (Rona, 1970) and north of the Cape Verde Islands at about long. 24° W, lat. 19° N (Schneider, 1969b). "Diapirs" of the first group are associated with magnetic anomalies exceeding 100 gammas. The second group occurs in a zone of oceanic volcanic features as shown in a seismic reflection profile (location, Fig. 9; profile, Fig. 10). Drilling on one of these "diapirs" produced a basement rock of altered basalt (DSDP site 141; Hayes *et al.*, 1971). It seems that deep-sea features off West Africa which appear diapiric are probably basaltic intrusions or volcanic features which have been draped with sediment.

Extensive outpourings of basic lava occurred during Triassic and possibly Permian throughout West Africa in Morocco (Ambroggi, 1963; Salvan, 1968; etc.), Spanish Sahara (Querol, 1966), Algeria (Demaison, 1965), Senegal, Guinea and Sierra Leone (Pollett, 1956), Mauritania (Villemur, 1967), Mal, (Dars, 1961; Lay and Reichelt, 1971), and Ivory Coast (Black and Girodi

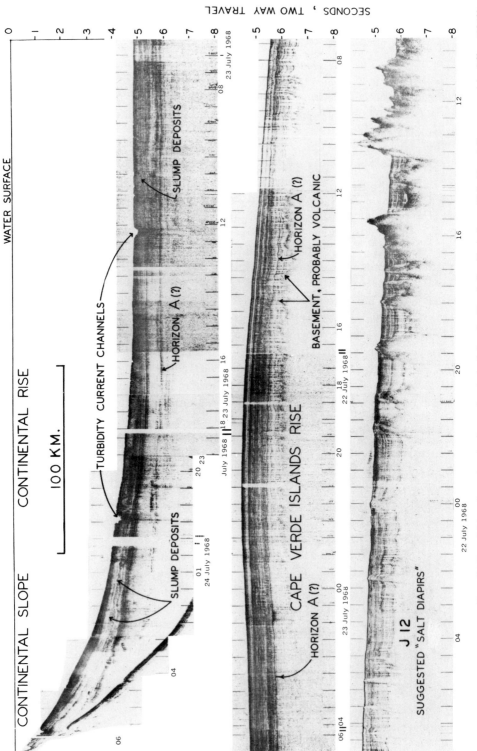

Fig. 10. Seismic reflection profile off West Africa. Location is shown by dashed line in Fig. 9. Profile is from U.S. Naval Oceanographic Office (1969).

1970). The tholeiitic diabase–granophyre association of Mauritania, Mali, Guinea, and Ivory Coast is considered to represent a major phase of igneous activity (Black and Girod, 1970). These rocks are actually dated as post-Visean to pre-Jurassic based on stratigraphic relationships but five dates of 275–230 m.y. have been obtained (Permian) (Lay and Reichelt, 1971). They occur as sheets, sills, and dike swarms, restricted to the West African craton, and cropping out around the Taoudeni basin. Aeromagnetic measurements have suggested the intrusion of Triassic dikes in other areas. Short-wavelength, low-amplitude (about 75 gamma) magnetic anomalies near the coast at about lat. 21° N probably represent such features (Rona, Brakl, and Heirtzler, 1970) and are probably similar to those identified by Behrendt and Wotorson (1970a) in Liberia. The Liberian dikes may be somewhat younger than those to the north, as they have been dated as 176–192 m.y. old, Late Triassic–Early Jurassic (Behrendt and Wotorson, 1970b). The massive Freetown igneous complex in Sierra Leone, located on the coast at about lat. 8.5° N, is also apparently of Late Triassic age, having been emplaced between 180 and 220 m.y. ago (Briden, Henthorn, and Rex, 1971) These age dates indicate a significant igneous event in Late Permian to Triassic. Other radiometric dates between 240 and 200 m.y. are known from rocks of Morocco, Senegal and Mauritania (Choubert et al., 1965; Bassot et al., 1963; Giraudon and Vachette, 1964), but their relationship with this igneous event is not clear.

2. *Jurassic*

During Jurassic to Eocene time sedimentation occurred in the western continental shelf areas and coastal basins, including the Senegal, Aaiun, and Essaouira basins and the western part of the present High Atlas Mountains (Fig. 1). The coastal basins and shelves formed a continuous continental margin zone of subsidence and deposition as suggested diagrammatically in Fig. 11. The Bové basin did not subside and receive sediment during Mesozoic–Tertiary time and is not considered one of these coastal basins.

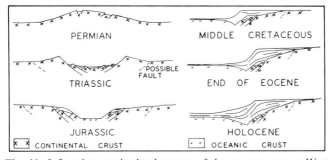

Fig. 11. Inferred stages in development of the present western West
African continental margin.

Jurassic sediments are known in all three west coastal basins but crop out only in the Essaouira basin. Lowermost Jurassic deposits are similar to Triassic sediments; that is, continental types. This type of continental deposition with evaporites continued until Early Cretaceous time in Algeria (Demaison, 1965) and other areas inland from the coastal basins (Faure *et al.*, 1956; Demaison, 1965). These rocks, the "Continental intercalaire," include red beds of sandstone and shale. During late Liassic (Early Jurassic) in Morocco, however, a truly marine transgression began (Ambroggi, 1963). By Late Jurassic time this marine incursion had spread southward to the southern end of the Senegal basin (Reyre, 1966). The marine transgression may have been assisted by a general rise of sea level in the Jurassic. At the beginning of Jurassic time seas covered less than 5% of the continents, whereas in Late Jurassic, 25% of the continents was covered (Hallam, 1969). Hallam (1969) suggests that this general transgression resulted in an increase in calcareous deposition at the expense of clastics. This does hold true for northwest Africa, as most of the Middle and Upper Jurassic sediments contain carbonates.

Essaouira basin sediments are dominantly carbonate. Evaporites which are present are thought to have formed in a back-reef environment (Reyre, 1966).

In the western Atlas the sediments are dominantly detrital, but dolomite and anhydrite are present. Apparently, there was a tendency for the central High Atlas to be upwarped as early as Jurassic and to shed clastics toward the west. A sedimentary break between Lias and Dogger has been attributed to a tectonic disturbance at that time (du Dresnay, 1964). The central part of the western High Atlas apparently began to rise by early Kimmeridgian (Late Jurassic), and major deposition locations in this area began to shift to troughs on either side of the present High Atlas Mountains (Société Chérifienne des Pétroles, 1966).

In the Aaiun basin conditions became progressively more marine during Jurassic (Querol, 1966). This is consistent with the idea of a north to south opening of the Atlantic and a general Jurassic transgression. Continental or evaporitic environments alternating with marine environments are found in Lower and Middle Jurassic, but marine facies become more and more predominant toward the top of the section. Upper Jurassic rocks are typically marine argillaceous limestones, best developed in northwest Spanish Sahara. Transgression in the Aaiun basin reached a maximum in Late Jurassic time.

The Senegal basin contains thick Jurassic limestones (Templeton, 1971) becoming sandy to the eastward (Spengler *et al.*, 1966). The schematic profile of the Senegal basin (Fig. 12) shows the general pattern of all of the coastal basins of the west coast of West Africa. However, the Senegal basin, near Dakar at the location of the profile, has undergone faulting, folding, and

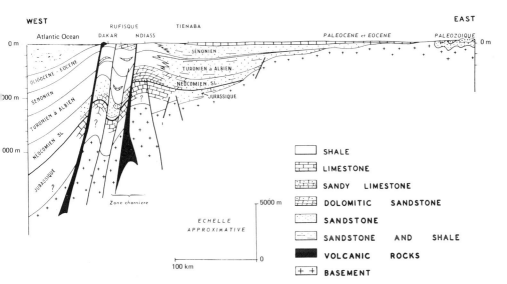

Fig. 12. Schematic cross section of the Senegal basin, from Spengler *et al.* (1966) with permission of Association des Services Géologiques Africains.

volcanic activity much more intensively than experienced by the other basins apparently, and thus shows a more complex structure than average.

On the southern coast of West Africa, the continental margin sedimentation areas, including the Abidjan basin (Fig. 1), and continental shelves apparently began to form later than those on the west coast, probably because rifting occurred later there. Schlee (1972) notes that sediment accumulation is thicker on the African margin west of long. 9° W and suggests that this may be due to a longer time of accumulation in the region of thicker sediments. The oldest deposits related to the present southern continental margin of West Africa are probably of Jurassic age. Over 2000 m of Jurassic continental conglomerate and other coarse terrigenous sediments were drilled in one probable graben (Arens *et al.*, 1971), and the Upper Jurassic and Lower Cretaceous section is thought to be over 5000 m thick (Martin, 1971). A schematic cross section of the Abidjan basin is shown in Fig. 13.

In general, the western continental margin of West Africa has a simple structure formed by normal faulting of the basement with subsequent subsidence (Fig. 11). The slope and rise generally form a fan structure (Fig. 10). However, at the southern continental margin of West Africa, the apparently simple structure of the western continental margin is replaced by a more complex ridge and basin structure. The Abidjan basin is very narrow onshore, and the continental shelf is also very narrow. The outer shelf is formed of a shallow, fault-bounded block of crystalline basement which is exposed on the shelf surface west of about long. 5° W (Fig. 9) (Arens *et al.*, 1971). Seaward

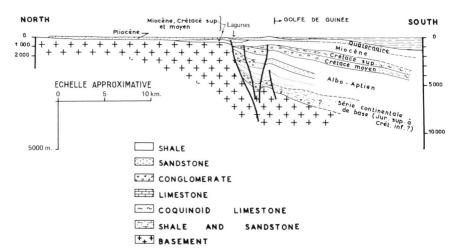

Fig. 13. Schematic cross section of Abidjan basin, from Spengler and Delteil (1966) with permission of Association des Services Géologiques Africains. Subsequent profiling appears to indicate a fault and possibly a basement block near the shelf edge (Arens *et al.*, 1971).

of the continental shelf, a thick sedimentary section is bounded by the shelf-edge basement block to the north and another basement ridge on the ocean floor to the south. This southern ridge forms an extension of the Romanche fracture zone (Fig. 9). It lies about 250–300 km south of the shelf edge off the central Abidjan basin and trends diagonally east-northeast into the shelf edge off Cape Three Points in western Ghana. The basin formed on the sea floor between the ridges has been called the "Ivory Coast basin" (Arens *et al.*, 1971). The Chain fracture zone, south of the Romanche fracture zone, also extends into the Guinea abyssal plain (Fig. 9) (Fail *et al.*, 1970), and its effects dominate sea-floor topography.

Off southern Liberia, another fracture zone modifies the structural and sedimentational patterns of the African continental margin. There, the St. Paul's fracture zone extends from the mid-Atlantic ridge to the African continent along the south side of the Ivory Coast rise (Fig. 9). Faulting on the extension of the fracture zone results in a region of block faulting and slumping on the eastern one-third of the Liberian continental shelf (Schlee, 1972; Robb *et al.*, 1973).

The Guinea fracture zone also extends into the African continental margin at about lat. 9° N (Krause, 1964). However, at this location, where continental breakup occurred with movement normal to the suture between North America and Africa, no major strike-slip movements would be anticipated. The impingement of a fracture zone's transform fault against the continent does not require that faulting extend into the continent, and the linearity of trend of the older Rokelides across the Guinea fracture zone trend shows that fault movements related to the fracture zone have not affected the continent. The nose extending seaward off Guinea, as well as that off Liberia (Ivory Coast rise) (Fig. 9), may have resulted in part from buildup of sediments on the ocean floor restricted by positions of the continents during early stages of drift.

3. Cretaceous

In general, sedimentation in Cretaceous time was characterized by gradients from marine deposition on the seaward sides of the west coast basins to continental deposition on the landward sides. Superimposed on this was a general emergent condition in Early Cretaceous and generally submergent condition in Late Cretaceous. However, rates of subsidence and infilling were roughly equivalent. A Late Cretaceous transgression may have been associated with a eustatic sea level change or epeirogenic subsidence of West Africa, as it was accompanied by development of a shallow epicontinental sea which separated West Africa from the rest of the continent in Turonian. The sea spread across a zone from Nigeria to northeast Algeria (Radier, 1959; Furon, 1964, 1968; Faure, 1966; King, 1967; Reyment, 1971; Kogbe, 1973).

In Morocco, Cretaceous sedimentation was quite variable but generally consisted of carbonates with silt and clay. No appreciable deposition occurred in the western Atlas after Coniacian, indicating that uplift was becoming effective there before the Tertiary orogenies which would fold the rocks (Société Chérifienne des Pétroles, 1966). During the Late Cretaceous transgression, gulfs formed at the Souss trough, near the coast in the western High Atlas and in the northern Essaouira basin. Phosphorites were deposited in these gulfs from Maestrichtian to Eocene (Visse, 1948; Choubert and Faure-Muret, 1962; Summerhayes, Nutter, and Tooms, 1971).

Aaiun basin sediments are generally dolomites, with shales and sandstones toward the ocean and dominantly sandstones to landward. In late Early Cretaceous an influx of coarse arkosic sands probably indicated elevation of a new sediment source to the east (Querol, 1966).

In the western part of the Senegal basin over 4000 m of Cretaceous deposits are present (Maugis, 1955, de Spengler, 1966; Templeton, 1971; Choubert and Faure-Muret, 1971f). In general these consist of a thick limestone section of Lower Cretaceous age with silts, shales, and some anhydrite in Albian. Above an erosional break, transgressive shales are dominant. The difference in facies between seaward and landward parts of the basins during Cretaceous is especially marked in the Senegal basin. To the west, the Cretaceous deposits are dominantly shales and limestones, whereas to the east, sandstone and some shale predominate. The eastern sediments were deposited in alternating continental, lagoonal, and shallow marine environments.

A regression seems to have taken place at the end of Cretaceous in both the Aaiun and Senegal basins. In Senegal the sands spread westward for the first time in Maestrichtian, and sediments became dark bituminous shales near Dakar. To the eastward the sandstones eventually become continental (Maugis, 1954). Late Cretaceous folding and faulting apparently occurred near Dakar (Maugis, 1955). At the very end of Cretaceous time or in Paleocene the sea transgressed to its easternmost limit and carbonates were deposited.

The basement of the West African west coast basins commonly slopes gently seaward in the landward parts of the basins, then near the center of the basin the basement slope increases and sediments thicken markedly to seaward (Fig. 12) (A.S.G.A.–U.N.E.S.C.O., 1968; oil company reports). This abrupt change in basement slope seems to be traceable through the Senegal and Aaiun basins and from the Aaiun basin northward, along the coast through the Souss trough and western High Atlas (G. Choubert, 1967, pers. comm.) and through the Essaouira basin. It is not clear whether this resulted from fault movement, as inferred in Fig. 12 and by Ambroggi (1963), or from a monoclinal flexure.

South of the Senegal basin, Cretaceous deposition continued on the continental shelf. Lower Cretaceous graywackes and conglomerates are present in Liberia (Behrendt and Wotorson, 1970) and 5 km of Cretaceous and Tertiary deposits reportedly have been drilled off Liberia (Schlee *et al.*, 1972).

The eastern part of the Abidjan basin contains coarse detrital sediments deposited throughout most of Cretaceous, whereas the western part received dominantly clayey deposits (Fail *et al.*, 1970; Arens *et al.*, 1971). These authors suggest that the western part of the basin was subsident while the eastern part was maintained at shallower depth.

4. Tertiary and Quaternary

During Paleocene and Eocene a thin layer of sediment accumulated in the coastal basins of West Africa. Deposition generally was ended by regression which occurred toward the end of Eocene or in Oligocene time, although a shallow sea separated West Africa from the rest of the continent through Algeria, Mali, Niger, and Nigeria.

Paleocene and Eocene deposition in the western High Atlas consisted of varied phosphatic and siliceous deposits, marly siltstones, and limestones in shallow embayments. In the Aaiun basin, sands to landward graded to shales toward the ocean (Querol, 1966) and phosphates were deposited. In the Senegal basin Paleocene and Eocene deposits were completely different from the Cretaceous terrigenous sands and clays which had preceded them. Early Tertiary deposits consisted of a variable suite, including thin carbonate deposits with glauconite, phosphorite, and chert, a starved basin assemblage similar to that produced in Morocco by the early Tertiary transgression (Tessier, 1952; Maugis, 1954; Spengler *et al.*, 1966; Templeton, 1971; Trenous and Michel, 1971; Flicoteaux, 1972).

A major regression affected the west coastal basins, probably beginning in late Eocene (Allard, Cochet, and Duffaud, 1958; Ambroggi, 1963; Monciardi, 1966; Querol, 1966). Some authors have suggested that it occurred later (Oligocene: Summerhayes, Nutter, and Tooms, 1971; early Miocene: Templeton, 1971) and gulfs probably existed in northern and southern Senegal until Miocene (Gorodiski, 1958; Brancart and Flicoteaux, 1971). The Abidjan basin also shows an Oligocene hiatus and apparently was affected by the regression (Spengler and Delteil, 1966).

An early pulse of the orogenies which formed the Atlas Range has been considered responsible for the regressions in the western High Atlas (Ambroggi, 1963) and in the Aaiun basin (Querol, 1966) as well as for uplift without folding of the Anti-Atlas (Choubert and Marcais, 1952). The Eocene–Oligocene regression was accompanied by large deformations of the continental shelf off the Ivory Coast (Arens *et al.*, 1971). Apparently the regression resulted from a general epeirogenic episode.

The Atlas Range was created during Tertiary by a series of orogenic pulses. Mesozoic sedimentary rocks were folded in the western High Atlas as well as in the Saharan Atlas and in the Tell Range (Fig. 1) (Caire, 1971). The High Atlas and Middle Atlas, which were also folded at this time, had been uplifted in Jurassic time, and in the central High Atlas, upwarped Hercynian structures are exposed, with trends perpendicular to the Atlas chain (DeSitter, 1956; Faure-Muret, Choubert, and Kornprobst, 1971). The main folding in the central High Atlas is Eocene (Laville, Lesage and Seguret, 1973). In the western High Atlas Ambroggi (1963) has distinguished 11 phases of orogenic and epeirogenic activity. The important ones were the Pyreneic phase in late Eocene time, and two of the Alpine phases in late Oligocene and Miocene (Choubert *et al.*, 1965). The Miocene phase produced the main folding in the Atlas (Allard, Cochet, and Duffaud, 1958). Main alpine tectonism ended at the end of Pliocene (Sitter, 1956), but uplift has continued.

The Rif Mountains (Fig. 1) consist of an "interior zone" to the north and an "exterior zone" to the south. "Interior" and "exterior" refer to positions relative to the entire horseshoe-shaped chain, which includes the Betic cordillera of southern Spain (Durand-Delga, 1969, 1972). The internal zone formed first, with principal folding probably at the end of Eocene or beginning of Oligocene (Faure-Muret, Choubert, and Kornprobst, 1971) and associated metamorphism dated at 34 m.y. (Choubert, Diouri, and Faure-Muret, 1965). This early orogenic pulse was approximately contemporaneous with the probable epeirogenic movements producing the general regression in West Africa. This folding formed arcuate structures paralleling the border of the chain. In middle to upper Miocene the external zones were folded, the internal zone was thrust over th external zone, and nappes slid toward the south in Africa (Durand-Delga *et al.*, 1962; Andrieux, 1971; Andrieux, Fontbote, and Mattauer, 1971). Some intermediate to basic volcanic activity occurred in the eastern Rif from upper Miocene to Quaternary. During the Tertiary orogenies, the Moroccan Meseta (Fig. 1) acted as a stable block. Volcanism apparently occurred during Miocene and Quaternary at Dakar, Senegal, on the coast at about lat. 15° N (Hébrard, Faure, and Elouard, 1969). This may be related to the volcanism of the Cape Verde and Canary Islands which reached a peak of activity at these times.

Northwest Africa has been emergent to a large extent since Eocene, and middle and late Tertiary deposits which crop out are generally continental. In northern West Africa (the Atlas, Rif, and High Plateau zones, Fig. 1), sediments are generally orogenic-type continental red beds and conglomerates (molasse) of Oligocene and younger age (Allard, Cochet, and Duffaud, 1958; Société Chérifienne des Pétroles, 1966; A.S.G.A.–U.N.E.S.C.O., 1968). During early Tertiary, West Africa experienced a hot, humid climate, resulting in strong chemical weathering, with development of residual laterite, bauxite and kaolinite deposits, while shale, phosphorite, limestone and chert formed in the marine environment (Millot, Radier and Bonifas, 1957; Millot, Elouard, Lucas and Slansky, 1960). After Oligocene, epeirogenic movements resulted in renewed erosion, providing material for the "Continental terminal," kaolinitic sandstone and shale, lignitic shale and ferruginous oolites (Radier, 1959; Faure, 1966).

Pliocene marine sediments are found in the Souss Trough (Fig. 1) south of the Atlas Mountains. This trough was downwarped as an estuary in Pliocene time (Ambroggi, 1963), and blue muds and calcareous shell sediments were deposited. In late Pliocene, uplift resulted in strong erosion of the Atlas Range, and brought a flood of coarser clastic sediment but also resulted in uplift of the trough.

Little deposition occurred during middle and late Tertiary in the Aaiun basin, which was gently uplifted from Pliocene to Quaternary according to Querol (1966). Covering much of Senegal are lignitic clays, sands, and gravels which represent the stream and lake deposits of the "Continental terminal" of mainly Miocene and Pliocene age (Tessier, 1952; Dieng, 1965; Spengler *et al.*, 1966; Templeton, 1971).

In Sierra Leone, the Bullom Series of red or buff sands and white and mottled clay is referred to as Eocene to Holocene in age (Pollett, 1956; Dixey and Willbourn, 1951).

Even the continental shelves have accumulated little sediment from middle Tertiary to present (McMaster and Lachance, 1968; Dillon, 1969; Summerhayes, Nutter, and Tooms, 1971). Miocene deposits form a thin layer over the outer shelf off northern and central Morocco (Robb, 1971; Summerhayes, Nutter, and Tooms, 1971). These deposits are formed of glauconitic, phosphatic conglomerates and may have resulted in some outbuilding of the shelf. The Souss trough extends beneath the continental shelf, and this extension of the trough continued to subside and probably received sediment through Tertiary (Dillon, 1969). A thick Miocene section has been reported for a well in the northwestern corner of Spanish Sahara (Querol, 1966; Anonymous, 1969, Fig. 6), but seismic reflection profiles (Dillon, in press) suggest that this sediment may have accumulated by progradation into a large slump scar. Off Guinea, McMaster *et al.* (1971) report development of a delta and suggest that some upbuilding and outbuilding of the shelves of Portuguese Guinea, Guinea, and Sierra Leone occurred in late Tertiary. A very strong uplift of the margins of much of the African block resulted in "many large captures of drainage" in Pliocene time (King, 1967, p. 296). This may be one of the uplifts experienced since Hercynian by the Fouta Djallon Mountains in Guinea, east of the Bové basin (Carrington da Costa, 1951). The uplift of the Fouta Djallon Range may have resulted in capture by the Niger River of much of the drainage of the southern part of West Africa. Prior to the uplift a great deal of the water could have drained westward and flowed to the sea, perhaps in the channels of the present rivers of Portuguese Guinea, which may have supplied sediment to the continental shelf.

The relatively small amount of middle and late Tertiary deposits results in a West African shelf consisting in large part of Cretaceous to Eocene beds which are level or dip gently seaward and which have been beveled by erosion during Tertiary to Pleistocene low stands of the sea (McMaster and Lachance, 1968; Tooms, Summerhayes, and McMaster, 1971). The Alpine orogenies did result in considerable folding and faulting of continental margin deposits off the western High Atlas (Robb, 1971).

As little sediment accumulated on the continental block after early Tertiary, sediment which was subsequently derived from the continent must have gone to build the continental slope and rise and to the deep sea. This seems to be confirmed by the level trend of horizon A beneath the wedge of sediments forming the continental rise, a trend which is seen in many seismic profiles (Fig. 10). Horizon A is formed of a series of Eocene chert beds in the Atlantic (Ewing, Windisch, and Ewing, 1970) and was identified in a core at JOIDES site 12, near the position of the profile shown in Fig. 10 (approximate location indicated on profile) (Maxwell *et al.*, 1970). Sediments above horizon A, and therefore younger than early Eocene, form the major part of the slope–rise accumulation. A rather similar circumstance is found on the western side of the North Atlantic (Emery *et al.*, 1970).

During the regressions resulting from the Pleistocene glacial advances northern West Africa experienced a cool, rainy climate, while that of southern

West Africa was hot and dry. During the interglacial periods of transgression the northern part was cool and dry, while the southern part was hot and humid (Beaudet, Maurer, and Ruellan, 1967; Michel, 1970a,b; Beaudet, 1971; Hébrard, 1972b). These alternations of climate resulted in a series of weathered surfaces (Tessier, 1952; Beaudet, Maurer, and Ruellan, 1967; Conrad, 1968; Grandin and Delvigne, 1969; Michel, 1970a,b; Nahon and Ruellan, 1972). Crusts on these weathered surfaces generally are calcareous to the north and ferruginous-kaolinitic to the south. Correlations between Quaternary marine deposits, produced during transgressions, and continental deposits are still not well known in much of West Africa (Faure and Elouard, 1967; Elouard, Faure, and Hébrard, 1969; Beaudet, 1971; Hébrard, 1972a; Martin and Delibrias, 1972; Martin and Tastet, 1972).

Holocene sediments on the continental shelves generally form only a thin layer upon the surface produced by Tertiary and Pleistocene erosion. Many areas are characterized by differential erosion features and relict deposits formed during the Pleistocene sea-level depression. Bezrukov and Senin (1971) distinguish two zones of different weathering types on the west coast of West Africa at present. North of lat. 15° N conditions are arid with mainly mechanical weathering, poorly developed drainage, and low supply of sediment to the shelves. South of lat. 15° N is an equatorial humid zone, where high temperatures and high rainfall produce a thick crust of chemical weathering and rivers carry much fine-grained, iron-enriched sediment. Calcarenites are the most common sediments in the northern section along with some terrigenous quartz sand. Placer phosphorite deposits off Morocco and Spanish Sahara are relict Pleistocene features, and phosphorite apparently is not forming at present (Summerhayes, 1972; Summerhayes, Nutter, and Tooms, 1972). McMaster and Lachance (1969), in a most complete examination of West African shelf sedimentation, note that carbonate sands are the dominant shelf sediments in tectonically stable areas. Mid-shelf deposits of silt occur, scattered along the western continental margin of West Africa, where the shelf is narrow and sediment supply is large. Where the shelf is wider, sediments are deposited on the inner shelf and in estuaries. Off Portuguese Guinea, Guinea, and Sierra Leone much of the shelf is covered by relict terrigenous sands, but river-transported silts and clays are presently building deltas, and one reaches the shelf edge near lat. 11° N (McMaster, Lachance, and Ashraf, 1970). Other drowned deltas near the shelf edge presumably were formed during the Pleistocene low sea level.

The south coast of West Africa is generally comparable to the southern west coast, with mainly fine materials transported to the sea. A considerable sediment load is carried by the Volta River, which enters the ocean in eastern Ghana and has built its delta to the shelf edge (Martin, 1971).

B. Development of West Africa during Formation of the Present Atlantic Ocean Basin

1. *Permian and Triassic*

Upper Permian–Triassic sediments were deposited on a block-faulted surface. They consisted of continental muds, sands and gravels, and evaporites, and deposition was accompanied by basaltic intrusions and extrusions. The faulting and igneous activity probably is evidence for extensional stresses which finally resulted in continental breakup to form the present ocean basin (May, 1971). Permian deposits seem to have been restricted to accumulations in grabens in northwestern West Africa.

The first homogeneous regularly subsident layer in the western High Atlas Range accumulated in Triassic time. The subsidence of the continental margin of West Africa was accompanied by faulting, both parallel and normal to the continental margin, and much of this block-faulting activity apparently persisted at least into Tertiary time (Alia and Medina, 1949; Alia, 1960; Rod, 1962; Querol, 1966; Spengler *et al.*, 1966; McMaster, de Boer, and Ashraf, 1970; Robb, 1971). THe fault zone ultimately subsided below sea level, and the long, restricted rift valley received ocean water. Evaporites began to be deposited from northern Morocco to at least as far south as southern Senegal (Templeton, 1971) because of the restricted circulation and excess of evaporation over rainfall plus runoff from the surrounding continents. Relatively arid conditions would be anticipated near the center of the huge landmass because much atmospheric moisture would have been lost before penetrating to the rifting zone. Perhaps at the end of Triassic the rift stage ended and the phase of drift began, with formation of new oceanic crust (Hallam, 1971; Pitman, Talwani, and Heirtzler, 1971; Le Pichon and Fox, 1971; Phillips and Forsyth, 1972). Kennedy (1965) has suggested that the oldest sediments of the northwest African coastal basins become progressively younger from north to south. This would imply that West Africa split away from North America from north to south, although, conversely Dietz and Holden (1970) have hypothesized that the continents split from south to north.

2. *Jurassic*

During the Jurassic, marine deposition became dominant on the western continental margins, including the Essaouira, Aaiun, and Senegal basins. At that time the spreading North Atlantic apparently had opened enough to allow relatively free circulation connected to the rest of the world ocean. It would seem that the southern margin of West Africa did not separate from northern South America until some time in Jurassic (Le Pichon and Hayes,

1971, suggest 140 m.y.—Late Jurassic). When this separation finally began it did not occur as a simple extensional rifting. Rather, the geometry of the plates requires that movements were dominantly transcurrent. The result was a ridge and basin structure to the continental margin as well as the nearby ocean floor, instead of the simple subsiding block pattern which is inferred to form the deep structure of the basins on the west coast. A distinct facies difference between the eastern and western parts of the Abidjan basin may be related to Early Cretaceous activity on the fracture zones which apparently extend into this area (Fail *et al.*, 1970).

3. *Cretaceous*

Cretaceous was the time when the major sediment accumulations occurred in the west coast basins and presumably on the continental shelves. These zones underwent major subsidence while mainly muds and carbonates were deposited. Progressively greater seaward dips of the older rocks indicate that the center of subsidence was seaward of the present shoreline. In the Aaiun basin Jurassic beds dip at about 35 m/km and Cretaceous at 7–17 m/km (oil company reports; Martinis and Visintin, 1966), while lower Tertiary beds are horizontal. The climate apparently changed rapidly from hot-dry to hot-humid during Cretaceous, which, along with some epeirogenic movements, may have produced the change from mixed detrital-chemical deposition on land to lateritic weathering with chemical sediments (Radier, 1959; Faure, 1966).

4. *Tertiary and Quaternary*

The great subsidences and accumulations of sediments were ended by the beginning of Tertiary. A minor transgression in Paleocene apparently was not accompanied by a strong influx of clastic terrigenous deposits. Probably in late Eocene a major epeirogenic phase caused a regression which seems to have affected much of west Africa. The continental bauxites and laterites of Eocene gave way to the detrital deposits of the Continental terminal after Oligocene as regression occurred and climate probably became drier, although several different laterite crusts have developed since Eocene.

From late Eocene through Miocene northern West Africa was subjected to complex orogenic pulses, presumably resulting from collisions of the African and European plates. A small subplate, which included the Rif zone in Africa, probably was shoved westward in early Tertiary, then squeezed between the European and African plates in the major Miocene pulse, producing the thrusts and nappes of the Rif Mountains according to Andrieux, *et al.* (1971).

Little sediment has accumulated on the African continental block including the continental shelves since the end of Eocene time. The probable reason for this is the lack of subsidence of the continental margins since then. Subsequently, detritus shed from the continent has largely gone to construct the continental slope and rise and to the deep sea. The shelves were eroded during Tertiary and Pleistocene regressions and only a thin discontinuous accumulation of Holocene deposits is present.

VI. VOLCANIC ISLANDS

A. Canary Islands

The visible parts of the Canary Islands have undergone a two-stage development. The first stage produced a basal complex and the second, the basaltic rocks which form most of the present surface of the islands.

1. Basal Complex

The basal complex crops out on Fuerteventura, Gomera, and La Palma (Fig. 14) (Cendrero, 1971; Fúster et al., 1958; Hausen, 1958; Gastesi, 1970; Ibbarola, 1970). It is probably represented by inclusions in the basalts of Lanzarote (Fúster, Páez, and Sagredo, 1970), and possibly by the high-density core modeled for Tenerife and Gran Canaria (MacFarlane and Ridley, 1968; Bosshard and MacFarlane, 1970). The complex consists of: (1) plutonic mafic and ultramafic rocks and syenites; (2) submarine volcanic rocks; and (3) sedimentary rocks.

The plutonic rocks include peridotites, pyroxenites, dunites, wehrlites, gabbros, olivine gabbros, and alkaline gabbros. They apparently are part of a large, mafic, layered intrusive body sometimes called the "stratiform complex" (Cendrero, 1971; Gastesi, 1970; Rothe and Schmincke, 1968). The chemical composition of the plutonic rocks suggests a different source from

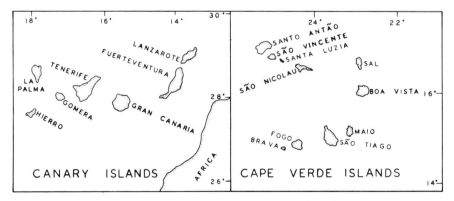

Fig. 14. The Canary and Cape Verde Islands.

that of the younger volcanic rocks which cover then (Fúster, Páez, and Sagredo, 1970). They may represent a former part of the upper mantle, similar to the Troodos massif (Gastesi, 1970) or part of the lower crust (Cendrero, 1971). The mafic and ultramafic rocks are cut by younger syenite, quartz syenite, and possibly nepheline syenite dikes, which caused alkalinization of the plutonic rocks (Cendrero, 1971; Rothe and Schmincke, 1968).

Associated with the plutonic rocks of the basal complex are submarine basic lavas with remnants of pillow structure (Cendrero, 1971) and marine sedimentary rocks, including a thick section of limestone, shale, chert, marl, siltstone, and sandstone on Fuerteventura (Fuster, Cendrero, et al., 1968). The sediments were probably deposited on an erosion surface on the plutonic rocks (Cendrero, 1971), although Rothe and Schmincke (1968) believe that they are older than the plutonics. Age of limestones on the basal complex require that it be older than Miocene (Gastesi, 1970). However, an age of Late Cretaceous or older, determined from foraminifera present in the sediments included in the complex (Rothe, 1968a; Rothe and Schmincke, 1968), shows that at least part of the basal complex is as old as Mesozoic. The sedimentary sequence is strongly folded.

A very extensive dike network commonly forms 80% of the basal complex (Cendrero, 1971; López Ruiz, 1970). The dikes generally trend north-northeast, parallel to faults and to the overall alignment of the islands as well as to faults on the African continent (the Zemmour fault zone) (Alia, 1960; Rod, 1962; Rothe and Schmincke, 1968). The dike swarm indicates considerable extension normal to the African continental margin and probably represents the roots of an old volcanic field which has been eroded away (Cendrero, 1971).

2. Basaltic Rocks

The oldest subaerial volcanic rocks of the Canary Islands are tabular basaltic flows commonly more than 1000 m thick, which are present on all of the islands. They are seen to rest on the eroded basal complex where it is visible, and are referred to as the Tableland Formation or Basaltic Series I (Hausen, 1958; Fuster, Arana, et al., 1968; Fuster, Cendrero, et al., 1968; Fuster, Hernandez-Pacheco, et al., 1968; Ibbarola, 1970) or the ältere plateau basalte (Rothe, 1966). The islands apparently formed as large composite shield volcanoes (Schmincke, 1967) which were built up above sea level and later collapsed with subsequent production of pyroclastics (Borley, 1966; Schmincke, 1968). The basaltic rocks display a spectrum of differentiates from mafic (oceanites and ankaramites) to salic (trachybasalts or hawaiites), but the variations can all be accounted for by differentiation from an initial magma with a composition of alkaline olivine basalt (Ibarrola, 1970; Schmincke, 1973). The major basaltic volcanism apparently began in early to middle Miocene and continued into Pliocene both on the basis of paleontologic and radiometric dating (Furon, 1963, 1968; Abdel-Monem, Watkins and Gast, 1967*b*; Fúster-Casas, *et al.*, 1967; Fúster, Fernandez-Santin and Sagredo, 1968; Fúster, Hernandez-Pacheco, *et al.*, 1968).

A second, less important peak in intensity of basaltic volcanism occurred in the Pleistocene after a period of quiescence (Schmincke, 1968; Tinkler, 1968; Frisch and Schmincke, 1969; Ibarrola, 1970). This period of activity may be considered to be continuing, as eruptions have occurred in 1704–1706, 1730–36, 1798, 1824, 1909, and 1949 (U.S. Naval Oceanographic Office, 1965; Tinkler, 1966, 1968; MacFarlane and Ridley, 1968). Ibarrola (1970, p. 774) suggests that, although magmatic evolution during the Late Tertiary pulse was "homogeneous," the Quaternary volcanism of the central Canarian chain was different from that on the eastern islands (Lanzarote and Fuerteventura). The central islands showed strong alkalic trends, with alkali mafic rocks (basanites and tephrites) progressing to more acid alkalic types (trachybasalts). This developmental pattern was repeated several times. However, on the eastern islands Quaternary basalts evolved toward less alkalic types with tholeiitic affinities.

3. *Summary of Late Cretaceous to Present Development of the Canary Islands*

The oldest exposed part of the Canary Islands may be the plutonic rocks of the basal complex which may represent former parts of the mantle or lower oceanic crust. The zone around the islands is thought to represent old oceanic crust, because oceanic-type magnetic anomalies pass through the area (Rona, Brakl, and Heirtzler, 1970). The early volcanism of the islands may be the result of activity along an oceanic fracture zone or perhaps eruptions produced as the oceanic crust drifted across a hot spot in the mantle (Le Pichon and Fox, 1971). The subsequent, very extensive Tertiary volcanism may suggest presence of some long-standing crustal weakness in the area, however.

The plutonic rocks were probably uplifted and eroded prior to Late Cretaceous, then were submerged and covered by marine sediments derived from the African continent. Submarine volcanism became important in Late Cretaceous or early Tertiary. Tectonism resulted in uplift of the island area in early to middle Tertiary, and erosion beveled the basal complex. An alternative hypothesis proposed by Rothe (1968*a*) and Rothe and Schmincke (1968) would require deposition of the Late Cretaceous sediments of the eastern islands on the continental margin of Africa. Subsequently, rifting would have moved the islands away from the continent and the plutonic rocks would have been intruded, then uplifted and eroded.

In late Tertiary (Miocene and Pliocene) a great outpouring of basaltic lava formed shield volcanoes, producing most of the present island areas.

Collapse structures formed, then after a quiet period, volcanism began again in the Quaternary.

4. *Deep Structure of the Canary Islands*

The deep structure and early development of the Canary Islands is controversial. The principal disagreement regards the presence (Rothe and Schmincke, 1968) or absence (Fúster, Cendrero, *et al.*, 1968) of continental crust beneath the inner islands (Fuerteventura and Lanzarote). Evidence cited for continental blocks beneath some islands is based on assumed prerift plate fits, composition of the volcanic rocks, crustal thicknesses, comparison of sediments to those on the African continent, and presence of some organisms for which a land bridge to the islands is considered necessary.

A sialic basement has been suggested in order to close a gap in prerift fits of Africa against North America (Dietz and Sproll, 1970; Rona and Nalwalk, 1970). Such fits are somewhat arbitrary and alternative means of eliminating gaps would be possible, thus this evidence for continental fragments is not definitive.

The difference in composition between basalts from the eastern islands and those from the other islands was cited above (Ibbarola, 1970) and might be accounted for by contamination by continental materials beneath the eastern islands. However, the source of contaminants, if they did in fact exist, also could be the great pile of continent-derived sediments of the continental rise through which the magma was intruded. The highly alkalic nature of the extrusive rocks also could suggest the presence of continental material beneath all of the islands (Hausen, 1958). However, it is pointed out by Dash and Bosshard (1968) that although alkali lavas are present in the islands and generally regarded as continental, tholeiitic lavas, which are typically oceanic, are also common. They conclude (*ibid.*, p. 251) that "alkali lavas could quite easily be secondary products formed by metamorphism and chemical processes in older material as a result of recent volcanism."

Carbonatites are present in the Canary Islands on Fuerteventura (Fuster, Cendrero, *et al.*, 1968, p. 209–211). Such rocks have also been identified in the Cape Verde Islands (de Assunção, Machado and Gomes, 1965; de Assunção, Machado, and Serralheiro, 1968; Allegre *et al.*, 1971) and in Jurassic limestones of the High Atlas in Morocco (Agard, 1960; Gittins, 1966, p. 435). About 90% of carbonatites are situated in continental shield areas (Verwoerd, 1966), and therefore the presence of these rocks might suggest continental material beneath the islands. However, carbonatites probably form as late-stage differentiates (Tuttle and Gittins, 1966), perhaps in gas traps formed by irregularities on the bottom of the crust (Dawson, 1966). Fuster, Cendrero, *et al.* (1968, p. 211) indicated that they believe the Fuerteventura carbonatites to have been "magmatic residual liquids proceeding from the alkaline syenitic intrusions," and thus not indicative of continental rocks beneath the islands.

Differences in lead isotope composition between the two eastern islands and the rest of the group probably result from small additions of lead from continental material beneath the islands according to Oversby, Lancelot, and Gast (1971). Whether the lead isotopes could have been contributed from continent-derived sediments is not clear.

The Mohorovicic discontinuity beneath the Canary Islands deepens fairly constantly from normal oceanic depth to the west of the islands to 21–25 km beneath the African shelf, with local increases beneath some of the islands (Bosshard and MacFarlane, 1970; Roeser, Hinz, and Plaumann, 1971). Presumably, the depression of the Moho is related to the presence of the transition zone from oceanic to continental crust and also to loading of the crust by the weight of the islands. The crustal thickening does not necessarily require the presence of continental rocks, as it could result from the piling of volcanic material on an oceanic crust or from tectonic compression, which probably is required if the plutonic rocks of the basal complex truly represent lower crust or mantle material.

A possible pattern of ocean-opening magnetic stripes may indicate the presence of oceanic crust in the islands area (Rona, Brakl, and Heirtzler, 1970; Roeser, Hinz, and Plaumann, 1971). Roeser, Hinz, and Plaumann (1971) calculate that depth to the Mohorovicic discontinuity is 21 km in the eastern part of this area near the African margin and that depth to magnetic basement in the same locality is 10–15 km. The difference, 6–11 km, is in the range of average oceanic crustal thickness. Thus they conclude that the crust consists of a deep section of normal oceanic crust covered by a thicker section of metamorphic and sedimentary (and possibly volcanic) rocks, a conclusion consistent with velocities determined by refraction techniques. Their evidence suggests a lack of rifted continental remnants in the area studied.

Rothe (1968a) has found that the sedimentary sequence on Fuerteventura is similar to that of the Aaiun basin. However, the presence of deep-water limestones, turbidites (Sauer and Rothe, 1972) graded beds, and intraformational pebbles (Rothe, 1968) in the Fuerteventura rocks, would seem to indicate that these rocks are not strictly comparable to a shelf facies, such as is present in the Aaiun basin, thus the correlation seems questionable. The presence of Miocene–Pliocene ostrich shells also has been cited as evidence that the islands were separated from the mainland relatively recently (Sauer and Rothe, 1972). Such conclusions seem unwarranted because other means exist to transport organisms to islands, as was pointed out by Rothe and Schmincke (1968, p. 1154).

None of the evidence cited is adequate to prove presence or absence of continental crustal fragments beneath any or all of the Canary Islands.

B. Cape Verde Islands

Most of the area around the Cape Verde Islands, including the Cape Verde Islands rise (Fig. 10), appears to be underlain by oceanic crust as indicated by the presence of linear, ocean-opening-type, remanent magnetic stripes (Rona, Brakl, and Heirtzler, 1970). The oldest part of the visible islands may be the palagonites of Brava (Fig. 14), which are thought to be ancient sea-floor rocks (Machado, Azeredo Leme, and Monjardino, 1967). The oldest dated rocks of the islands probably are the fossiliferous Upper Jurassic to Cretaceous limestones, argillaceous limestones, and cherts of Maio (Stahlecker, 1935; Bourcart, 1946; Furon, 1968; Klerkx and dePaepe, 1971) which are probably deep-water deposits (de Assunção et al., 1968; Templeton, 1971). These beds, which dip east and west away from the center of the island, were apparently warped by an uplift of the central part of Maio and are covered by horizontal Tertiary and Quaternary sediments with many volcanic interbeds. Fossiliferous marly limestone interlayered with basalt on São Nicolau is of Eocene and probably Cretaceous (Senonian) age (Bebiano and Soares, 1951). The Cape Verde Islands, like the Canary Islands, are notable for the presence of alkalic rocks including nepheline syenites (Machado, Azeredo Leme, and Monjardino, 1967). Carbonatites are present (de Assunção et al., 1965; Machado et al., 1967; de Assunção et al., 1968) which are true carbonatites derived from a magma rather than hydrothermal limestones (Allègre et al., 1971).

The Cape Verde Islands probably began to form by submarine volcanic processes prior to the end of Jurassic. Intrusion of nepheline syenites was followed by emplacement of carbonatites (Machado, Azeredo Leme, and Monjardino, 1967). Deep-water deposition of limestone and argillaceous sediments continued into Late Cretaceous. Between middle Cretaceous and upper Miocene times a major tectonic disturbance caused upwarping of the islands (Machado, Azeredo Leme, and Monjardino, 1967), and they were eroded at sea level. The basement so formed, which is comparable to the basal complex of the Canary Islands, is now covered mainly by phonolitic tuff as well as late Tertiary and Quaternary coastal sediments. The volcanic rocks were probably erupted in Miocene and Quaternary times and Fogo last erupted in 1951 (Ribeiro, 1954).

The similarities in development of the Canary and Cape Verde Island groups are striking, although relatively little work has been carried out in the Cape Verde Islands and therefore correlations are less than certain. Both groups have strongly alkalic composition with differentiates including phonolites and nepheline syenites similar to the nepheline syenites of the coast of Guinea (La Croix, 1911; Millot and Dars, 1959). Carbonatites are present and observed in ring structures in both sets of islands (Fúster and Aguilar, 1965; Muñoz, 1970). Both seem to have a Mesozoic basal complex, formed of

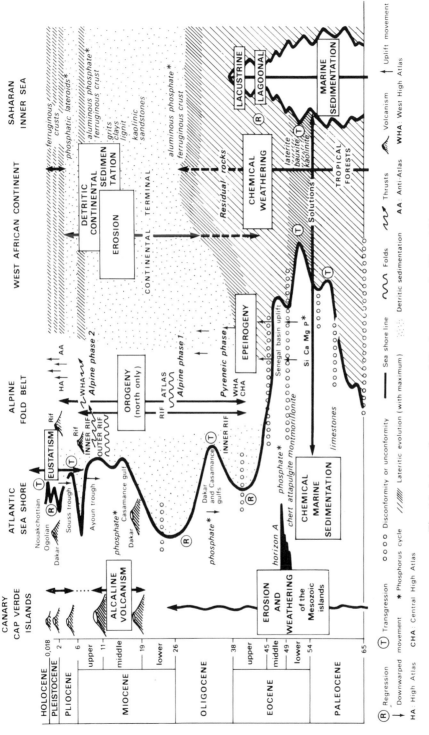

Fig. 15. Interpretive summary of Tertiary events in West Africa.

volcanic, plutonic, and sedimentary rocks, which was eroded, then covered by Miocene and Quaternary volcanic rocks. There is some indication that Miocene and Quaternary volcanism was important not only in the Cape Verde and Canary Islands but also in the Anti-Atlas Mountains (Choubert, 1963; Choubert *et al.*, 1968) and near Dakar on the continent east of the Cape Verde Islands.

The time equivalence of the two main Tertiary–Quaternary periods of volcanism suggests a relationship in causative factors. During part of Tertiary it appears that Africa and Europe were being shoved one against the other. The compressive stresses which resulted produced folding in North Africa (Pyrenean and Alpine tectonism). It seems likely that volcanism takes place when the crust is extended. Extension of the crust on the west side of the African block while the north side was being compressed suggests a rotation of the block in a counterclockwise manner, around a pivot point near its northwest corner. Some reorganization of plate movements might be anticipated for late Miocene, if, as Le Pichon (1968, p. 3693) suggests, spreading of the ocean floor developed a renewed intensity at that time.

VII. SUMMARY AND CONCLUSIONS

The African shield appears in three massifs, the West African, Congo, and Kalahari cratons (Fig. 4). In West Africa the craton crops out as three east-northeast-trending belts, the Anti-Atlas, Reguibat, and Léo uplifts (Fig. 1). In general, the oldest rocks of these uplifts are found in the western parts and are about 2700 m.y. old (Figs. 3 and 5). The eastern parts of the uplifts are about 1900 m.y. old. The West African craton probably has reacted as a stable block since the last main granitic intrusions of about 1700 m.y.

The cratonized basement was beveled by erosion, and a sedimentary blanket was laid down (Fig. 6). Deposits of upper Precambrian to Upper Carboniferous age are present in a series of basins—the Tindouf, Taoudeni, Bové, and Volta basins (Fig. 1). Basin sediments are generally epicontinental marine and continental types. From about 1000 to 650 m.y. ago shales, sandstones, and stromatolite limestones of the first series were laid down. The area north of the Anti-Atlas zone (north of the "major fault," Fig. 5) apparently formed as a continental slope and rise in Precambrian and has been accreted to Africa since 1000 m.y. ago.

The interval of deposition from about 650 to 440 m.y. ago was bounded by major glacial episodes in Eocambrian(?) at about 620–650 m.y. ago and at the end of Ordovician (about 440 m.y. ago). This second series begins with withspread tillites, followed by green shales, then red beds, and ends with Lower Ordovician sandstone. The third series begins above a glacial disconformity with Late Ordovician tillite and sandstone succeeded by a Silurian black graptolitic shale deposited during a eustatic transgression, perhaps related to the melting of the ice sheet.

Pan-African orogenies at about 500–650 m.y. affected the borders of the previously cratonized West African massif and folded some of the sedimentary cover. Since the Pan-African events, much of Africa has reacted as a stable block of the earth's lithosphere. During the Pan-African orogeny, Africa, North America, and South America may have formed a large continental mass, but an early Paleozoic paleo-Atlantic ocean basin soon must have formed, with North America separated from Africa. The ocean-opening suture must have been located somewhat west of the present North Atlantic. That is, some continental fragments which were on the African side during the early Paleozoic opening of the Atlantic seem to be found on the North American side in this present opening. Closing of the early Paleozoic Atlantic probably resulted in the Caledonian orogeny, due to forces related to subduction of the oceanic lithosphere, which began in Ordovician time.

The early-stage Hercynian orogenies at the end of Devonian time, which resulted in major folding and thrusting toward the east as well as uplift, may have been caused by continental collision of Africa and North America. Devonian sediments of the West African Paleozoic basins thicken from the southeast toward the old continental margin in the northwest. On the other hand, the coarser Carboniferous deposits thicken eastward, probably indicating the rise of the Mauritanides chain due to the Devonian–Carboniferous orogenies. The remnants of the Mauritanides chain crop out in two sections in west-central West Africa (Fig. 8). The rocks were affected by Precambrian and Pan-African orogenies as well as the Caledonian and Hercynian orogenies which formed the presently exposed orogenic zone. Rocks as young as Upper Devonian are folded at least in the eastern part of the chain. Younger (Carboniferous) Hercynian folding occurred in the Moroccan Meseta, and nappes were formed during Carboniferous uplift.

After Early Carboniferous, deposition in most of West Africa was continental until middle Mesozoic time as the Mauritanide mountains were eroded. In Late Permian and Triassic, the extension began which resulted in rifting and, ultimately, in opening of the present North Atlantic basin. Normal faulting produced grabens, and continental clastic sediments accumulated in them. The extensional movements also resulted in volcanic activity in West Africa. The rift zone subsided below sea level, and sea water invaded it bringing about deposition of salt. Relatively unrestricted oceanic circulation finally became established in Jurassic, and true marine deposits began to accumulate. Mesozoic sedimentation took place on the present continental shelf area and the coastal basins (Senegal, Aaiun, and Essaouira basins and western High Atlas—Fig. 1).

The southern margin of West Africa probably did not begin to form until Jurassic, although Devonian–Carboniferous marine deposits along the south coast of Ghana may indicate an ancient zone of weakness. Direction of movement, dictated by the geometry of plates, requires a right-lateral transcurrent movement along this margin. The result of this dominantly strike-slip activity was formation of a ridge and basin topography. Subsequently, sediments were trapped by basement ridges both in a narrow shelf basin (Albidjan basin— Fig. 1), and in a basin on the deep-sea floor.

Most subsidence and most sediment accumulation occurred in Cretaceous in the west coast basins and shelf. During a Late Cretaceous transgression, an epicontinental sea extending from Algeria to Nigeria isolated West Africa from the rest of the continent. In early Tertiary, marine phosphatic deposits accumulated in coastal basins, but in late Eocene time, an epeirogenic period apparently caused cessation of sedimentation in the coastal basins. Subsequently, the major locus of sedimentation shifted to the slope and rise as the margin built into the deep sea. Continental deposition of laterite and bauxite during Cretaceous and early Tertiary may have been due to hot and humid climatic conditions.

From late Eocene to Pliocene West Africa was in collision with Europe and a series of mountain chains were created, the Atlas, Tell, and Rif (Fig. 1). The major orogenic pulse was in Miocene.

The Cape Verde and Canary Islands are oceanic volcanoes, probably dating almost from time of development of the earliest oceanic crust. Continental crust may exist beneath the two inner Canary Islands, however. A great outpouring of lava (presumably related to extension in the crust) took place in Miocene along the West African coast. At the same time major folding occurred in northern Africa. These two occurrences may be related. Perhaps collision of Africa and Europe caused Africa to rotate in a counterclockwise manner. In that case, compression of North Africa and tensional effects along its western coast could have resulted from a single set of plate movements.

The west coast shelves of West Africa were eroded during late Tertiary and Pleistocene. The present shelves are covered mainly by relict sands with a few areas of Holocene silts on the midshelf and some Holocene deltas where silt and clay accumulate in the southern, tropical zone.

ACKNOWLEDGMENTS

Thanks are due Dr. Robert McMaster of the University of Rhode Island, who first kindled the interest of the senior author in African problems. The senior author appreciated the opportunity to publish this work provided by Chevron Overseas Petroleum Incorporated, as much of the research for the paper was done while working for this organization. Discussions with Malcolm Boyce, Arthur Martinez, and Anthony Paap of Chevron and James Robb of the United States Geological Survey were very fruitful.

The second author thanks Prof. A. E. M. Nairn for help in developing this cooperation between a West African onshore geologist, who has worked in the area for 25 years, with an American marine geologist, interested in African problems. He is indebted for the comments of MM. R. du Dresnay and H. Hollard, with their knowledge of Morocco; to Prof. G. Young for advice and help with English translation; to MM. R. Flicoteaux and D. Nahon for their useful suggestions concerning the Tertiary and Quaternary geology. His very special thanks are expressed to Dr. R. Trompette for his sustained interest in the work and his help, with the St. Jerôme Library staff, in gathering reference material; and to R. Dassulle for the drawings. Part of the work has been made possible by material assistance of the C.N.R.S. Associate Laboratory no. 132 "Études Géologiques Ouest-Africaines," Marseille St. Jerôme and its staff.

Finally, we both must acknowledge the patience shown by the editors of this volume.

REFERENCES

Abdel-Monem, A., Watkins, N. D., and Gast, P. W., 1967a, K-Ar geochronology and paleomagnetic studies of the volcanism on the Canary Islands, (abs.): *Geol. Soc. Am.*, 1967 Meeting, New Orleans.

Abdel-Monem, A., Watkins, N. D., and Gast, P. W., 1967b, Volcanic history of the Canary Islands, (abs.): *Am. Geophys. Union Trans.*, v. 48, p. 226–227.

Abdel-Monem, A., Watkins, N. D., and Gast, P. W., 1968, Volcanic stratigraphy and magnetic history of the Canary Islands and Madeira, (abs.): *Intern. Sympos. Volcanol.*, Tenerife, Sept. 1968, p. 1.

Abdel-Monem, A., Watkins, N. D., and Gast, P. W., 1971, Potassium-argon ages, volcanic stratigraphy, and geomagnetic polarity history of the Canary Islands: Lanzarote, Fuerteventura, Gran Canaria, and La Gomera: *Am. J. Sci.*, v. 271, p. 490–521.

Agard, J., 1960, Les carbonatites et les roches à silicates et carbonates associés du massif de roches alcalines du Tamazert (Haut-Atlas de Midelt, Maroc) et les problèmes de leur genèse: *Rep. 21th Sess. Intern. Geol. Congr.*, Copenhagen, v. 13, p. 293–309.

Ali, O., 1973, Stratigraphy of Lower Triassic sandstone of northwest Algerian Sahara, Algeria: *Am. Assoc. Petrol. Geol. Bull.* v. 57, n. 3, p. 528–540.

Alía Medina, M., 1949, Contribución al conocimiento geomorfologico de las zonas centrales del Sahara Español: Consejo sup. Investig. cient., Inst. Estud. Afr., Madrid, 234 p.

Alía Medina, M., 1960, La téctonica del Sahara español: *Rep. 21st Sess. Intern. Geol. Congr.*, Copenhagen, v. 18, p. 193–202.

Allard, P. L., Cochet, E. and Duffaud, F., 1958, L'Oligocène dans le Haut Atlas Occidental: *Notes Mém. Serv. Géol. Maroc*, No. 143, *Notes*, t. 16, p. 7–16.

Allary, A., Andrieux, J., Lavenu, A., and Ribeyrolles, M., 1972, Présence de décrochements dans la Meseta sud-orientale du Maroc central: *C.R. Acad. Sci.*, Paris, Ser. D, v. 274, p. 653–656.

Allary, A., Andrieux, J., Lavenu, A., and Ribeyrolles, M., 1973, La chaîne hercynienne anté-viséenne du Maroc central: *Ière Réun. Ann. Sci. Terre*, Paris, 19–22 mars, p. 40.

Allègre, C. J., and Caby, R., 1972, Chronologie absolue du Précambrien de l'Ahaggar occidental: *C.R. Acad. Sci.*, Paris, Sér. D, v. 275, p. 2095–2098.

Allègre, C. J., Pineau, F., Bernat, M., and Javoy, 1971, Evidence for the occurrence of carbonatites on the Cape Verde and Canary Islands: *Nature (Phys. Sci.)*, Oct. 4, v. 233, p. 103–104.

Allen, P. M., 1965, A preliminary note on the Rokel River series, Sierra Leone: *9th Ann. Rep. Res. Inst. Afr. Geol.*, 1963–64: Leeds, U.K., p. 34–36.

Allen, P. M., 1968, The stratigraphy of a geosynclinal succession in western Sierra Leone, West Africa: *Geol. Mag.*, v. 105, n. 1, p. 62–73.

Allen, P. M., 1969, The geology of part of an orogenic belt in western Sierra Leone, West Africa: *Geol. Rundschau*, v. 58, n. 2, p. 588–620.

Allen, P. M., Snelling, N. L., and Rex, D. C., 1967, Age determinations from Sierra Leone, in: *Variations in isotopic abundances of strontium, calcium and argon and related topics*: Mass. Inst. Technol., Cambridge, M.I.T. 1381-15, *15th Ann. Progress Rep.*, p. 25–26.

Ambroggi, R., 1963, Etude géologique du versant méridional du Haut Atlas occidental et de la plaine du Souss: *Notes Mém. Serv. Géol. Maroc*, No. 157, 321 p.

Andrews-Jones, D. A., 1971, Structural history of Sierra Leone, in: *Tectonique de l'Afrique*: U.N.E.S.C.O., Paris, *Sci. Terre* No. 6, p. 205–207.

Andrieux, J., 1971, La structure du Rif central: Montpellier University thesis, France, *Notes Mém. Serv. Géol. Maroc*, v. 235, 155 p.

Andrieux, J., Fontbote, J.-M., and Mattauer, M., 1971, Sur un modèle explicatif de l'arc de Gibraltar: *Earth Planet. Sci. Let.*, v. 12, p. 191–198.

Apostolescu, V., 1963, Essai de zonation par les Ostracodes dans le Crétacé du bassin du Sénégal: *Rev. Inst. Fr. Pétrole*, v. 18, n. 12, p. 1675–1694.

Applin, P. L., 1951, Preliminary report on buried pre-Mesozoic rocks in Florida and adjacent states: *U.S. Geol. Surv. Circ.*, 91, 28 p.

Applin, P. L., 1952, Sedimentary volumes in Gulf Coastal Plain of the United States and Mexico, Part I: Volume of Mesozoic sediments in Florida and Georgia: *Bull. Geol. Soc. Am.*, v. 63, p. 1159–1163.

Arens, G., Delteil, J. R., Valéry, P., Damotte, B., Montadert, L., and Patriat, P., 1971, The continental margin off the Ivory Coast and Ghana, in: *The geology of the East Atlantic continental margin*, F. M. Delaney, ed.: G. Brit. Inst. Geol. Sci. Rep., 70/16, p. 61–78.

Arnould, A., 1972, Rapport scientifique récapitulatif 1969-1972, in: Rapport scientifique pour la période 1969-1972 du Laboratoire Associé au C.N.R.S. No. 132: *Etudes géologiques ouest-africaines*, par J. Sougy: *Trav. Lab. Sci. Terre St. Jérôme*, Marseille, Ser. E, v. 10, p. 75–76.

Arnould, M., 1961, Etude géologique des migmatites et des granites précambriens du NE de la Côte d'Ivoire et de la Haute-Volta méridionale. Cadre géologique, classification, principaux types: *Mém. Bur. Rech. Géol. Min.*, Paris, v. 3, p. 174.

Arribas, A., 1968, El Precámbrico del Sahara español y sus relaciones con las series sedimentarias más modernas: *Bol. Geol. Minero*, v. 79, n. 5, p. 445–480.

Assémien, P., Filleron, J. C., Martin, L., and Tastet, J. P., 1970, Le Quaternaire de la zone littorale de Côte d'Ivoire: *Bull. Liaison Assoc. Sénégalaise Et. Quatern. Ouest Afr.*, Dakar, v. 25, p. 65–78.

Association Services Géologiques Africains - U.N.E.S.C.O., 1963, Carte géologique de l'Afrique au 1:5,000,000: U.N.E.S.C.O., Paris.

Association Services Géologiques Africains - U.N.E.S.C.O., 1968, Carte tectonique internationale de l'Afrique au 1:5,000,000: U.N.E.S.C.O., Paris.

Assunção, C. F. T. de, Machado, F., and Conceição Silva, L., 1967, Petrologia e vulcanismo da Ilha do Fogo (Cabo Verde): *Garcia de Orta*, Lisboa, v. 15, n. 1, p. 99–110.

Assunção, C. T. de, Machado, F., and Gomes, D. R. A., 1965, On the occurence of carbonatites in the Cape Verde islands: *Bol. Soc. Geol. Port.*, v. 16, p. 179–188.

Assunção, C. T. de, Machado, F., and Serralheiro, A., 1968, New investigations on the geology and volcanism of the Cape Verde Islands: *Rep. 23rd Session Intern. Geol. Congress*, Czechoslovakia, Proc. Section 2, p. 9–16.

Aymé, J.-M., 1965, The Senegal salt basin, in: *Salt basins around Africa*: Elsevier Publ. Co., Amsterdam, p. 83–90.

Bailey, D. K., 1966, Carbonatite volcanoes and shallow intrusions in Zambia, in: *Carbonatites*, O. F. Tuttle and J. Gittins, eds.: Interscience–Wiley, New York, p. 127–154.

Barrère, J., 1967, Le groupe précambrien de l'Amsaga entre Atar et Akjoujt (Mauritanie). Etude d'un métamorphisme profond et de ses relations avec la migmatisation: *Mém. Bur. Rech. Géol. Min.*, Paris, v. 42, 275 p.

Barrère, J., 1969, Aperçu sur le métamorphisme et la migmatisation dans les séries précambriennes de l'Amsaga (Mauritanie sud-occidentale): *Bull. Soc. Géol. France*, Sér. 7, v. 11, p. 150–159.

Bass, M. N., 1969, Petrography and ages of crystalline basement rocks of Florida—Some extrapolations: *Am. Assoc. Petrol. Geol. Mem.*, v. 11, p. 283–310.

Bassot, J.-P., 1966, Etude géologique du Sénégal oriental et de ses confins guinéo-maliens: *Mém. Bur. Rech. Géol. Min.*, Paris, v. 40, 322 p.

Bassot, J.-P., 1969, Aperçu sur les formations précambriennes et paléozoïques du Sénégal oriental: *Bull. Soc. Géol. France*, Sér. 7, v. 11, p. 160–169.

Bassot, J.-P., Bonhomme, M., Roques, M., and Vachette, M., 1963, Mesures d'âges absolus sur les séries précambriennes et paléozoïques du Sénégal oriental: *Bull. Soc. Géol. France*, Sér. 7, v. 5, p. 401–405.

Beaudet, G., 1971, Le Quaternaire marocain: état des études: *Rev. Géograph. Maroc*, v. 20, p. 3–56.

Beaudet, G., Maurer, G., and Ruellan, A., 1967. Le Quaternaire marocain, Observations et hypothèses nouvelles: *Rev. Géograph. Phys. Géol. Dynam.*, Paris, Sér. 2, v. 9, n. 4, p. 269–309.

Bebiano, J. B., and Soares, J. M. P., 1951, Note on some supposed Senonian fossils from São Nicolau Island (Cape Verde Islands): *Rep. 18th Sess. Intern. Geol. Congr.*, London, 1948, v. 14, p. 186–189.

Behrendt, J. C., and Wotorson, C. S., 1970a, Aeromagnetic and gravity investigations of the coastal area and continental shelf of Liberia, West Africa and their relation to continental drift: *Geol. Soc. Am. Bull.*, v. 81, p. 3563–3574.

Behrendt, J. C., and Wotorson, C. S., 1970b, Aeromagnetic survey unveils Liberian coastal basins: *Oil Gas J.*, June 29, p. 160–164.

Behrendt, J. C., and Wotorson C. S., 1971, An aeromagnetic and aeroradioactivity survey of Liberia, West Africa: *Geophys.*, v. 36, p. 590–604.

Berry, W. B. N., and Boucot, A. J., 1973, Glacio-eustatic control of late Ordovician-early Silurian platform sedimentation and faunal changes: *Geol. Soc. Am. Bull.*, v. 84, p. 275–284.

Bertrand-Sarfati, J., and Raaben, M. E., 1970, Comparaison des ensembles stromatolitiques du Précambrien supérieur du Sahara occidental et de l'Oural: *Bull. Soc. Géol. France*, Sér. 7, v. 12, p. 364–371.

Beuf, S., Biju-Duval, B., de Charpal, O., Rognon, P., Gariel, O., and Bennacef, A., 1971, Les grès du Paléozoïque inférieur au Sahara. Sédimentation et discontinuités. Evolution structurale d'un craton: Publ. Inst. Fr. Pétrole, Technip edit., 480 p.

Beuf, S., Biju-Duval, B., Stevaux, J., and Kulbicki, G., 1966, Ampleur des glaciations "silu-riennes" au Sahara: leurs influences et leurs conséquences sur la sédimentation: *Rev. Inst. Fr. Pétrole*, v. 21, n. 3, p. 363–381.

Bezrukov, P. L., and Senin, K. M., 1971, Sedimentation on the west African shelf, in: *The geology of the East Atlantic continental margin*, F. M. Delaney, ed., G. Brit. Inst. Geol. Sci. Rep., 70/16, p. 1–7.

Bird, J. M. and Dewey, J. F., 1970, Lithosphere plate-continental margin tectonics and the evolution of the Appalachian orogen: *Geol. Soc. Am. Bull.*, v. 81, p. 1031–1060.

Black, R., 1966, Sur l'existence d'une orogénie riphéenne en Afrique occidentale: *C. R. Acad. Sci.*, Paris, v. 262, Sér. D, p. 1046–1049.

Black, R., 1967, Sur l'ordonnance des chaînes métamorphiques en Afrique occidentale: *Chron. Mines Rech. Min.*, Paris, v. 364, p. 225–238.

Black, R., and Girod, M., 1970, Late Palaeozoic to Recent igneous activity in West Africa and its relationship to basement structure, in: *African magmatism and tectonics*, T. N. Clifford and I. G. Gass, eds.: Oliver and Boyd, Edinburgh, v. 9, p. 185–210.

Blanchot, A., 1955, Le Précambrien de Mauritanie occidentale (esquisse géologique): *Bull. Dir. Fédér. Mines Géol. Afr. Occ. Fr.*, Dakar, v. 17, 308 p.

Blot, C., Crenn, Y., and Rechenmann, J., 1962, Eléments apportés par la gravimétrie à la connaissance de la tectonique profonde du Sénégal: *C. R. Acad. Sci.*, Paris, v. 254, p. 1131–1133.

Bongrand, M.-O., Dars, R., and Sougy, J., 1961, Sur la présence de galets éoliens dans le complexe de base de l'Adrar mauritanien: *Bull. Soc. Géol. France*, Sér. 7, v. 3, p. 210–215.

Bonhomme, M., 1962, Contribution à l'étude géochronologique de la plate-forme de l'Ouest-africain: *Ann. Fac. Sci. Univ. Clermont-Ferrand*, France, v. 5, 62 p.

Borley, G. D., 1966, The geology of Tenerife, Volcanic studies group meeting: *Geol. Soc. London Proc.*, 2 June 1966, v. 1635, p. 173–176.

Bosshard, E., and MacFarlane, D. J., 1970, Crustal structure of the western Canary Islands from seismic refraction and gravity data: *J. Geophys. Res.*, v. 75, n. 26, p. 4901–4918.

Boucot, A. J., Johnson, J. G., and Talent, J. A., 1969, Early Devonian brachiopod zoo-geography: Geological Society of America, Special Paper, 119, 107 p.

Bourcart, J., 1946, Géologie des îles Atlantides, in: *Contribution à l'étude du peuplement des îles Atlantides*, P. Chevalier, ed.: *Mém. Soc. Biogéogr.*, Paris, v. 8, p. 9–40.

Brancart, R., and Flicoteaux, R., 1971, Age des formations phosphatées de Lam Lam et de Taïba (Sénégal occidental). Données micropaléontologiques, conséquences stratigra-phiques et paléogéographiques: *Bull. Soc. Géol. France*, Sér. 7, v. 13, p. 399–408.

Briden, J. C., Henthorn, D. I., and Rex, D. C., 1971, Palaeomagnetic and radiometric evidence for the age of the Freetown igneous complex, Sierra Leone: *Earth Planet. Sci. Let.*, v. 12, p. 385–391.

Bridge, J., and Berdan, J. M., 1952, Preliminary correlation of the Paleozoic rocks from test wells in Florida and adjacent parts of Georgia and Alabama: (Florida Geol. Surv., ed.). Assoc. Am. State Geol., 44th Ann. Meeting, Tallahassee, 1952, guide-book, p. 29–38.

Bronner, G., 1972, Rapport scientifique récapitulatif 1969-1972, in: Rapport scientifique pour la période 1969-1972 du Laboratoire associé au C.N.R.S. Etudes géologiques W-africaines, par J. Sougy: *Trav. Lab. Sci. Terre Marseille Saint-Jérôme*, France, Sér. E, v. 10, p. 27–28.

Bronner, G., and Sougy, J., 1969a, Etude structurale de la région de Tourarine (Mauritanides, NW d'Akjoujt, République islamique de Mauritanie): 5e Colloque Géol. afr., Clermont-Ferrand, 9-11 avril, *Ann. Fac. Sci. Univ. Clermont-Ferrand, Géol. Minéral.*, v. 41, n. 19, p. 77–78.

Bronner, G., and Sougy, J., 1969b, Extension de la glaciation fini-ordovicienne à la région d'Aoucert (Sahara espagnol méridional): *Ann. Fac. Sci. Univ. Clermont-Ferrand*, France, *Géol. Minéral.*, v. 41, n. 19, p. 79–80.

Bullard, S. C., Everett, J., and Smith, A., 1965, The fit of the continents around the Atlantic, in: *A symposium on continental drift*: Royal Soc., London, *Philos. Trans.*, v. 1088, 258A, p. 41–51.

Burri, C., 1960, Petrochemie der Capverden und Vergleich des Capverdischen Vulkanismus mit demjenigen des Rheinlandes: *Schweiz. Min. Petr. Mitt.*, v. 40, p. 115–161.

Caby, R., 1970, La chaîne pharusienne dans le Nord-Ouest de l'Ahaggar (Sahara central, Algérie); sa place dans l'orogenèse du Précambrien supérieur en Afrique: Thèse, Université de Montpellier, France, 336 p.

Caby, R., and Leblanc, M., 1973, Les ophiolites précambriennes sur les bords est et nord du craton ouest-africain: *1ère Réun. Ann. Sci. Terre*, Paris, 19-22 mars, p. 112.

Caire, A., 1971, Chaînes alpines de la Méditerranée centrale (Algérie et Tunisie septentrionale, Sicile, Calabre et Apennin méridional), in: *Tectonique de l'Afrique*: *Sci. Terre*, U.N.E.S.C.O., Paris, No. 6, p. 61–90.

Cann, J. R., 1967, A second occurrence of dalyite and the petrology of some ejected syenite blocks from São Miguel, Azores: *Mineral. Mag.*, v. 36, p. 227–232.

Cariou-Ogundaré, H., 1972, Analyse micropaléontologique de carottes sur le plateau continental ivoirien: Thesis 3rd cycle, Univ. Paris Sud, Orsay, 106 p.

Carrington da Costa, J., 1951, Notes on the stratigraphy and tectonics of Portuguese Guinea: *Rep. 18th Sess. Intern. Geol. Congr.*, London, Serv. géol. afr., v. 14, p. 84–86.

Castelain, J., 1965, Aperçu stratigraphique et micropaléontologique du bassin du Sénégal, Historique de la découverte paléontologique, Colloque inter. Micropaléont.: Dakar, 6–11 mai 1963, *Mém. Bur. Rech. Géol. Min.*, Paris, v. 32, p. 135–159.

Cendrero, A., 1971, The volcano-plutonic complex of La Gomera (Canary Islands): *Bull. Volcanol.* (Napoli), v. 34, n. 2, p. 537–561.

Charlot, R., Choubert, G., Faure-Muret, A., and Tisserant, D., 1970, Etude géochronologique du Précambrien de l'Anti-Atlas (Maroc): *Notes Mém. Serv. Géol. Maroc*, 225, *Notes*, v. 30, p. 99–134.

Chételat, E. (de), 1938, Sur l'extension du Gothlandien en Guinée française: *C. R. Acad. Sci.*, Paris, v. 207, p. 371–372.

Chiron, J.-C., 1969, Esquisse géologique de la chaîne des Mauritanides entre M'Bout et Moudjéria (Mauritanie occidentale): *Bull. Soc. Géol. France*, Sér. 7, v. 11, p. 170–184.

Chiron, J.-C., 1973, Etude géologique de la chaîne des Mauritanides entre le parallèle de Moudjéria et le fleuve Sénégal (Mauritanie). Un exemple de ceinture plissée précambrienne reprise à l'Hercynien: Thesis Univ. Lyon, France, 466 p.

Choubert, B., 1969, Les Guyano-éburnéides de l'Amérique du Sud et de l'Afrique occidentale: *Bull. Bur. Rech. Géol. Min.*, Paris, Sér. 2, Sect. 4, v. 4, p. 39–69.

Choubert, G., 1951, La limite du Pliocène et du Quaternaire au Maroc, résumé: *Rep. 18th Sess. Intern. Geol. Congr.*, London, 1948, v. 14, p. 176.

Choubert, G., 1952, Histoire géologique du domaine de l'Anti-Atlas, in: *Géologie du Maroc*, fasc. 1: *Notes Mém. Serv. Géol. Maroc*, v. 100, p. 75–195.

Choubert, G., 1955, Vue d'ensemble sur l'Infracambrien et le Précambrien de l'Anti-Atlas (Maroc): Assoc. Serv. Géol. Afr., réunion de Nairobi, 1954, C. R., p. 105–116.

Choubert, G., 1963, Histoire géologique du Précambrien de l'Anti-Atlas. Tome I: *Notes Mém. Serv. Géol. Maroc*, v. 162, 352 p.

Choubert, G., Charlot, R., Faure-Muret, A., Hottinger, L., Marçais, J., Tisserant, D. and Vidal, P., 1968, Note préliminaire sur le volcanisme messinien-"pontien" au Maroc: *C. R. Acad. Sci.*, Paris, Sér. D, v. 266, p. 197–199.

Choubert, G., and Debrenne, F., 1964, Sur la paléogéographie des calcaires à Archéocyathes dans l'Anti-Atlas occidental: *C. R. Acad. Sci.*, Paris, v. 258, p. 2616–2618.

Choubert, G., Diouri, M. and Faure-Muret A., 1965, Mesures géochronologiques récentes par la méthode A^{40}/K^{40} au Maroc: *Notes Mém. Serv. Géol. Maroc*, v. 183, *Notes*, 24, p. 53–62.

Choubert, G., and Faure-Muret, A., 1962, Evolution du domaine atlasique marocain depuis les temps paléozoïques, in: *Livre à la mémoire du Professeur Paul Fallot*, M. Durand-Delga ed.: Soc. Géol. France, v. 1, p. 447–527.

Choubert, G., and Faure-Muret, A., 1965, Manifestations tectoniques au cours du Quaternaire dans le sillon préafricain (Maroc): *Notes Mém. Serv. Géol. Maroc*, no. 185, *Notes* 25, p. 57–62.

Choubert, G., and Faure-Muret, A., 1969, Sur la série stratigraphique précambrienne de la partie sud-ouest du massif du Bas Dra (Tarfaya, Sud marocain): *C. R. Acad. Sci.*, Paris, v. 269, Sér. D, p. 759–762.

Choubert, G., and Faure–Muret, A., 1970, Principales caractéristiques du Précambrien de l'Anti-Atlas: *Notes Mém. Serv. Géol. Maroc*, v. 229, p. 7–42.

Choubert, G. and Faure-Muret, A., 1971a, Anti-Atlas (Maroc), in: *Tectonique de l'Afrique*: *Sci. Terre*, U.N.E.S.C.O., Paris no. 6, p. 163–175.

Choubert, G. and Faure-Muret, A., 1971b, Dorsale Reguibat (Sahara occidental), in: *Tectonique de l'Afrique*: *Sci. Terre*, U.N.E.S.C.O., Paris, no. 6, p. 177–184.

Choubert, G. and Faure-Muret, A., 1971c, Bouclier éburnéen (ou libéro-ivoirien), in: *Tectonique de l'Afrique*: *Sci. Terre*, U.N.E.S.C.O., Paris, no. 6, p. 185–200.

Choubert, G. and Faure-Muret, A., 1971d, Vue d'ensemble sur l'histoire orogénique du craton de l'Afrique occidentale, in: *Tectonique de l'Afrique*: *Sci. Terre*, U.N.E.S.C.O., Paris, no. 6, p. 201–204.

Choubert, G. and Faure-Muret, A., 1971e, Grands bassins sédimentaires de l'Afrique occidentale, in: *Tectonique de l'Afrique*, *Sci. Terre*, U.N.E.S.C.O., Paris, no. 6, p. 267–293.

Choubert, G. and Faure-Muret, A., 1971f, Bassins récents du Maroc au Dahomey, in: *Tectonique de l'afrique*: *Sci. Terre*, U.N.E.S.C.O., Paris, no. 6, p. 391–402.

Choubert, G., Faure-Muret, A., Alía, M., Menchikoff, N., and Burollet, P. F., 1966, Note explicative de la carte tectonique de l'Afrique nord-occidentale: *22nd Sess. Intern. Geol. Congr.*, New-Delhi, 1964, Commun. Sci. Comm. Carte géol. Monde, p. 61–75.

Choubert, G., Faure-Muret, A., and Charlot, R., 1968, Le problème du Cambrien en Afrique nord-occidentale: *Rev. Géogr. Phys. Géol. Dynam.*, Paris, Sér. 2, v. 10, n. 4, p. 289–309.

Choubert, G., Faure-Muret, A., and Hottinger, L., 1971, La série stratigraphique de Tarfaya (Maroc méridional) et le problème de la naissance de l'océan Atlantique: *Notes Mém. Serv. Géol. Maroc*, 237, *Notes*, t. 31, p. 29–40.

Choubert, G., and Marçais, J., 1952, Aperçu structural, in: *Géologie du Maroc, fascicule I*: *Notes Mém. Serv. Géol. Maroc*, v. 100, p. 7–73.

Chudeau, R., 1911, Note sur la géologie de la Mauritanie: *Bull. Soc. Géol. France*, Sér. 4, v. 11, p. 413–428.

Clauer, N., Utilisation de la méthode rubidum-strontium pour la datation de niveaux sédimentaires du Précambrien supérieur de l'Adrar mauritanien (Sahara occidental) et la mise en évidence de transformations précoces des minéraux argileux: *Ière Réun. Ann. Sci. Terre*, Paris, 19–22 mars, p. 134.

Clifford, T. N., 1966, Tectono-metallogenic units and metallogenic provinces of Africa: *Earth Planet. Sci. Lett.*, v. 1, n. 6, p. 421–434.

Clifford, T. N., 1968, Radiometric dating and the pre-Silurian geology of Africa, in: *Radiometric dating for geologists*, E. I. Hamilton and R. M. Farquhar, eds.: Interscience Publ., London, p. 299–416.

Clifford, T. N., 1970, The structural framework of Africa, in: *African magmatism and tectonics*, T. N. Clifford and I. G. Gass, eds.: Oliver and Boyd, Edinburgh, v. 1, p. 1–26.

Collignon, M., 1967, Les Ammonites crétacées du bassin côtier de Tarfaya, Sud marocain: *C. R. Acad. Sci.*, Paris, Ser. D, v. 264, p. 1390–1392.

Colom, G., 1955, Jurassic-Cretaceous pelagic sediments of the western Mediterranean and the Atlantic area: *Micropaleontology*, v. 1, p. 109.

Conrad, G., 1969, L'évolution continentale post-hercynienne du Sahara algérien (Saoura, Erg Chech-Tanezrouft, Ahnet-Mouydir): C.N.R.S. ed., *Publ. Centre Rech. Zones Arides*, Sér. Géol., Paris, v. 10, p. 527.

Cooper, W. G. G., 1927, Report on a rapid geological survey of the Gambia, British West Africa: *Gold Coast Geol. Surv. Bull.*, No. 3, 36 p.

Cox, L. R., 1946, The fossils and age of the Sekondi series (Upper Palaeozoic) of the Gold Coast: *Ann. Mag. Natur. Hist.*, Ser. 11, v. 13, n. 105, p. 602–624.

Cox, L. R., 1948, Report on fossils from the Sekondian system of the Gold Coast: *Ann. Rep. Gold Coast Geol. Surv.*, 1946–47, p. 5–6.

Cramer, F. H., 1971, Position of the North Florida lower Paleozoic block in Silurian time: phytoplankton evidence: *J. Geophys. Res.*, v. 76, p. 4754–4757.

Crow, A. T., 1952, The rocks of the Sekondi series of the Gold Coast: *Bull. Gold Coast Geol. Surv.*, v. 18, p. 1–68.

Dars, R., 1961, Les formations sédimentaires et les dolérites du Soudan occidental (Afrique de l'Ouest): *Mém. Bur. Rech. Géol. Min.*, Paris Univ. Thesis, Paris, no. 12, 329 p.

Dars, R. and Le Page, A., 1973, Tectonique de la "série de Bakel" en rive gauche du fleuve Sénégal (Sénégal, Afrique de l'Ouest): *1ère Réun. Ann. Sci. Terre*, Paris, 19–22 mars, p. 151.

Dash, B. P., 1968, Crustal studies around the Canary Islands: *Rep. 23rd Sess. Intern. Geol. Congr.*, Prague, 1968, p. 15, (abs.).

Dash, B. P. and Bosshard, E., 1968, Crustal studies around the Canary Islands, *Rep. 23rd Sess. Intern. Geol. Congr.*: Czechoslovakia, Section 1, p. 249–260.

Dash, B. P., and Bosshard, E., 1969, Seismic and gravity investigations around the western Canary Islands: *Earth Planet. Sci. Let.*, v. 7, p. 169–177.

Dawson, J. B., 1966, Oldoinyo Lengai—An active volcano with sodium carbonatite lava flows, in: *Carbonatites*, O. F. Tuttle and J. Gittins, eds.: Interscience–Wiley, New York, p. 155–168.

Delaire, L., and Renaud, L., 1955, Notice explicative sur la feuille Conakry-Ouest: *Carte Géol. Reconnaissance au 500.000e A.O.F.*, Dakar, no. NC. 28–SE.-O.11, 1 colored map, 21 p.

Demaison, G. J., 1965, The Triassic salt in the Algerian Sahara, in: *Salt basins around Africa*: Inst. Petrol., London, Elsevier Publ. Co., Amsterdam, p. 91–100.

Denison, R. E., Raveling, H. P., and Rouse, J. T., 1967, Age and descriptions of sub-surface basement rocks, Pamlico and Albemarle Sound areas, North Carolina: *Am. Assoc. Petrol. Geol. Bull.*, v. 51, n. 2, p. 268–272.

Departamento de Investigaciones Petroliferas de AUXINI, 1969, Correlación estratigráfica de los sondeos perforados en el Sahara español; *Bol. Geol. Min.*, Madrid, v. 80, n. 3, p. 235–251.

De Sitter, L. U., 1956, *Structural geology*: McGraw-Hill Publ. Co., New York, 552 p.

Destombes, J., 1968a, Sur la présence d'une discordance générale de ravinement d'âge Ashgill supérieur dans l'Ordovicien terminal de l'Anti-Atlas (Maroc): *C. R. Acad. Sci.*, Paris, v. 267, p. 565–567.

Destombes, J., 1968b, Sur la nature glaciaire des sédiments du groupe du 2ème Bani, Ashgill supérieur de l'Anti-Atlas, Maroc: *C. R. Acad. Sci.*, Paris, v. 267, p. 684–686.

Destombes, J., 1970, Cambrien moyen et Ordovicien, in: *G. Choubert et A. Faure-Muret, Colloque international sur les corrélations du Précambrien*: Notes Mém. Serv. Géol. Maroc, v. 229, p. 161–170.

Destombes, J., 1971, L'Ordovicien au Maroc. Essai de synthèse stratigraphique, in: *Colloque Ordovicien-Silurien*, Brest, Sept.: *Mém. Bur. Rech. Géol. Min.*, Paris, v. 73, p. 237–263.

Destombes, J. and Jeannette A., 1955, Etude pétrographique et sédimentologique de la série acadienne de Casablanca; présence de glissements sous-marins (slumpings): *Notes Mém. Serv. Géol. Maroc*, v. 123, p. 75–98.

Destombes, J., Sougy, J., and Willefert, S., 1969, Révisions et découvertes paléontologiques (Brachiopodes, Trilobites et Graptolites) dans le Cambro-Ordovicien du Zemmour (Mauritanie septentrionale): *Bull. Soc. Géol. France*, Sér. 7, v. 11, p. 185–206.

Dewey, J. F., and Bird, J. M., 1970, Mountain belts and the new global tectonics: *J. Geophys. Res.*, v. 75, p. 2625–2647.

Deynoux, M., 1971, Essai de synthèse stratigraphique du bassin de Taoudéni (Précambrien supérieur et Paléozoïque d'Afrique occidentale): *Trav. Lab. Sci. Terre, Marseille St-Jérôme*, Sér. B, v. 3, 71 p.

Deynoux, M., Dia, O., Sougy, J., and Trompette, R., 1972, La glaciation "finiordovicienne" en Afrique de l'Ouest, Colloque Ordovicien-Silurien, Brest, Sept. 1971: *Bull. Soc. Géol. Minéral. Bretagne*, Rennes, France, Sér. C, t. 4, n. 1, p. 9–16.

Dia, O., 1972, Rapport scientifique récapitulatif 1969-1972, in: *Rapport scientifique pour la période 1969-1972 du Laboratoire associé au C.N.R.S.* n° 132, *Etudes géologiques W-africaines*, par J. Sougy: *Trav. Lab. Sci. Terre, Marseille St-Jérôme*, Sér. E, v. 10, p. 33–37.

Dia, O., Sougy, J., and Trompette, R., 1969, Discordances de ravinement et discordance angulaire dans le "Cambro-Ordovicien" de la région de Mejeria (Taganet occidental, Mauritanie): *Bull. Soc. Géol. France*, Sér. 7, v. 11, p. 207–221.

Dieng, M., 1965, Contribution à l'étude géologique du Continental terminal du Sénégal: *Rept. Bur. Rech. Géol. Min.*, Dakar, DAK65-A27, 181 p.

Dietz, R. S., and Holden, J. C., 1970, Reconstruction of Pangea: break up and dispersion of continents, Permian to Present: *J. Geophys. Res.*, v. 75, n. 26, p. 4939–4956.

Dietz, R. S., Holden, J. C., and Sproll, W. P., 1970, Geotectonic evolution and subsidence of Bahama platform: *Geol. Soc. Am. Bull.*, v. 81, p. 1915–1927.

Dietz, R. S., Knebel, H. J., Somers, L. H., 1968, Cayar submarine canyon: *Geol. Soc. Am. Bull.*, v. 79, p. 1821–1828.

Dietz, R. S. and Sproll, W. P., 1970, East Canary Islands as a micro-continent within the Africa-North America continental drift fit: *Nature*, v. 266, p. 1043–1045.

Dillon, W. P., 1969, Structural geology of the southern Moroccan continental margin: Thesis, Univ. of Rhode Island, Kingston, Rhode Island, 82 p.

Dillon, W. P., 1974, Structure and development of the southern Moroccan continental shelf: *Marine Geol.* (in press).

Dixey, F., and Willbourn, E. S., 1951, The geology of the British African colonies: *Rep. 18th Sess. Intern. Geol. Congr.*, London, Part 14, p. 87–109.

Drake, C. L., and Ewing, J. I., 1968, The continental margin of the eastern United States: *Can. J. Earth Sci.*, v. 5, p. 993–1010.

Dresnay, R. (du), 1964, Les discontinuités de sédimentation pendant le Jurassique dans la partie orientale du domaine atlasique marocain, leurs conséquences stratigraphiques et leurs relations avec l'orogenèse atlasique, in: *Colloque du Jurassique* (Luxembourg, 1962): Inst. Grand-Ducal, Sect. Sci. Natur. Phys. Math., p. 899–912.

Drot, J., 1966, Présence du genre *Amphigenia* (Brachiopode, Centronellidae) dans le bassin de Taoudeni (Sahara occidental): *C. R. Somm. Séances Soc. Géol. France*, v. 9, p. 373.

Duffaud, F., Brun, L., and Plauchut, B., 1966, Le bassin du Sud-Ouest marocain, in: *Bassins sédimentaires du littoral africain*, D. Reyre, éd.: Symposium, 1e partie: Littoral Atlantique (New-Delhi, 1964), *Assoc. Serv. Géol. Afr.*, Paris, p. 5–12.

Durand-Delga, M., 1969, Mise au point sur la structure du Nord-Est de la Berbérie: *Bull. Serv. Géol. Algérie*, New Ser., v. 39, p. 89–131.

Durand-Delga, M., 1972, La courbure de Gibraltar, extrémité occidentale des chaînes alpines, unit l'Europe et l'Afrique: *Eclogae Geol. Helv.*, Bâle, v. 65, n. 2, p. 267–278.

Durand-Delga, M., Hottinger, L., Marçais, J., Mattauer, M., Milliard, Y., and Suter, G., 1962, Données actuelles sur la structure du Rif, in: *Livre à la Mémoire du Professeur Paul Fallot*, M. Durand-Delga ed.: *Soc. Géol. France*, Paris, v. 1, p. 399–422.

Eardley, A. J., 1962, *Structural geology of North America*, 2nd ed.: Harper and Row, New York, 743 p.

Egloff, J., 1972, Morphology of ocean basin seaward of northwest Africa: Canary Islands to Monrovia, Liberia: *Am. Assoc. Petrol. Geol. Bull.*, v. 56, n. 4, p. 694–706.

Elouard, P., 1966, Le Quaternaire du bassin du Sénégal, in: *Sedimentary basins of the African coasts, 1 - Atlantic coast*, D. Reyre, ed.: *Assoc. Afr. Geol. Surv.*, Paris, p. 95–97.

Elouard, P., 1973, Formations sédimentaires de Mauritanie atlantique, in: *Notice Carte géologique de Mauritanie au 1/1.000.000*: (in press).

Élouard, P. and Deynoux, M., 1969, Bassin sédimentaire secondaire et tertiaire sénégalo-mauritanien. Bassin quaternaire de Mauritanie. Bibliographie: *Rep. Lab. Géol. Fac. Sci. Univ. Dakar*, v. 30, 124 p.

Elouard, P., Faure, H., and Hébrard, L., 1969, Quaternaire du littoral mauritanien entre Nouakchott et Port-Etienne (18°-21° Latitude Nord): *Bull. Liaison Assoc. Sénégalaise Et. Quatern. Ouest Afr.*, Dakar, v. 23, p. 15–24.

Emery, K. O., Uchupi, E., Phillips, J. D., Bowin, C. O., Bunce, E. T., and Knott, S. T., 1970: Continental rise off eastern United States: *Am. Assoc. Petrol. Geol. Bull.*, v. 54, n. 1, p. 44–108.

Ewing, J., Hollister, C., Hathaway, J., Paulus, F., Lancelot, Y., Habib, D., Poag, C. W., Luterbacher, H. P., Worstell, P., and Wilcoxon, J. A., 1970, Deep sea drilling project: Leg 11: *Geotimes*, v. 15, p. 14–16.

Ewing, J., Windisch, C., and Ewing, M., 1970, Correlation of horizon A with Joides bore-hole results: *J. Geophys. Res.*, v. 75, p. 5645–5653.

Fabre, J. and Moussine-Pouchkine, A., 1971, Régressions et transgressions permocarbonifères sur le Nord-Ouest de la plate-forme africaine: épirogenèse ou variation eustatique?: *Bull. Soc. Géol. France*, Sér. 7, v. 13, p. 140–145.

Fabre, J., and Villemur, J. R., 1959, Le Carbonifère continental du bassin de Taoudenni: *Ann. Soc. Géol. Nord*, Lille, France, v. 79, p. 89–113.

Fail, J. P., Montadert, L., Delteil, J. R., Valéry, P., Patriat, P., and Schlich, R., 1970, Prolongation des zones de fractures de l'Océan atlantique dans le golfe de Guinée: *Earth Planet. Sci. Let.*, v. 7, p. 413–419.

Faure, H., 1966, Reconnaissance géologique des formations sédimentaires post-paléozoïques du Niger oriental: *Mém. Bur. Rech. Géol. Min.*, Univ. Paris thesis, Paris, v. 47, 630 p.

Faure, H., 1973, Evolution structurale du craton africain depuis le Jurassique: *1ère Réun. Ann. Sci. Terre*, Paris, 19-22 mars, p. 185.

Faure, H., Démoulin, D., Hébrard, L., and Nahon, D., 1971, Données sur la néotectonique de l'extrême ouest de l'Afrique: *Proc. Confer. Afr. Geol.*, Ibadan, Nigeria, Dec. 1970.

Faure, H. and Élouard, P., 1967, Schéma des variations du niveau de l'océan Atlantique sur la côte de l'Ouest de l'Afrique depuis 40 000 ans: *C. R. Acad. Sci.*, Paris, Sér. D, v. 265, p. 784–787.

Faure, H., Furon, R., Lelubre, M., Monod, T., Pires Soares, J. M., Sougy, J., and Tessier, F., Sahara, Afrique Occidentale Française et Portugaise, (R. Furon ed.): Lexique Stratigraphique International, Paris, C.N.R.S., v. 4, fasc. 2, 77 p.

Faure-Muret, A., Choubert, G., and Kornprobst, J., 1971, Le Maroc. Domaine rifain et atlasique, in: *Tectonique de l'Afrique*: *Sci. Terre*, U.N.E.S.C.O., Paris, no. 6, p. 17–46.

Fitches, W. R., 1970, "Pan-African orogeny" in the coastal region of Ghana: *Nature*, v. 226, p. 744–746.

Flicoteaux, R., 1972, Rapport scientifique récapitulatif 1969-1972, in: Rapport scientifique pour la période 1969-1972 du Laboratoire Associé au C.N.R.S. n° 132, *Etudes géologiques ouest-africaines*: *Trav. Lab. Sci. Terre St-Jérôme*, Marseille, Sér. E, v. 10, p. 77–78.

Flicoteaux, R., and Tessier, F., 1971, Précisions nouvelles sur la stratigraphie des formations du plateau de Thiès (Sénégal occidental) et sur leurs altérations. Conséquences paléogéographiques: *C. R. Acad. Sci.*, Paris, Sér. D, v. 272, p. 364–367.

Frisch. T., and Schmincke, H. U., 1970, Petrology of clinopyroxene-amphibole inclusions from the Roque Nublo volcanics. Gran Canaria, Canary Islands: *Bull. volcanolog.*, v. 33, p. 1073–1088.

Funnell, B. M., and Smith, A. G., 1968, Opening of the Atlantic Ocean: *Nature*, v. 219, p. 1328–1333.

Furon, R., 1963, *Geology of Africa*: Hafner Publ. Co., New York, 377 p.

Furon, R., 1964, *Le Sahara, Géologie, Ressources minérales*: Payot, ed., Paris, 313 p.

Furon, R., 1968, *Géologie de l'Afrique*, 3e éd.: Payot, ed., Paris, 374 p.

Furon, R., and Nicklès, M., 1956, Afrique équatoriale française, Cameroun français, Guinée espagnole et San Tomé: *Lexique Stratigraph. Internat.*, Centre Nat. Rech. Sci., Paris, v. IV, n. 6, 58 p.

Fúster, J. M., 1957, Graciosa, Mapa geologico de España 1:50,000, Mapa nacional 1.082: Inst. Geol. Min. España.

Fúster, J. M. and Aguilar, T., 1965, Nota previa sobre la geologia del macizo de Betancuria, Fuerteventura (Islas Canarias): Inst. "Lucas Mallada", Estudios Geol., v. 21, p. 181–197.

Fúster, J. M., Araña, V., Brandle, J. L., Navarro, M., Alonso, U., and Aparicio, A., 1968, Geologia y volcanologia de las Islas Canarias - Tenerife: Inst. "Lucas Mallada", Consejo Sup. Investig. Cient., Madrid, 218 p.

Fúster, J. M., Cendrero, A., Gastesi, P., Ibarrola, E., and López Ruiz, J., 1968, Geologia y volcanologia de las Islas Canarias - Fuerteventura: Inst. "Lucas Mallada", Consejo Sup. Investig. Cient. Madrid, 239 p.

Fúster, J. M., Fernandez Santin, A., and Sagredo, J., 1968, Geologia y volcanologia de las Islas Canarias - Lanzarote: Inst. "Lucas Mallada", Consejo Sup. Investig. Cient., Madrid, 177 p.

Fúster, J. M., Hernandez-Pacheco, A., Muñoz, M., Rodriguez Badiola, E., and Garcia Cacho, L., 1968, Geologia y volcanologia de las Islas Canarias - Gran Canaria: Inst. "Lucas Mallada", Consejo Sup. Investig. Cient., Madrid, 243 p.

Fúster, J. M., Páez, A., and Sagredo, J., 1970, Significance of basic and ultramafic rock inclusions in the basalts of the Canary Islands: *Bull. volcanolog.*, v. 33, p. 665–693.

Fúster, J. M., *et al.*, 1958, Tuineje, Mapa geologico de España 1:50,000, Mapa nacional 1.115: Inst. Geol. Min. España.

Fúster Casas, J. M., *et al.*, 1967, Alegranza, Mapa geologico de Espana 1:50,000, Mapa nacional 1079-1080: Inst. Geol. Min. España.

Gass, I. G., 1967, Geochronology of the Tristan da Cunha group of Islands: *Geol. Mag.*, v. 104, p. 160–170.

Gastesi, P., 1970, Petrology of the ultramafic and basic rocks of Betancuria massif, Fuerteventura Island (Canarian archipelago): *Bull. volcanolog.*, v. 33, p. 1008–1038.

Gevin, P., 1960, Etudes et reconnaissances géologiques sur l'axe cristallin Yetti-Eglab et ses bordures sédimentaires: Thèse Paris, 1958, *Publ. Serv. Carte Géol. Algérie Bull.*, Nouv. Sér., v. 23, 328 p.

Gibson, T. G., 1970, Late Mesozoic-Cenozoic tectonic aspects of the Atlantic coastal margin: *Geol. Soc. Am. Bull.*, v. 81, p. 1813–1822.

Gigout, M., 1956, Recherches sur le Pliocène et le Quaternaire atlantiques marocains: *Trav. Inst. Sci. Chérifien*, Rabat, Sér. Géol. Géogr. Phys., v. 5, 94 p.

Giraudon, R. and Sougy, J., 1963, Position anormale du socle granitisé des Hajar Dekhen sur la série d'Akjoujt et participation de ce socle à l'édification des Mauritanides hercyniennes (Mauritanie occidentale): *C. R. Acad. Sci.*, Paris, v. 257, n. 9, p. 937–940.

Giraudon, R., and Vachette, M., 1964, Mesures d'âges absolus sur des formations de Mauritanie: *C. R. Acad. Sci.*, Paris, v. 258, n. 9, p. 3520–3523.

Gittins, J., 1966, Summaries and bibliographies of carbonatite complexes, in: *Carbonatites*, O. F. Tuttle and J. Gittins, eds.: Interscience–Wiley, New York, p. 417–540.

Gorodiski, A., 1958, Miocène et indices phosphatés de Casamance (Sénégal): *C. R. Somm. Séances Soc. Géol. Fr.*, v. 1958, n. 13, p. 293–297.

Gougenheim, A., 1959, Levé bathymétrique de la Côte du Maroc: *C. R. Acad. Sci.*, Paris, v. 249, p. 2599–2601.

Gough, D. I., Opdyke, N. D., and McElhinny, M. W., 1964, The significance of paleomagnetic results from Africa: *J. Geophys. Res.*, v. 69, p. 2509–2519.

Grandin, G., and Delvigne, J., 1969, Les cuirasses de la région birrimienne volcano-sédimentaire de Toumodi: jalons de l'histoire morphologique de la Côte d'Ivoire: *C. R. Acad. Sci.*, Paris, Sér. D, v. 269, p. 1474–1477.

Grant, N. K., 1967, Complete late Pre-Cambrian to early Palaeozoic orogenic cycle in Ghana, Togo and Dahomey: *Nature*, v. 215, n. 5101, p. 609–610.

Grant, N. K., 1969a, The late Precambrian to early Paleozoic Pan-African orogeny in Ghana, Togo, Dahomey and Nigeria: *Geol. Soc. Am. Bull.*, v. 80, p. 45–56.

Grant, N. K., 1969b, The nature of the Pan-African orogeny in Nigeria: *13th Ann. Rep. Res. Inst. Afr. Geol. Leeds*, Great Britain, Dec., p. 20–21.

Grant, N. K., 1970, Geochronology of Precambrian basement rocks from Ibadan, southwestern Nigeria: *Earth Planet. Sci. Let.*, v. 10, p. 29–38.

Guilcher, A., 1963, Continental shelf and slope (continental margin), in: The sea, M. N. Hill, ed.: Interscience, New York, v. 3, p. 281–311.

Guilcher, A., and Joly, F., 1954, Recherches sur la morphologie de la côte atlantique du Maroc: *Trav. Inst. Sci. Chérifien*, Rabat, Sér. Géol. Géogr. Phys., no. 2, 140 p.

Gutenberg, B., and Richter, C. F., 1954, *Seismicity of the Earth*: Hafner Publ. Co., New York, (re-ed. 1965), 310 p.

Hailwood, E. A., and Mitchell, J. G., 1971, Palaeomagnetic and radiometric dating results from Jurassic intrusions in South Morocco: *Geophys. J. R. Astr. Soc.*, v. 24, p. 351–364.

Hallam, A., 1969, Tectonism and eustasy in the Jurassic: *Earth-Sci. Rev.*, v. 5, n. 1, p. 45–68.

Hallam, A., 1971, Mesozoïc geology and the opening of the North Atlantic: *J. Geol.*, v. 79, p. 129–157.

Hausen, H., 1956, Contributions to the geology of Tenerife (Canary Islands): *Soc. Sci. Fennica, Comment. Phys.-Math.*, v. 18, n. 1, 271 p.

Hausen, H., 1958, On the geology of Fuerteventura (Canary Islands): *Soc. Sci. Fennica, Comment. Phys.-Math.*, 1 geol. map 1:300,000, 1959, v. 22, n. 1, 211 p.

Hausen, H., 1959, On the geology of Lanzarote, Graciosa and the Isletas (Canarian Archipelago): *Soc. Sci. Fennica, Comment. Phys.-Math.*, v. 23, n. 4, 116 p.

Hausen, H., 1962, New contributions to the geology of Grand Canary (Gran Canaria, Canary Islands): *Soc. Sci. Fennica, Comment. Phys.-Math.*, v. 27, n. 1, 418 p.

Hayes, D. E., Pimm, A. C., Benson, W. E., Berger, W. H., von Rad, U., Supko, P. R., Beckmann, J. P., Roth, P. H., and Musich, L. F., 1971, Deep sea drilling project, Leg 14: *Geotimes*, v. 16, n. 2, p. 14–17.

Hazzard, J. C., 1961, Bioherms in Middle Devonian of Northeastern Spanish Sahara, Northwest Africa: *Bull. Am. Assoc. Petrol. Geol.*, v. 45, n. 1, p. 129.

Hébrard, L., 1972a, Un épisode quaternaire en Mauritanie "Afrique occidentale" à la fin du Nouakchottien: le Tafolien, 4000–2000 ans avant le présent, *Bull. Liaison Assoc. Sénégalaise Et. Quatern. Ouest Afr.*: Dakar, v. 33–34, p. 5–15.

Hébrard, L., 1972b, Contribution à l'étude géologique du Quaternaire du littoral mauritanien entre Nouakchott et Nouadhibou, 18°-21° latitude nord: Lyon Univ. thesis, Rep. Lab. Géol. Univ. Dakar, 549 p.

Hébrard, L., Faure, H., and Elouard, P., 1969, Age absolu du volcanisme quaternaire de Dakar (Sénégal): *Bull. Liaison Assoc. Sénégalaise Et. Quatern. Ouest Afr.*, Dakar, v. 22, p. 15–19.

Hedberg, H. D., 1961, Petroleum developments in Africa in 1960: *Bull. Am. Assoc. Petrol. Geol.*, v. 45, p. 1143–1185.

Heezen, B. C., and Johnson, G. L., 1963, A moated knoll in the Canary passage: *D. Hydrograph. Z.*, v. 16, n. 6, p. 269–272.

Heezen, B. C., and Menard, H. W., 1963, Topography of the deep sea floor, in: *The Sea*, M. N. Hill, ed.: Interscience, New York, v. 3, p. 233–280.

Heezen, B. C., and Tharp, M., 1968, Physiographic diagram of the North Atlantic ocean: Geol. Soc. Am., spec. pap. 65.

Heezen, B. C., Tharp, M., and Ewing, M., 1959, The floors of the oceans, I. North Atlantic: Geol. Soc. Am., spec. pap. 65, 122 p.

Heirtzler, J. R. and Hayes, D. E., 1967, Magnetic boundaries in the North Atlantic Ocean: *Science*, v. 157, p. 185–187.

Hernández-Pacheco, A., 1970, The tahitites of Gran Canaria and haüynitization of their inclusions: *Bull. volcanolog.*, v. 33, p. 701–728.

Hernández-Pacheco, E., 1950, Morfología y evolución de las zonas litorales de Ifni y del Sahara Español: *C. R. 7e Congr. Intern. Géogr.*, Lisbonne 1949, Union Géogr. Intern., v. 2, p. 487–505.

Hollard, H., 1967a, Le Dévonien du Maroc et du Sahara nord-occidental, in: *International Symposium on the Devonian System, Calgary*: D. H. Oswald, ed.: *Alberta Soc. Petrol. Geol.*, v. 1, p. 203–244.

Hollard, H., 1967b, Sur la transgression dinantienne au Maroc présaharien: *C. R. 6e Congr. Intern. Stratigr. Géol. Carbonifère*, p. 923–936.

Hollard, H. and Jacquemont, P., 1958, Le Gothlandien, le Dévonien et le Carbonifère des régions du Dra et du Zemoul: *Notes Mém. Serv. Géol. Maroc*, v. 135, *Notes* 15, p. 7–33.

Hurley, P. M., *et al.*, 1966, Continental drift investigations: *14th Ann. Progr. Rep. Mass. Inst. Technol.*, Cambridge, M.I.T., 1381–14, p. 3–15.

Hurley, P. M., de Almeida, F. F. M., Melcher, G. C., Cordani, U. G., Rand, J. R., Kawashita, K., Vandoros, P., Pinson, W. H. Jr., and Fairbairn, H. W., 1967, Test of continental drift by comparison of radiometric ages. A predrift reconstruction shows matching geologic age provinces in West Africa and North Brazil: *Science*, v. 157, n. 3788, p. 495–500.

Hurley, P. M., Laing, E. M., Fairbairn, H. W., and Pinson, W. H., 1968, Age determinations in the older Precambrian basement rocks of Sierra Leone, in: *Variations in isotopic abundances of strontium, calcium and argon and related topics*: *16th Ann. Progr. Rep.*, Mass. Inst. Technol., Cambridge, M.I.T., 1381-16, p. 83–86.

Hurley, P. M., White, R. W., and Leo, G. W., 1969, The location of the boundary between the Eburnean and Liberian orogenic provinces in eastern Liberia, in: *Variations in isotopic abundances of strontium, calcium and argon and related rocks*: *17th Ann. Progr. Rep.*, M.I.T.-1381-17, Dept. Geol. Geophys., M.I.T., Cambridge, p. 25–26.

Huvelin, P., 1967, Nappe de glissement précoce hercynienne dans les Jebilet, (Maroc): *C. R. Acad. Sci.*, Paris, Sér. D, v. 265, p. 1039–1042.

Huvelin, P., 1970a, Chevauchements et écaillages précoces hercyniens des terrains anté-viséens dans le domaine atlasique (Maroc): *C. R. Acad. Sci.*, Paris, Sér. D, v. 270, p. 2760–2763.

Huvelin, P., 1970b, Mouvements hercyniens précoces dans la région de Mrirt (Maroc): *C. R. Acad. Sci.*, Paris, Sér. D, v. 271, p. 953–955.

Ibarrola, E., 1970, Variation trends in basaltic rocks of the Canary Islands: *Bull. Volcanolog.*, v. 33, p. 729–777.

Isacks, B., Oliver, J., and Sykes, L. R., 1968, Seismology and the new global tectonics: *J. Geophys. Res.*, v. 73, n. 18, p. 5855–5899.

Jacobson, R. R. E., Snelling, N. J., and Truswell, J. F., 1963, Age determinations in the geology of Nigeria, with special reference to the Older and Younger granites: *Overseas Geol. Mineral Res.*, v. 9, p. 162–182.

Jérémine, E., 1935, Contribution à l'étude des îles Hierro et Gomera (archipel canarien): *Bull. Soc. Fr. Minéral.*, v. 58, p. 350–363.

J.O.I.D.E.S., 1965, Ocean drilling on the continental margin: *Science*, v. 150, n. 3697, p. 709–716.

Kennedy, W. Q., 1964, The structural differentiation of Africa in the Pan-African ($\pm$500 million years) tectonic episode: *8th Ann. Rep. Res. Inst. Afr. Geol.* (1962–63), Leeds Univ., (U.K.), p. 48–49.

Kennedy, W. Q., 1965, The influence of basement structure on the evolution of the coastal (Mesozoic and Tertiary) basins of Africa, in: *Salt basins around Africa*: Institute of Petroleum, London, Elsevier Publishing Co., Amsterdam, p. 7–16.

Kilian, C., 1926, Sur la mesure dans laquelle la période silurienne est représentée par ses formations de l'Enceinte tassilienne et sur la présence de l'Ordovicien au Sahara: *C. R. Acad. Sci.*, Paris, v. 182, p. 146–148.

King, L. C., 1967, *Morphology of the earth*, 2nd ed.: Hafner Publ. Co., New York, 726 p.

King, P. B., 1959, *The evolution of North America*: Princeton Univ. Press, Princeton, New Jersey, 190 p.

King, P. B., 1969, The tectonics of North America, a discussion to accompany the tectonic map of North America, scale 1:5,000,000: *U.S. Geol. Surv. Prof. Paper*, 628, 95 pp.

Klemme, H. D., 1958, Regional geology of circum-mediterranean region: *Bull. Am. Assoc. Petrol. Geol.*, v. 42, n. 3, p. 477–512.

Klerkx, J., and Paepe (de), P., 1971, Cape Verde Islands: evidence for a Mesozoic oceanic ridge: *Nature* (Phys. Sci.), Oct. 11, v. 233, p. 117–118.

Kogbe, C. A., 1973, Geology of the Upper Cretaceous and Tertiary sediments of the Nigerian sector of the Iullemeden basin (West Africa): *Geol. Rundschau*, v. 62, n. 1, p. 197–211.

Koning (de), G., 1957, Géologie des Ida ou Zal (Maroc): *Leidse Geol. Mededelingen*, v. 23, 209 p.

Krause, D. C., 1963, Seaward extension and origin of the Freetown layered basic complex of Sierra Leone: *Nature*, v. 200, p. 1280–1281.

Krause, D. C., 1964, Guinea fracture zone in the equatorial Atlantic: Science, v. 146, p. 57–59.

Lacroix, A., 1911, Les syénites néphéliniques de l'archipel de Los et leurs minéraux: *Nouv. Archiv. Mus. Hist. Natur.*, Paris, Sér. 5, v. 3.

Lasserre, M., Lameyre, J., and Buffière, J.-M., 1970, Données géochronologiques sur l'axe précambrien Yetti-Eglab en Algérie et en Mauritanie du Nord: *Bull. Bur. Rech. Géol. Min.*, Sér. 2, Section 4, v. 2, p. 5–13.

Laville, E., Lesage, J.-L., and Seguret, M., 1973, Résultats préliminaires sur la tectonique hercynienne et atlasique de l'Atlas central et de l'"accident sud-atlasique": *1ère Réun. Ann. Sci. Terre*, Paris, 19–22 mars, p. 260.

Lay, C. and Reichelt, R., 1971, Sur l'âge et la signification des intrusions de dolérites tholéitiques dans le bassin de Taoudenni (Afrique occidentale): *C. R. Acad. Sci.*, Paris, Sér. D, v. 272, p. 374–376.

Leblanc, M., 1968, Chevauchements dans la boutonnière précambrienne de Bou-Azzer El Graara (Anti-Atlas, Maroc): *Bull. Soc. Géol. France*, 7e Sér., v. 10, n. 1, p. 93–96.

Lecointre, G., 1952, Recherches sur le Néogène et le Quaternaire marins de la côte atlantique du Maroc: *Notes Mém. Serv. Géol. Maroc*, no. 99, Notes, v. 1, 198 p.

Lecointre, G., 1963, Sur les terrains sédimentaires de l'île de Sal: *Garcia de Orta*, v. 11, p. 275–289.

Lecointre, G., 1966, Néogène et Quaternaire du Rio de Oro (Maroc espagnol): *C. R. Somm. Séances Soc. Géol. France*, v. 10, p. 404–405.

Lécorché, J.-P., and Sougy, J., 1969, Relations des formations d'Akjoujt (Mauritanides) avec le Paléozoïque de l'Adrar (bassin de Taoudeni) dans la région d'Iriji (Est d'Akjoujt, Mauritanie occidentale): *Bull. Soc. Géol. France*, 7e Sér., v. 11, p. 233–250.

Legoux, P., 1939, Esquisse géologique de l'Afrique Occidentale Française: *Bull. Serv. Min. A. O. F.*, Dakar, v. 4, 134 p.

Legrand, P., 1969, Description de *Westonia chudeaui* nov. sp., Brachiopode inarticulé de l'Adrar mauritanien (Sahara occidental): *Bull. Soc. Géol. France*, Sér. 7, v. 11, p. 251–256.

Le Maître, D., 1950, Nouveaux éléments communs avec l'Amérique dans la faune dévonienne de l'Afrique du Nord: *C. R. Sommaire Séances Soc. Géol. France*, v. 14, p. 253–256.

Le Maître, D., 1961, Découverte de nouveaux gisements africains de *Pustulatia pustulosa* (Hall), Répartition stratigraphique de ce Brachiopode en Afrique: *C. R. Sommaire Soc. Géol. France*, v. 7, p. 190–191.

Leo, G. W. and White, R. W., 1968, Age investigation in Liberia, in: *Variations in isotopic abundances of strontium, calcium, and argon and related topics*: *16th Ann. Progr. Rep.*, Mass. Inst. Technol., Cambridge, M.I.T. 1381–16, p. 87–93.

Le Page, A., 1972, Rapport scientifique récapitulatif 1969-1972, in: Rapport scientifique pour la période 1969-1972 du Laboratoire associé au C.N.R.S. n° 132, *Etudes géologiques W-africaines*, par J. Sougy: *Trav. Lab. Sci. Terre Marseille St-Jérôme*, Sér. E, v. 10, p. 41–43.

Le Pichon, X., 1968, Sea floor spreading and continental drift: *J. Geophys. Res.*, v. 73, p. 3661–3697.

Le Pichon, X., and Fox, P. J., 1971, Marginal offsets, fracture zones and the early opening of the North Atlantic: *J. Geophys. Res.*, v. 76, p. 6294–6308.

Le Pichon, X., and Hayes, D. E., 1971, Marginal offsets, fracture zones and the early opening of the South Atlantic: *J. Geophys. Res.*, v. 76, p. 6283–6293.

Leprun, J.-C., and Trompette, R., 1970, Subdivision du Voltaïen du massif de Gobnangou (République de Haute-Volta) en deux séries discordantes séparées par une tillite d'âge éocambrien probable: *C. R. Acad. Sci.*, Paris, Sér. D, v. 269, p. 2187–2190.

Lille, R., 1969, Précambrien et Cambro-Ordovicien du Guidimaka (Mauritanie méridionale). Tectoniques, métamorphismes syntectoniques et volcanisme dans un secteur des Mauritanides: *Bull. Soc. Géol. France*, Sér. 7, v. 11, p. 257–267.

Loczy, L. de, 1951, Sur le problème du Trias-salifère et sur l'existence du Trias-alpin dans la partie septentrionale du Maroc: *Rep. 18th. Session Intern. Geol. Congr.*, London, 1948, part 14, p. 164–174.

López Ruiz, J., 1970, Le complexe filonien de Fuerteventura (Iles Canaries): *Bull. volcanolog.*, v. 33, p. 1166–1185.

Loubet, M., and Allègre, C. J., 1970, Analyse des terres rares dans les échantillons géologiques par dilution isotopique et spectrométrie de masse. Application à la distinction entre carbonatites et calcaires: *C. R. Acad. Sci.*, Paris, Sér. D, v. 270, p. 912–915.

Louis, P., 1969, Les anomalies gravimétriques régionales et le bâti structural de l'Afrique occidentale et centrale: *5e Colloque Géol. Afr.*, Clermont-Ferrand, 9-11 Avril, *Ann. Fac. Sci. Univ. Clermont-Ferrand*, 41, *Géol. Minéral.*, v. 19, p. 35–39.

MacFarlane, D. J., and Ridley, W. I., 1968, An interpretation of gravity data for Tenerife, Canary Islands: *Earth Planet. Sci. Let.*, v. 4, p. 481–486.

MacFarlane, D. J., and Ridley, W. I., 1969, An interpretation of gravity data for Lanzarote, Canary Islands: *Earth Planet. Sci. Let.*, v. 6, p. 431.

Machado, F., 1965, Mechanism of Fogo Volcano, Cape Verde Islands: *Garcia de Orta*, v. 13, n. 1, p. 51–56.

Machado, F., Azeredo Leme, J., and Monjardino, J., 1967, O complexo sienito-carbonatítico da ilha Brava, Cabo Verde: *Garcia de Orta*, Lisboa, v. 15, n. 1, p. 93–98.

Maluski, H., 1973, Datation par la méthode K-Ar des granites précambriens du Centre de la Côte d'Ivoire: *1ère Réun. Ann. Sci. Terre*, Paris, 19-22 mars, p. 284.

Marchand, J., Sougy, J., Rocci, G., Caron, J.-P. H., Deschamps, M., Simon, B., Deynoux, M., Tempier, C., and Trompette, R., 1971. Etude photogéologique de la partie orientale de la dorsale reguibat et de sa couverture sud (Mauritanie), tome I, synthèse géologique: *Trav. Lab. Sci. Terre St-Jérôme*, Marseille, Sér. X inéd., No. 11, 167 p.

Marchand, J., Sougy, J., and Trompette, R., 1972, Rapport scientifique de tournée sur la bordure nord-ouest du bassin de Taoudeni dans la région de Mejahouda-Tinioulig-El Zerem (Nord de la Mauritanie), in: Rapport annuel d'activité 1971-1972 du Laboratoire Associé au C.N.R.S. no. 132, *Etudes géologiques ouest-africaines*: *Trav. Lab. Sci. Terre St-Jérôme*, Marseille, Sér. E, v. 9, p. 53–55.

Martin, L., 1971, The continental margin from Cape Palmas to Lagos: bottom sediments and submarine morphology, in: *The geology of the East Atlantic continental margin*, F. M. Delaney, ed.: *G. Brit. Inst. Geol. Sci. Rep.*, 70/16, p. 79–95.

Martin, L., 1973, Carte sédimentologique du plateau continental de Côte d'Ivoire: Edit. Off. Rech. Sci. technol. Outre-mer, Paris, Collection Notices, No. 48, 20 p.

Martin, L. and Delibrias, G., 1972, Schéma des variations du niveau de la mer en Côte d'Ivoire depuis 25 000 ans: *C. R. Acad. Sci.*, Paris, Sér. D, v. 274, p. 2848–2851.

Martin, L. and Tastet, J. P., 1972, Le Quaternaire du littoral et du plateau continental de Côte d'Ivoire. Rôle des mouvements tectoniques et eustatiques: *Bull. Liaison. Assoc. Sénégal. Et. Quatern. Ouest Afr.*, Dakar, v. 33-34, p. 17–32.

Martinis, B. and Visintin, V., 1966, Données géologiques sur le bassin sédimentaire côtier de Tarfaya (Maroc méridional), in: *Bassins sédimentaires du littoral africain, Symposium*, D. Reyre, éd.: 1e partie: Littoral atlantique (New Delhi, 1964), Assoc. Serv. Géol. Afr., Paris, p. 13–26.

Masoli, M., 1965, Sur quelques Ostracodes fossiles mésozoïques (Crétacé) du bassin côtier de Tarfaya (Maroc méridional), in: Colloque intern. Micropaleont.: *Mém. Bur. Rech. Géol. Min.*, Dakar, 6-11 mai 1963, Paris, v. 32, p. 119–134.

Massacrier, P., 1972, Carte géologique au 50 000e et notice explicative de la boutonnière des Aït Abdallah (Précambrien de l'Anti-Atlas occidental): *Trav. Lab. Sci. Terre Marseille St-Jérôme*, Sér. X, v. 20, 7 p.

Mattauer, M., Proust, F., and Tapponnier, P., 1972, Major strike-slip fault of late Hercynian age in Morocco: *Nature*, May 19, No. 5351, v. 237, p. 160–162.

Maugis, P., 1954, L'activité de la mission de préreconnaissance du Bureau de Recherches de Pétrole dans les bassins sédimentaires de l'A.O.F.: *Rev. Inst. Fr. Pétrole*, v. 9, n. 1, p. 3–17.

Maugis, P., 1955a, Compte rendu des études de préreconnaissance pétrolière dans le bassin sédimentaire du Sénégal: *Rev. Inst. Fr. Pétrole*, v. 10, n. 5, p. 269–282.

Maugis, P., 1955b, Etudes de préreconnaissance pétrolière dans le bassin sédimentaire du Sénégal: *Bull. Dir. Fédér. Min. Géol. A.O.F.*, Dakar, v. 19, p. 99–128.

Maxwell, A. E., von Herzen, R. P., Andrews, J. E., Boyce, R. E., Milow, E. D., Hsu, K. J., Percival, S. F., and Saito, T., 1970, Initial reports of the deep sea drilling project: U.S. Govt. Print. Off., Washington, vol. III, 806 p.

May, P. R., 1971, Pattern of Triassic-Jurassic diabase dikes around the North Atlantic in the context of predrift position of the continents: *Geol. Soc. Am. Bull.*, v. 82, p. 1285–1291.

Mazéas, J.-P., and Pouit, G., 1968, Marques de mouvements hercyniens à composante tangentielle de grande amplitude dans la boutonnière précambrienne et infracambrienne du bas Oued Dra (Maroc méridional): *C. R. Acad. Sci.*, Paris, Sér. D, v. 267, p. 1549–1552.

McBirney, A. R., and Gass I. G., 1967, Relations of oceanic volcanic rocks to mid-oceanic rises and heat flow: *Earth Planet. Sci. Let.*, v. 2, n. 4, p. 265–276.

McConnell, R. B., 1969, Fundamental fault zones in the Guiana and West African shields in relation to presumed axes of Atlantic spreading: *Geol. Soc. Am. Bull.*, v. 80, p. 1775–1782.

McCurry, P., 1971, Pan-African orogeny in northern Nigeria: *Geol. Soc. Am. Bull.*, v. 82, p. 3251–3262.

McKie, D., 1966, Fenitization, in: *Carbonatites*, O. F. Tuttle and J. Gittins, eds.: Interscience–Wiley, New York, p. 261–294.

McMaster, R. L., De Boer, J., and Ashraf, A., 1970, Magnetic and seismic reflection studies on continental shelf off Portuguese Guinea, Guinea, and Sierra Leone, West Africa: *Am. Assoc. Petrol. Geol. Bull.*, v. 54, p. 158–167.

McMaster, R. L., and Lachance, T. P., 1968, Seismic reflectivity studies on north-western African continental shelf: strait of Gibraltar to Mauritania: *Am. Assoc. Petrol. Geol. Bull.*, v. 52, p. 2387–2395.

McMaster, R. L. and Lachance, T. P., 1969, Northwestern African continental shelf sediments: *Mar. Geol.*, v. 7, p. 57–67.

McMaster, R. L., Lachance, T. P., and Ashraf, A., 1970, Continental shelf geomorphic features off Portuguese Guinea, Guinea and Sierra Leone (West Africa): *Mar. Geol.*, v. 9, p. 203–213.

McMaster, R. L., Lachance, T. P., Ashraf, A., and de Boer, J., 1971, Geomorphology, structure and sediments of the continental shelf and upper slope off Portuguese Guinea, Guinea and Sierra Leone, in: *The geology of the East Atlantic continental margin*, F. M. Delaney, ed.: G. Brit. Inst. Geol. Sci. Rep., 70/16, p. 105–119.

Menchikoff, N., 1930, Recherches géologiques et morphologiques dans le Nord du Sahara occidental: *Rev. Géogr. Phys. Géol. Dynam.*, v. 3, n. 2, p. 147.

Michel, P., 1970a, Chronologie du Quaternaire des bassins des fleuves Sénégal et Gambie, Essai de synthèse (1ère partie): *Bull. Liaison Assoc. Sénégalaise Et. Quatern. Ouest Afr.*, Dakar, v. 25, p. 53–64.

Michel, P., 1970b, Chronologie du Quaternaire des bassins des fleuves Sénégal et Gambie, Essai de synthèse (2ème partie): *Bull. Liaison Assoc. Sénégalaise Et. Quatern. Ouest Afr.*, Dakar, v. 26, p. 25–37.

Millot, G. and Dars, R., 1959, L'archipel des îles de Los: une structure annulaire subvolcanique (République de Guinée): *Notes Serv. Géol. Prospect. Min.*, Dakar, v. 2, p. 47–56.

Millot, G., Elouard, P., Lucas, J., and Slansky, M., 1960, Une séquence sédimentaire et géochimique de minéraux argileux: montmorillonite, attapulgite, sépiolite: *Bull. Gr. Fr. Argiles*, v. 12, n. 7, p. 77–82.

Millot, G., Radier, H., and Bonifas, M., 1957, La sédimentation argileuse à attapulgite et montmorillonite: *Bull. Soc. Géol. France*, Sér. 6, v. 7, p. 425–433.

Monciardini, C., 1966, La sédimentation éocène du Sénégal, *Mém. Bur. Rech. Géol. Min.*, Paris, v. 43, 65 p.

Monod, T., 1952, L'Adrar mauritanien (Sahara occidental): Esquisse géologique, *Bull. Dir. Min. Afr. Occ. Fr.*, Dakar, v. 15, n. 1, p. 1–285.

Montadert, L., 1969, Données nouvelles sur la structure géologique du golfe de Guinée, 5e Colloque Géol. afr., *Ann. Fac. Sci. Univ. Clermont-Ferrand*, Clermont-Ferrand, 9-11 avr., 41, *Géol. Minéral.*, v. 19, p. 71–72.

Muñoz, M., 1970, Ring complexes of Pajara in Fuerteventura Island: *Bull. Volcanolog.*, v. 33, p. 840–861.

Nahon, D., 1971, Genèse et évolution des cuirasses ferrugineuses quaternaires sur grès: exemple du Massif de Ndias (Sénégal occidental), *Bull. Serv. Carte Géol. Alsace-Lorraine*, Strasbourg, France, v. 24, n. 4, p. 219–241.

Nahon, D., 1972, Rapport scientifique récapitulatif 1969-1972, in: Rapport scientifique pour la période 1969-1972 du Laboratoire Associé au C.N.R.S. n° 132, *Etudes géologiques ouest-africaines*, par. J. Sougy: *Trav. Lab. Sci. Terre St-Jérôme*, Marseille, Sér. E, v. 10, p. 81–83.

Nahon, D., and Démoulin, D., 1970, Essai de stratigraphie relative des formations cuirassées du Sénégal occidental: *C. R. Acad. Sci.*, Paris, Sér. D, v. 270, p. 2764–2767.

Nahon, D. and Démoulin, D., 1971, Contribution à l'étude des formations cuirassées du Sénégal occidental (pétrographie, morphologie et stratigraphie relative): *Rev. Géogr. Phys. Géol. Dynam.*, Sér. 2, v. 13, n. 1, p. 35–54.

Nahon, D. and Ruellan, A., 1972, Encroûtements calcaires et cuirasses ferrugineuses dans l'Ouest du Sénégal et de la Mauritanie: *C. R. Acad. Sci.*, Paris, Sér. D, v. 274, p. 509–512.

Navarro, F. de P., 1947, Exploracion oceanografica del Africa Occidental desde el Cabo Ghir al Cabo Juby: *Inst. España Oceanogr. Trabajos*, v. 20, 40 p.

Naylor, R. S., 1969, Lower Paleozoic continental accretion in the Northern Appalachians, (abs.): *Trans. Am. Geophys. Union*, v. 50, n. 4, p. 313.

Nota, D. J. G., 1958, Sediments of the western Guinea shelf: *Mededel. Landbouwhogeschool Wageningen*, v. 58, n. 2, 98 p.

Nutter, A. H., 1969, The origin and distribution of phosphate in marine sediments from the Moroccan and Portuguese continental margins: Thesis Univ. London, Great Britain, 158 p. (unpublished).

Oversby, V. M., Lancelot, J., and Gast, P. W., 1971, Isotopic composition of lead in volcanic rocks from Tenerife, Canary Islands: *J. Geophys. Res.*, v. 76, n. 14, p. 3402–3413.

Papon, A., Roques, M., and Vachette, M., 1968, Age de 2700 millions d'années, déterminé par la méthode au strontium, pour la série charnockitique de Man, en Côte d'Ivoire: *C. R. Acad. Sci.*, Paris, Sér. D, 266, p. 2046–2048.

Pareyn, C., 1962, Les massifs carbonifères du Sahara Sud-oranais: *Publ. Centre Rech. Sahariennes*, Paris, Géol. v. 1, 325 p.

Part, G. K., 1950, Volcanic rocks from the Cape Verde Islands: *Bull. Brit. Mus.* (Natur. Hist.), Mineralogy, v. 1, p. 25–72.

Peterson, M. N. A., Edgar, N. T., Cita, M., Gartner, S., Jr., Goll, R., Nigrini, C., and von der Borch, C., 1970, Initial reports of the deep sea drilling project: U.S. Govt. Print. Off., Washington, v. 2, 501 p.

Phillips, J. D., 1967, Magnetic anomalies over the mid-Atlantic ridge near 27°N: *Science*, v. 157, p. 920–923.

Phillips, J. D. and Forsyth, D., 1972, Plate tectonics, paleomagnetism, and the opening of the Atlantic: *Geol. Soc. Am. Bull.*, v. 83, p. 1579–1600.

Pineau, F. and Javoy, M., 1969, Détermination des rapports isotopiques $^{18}O/^{16}O$ et $^{13}C/^{12}C$ dans diverses carbonatites, implications génétiques: *C. R. Acad. Sci.*, Paris, Ser. D, v. 269, p. 1930–1933.

Pineau, F. and Javoy, M., 1973, Composition isotopique de l'oxygène dans les roches volcaniques de Tenerife et le complexe plutonique de Fuerteventura (îles Canaries): *Ière Réun. Ann. Sci. Terre*, Paris, 19-22 mars, p. 340.

Pires Soares, J. M., 1953, A proposito dos "Aptychi" da Ilha de Maio: *Publ. Junta Invest. Ultramar*, 3 p.

Pitman, W. C.III, Talwani, M., and Heirtzler, J. R., 1971, Age of the North Atlantic ocean from magnetic anomalies, *Earth Planet. Sci. Let.*, v. 11, p. 195–200.

Pollett, J. D., 1956, Sierra Leone et Gambie, in: *Lexique stratigraphique international*, R. Furon, ed.: Paris, C.N.R.S., fasc. 3: Afrique Occidentale Anglaise, v. 4, p. 3–11.

Proust, F. and Tapponnier, P., 1973, L'accident du Tizi n'Test (Haut Atlas-Maroc): décrochement tardi-hercynien repris par la tectonique alpine: *Ière Réun. Ann. Sci. Terre*, Paris, 19-22 mars, p. 350.

Puri, H. S. and Vernon, R. O., 1964, Summary of the geology of Florida and a guide-book to the classic exposures, Florida Geol. Surv. Spec. Publ., v. 5 (revised), 312 p.

Querol, R., 1966, Regional geology of the Spanish Sahara, in: *Sedimentary basins of the African coasts, 1st part: Atlantic Coast*, D. Reyre, ed.: Assoc. Afr. Geol. Surv., Paris, p. 27–39.

Radier, H., 1959, Contribution à l'étude du Soudan oriental (A.O.F.), tome I: Le Précambrien saharien au Sud de l'Adrar des Iforas; tome II: Le bassin crétacé et tertiaire de Gao - Le détroit soudanais: *Bull. Serv. Géol. Prospect. Min. A.O.F.*, Dakar, v. 26, 550 p.

Rancurel, P., 1965, Topographie générale du plateau continental de la Côte d'Ivoire et du Libéria: *Off. Rech. Sci. Techn. Outre-mer*, Paris (1968).

Ratschiller, L. K., 1966-67, Sahara: Correlazioni geologico-litostratigrafiche fra Sahara centrale ed occidentale: *Mem. Mus. Tridentino Sci. Natur.*, Trento, Anno 29-30, v. 16, n. 1, p. 53–293.

Ratschiller, L. K., 1970-71, Lithostratigraphy of the northern Spanish Sahara, *Mem. Museo Tridentino Sci. Natur.*, Trento, Anno 33-34, v. 18, n. 1, p. 1–80.

Reichelt, R., 1972, Géologie du Gourma (Afrique occidentale) - Un "seuil" et un bassin du Précambrien supérieur - Stratigraphie, tectonique, métamorphisme: Thesis Univ. Clermont-Ferrand, France, March 1971, *Mém. Bur. Rech. Géol. Min.*, Paris, v. 53, p. 1–213.

Reid, P. C. and Tucker, M. E., 1972, Probable late Ordovician glacial marine sediments from northern Sierra Leone: *Nature* (Phys. Sci.), v. 238, no. 81, p. 38–40.

Remack-Petitot, M.-L., 1960, Contribution à l'étude du Gothlandien du Sahara. Bassins d'Adrar Reggane et de Fort Polignac: *Bull. Soc. Géol. France*, Sér. 7, v. 2, p. 230–239.

Renaud, L. and Delaire, L., 1955, Notice explicative sur la feuille Conakry-Est: *Carte Géol. Reconnaissance 500 000e A.O.F.*, Dakar, no. NC 28-SE.-E.12, 1 colored map, 19 p.

Renaud, L., Delaire, L., and Lajoinie, J.-P., 1959, Notice explicative sur la feuille Kindia-Est, *Carte Géol. Reconnaissance 500.000e A.O.F.*, Dakar, no. NC 28-NE.-E.24, 1 colored map, 26 p.

Reyment, R. A., 1971, Experimental studies of Cretaceous transgressions for Africa: *Geol. Soc. Am. Bull.*, v. 82, p. 1063–1072.

Reyre, D., 1966, Particularités géologiques des bassins côtiers de l'Ouest africain. Essai de récapitulation, in: *Bassins sédimentaires du littoral africain, 1e partie, Littoral atlantique*, D. Reyre, ed.: Assoc. Serv. Géol. Afr., Paris, p. 253–304.

Ribeiro, O., 1954, A Ilha do Fogo e as suas erupções: *J. Inv. Ultramar*, Lisboa.

Ridley, W. I., 1969, The abundance of rock types on Tenerife and its petrogenetic significance: *Eos*, v. 50, n. 4, p. 341 (abs.).

Ridley, W. I., 1970, The petrology of the Las Canadas volcanoes, Tenerife, Canary Islands: *Contrib. Miner. Petrol.*, v. 26, p. 124–160.

Rippert, J.-C., 1973, Le Tamkarkart: la chaîne des Mauritanides contre la bordure du bassin de Taoudeni (République Islamique de Mauritanie). Etude structurale d'un bord de craton: Third cycle thesis, Marseille St-Jérôme Univ., 194 p.

Robb, J. M., 1971, Structure of continental margin between Cape Rhir and Cape Sim, Morocco, Northwest Africa: *Am. Assoc. Petrol. Geol. Bull.*, v. 55, p. 643–650.

Robb, J. M., Schlee, J., and Behrendt, J. C., 1973, Bathymetry of the continental margin off Liberia, West Africa: *J. Res. U. S. Geol. Surv.*, v. 1, n. 5, p. 563–567.

Rocci, G., 1957, Formations métamorphiques et granitiques de la partie occidentale du Pays Reguibat (Mauritanie du Nord), thesis, Nancy, 1955: *Bull. Dir. Fédér. Min. Géol. Afr. Occ. Fr.*, v. 21, 484 p.

Roch, E., 1931, Cartes géologiques provisoires des Abda et des Djebilet occidentales de la zone synclinale de Mogador et de l'Atlas occidental: Serv. Min. Carte Géol., Protectorat Républ. Fr. au Maroc, 29 p.

Roch, E., 1950, Histoire stratigraphique du Maroc: *Notes Mém. Serv. Géol. Maroc*, No. 80, 437 p.

Rod, E., 1962, Fault pattern, northwest corner of Sahara shield: *Bull. Am. Assoc. Petrol. Geol.*, v. 46, p. 529–534.

Rodgers, J., 1970, *The tectonics of the Appalachians*, Regional Geology series, L. U. de Sitter, ed.: Interscience–Wiley, New York, 271 p.

Rodgers, J., 1971, The Taconic orogeny: *Geol. Soc. Am. Bull.*, v. 82, p. 1141–1177.

Roeser, H. A., Hinz, K., and Plaumann, S., 1971, Continental margin structure in the Canaries, in: *The geology of the East Atlantic continental margin*, F. M. Delaney, ed.: G. Brit. Inst. Geol. Sci. Rep., 70/16, 27–36.

Rona, P. A., 1968, Northwest African continental margin between Canary and Cape Verde islands: Geol. Soc. Am. Abs. 1968 Ann. Mtg., p. 253–254.

Rona, P. A., 1969, Possible salt domes in the deep Atlantic off northwest Africa: *Nature*, v. 224, n. 5215, p. 141–143.

Rona, P. A., 1970, Comparison of continental margins of eastern North America at Cape Hatteras and northwestern Africa at Cap Blanc: *Am. Assoc. Petrol. Geol. Bull.*, v. 54, p. 129–157.

Rona, P. A., 1971, Bathymetry off central northwest Africa: *Deep-Sea Res.*, v. 18, p. 321–327.

Rona, P. A., Brakl, J., and Heirtzler, J. R., 1970, Magnetic anomalies in the northeast Atlantic between the Canary and Cape Verde Islands: *J. Geophys. Res.*, v. 75, n. 35, p. 7412–7420.

Rona, P. A., and Nalwalk, A. J., 1970, Post-Early Pliocene unconformity on Fuerteventura, Canary Islands: *Geol. Soc. Am. Bull.*, v. 81, p. 2117–2121.

Roques, M., 1949, Géologie de l'Afrique Occidentale Française, Encyclopédie coloniale et maritime: Paris, v. 5, p. 197–204.

Rothe, P., 1964, Zur geologischen Geschichte der Insel Gran Canaria: *Natur. Mus.*, Frankfurt a. M., v. 94, n. 1, p. 1–9.

Rothe, P., 1966, Zum Alter des Vulkanismus auf den östlichen Kanaren: *Commentationes Phys. Math. Soc. Sci. Fenn.*, v. 31, n. 13, p. 1–80.

Rothe, P., 1968a, Mesozoische Flysch-Ablagerungen auf der Kanareninsel Fuerteventura: *Geol. Rundschau*, v. 58, n. 1, p. 314–332.

Rothe, P., 1968b, Die Ostkanaren gehört zum afrikanischen Kontinent: *Umschau Wissenschaft Technik*, Frankfurt a. M., 1968, n. 4, p. 116–117.

Rothe, P. and Schmincke, H.-U., 1968, Contrasting origins of the eastern and western islands of the Canarian Archipelago: *Nature*, v. 218, n. 5147, p. 1152–1154.

Rothpletz, A. and Simonelli, V., 1890, Die marinen Ablagerungen auf Gran Canaria: *Z. Deutsch. Geol. Ges.*, Berlin, v. 42, p. 677–736.

Saito, T., Burckle, L. H., and Ewing, M., 1966, Lithology and paleontology of the reflective layer horizon A: *Science*, v. 154, n. 3753, p. 1173–1176.

Salvan, H. M., 1968, L'évolution du problème des évaporites et ses conséquences sur l'interprétation des gisements marocains: *Min. Géol.*, Rabat, v. 27, p. 5–30.

Sauer, F. E. G. and Rothe, P., 1972, Ratite eggshells from Lanzarote, Canary Islands: *Science*, v. 176, n. 4030, p. 43–45.

Schaer, J.-P., 1964, Aspects de la tectonique dans le bloc occidental du massif ancien du Haut-Atlas: *C. R. Acad. Sci.*, Paris, v. 258, n. 9, p. 2353–2356.

Schenk, P. E., 1971, Southeastern Atlantic Canada, Northwestern Africa, and continental drift: *Can. J. Earth Sci.*, v. 8, n. 10, p. 1218–1251.

Schenk, P. E., 1972, Possible Late Ordovician glaciation of Nova Scotia: *Can. J. Earth Sci.*, v. 9, n. 1, p. 95–107.

Schlee, J. C., 1972, Shallow structure of the Liberian continental margin (abs.): *Geol. Soc. Am.*, Abstracts with programs, v. 4, p. 655–656.

Schlee, J., Behrendt, J. C., Robb, J. M., O'Hara, C. J., Jennings, F. W., Martin, R. G., Hill, H. R., Nicholson, J. R., Todd, W. C., Pearl, J. E., Dunbar, J., and Garcia-Rodriguez, J., 1972, U.S.G.S. - I.D.O.E. 5: *Geotimes*, v. 17, n. 8, p. 16–17.

Schmincke, H.-U., 1967, Cone sheet swarm, resurgence of Tejeda caldera, and the early geologic history of Gran Canaria: *Bull. Volcanol.*, v. 31, p. 153–162.

Schmincke, H.-U., 1968, Faulting versus erosion and the reconstruction of the mid-Miocene shield volcano of Gran Canaria: *Geol. Mitt.*, 1967, v. 8, n. 1, p. 23–50.

Schmincke, H.-U., 1971a, Comments on paper by E. Bosshard and D. J. Macfarlane, Crustal structure of the western Canary Islands from seismic refraction and gravity data: *J. Geophys. Res.*, v. 76, n. 29, p. 7304–7305.

Schmincke, H.-U., 1971b, Tektonische Elemente auf Gran Canaria: *N. Jb. Geol. Paläont. Mh.*, Nov. 1971, n. 11, p. 697–700.

Schmincke, H.-U., 1973, Magmatic evolution and tectonic regime in the Canary, Madeira, and Azores Island groups: *Geol. Soc. Am. Bull.*, v. 84, p. 633–648.

Schmincke, H.-U., and Swanson, D. A., 1966, Eine alte Caldera auf Gran Canaria: *N. Jb. Geol. Paläont. Mh.*, 1966, p. 260–269.

Schmincke, H.-U., and Swanson, D. A., 1967, Laminar viscous flowage structures in welded ash flow tuffs, Gran Canaria, Canary Islands: *J. Geol.*, v. 25, p. 641–663.

Schneider, E. D., 1969a, The deep sea—A habitat for petroleum?: *Under Sea Technology*, v. 10, n. 10, p. 32–34 and 54, 56, 57.

Schneider, E. D., 1969b, The evolution of the continental margins and possible long term economic resources: Offshore Technology Conference, Dallas, paper no. OTC 1027, 8 p.

Shepard, F. P., 1963, *Submarine geology*, 2nd ed.: Harper and Row Publ., New York, 557 p.

Sheridan, R. E., Drake, C. L., Nafe, J. E., and Hennion, J., 1966, Seismic-refraction study of continental margin east of Florida: *Bull. Am. Assoc. Petrol. Geol.*, v. 50, p. 1972–1991.

Sheridan, R. E., Houtz, R. E., Drake, C. L., and Ewing, M., 1969, Structure of continental margin off Sierre Leone, West Africa: *J. Geophys. Res.*, v. 74, p. 2512–2530.

Sinclair, J.-H., 1920, Discovery of Silurian fossils in French Guinea: *J. Geol.*, v. 36, p. 475–478·

Smith, A. G. and Hallam, A., 1970, The fit of the southern continents: *Nature*, v. 225· p. 139–144.

Société Chérifienne des Pétroles, 1966, Le bassin du Sud-Ouest marocain, in: *Bassins sédimentaires du littoral africain, Symposium*, 1e partie: Littoral Atlantique, D. Reyre, éd.: New Delhi, 1964, Assoc. Serv. Géol. Afr., Paris, p. 5–12.

Sougy, J., 1955, Nouvelles observations sur le "Cambro-Ordovicien" du Zemmour (Sahara occidental): *Bull. Soc. Géol. France*, Sér. 6, v. 6, p. 99–113.

Sougy, J., 1956, Le Zemmour, point de jonction des types marocain et saharien du "Cambro-Ordovicien": *Rep. 20th Sess. Intern. Geol. Cong.*, Mexico, Assoc. Serv. Geol. Afr., p. 275–280.

Sougy, J., 1959, Les formations crétacées du Zemmour noir (Mauritanie septentrionale): *Bull. Soc. Géol. France*, Sér. 7, v. 1, p. 166–182.

Sougy, J., 1960, Les séries précambriennes de la Mauritanie nord-orientale (A.O.F.): *Rep. 21st Sess. Intern. Geol. Congr.*, Norden, Copenhagen, v. 9, n. 9, p. 59–68.

Sougy, J., 1962a, West African fold belt: *Geol. Soc. Am. Bull.*, v. 73, p. 871–876.

Sougy, J., 1962b, Contribution à l'étude géologique des Guelbs Bou Leriah (région d'Aoucert, Sahara espagnol): *Bull. Soc. Géol. France*, 7e Sér., v. 4, p. 436–445.

Sougy, J., 1964, Les formations paléozoïques du Zemmour noir (Mauritanie septentrionale). Etude stratigraphique, pétrographique et paléontologique, thesis Nancy 1961: *Ann. Fac. Sci. Univ. Dakar*, v. 15, n. XII, 695 p.

Sougy, J., 1969, Grandes lignes structurales de la chaîne des Mauritanides et de son avant-pays (socle précambrien et sa couverture infracambrienne et paléozoïque), Afrique de l'Ouest: *Bull. Soc. Géol. France*, Sér. 7, v. 11, p. 133–149.

Sougy, J., 1970, Rapport annuel d'activité 1969-1970 du Laboratoire associé au Centre National de la Recherche Scientifique n° 132 "Etudes géologiques W-africaines": *Trav. Lab. Sci. Terre St-Jérôme*, Marseille, Sér. E, v. 5, 106 p.

Sougy, J., 1971, Remarques sur la stratigraphie du Protérozoïque supérieur du bassin voltaïen; influence de la paléosurface d'érosion glaciaire de la base du groupe de l'Oti sur le tracé sinueux des Volta et de certains affluents: *C. R. Acad. Sci.*, Paris, Sér. D, v. 272, p. 800–803.

Sougy, J., 1972 (1970), Etat des connaissances géologiques sur la partie mauritanienne de la Dorsale réguibat précambrienne, Colloque intern. sur les corrélations du Précambrien, Agadir-Rabat, 3–23 mai 1970: *Notes Mém. Serv. Géol. Maroc*, v. 236, p. 95–103.

Sougy, J., 1973, Evolution structurale de l'Ouest africain: *1ère Réun. Ann. Sci. Terre*, Paris, 19–22 mars, p. 385.

Sougy, J., and Bronner, G., 1969, Nappes hercyniennes au Sahara espagnol méridional (tronçon nord des Mauritanides), 5e Colloque Géol. Afr., Clermont-Ferrand, 9-11 avril: *Ann. Fac. Sci. Univ. Clermont-Ferrand*, 41, *Géol. Minéral.*, v. 19, p. 75–76.

Sougy, J. and Lécorché, J.-P., 1963, Sur la nature glaciaire de la base de la série de Garat el Hammoueïd (Zemmour, Mauritanie septentrionale): *C. R. Acad. Sci.*, Paris, v. 256, p. 4471–4474.

Sougy, J. and Pouit, G., 1968, Etats des connaissances sur la géologie et les minéralisations de la République Islamique de Mauritanie: Assoc. Serv. Géol. Afr., *23th Sess. Intern. Geol. Congr.*, Prague, 19 p.

Sousa Torres, A. and Pires Soares, J. M., 1946, Formacoes sedimentares do Arquipelago do Cabo Verde: *Mem. Minist. Colon.*, Ser. Geol., v. 3, p. 1–397.

Spengler, A. de, Castelain, J., Cauvin, J., and Leroy, M., 1966, Le bassin secondaire-tertiaire du Sénégal, in: *Bassins sédimentaires du Littoral africain, 1e partie: Littoral atlantique, Symposium*, D. Reyre, ed.: Assoc. Serv., Géol. Afr., Paris, p. 80–94.

Spengler, A. de and Delteil, J. R., 1966, Le bassin secondaire-tertiaire de Côte d'Ivoire (Afrique occidentale), in: *Bassins sédimentaires du Littoral africain, 1e partie: Littoral atlantique, Symposium,* D. Reyre, ed.: Assoc. Serv. Géol. Afr., Paris, p. 99–113.

Stahlecker, R., 1935, Neokom auf der Kapverden - Insel Maio: *Neues Jahrb. Miner. Geol. Paleont.,* B, v. 73, p. 302–311.

Stearns, C. E., and Thurber, D. L., 1965, Th230-U234 dates of late Pleistocene marine fossils from the Mediterranean and Moroccan littorals: *Quaternaria,* v. 7, p. 29–42.

Strangway, D. W. and Vogt, P. R., 1970, Aeromagnetic tests for continental drift in Africa and South America: *Earth Planet. Sci. Let.,* v. 7, p. 429–435.

Summerhayes, C. P., 1970, Phosphate deposits on the northwest Africa continental shelf and slope: Thesis Univ. London, 282 p. (unpublished).

Summerhayes, C. P., 1972, Geochemistry of continental margin sediments from northwest Africa: *Chem. Geol.,* v. 10, p. 137–156.

Summerhayes, C. P., Nutter, A. H., and Tooms, J. S., 1971, Geological structure and development of the continental margin of northwest Africa: *Mar. Geol.,* v. 11, p. 1–25.

Summerhayes, C. P., Nutter, A. H., and Tooms, J. S., 1972, The distribution and origin of phosphate in sediments off northwest Africa: *Sediment. Geol.,* v. 8, p. 3–28.

Suter, G., 1965, La région du Moyen Ouerrha (Rif, Maroc): étude préliminaire sur la stratigraphie et la tectonique: *Notes Mém. Serv. Géol. Maroc,* 183, *Notes,* t. 24, p. 7–17.

Tagini, B., 1971, Esquisse structurale de la Côte d'Ivoire. Essai de géotectonique régionale: SODEMI, Abidjan, 302 p.

Teixeira, C., 1948, Notas sobre a geologia das Ilhas Atlântidas: *An. Fac. Ciên. Porto,* Portugal, v. 33, p. 193–233.

Templeton, R. S. M., 1971, The geology of the continental margin between Dakar and Cape Palmas, in: *The Geology of the East Atlantic Continental Margin,* F. M. Delaney, ed.: G. Brit. Inst. Geol. Sci. Rep., 70/16, p. 43–60.

Tenaille, M. M. A., Nicod and de Spengler A., 1960, Petroleum exploration in Senegal-Mauritania and Ivory Coast coastal basins (West Africa): *45th Ann. Mtg. Am. Assoc. Petrol. Geol.,* Atlantic City, New Jersey.

Tessier, F., 1952, Contributions à la stratigraphie et à la paléontologie de la partie ouest du Sénégal (Crétacé et Tertiaire): *Bull. Dir. Min. Afr. Occ. Fr.,* Dakar, no. 14, p. 7–573.

Tessier, F., Dars, R., and Sougy, J., 1961, Mise en évidence de charriages dans la "série d'Akjoujt" (République Islamique de Mauritanie): *C. R. Acad. Sci.,* Paris, v. 252, p. 1186–1188.

Tessier, F., Hébrard, L., and Lappartient, J.-R., 1971, Découverte de fragments d'oeufs de *Psammornis* et de *Struthio* dans le Quaternaire de la presqu'île du Cap Blanc (République Islamique de Mauritanie): *C. R. Acad. Sci.,* Paris, Sér. D, v. 273, p. 2418–2421.

Tinkler, K. J., 1966, Volcanic chronology of Lanzarote (Canary Islands): *Nature,* v. 209, n. 5028, p. 1122–1123.

Tinkler, K., 1968, Volcanic chronology of Lanzarote, Canary Islands: *Proc. Geol. Soc. London,* 24 Jan., n. 1649, p. 118–119.

Tooms, J. S., and Summerhayes, C. P., 1968, Phosphatic rocks from the north west African continental shelf: *Nature,* v. 218, p. 1241–1242.

Tooms, J. S., Summerhayes, C. P., and McMaster, R. L., 1971, Marine geological studies on the northwest African margin: Rabat-Dakar, in: *The geology of the East Atlantic continental margin,* F. M. Delaney, ed.: G. Brit. Inst. Geol. Sci. Rep., 70/16, p. 9–25.

Torres, A. S., and Soares, J. M. P., 1946, Formações sedimentares do Arquipélago de Cabo Verde, I-Actualização de conhecimentos: *Mem. Ser. Geol. III, Ed. Junta Missões Geogr. Invest. Coloniais,* p. 204, 209, 210.

Tougarinov, A. I., Knorre, K. G., Shanin, L. L., and Prokofieva, L. N., 1968, The geochronology of some Precambrian rocks of southern West Africa: *Can. J. Earth Sci.*, v. 5, p. 639–642.

Trénous, J.-Y., and Michel, P., 1971, Etude de la structure du dôme de Guier (Sénégal nord-occidental): *Bull. Soc. Géol. France*, Sér. 7, v. 13, p. 133–139.

Trompette, R., 1969*a*, Etude de la série 1, Précambrien supérieur ou Infracambrien, en Adrar de Mauritanie (Sahara occidental): Serv. Min. Géol. Mauritanie, Nouakchott, 248 p., (unpublished report).

Trompette, R., 1969*b*, Les Stromatolites du "Précambrien supérieur" de l'Adrar de Mauritanie (Sahara occidental): *Sedimentology*, v. 13, p. 123–154.

Trompette, R., 1970, Etude de la série 2 "Eocambrien et Cambro-Ordovicien" en Adrar de Mauritanie (Sahara occidental): Serv. Min. Géol. Mauritanie, Nouakchott, 260 p., (unpublished report).

Trompette, R., 1972, Présence, dans le bassin voltaïen, de deux glaciations distinctes à la limite Précambrien supérieur - Cambrien. Incidences sur l'interprétation chronostratigraphique des séries de bordure du craton ouest-africain: *C. R. Acad. Sci.*, Paris, v. 275, p. 1027–1030.

Trompette, R., 1973*a*, Stratigraphie et sédimentologie du Précambrien supérieur et du Paléozoïque inférieur de l'Adrar de Mauritanie (Afrique de l'Ouest). Contribution à l'évolution structurale de l'Ouest africain: *1ère Réun. Ann. Sci. Terre*, Paris, 19-22 mars, p. 404.

Trompette, R., 1973*b*, Le Précambrien supérieur et le Paléozoïque inférieur de l'Adrar de Mauritanie (bordure occidentale du bassin de Taoudeni, Afrique de l'Ouest): un exemple de sédimentation de craton. Etude stratigraphique et sédimentologique: Marseille Univ. thesis, France, 702 p.

Tuttle, O. F. and Gittins, J., 1966, Introduction, in: *Carbonatites*, O. F. Tuttle and J. Gittins, eds.): Interscience–Wiley, New York, p. XI–XIX.

Uchupi, E., 1971, Bathymetric atlas of the Atlantic, Caribbean and Gulf of Mexico: Woods Hole Oceanog. Inst., Ref. 71–72, (unpublished manuscript).

Uchupi, E. and Emery, K. O., 1967, Structure of continental margin off Atlantic coast of United States: *Am. Assoc. Petrol. Geol. Bull.*, v. 51, p. 223–234.

U.S. Naval Oceanographic Office, 1965, Oceanographic atlas of the North Atlantic, Section V, Marine Geology, Pub. No. 700.

U.S. Naval Oceanographic Office, 1969, Global ocean floor analysis and research data series: U.S. Govt. Print. Off., Washington, 971 p.

Vachette, M., 1964, Essai de synthèse des déterminations d'âges radiométriques de formations cristallines de l'Ouest africain (Côte d'Ivoire, Mauritanie, Niger): *Ann. Fac. Sci. Univ. Clermont-Ferrand*, France, 25, Géol. Minéral., v. 8, p. 7–29.

Vachette, M., Rocci, G., Sougy, J., Caron, J.-P. H., Marchand, J., Simon, B., and Tempier, C., 1973, Ages radiométriques Rb/Sr, de 2000 à 1700 MA, de séries métamorphiques et granites intrusifs précambriens dans la partie N et NE de la dorsale réguibat (Mauritanie septentrionale): *C. R. 7ème Colloque Géol. Afr.*, Florence, 25–27 avr., 3 p., (in press).

Verwoerd, W. J., 1966, Fenitization of basic igneous rocks, in: *Carbonatites*, O. F. Tuttle and J. Gittins, eds.: Interscience–Wiley, New York, p. 295–308.

Villemur, J. R., 1967, Reconnaissance géologique et structurale du Nord du bassin de Taoudenni: *Mém. Bur. Rech. Géol. Min.*, Paris, v. 51, 151 p.

Villemur, J. R. and Drot, J., 1957, Contribution à la fauna dévonienne du bassin de Taoudeni: *Bull. Soc. Géol. France*, Sér. 6, v. 7, p. 1077–1082.

Viotti, C., 1965, Microfaunes et microfaciès du sondage Puerto Cansado 1 (Maroc méridional - Province de Tarfaya): Colloque Intern. Micropaléont., Dakar, 6–11 mai 1963, *Mém. Bur. Rech. Géol. Min.*, Paris, v. 32, p. 29–60.

Visse, L., 1948, Contribution à l'étude pétrographique des phosphates marocains: *Bull. Soc. Géol. France*, Sér. 5, v. 18, p. 675–684.

Viterbo, I., 1965, Examen micropaléontologique du Crétacé du Maroc méridional (bassin côtier de Tarfaya): Colloque Intern. Micropaléont., Dakar, 6–11 mai 1963, *Mém. Bur. Rech. Géol. Min.*, Paris, v. 32, p. 61–100.

White, R. W., 1969, Sedimentary rocks of the coast of Liberia: U.S. Geol. Surv. Open-File rep., Liberian Geol. Surv. Memo Rep. 39.

White, R. W. and Leo, G. W., 1969, Geological reconnaissance in western Liberia: Liberian Geol. Surv. Spec. Pap. 1.

Whittington, H. B., 1966, Presidential address - Phylogeny and distribution of Ordovician Trilobites: *J. Paleont.*, v. 40, n. 3, p. 696–737.

Wilson, J. T., 1965, Evidence from ocean islands suggesting movement in the earth, in: *A symposium on continental drift*, P. M. S. Blackett, E. Bullard and S. K. Runcorn, eds.: Royal Soc. London, Philos. Trans., No. 1088, p. 145–167.

Wilson, T. J., 1966, Did the Atlantic close and then re-open?: *Nature*, v. 211, n. 5050, p. 676–681.

Windisch, C. C., Leyden R. J., Worzel, J. L., Saito, T., and Ewing, J., 1968, Investigation of horizon beta: *Science*, v. 162, n. 3861, p. 1473–1479.

Worzel, J. L., 1968, Advances in marine geophysical research of continental margins: *Can. J. Earth Sci.*, v. 5, p. 963–983.

Wyllie, P. J., 1966, Experimental studies of carbonatite problems: the origin and differentiation of carbonatite magmas, in: *Carbonatites*, O. F. Tuttle and J. Gittins, eds.: Interscience–Wiley, New York, p. 311–352.

Zimmermann, M., 1960, Nouvelle subdivision des séries antégothlandiennes de l'Afrique occidentale (Mauritanie, Soudan, Sénégal): *Rep. 21st Sess. Intern. Geol. Congr., Norden*, Copenhagen, v. 8, n. 8, p. 26–36.

Chapter 11

CENOZOIC TO RECENT VOLCANISM IN AND AROUND THE NORTH ATLANTIC BASIN

Arne Noe-Nygaard*

Mineralogisk Museum
Copenhagen, Denmark

I. INTRODUCTION

Our knowledge of the sea bottom of the North Polar Sea, the Greenland–Norwegian Sea, and the North Atlantic was for a long period based on bathymetric maps by Nansen (1904) and bottom-sample investigations by Bøggild (1900, 1907).

The almost explosive development within the last decade of research activity in the oceans, including the North Atlantic, has brought us a steadily increasing wealth of new geophysical and geological data, of new concepts, and of new names for peaks and depressions and linear structures, whether ridges or furrows (Johnson and Eckhoff, 1966; Johnson and Heezen, 1967; Johnson, Closuit, and Pew, 1969; Vogt, Ostenso, and Johnson, 1970; Johnson, Vogt, and Schneider, 1971).

Against this background it seems justified to try to summarize what we know today of the Cenozoic volcanic areas along the borders of the ocean and on the volcanic islands which lie in it.

* With contributions by Kent Brooks, Sveinn Jakobsson, and Asger Ken Pedersen.

There is a marked contrast between the volcanic islands of the South Atlantic (Darwin, 1844; Daly, 1926, 1942; Baker, 1973) and corresponding volcanic regions in and around the North Atlantic. The volcanic islands in the South Atlantic constitute the tops of large submarine cones, and thus mainly show us the latest results of volcanic activity in any given place and the latest products of magmatic differentiation. Several of the regions in the North Atlantic which have been affected by early Cenozoic to Recent volcanism lie along the margins of continental masses, such as northwest Scotland and West and East Greenland, or on part of an aseismic ridge, the Faeroe Islands.

Due to upheaval after their formation the basal parts of lava piles and the sedimentary sequences underneath them, which have been widely intruded by dolerite sills (Skye, East Greenland), have been laid bare, as have a large number of subvolcanic plutonic bodies (northern Ireland, northwestern Scotland, East Greenland). In the North Atlantic we therefore have good access to the volcanic products and to part of the plutonic realm underneath them.

The hypothesis of sea-floor spreading and plate tectonics has given us an excellent model on which geologists and geophysicists may concentrate their work in a common effort. This model is not commented upon here but is in principle accepted.

The formation of the North Atlantic Ocean may have begun with a fissuring and breaking up of a Laurasian landmass (Wegener, 1922; Brooks, 1973). We do not know exactly when this breakup began, but mid-Mesozoic is a likely estimate.

Tholeiitic basalts of Upper Jurassic and Lower Cretaceous age are found in Spitzbergen (Hamberg, 1899; Tyrrell and Sandford, 1933); rocks of the same type and age have been reported from northeast Canada (Blackadar, 1964) and southwest Greenland (Watt, 1969). Perhaps also some master dikes in central West Greenland are of Jurassic age (p. 396).

The first-formed continental slope deposits of the European side of the North Atlantic appear to be Cretaceous in age (Stride *et al.*, 1969).

The intense magmatic activity, however, which can be traced in a belt from Baffin Land over Greenland and the Faeroe Islands to northern Britain, did not reach its culmination until lower Tertiary (Fig. 1).

Of the same age is also the volcanism in the Norwegian deep as shown by the Eocene ash layers in northwest Denmark (Bøggild, 1918; Norin, 1934; Noe-Nygaard, 1965).

Volcanism related to the Fennoscandian border zone in Scania began in the Cretaceous and went on in lower Tertiary (Larsen and Printzlau, 1973).

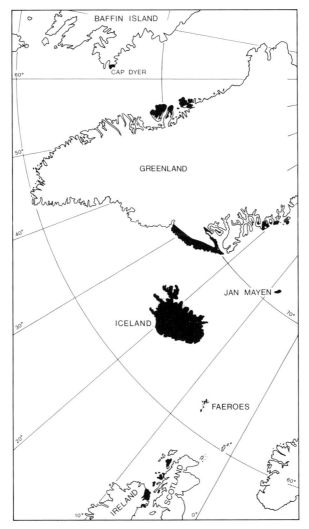

Fig. 1. The Thulean basalt province.

II. REGIONAL DESCRIPTION OF VOLCANISM IN AND AROUND THE NORTH ATLANTIC BASIN

The present account first deals with the lower Tertiary volcanism caused by the rifting in the Davis Strait, that is, Baffin Island and West Greenland, then with East Greenland, the Faeroe Islands, and the Hebridean province which are related to a rifting east of Greenland. In early Tertiary these landmasses were much closer together than today.

Thereafter follows the description of Iceland, where the oldest volcanic rocks formed in Miocene, and of Jan Mayen, which is only Pleistocene in age. These two islands are both related to the present mid-Atlantic ridge, Iceland being the only larger landmass in the world which lies across an active oceanic ridge; Jan Mayen is influenced by a cross-cutting fracture zone which cuts the submarine plateau north of Iceland from west-northwest to east-southeast between lat. 71° and 72° N.

A. The Davis Strait

An extensive pile of sediments and lavas has recently been disclosed beneath the waters of Baffin Bay, the Davis Strait, and the Labrador Sea, of which the onshore Cretaceous–Tertiary sediments and lavas of central West Greenland only form the eastern marginal part (Henderson, 1972; Park et al., 1972). A corresponding, though very narrow, western marginal part lies on the east coast of Baffin Island (Clarke, 1970).

Canadian geologists have shown the presence of a wide submarine graben parallel to the Greenland coast in Melville Bugt; it contains about 6000 m of sediments. In the offshore area from central West Greenland an approximation of the extension on the sea bottom of the basalts has been achieved by bottom sampling and geophysical work.

It is held by Henderson (1972) that the sediments of the West Greenland basin were laid down in a continuation of the Melville Bugt graben that terminates southward as an embayment bounded by faults along the south end of Disko Bugt, and that a second graben, or at least a coast-parallel fault, developed west of Disko. The continuation of the Melville Bugt graben and the graben or fault on the west side of Disko became linked by a subsidiary graben structure between Ubekendt Ejland and Disko (Fig. 2).

It is suggested that an almost continuously subsiding graben would account for the very large thickness of basalt in the western part of the onshore area (p. 398) and that the graben structure itself was the source of the basalts (Henderson, 1972).

B. Baffin Island

On the east coast of Baffin Island, around Cape Dyer, limnic sedimentary rocks with Paleocene plant remains crop out along the coast, whereas marine sedimentary strata have not been recorded from the area.

The limnic sedimentary rocks are overlain by subaqueous basalt breccia and picritic lavas which form scattered outcrops also farther inland; they rest here directly on Precambrian gneisses. No trace of plutonic activity has been recorded (Wilson and Clarke, 1965; Clarke, 1970; Andrews et al., 1972).

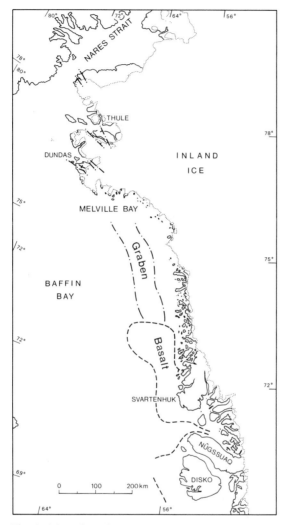

Fig. 2. Map of northern West Greenland and Baffin Bay showing the graben in Melville Bay and the probable offshore extension of basalts (after Henderson, 1972).

C. West Greenland

1. Pre-Cenozoic Geology of the Disko–Svartenhuk Region

West Greenland is dominated by Precambrian complexes, and only between Disko and Svartenhuk do younger strata occur. On Nûgssuaq peninsula mountains consisting of Cretaceous–Tertiary sedimentary rocks covered by basalts approach 2000 m in height.

The history of research of this region goes back to Giesecke 1806–1813 (1910), followed by Steenstrup (1883), Heim (1910), Ravn (1918), and Krueger (1928). It is, however, mainly the 15 expeditions conducted by Prof. A. Rosenkrantz since 1938 that have established our present detailed knowledge of the area (Rosenkrantz et al., 1940; Rosenkrantz et al., 1942; Rosenkrantz, 1952, 1964; Rosenkrantz and Pulvertaft, 1969). Recent reviews of the volcanic province are given by Clarke and Pedersen (1973) and Henderson (1972) to which the reader is referred; the present treatment is therefore only a short one.

The pre-Cretaceous relief was considerable and in many places the Precambrian basement was deeply weathered before the Cretaceous sedimentary strata were deposited. These deposits are psammitic limnic in the south and east, and become increasingly pelitic and marine toward the northwest; in central and southwest Nûgssuaq transitional facies occur (Rosenkrantz and Pulvertaft, 1969).

On Disko and southeast Nûgssuaq most of the Cretaceous–Tertiary sedimentary sequence is fluviatile; this is also valid for the sedimentary strata along the northeastern boundary of the area on Upernavik Ø and Qeqertarssuaq. Marine strata dominate on Svartenhuk and in the central and northern part of Nûgssuaq, and taken together they comprise an almost complete succession from upper Turonian to Maestrichtian, 400–500 m in thickness. Following upon an unconformity come ca. 600 m of Danian strata (lowermost Tertiary), in which the lower and upper Danian are separated by another unconformity. The main lithologic rock type in the marine Cretaceous and Tertiary strata is dark, bituminous shale.

The sea which first transgressed outer Svartenhuk in late Turonian time, reached Nûgssuaq during the Conacian; it carried a North American ammonite fauna (Birkelund, 1965). Later, the sea must have linked with that covering western Europe as indicated by the European affinities of the molluscs, a feature particularly marked in the early Tertiary (Rosenkrantz, 1970).

The principal movements occurred in three phases: at the end of the Maestrichtian; in mid-Danian time; and after the extrusion of the basalts. Sinking and faulting occurred continuously during the whole volcanic episode.

The oldest magmatic rocks in the province were probably those which occupy a few coast-parallel, basaltic master dikes (Fig. 3) of supposed Jurassic age which can be traced in the Precambrian east of the area now occupied by sediments and volcanics (Rosenkrantz and Pulvertaft, 1969, Henderson, 1973).

The beginning of volcanism in the areas which are made up of Cretaceous–Tertiary sedimentary strata has been dated by means of marine fossils in

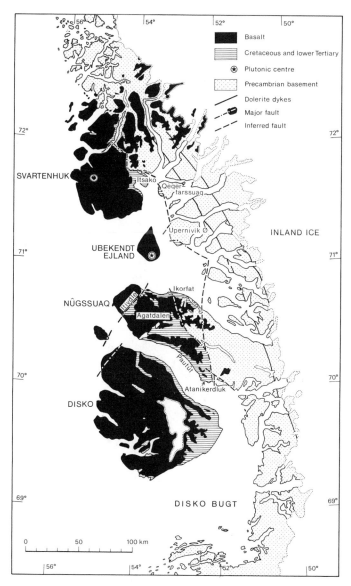

Fig. 3. Geological map of the basalt province in West Greenland
(slightly modified after Rosenkrantz and Pulvertaft, 1969).

Nûgssuaq (Rosenkrantz *et al.*, 1942; Munck and Noe-Nygaard, 1957; Rosen-krantz and Pulvertaft, 1969). The first tuff layers occur in lower Danian, where the uppermost part of the bituminous shale sequence is interstratified with fossiliferous tuffs, equivalent to about 1 km of early lavas and breccias on Disko southwest of the marine basin.

2. *The Tertiary Volcanic Sequence*

The volcanic strata are divided into two formations. A lower lava formation which dominantly consists of olivine-rich, subaquatic and subaeric volcanics may attain a thickness of 5 km or may be missing altogether. The upper lava formation mainly consists of plagioclase porphyritic tholeiitic flows but may also comprise an upper group of olivine porphyritic, alkalic basalts which rest discordantly on the tholeiitic lavas. As the top is often eroded away the thickness of the upper lava formation is difficult to estimate, but a thickness of 3 km has been measured locally.

The oldest exposed volcanics are normally subaquatic breccias made up of pillow fragments in a matrix of shattered basaltic glass and zeolites. The pillows are of olivine and picrite basalt composition. The breccia may exceed 500 m in thickness but is generally thinner. Their subaqueous, shallow-water origin was first recognized by Munck (in Rosenkrantz *et al.*, 1942). Large-scale cross-bedding is found in the breccias, in which single beds represent single eruptions. Nonbasaltic material is uncommon.

Once depressions were filled by pyroclasts, they were covered again by subaerial predominantly olivine and picrite lavas. Flows are typically a few meters thick and may show considerable internal olivine fractionation. Pyroclastics only formed when magma was brought into contact with water. Some thin red bole layers may occur, but on the whole volcanic activity seems to have been both intense and continuous.

In the northern part of the province these lower lavas form a monotonous sequence of olivine and picrite basalts, with locally plagioclase and olivine porphyritic lavas in the oldest breccias. In the southern part there are some aphyric or only slightly olivine porphyritic flows, some sediment-contaminated bronzite bearing lavas and tuffs, and some mildly alkaline olivine basalts. Occasionally uneven subsidence of the lava plateau formed local basins which were subsequently filled with pillow breccias. The tuffs spread into early Tertiary sedimentary basins to the east and northeast, permitting the establishment of a tephrochronology.

The upper lava sequence is formed of voluminous plagioclase porphyritic basalts. In some areas there is a sharp compositional break, in others there is a transitional sequence of olivine-rich basalts, plagioclase porphyritic and nearly aphyric basalts from the picritic to the plagioclase porphyritic lavas. This upper sequence is more extensive and rests in places directly on early Tertiary sediments. In western Nûgssuaq an erosional unconformity and a series of limnic sediments (Henderson, 1969; Hald, 1973) divide the upper lavas into two series. Resting discordantly upon the plagioclase porphyritic basalts on the island of Hareøen, northwest of Disko (Hald, 1971) are a series of mildly alkaline olivine basalts suggesting that a top series with this composition may have formerly been much more extensive.

The West Greenland basalt province is characterized by the low proportion of nonbasaltic volcanics. The greatest diversity is found on Ubekjendt Ejland (Steenstrup, 1883; Drever and Game, 1948; Drever, 1958). The plagioclase porphyritic basalts of the upper lava formation here include pitchstones, acid tuffs, biotite trachyte, and highly undersaturated lamphrophyric rocks, and several volcanic necks are found on the western side of the island. Other extreme differentiates from the upper lava formation include peralkaline anorthoclase trachytes on Svartenhuk (Nieland, 1931; Noe-Nygaard, 1942), and peralkaline acid tuffs on western Nûgssuaq (Henderson, 1969, Hald, 1973).

The telluric iron-bearing rocks form a special group of volcanics described in more than 80 papers since their first discovery by Nordenskjöld (1870). The native iron and associated sulfides on Disko occur in dikes and lavas in both the lower and upper lava formations.

Composite dikes and lavas have also been found which show that the basaltic magmas were contaminated with bituminous, sulfur-rich shales and with sandstones.

The most extreme contamination occurs in northwest Disko where plagioclase porphyritic basalt through a coupled crystal fractionation and a contamination process developed into telluric iron-bearing basalt, ignimbritic andesite, and sanidine–biotite rhyolite (Pedersen, 1970).

Central volcanic vents are scarce, although some are found in Ubekjendt Ejland where necks filled with carbonatized breccias occur (Drever and Game, 1948; Drever, 1958). Some craters also occur on Disko. The eruption forms are discussed by Rosenkrantz and Pulvertaft (1969) and by Henderson (1973). All over the province eruption was overwhelmingly from fissure volcanism (Noe-Nygaard, 1942). Dikes which grade into lava flows have been observed in several parts of the province.

A coast-parallel dike swarm comparable in magnitude to that of southeast Greenland (p. 408) does not occur, but several smaller swarms of dikes have been mapped. In such swarms the dike intensity increases near fault zones, which suggests that a major upwelling of magma took place in these areas (Henderson, 1973; Hald, 1973) (Fig. 4).

Some swarms of parallel dikes contain both picritic basaltic and plagioclase porphyritic dikes, which demonstrates that some magma upwelling took place in the same zones of weakness during the whole volcanic epoch.

Following a major fault zone which runs along the sediment basin in the east bordering on the Precambrian basement, a number of prominent dolerite dikes and sills occur, some of which were described in detail by Munck (1945).

A group of alkaline picritic intrusions occurs on the north coast of Nûgssuag near the settlement Qaersut (Steenstrup, 1883; Heim, 1910; Drescher and Krueger, 1928). The best described body is a 50-m-thick sill which has intruded into nonmarine sediments.

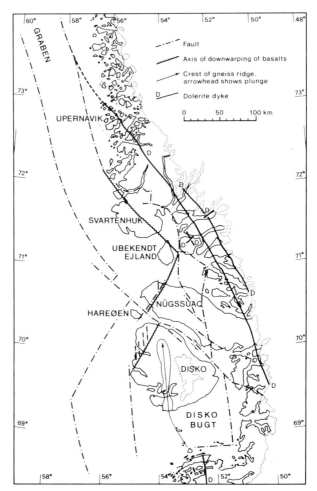

Fig. 4. Tectonic map of the West Greenland basalt province.
Note the two master dikes in the basement toward the east.
For further comments see text. (After Henderson, 1973.)

The only central intrusion, Sarqata qaqa on Ubekjendt Ejland (Steenstrup, 1883; Drever and Game, 1948; Thompson and Patrick, 1968), covers an area of about 15 sq. km. It consists of a lower unit of gabbro with igneous layering and an upper unit consisting of granophyre; a cone sheet swarm is connected with the complex.

In northwest Disko basaltic dikes may contain numerous rounded xenoliths of gabbro, some of which are igneous cumulates pointing at the existence of another Tertiary layered intrusion beneath the present erosion level of this area, and traces of further intrusive bodies may still be found onshore and offshore.

Early petrogenetic schemes based on the scarce chemical data then existing were given by von Wolff (1932) and Noe-Nygaard (1942). A complete picture of the chemistry and petrogenesis of the province is still not yet available, but much work is in progress.

Compared with the other regions of Tertiary volcanics in the North Atlantic, the occurrence of olivine and picrite basalts on a regional scale is most remarkable. These basalts are not merely olivine cumulative rocks, for studies by Drever (1952, 1956) of the intrusive picrites indicated that the rocks crystallized from highly magnesian melts. As experimental evidence increased Drever and Johnston (1966) could point out that the magnesian melts were likely to have been formed at high pressures in the mantle through partial melting of ultrabasic source rocks.

New chemical data were presented by Clarke (1970), who argued that the olivine basalts represent melts which equilibrated at about 30 kbar in equilibrium with garnet peridotite and that olivine basalts from Svartenhuk represent eclogite fractionated magmas evolved from a primary magma represented by the Baffin Island olivine basalts. Further petrogenetic reasoning is presented by Clarke and Pedersen (in press). The later evolution of the magmas in West Greenland was complex. The presence of magmas which equilibrated at high, intermediate, and low pressures is indicated by the rock compositions, and further complexities have arisen through reactions between the pre-volcanic pile of sediments and the basic magmas.

D. East Greenland

In East Greenland the Cenozoic volcanic province lies along a continental border, directly facing the ocean.

The gross structure and geological history of the East Greenland fjord belt from lat. 70 to 75° N differs markedly from that of the coastal region farther south (Fig. 5). The two regions are also separated by the huge fjord complex of Scoresby Sund and will therefore be treated separately in this account.

1. The Fjord Zone North of Scoresby Sund

The area is divided into two main units separated by a fault established in Early Devonian times which can be traced across through almost 5° of latitude extending northward from the inner parts of the Scoresby Sund fjord complex. West of this line, which is located approximately along the 24° W meridian, are rocks of the Caledonian complex, while to the east upper Paleozoic and Mesozoic sedimentary rocks predominate. The Caledonian complex, to which was added strips of Devonian and lowermost Carboniferous rocks,

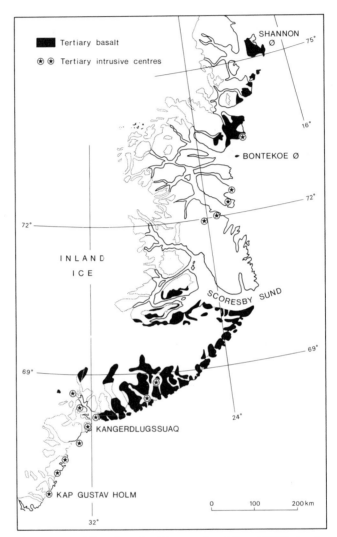

Fig. 5. Map of East Greenland between lat. 66° and 76°N showing
the three plateau lava fields and the plutonic centers of the coastal
region.

formed a steadily rising cratonic block contrasting with the subsiding belt
toward the coast in which great thicknesses of younger sediments accumulated.

    To begin with, the eastern belt may have formed only a few plates between
the main fault and the continental border. Later it was split up in antithetic
fault steps along auxiliary faults, some of which branched off from the main
fault, while others arose parallel to it. The net result of the faulting was a
crustal widening from west to east.

The single fault steps form escarpments towards the east and possess gently westward-inclined upper surfaces. Sedimentary strata are found on the western parts of the inclined steps, while on the higher eastern parts they may be missing altogether, either because they were never deposited there, or due to later erosion. The formation of fault steps has taken place periodically ever since the Devonian and has even continued in postbasaltic time.

Due to a repeated flooding of the low-lying western parts of the inclined steps, these were successively filled up with sediments, which become younger the farther east we go.

After the deposition of the youngest marine strata of Cretaceous age uplift occurred and the eruption of plateau basalts marked the beginning of the volcanic activity in the coastal belt.

These basalts often rest on a deeply eroded land surface, although occasionally thin, limnic or continental sandstones of Paleocene to Eocene age may be present. Only in the extreme east, on Sabine Ø, have marine sediments been reported (Vischer, 1943). The thin-flowing lavas spread across a roughly leveled land surface filling first the valleys and lowlands and then flooding the whole region. No pillow lavas are recorded, but Vischer (1943) believes that some of the lavas near the present coastline were erupted subaqueously.

The remnants of a dissected lava plateau, which is thickest in the east, i.e., at the present coast, and becomes thinner westwards, lie between lat. 73° and 75° N. Its total coverage is estimated at about 16,000 sq. km (Wenk, 1961) and its thickness at 700 m (Bütler, 1957). Plateau lavas are known to occur on Shannon Ø, Wollaston Forland, Sabine Ø, Hvalros Ø, Lille Pendulum Ø, Bass Rock, Clavering Ø, Jackson Ø, Hold-with-Hope, Hudson Land, Gauss Halvø, and Bontekoe Ø (Hochstetter, 1874; Lenz, 1874; Nathorst, 1901; Nordenskjöld, 1908; Wordie, 1926; Koch, 1929; Orvin, 1931; Backlund and Malmquist, 1932; Tyrrell, 1932; Nielsen, 1935; Vischer, 1943; Stern, 1964) (Fig. 6). South of Fosters Bay plateau basalts disappear, but at the same time the number of sills increases markedly (p. 404).

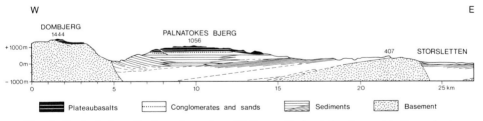

Fig. 6. Geological profile through northern Wollaston Forland. To the west is the cratonic region with a thin capping of basalt lavas; east hereof follows a downfaulted step covered by young Paleozoic and Mesozoic sediments and in Palnatokes Bjerg covered also by Paleocene–Eocene conglomerates and sands of limnic origin and plateau basalts. Toward the east the underlying basement is uncovered facing the next fault step at Storsletten.

Several authors give good descriptions of individual samples from widely separated parts of the region, and their composition is apparently tholeiitic; but little is known about the petrographic succession of lavas. Many lavas are described as aphyric, but some carry plagioclase phenocrysts, and occasionally lavas with phenocrystic olivine have been recorded.

Thin interbasaltic sedimentary layers are reported from several localities; they generally contain a considerable amount of nonbasaltic material, derived from local areas not yet flooded by the lavas or from the uplifted block to the west. The proportion of volcanic ash appears to be small.

Another region with plateau basalts, covering at least 2000 sq. km, exists at lat. 74° N in the innermost zone of nunataks in the elevated Caledonian belt, to the west of the main fault, and is quite isolated from the coastal occurrences. The base of this lava series rises from 2000 m in the east to 2300 m in the west (Katz, 1952). Haller (1956) is of the convinction that these basalts are of the same age as those of the coastal region.

Katz found feeder dikes as well as intercalated tuff layers in the lava sequence indicating that volcanic activity took place within the area itself. The lavas of the nunatak region are alkali basaltic to nephelinitic (p. 405).

Doleritic sills occur all along the coast. The sills are mainly intruded into the sedimentary strata below the plateau basalts. Their age must be the same as that of the basalt flows on top, for they cut off the sills in some places and are intruded by them in others.

One of the most fascinating sceneries in East Greenland is the region with the large sills. On the geological map it shows a striking similarity with random sections on the map of the dolerites of the Karoo System in South Africa.

From the eastern tip of Ymer Ø and southward the sills increase markedly in number and volume and spread out westward to form a broad belt, roughly parallel to the coast (Fig. 7). They dominate more than half of Geographical Society Ø (Stauber, 1938) and two-thirds of Traill Ø where they form the most typical sill country in East Greenland. There are also many sills in the northern part of Jameson Land; they even occur farther south in this peninsula, but here they are thinner. The maximum thickness of sills on Traill Ø is 250 m (Swinhufvud Bjerge), but this is reduced to 70 m near Sorte Fjelde and to 30 m at Mols Bjerge (Frebold and Noe-Nygaard, 1938; Putallaz, 1961).

There are no conspicuous dike swarms comparable with that south of Scoresby Sund, although locally small groups of dikes with a similar trend are found, e.g., Wollaston Forland, and a radial dike set occurs around the Werner pluton.

Knowledge of the petrology of these rocks is still very uneven; in some areas the best available results are observations made a century ago, while in others detailed petrological work has been published in the last decade. In

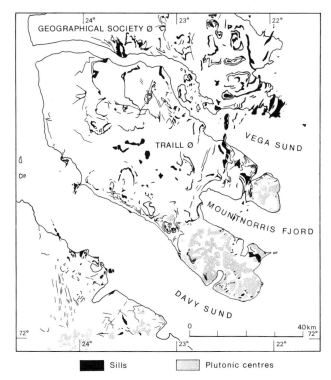

Fig. 7. The sill country of Geographical Society Ø and Traill Ø,
East Greenland. The plutonic center north of Mountnorris Fjord
is that of Cap Parry. The Forchhammer pluton lies to the south of
the same fjord.

many cases it is not possible to decide whether a given rock described was
extrusive or intrusive.

Tholeiitic basalt and dolerite (Backlund and Malmquist, 1932), basalts
s.s., and augite porphyrite basalts (Bearth, 1959; Kapp, 1960) are widespread.
Plagioclase porphyritic basalts (Krokström, 1944; Kapp, 1960) are most
common south of Kong Oscars Fjord, while the basalts with tabular felspars
(the Fame Ø porphyrite) find their main distribution in northern Jameson Land.
Olivine tholeiites and olivine basalts, ankaramite (Wordie, 1926; Tyrrell,
1932); alkaline olivine basalts (Tyrrell, 1932), and a related olivine trachy-
basalt (Backlund and Malmquist, 1932), alkali basalts (nephelinites, etc.)
(Katz, 1952; Haller, 1956; Stern, 1964) are found to a more limited extent.

A number of plutonic centers are found north of Scoresby Sund, showing
a general alignment from northeast to southwest between Kap Broyer Ruys
and Werner Bjerge in Scoresby Sund. The most important of these units are
the central volcanic complex of Kap Broer Ruys with quartzporphyries and

dikes (Lenz, 1874; Nathorst, 1901), the Kap Parry alkaline complex with syenites and alkalic dikes (Jameson, in Scoresby, 1823; Bäckström, in Nathorst, 1901; Tyrrell, in Wordie, 1932), the Forchhammer pluton (Callisen, in Koch, 1929; Backlund, 1937; Stauber, 1938, 1942; Reinhard, in Stauber, 1942), the alkaline massif of Werner Bjerge in Scoresby Sund with gabbro, syenite, and nepheline syenite (Nordenskjöld, 1908; Bearth, 1959), and a group of four small subvolcanoes between Antarctic Havn and Mesters Vig (Koch, 1929; Noe-Nygaard, 1937, 1940; Kapp, 1960) with rocks ranging from gabbro to monzonite, syenite, and alkali granite.

The course of geological events affecting the area north of Scoresby Sund can be summarized as follows. After an uplift had brought to an end Mesozoic sedimentation in all but the extreme easterly areas of the northern coastal region, flooding by tholeiitic basalt lava flows occurred from the oceanic border zone. The sills of Traill Island, however, are thickest farthest from the present coast, and it is therefore probable that rise of magma also occurred along the zones of weakness represented by the pre-existing fault pattern. Similarly, the alkalic flows of the nunatak region were probably also released along old, deep-going faults in the cratonic block, and a number of the alkalic intrusions may also have arisen in this way. The volcanic activity of this region, which can be shown to have undergone crustal extension, can be explained as being due to the sagging of a coastal strip along the edge of the westward-moving landmass. A postbasaltic flexure which occurs on the east coast of Wollaston Forland and Sabine Ø is consistent with this mechanism.

The plutonic centers are younger than the extrusives and follow a line approximately parallel to that of the plutons along the coast south of Scoresby Sund but displaced inland apparently west of the Jameson Land–Liverpool Land block. The basic parts of these plutons in all cases predate the more salic types.

The disappearance of the plateau basalts south of Fosters Bugt, where they are replaced by the numerous and vast doleritic sills invading the Mesozoic strata, and the gradually increasing depth to which the plutonic complexes have been eroded from Kap Broer Ruys to Werner Bjerge, suggests a differential uplift of the land, that is greater toward the south.

2. The Region South of Scoresby Sund

Between Scoresby Sund (lat. 70° N) and Kap Gustav Holm (lat. 67° N) the basement rocks which underlie the basalts consist mainly of Precambrian gneisses with ages ranging from 1600 to 2360 m.y. (Wager and Hamilton, 1964).

Northeast of Kangerdlugssuaq the metamorphic complex is overlain by a thin sedimentary series often intruded by basic sills. This so-called Kangerdlugssuaq sedimentary series rests on a fairly even surface of gneisses; it is shallow water and estuarine in origin and consists of conglomerates, sandstones, and sandy shales with seams of lignite. At one locality, Kulhøje, conglomerates containing basaltic pebbles show that igneous activity had already been in progress nearby (Wager, 1934; Deer, in press). The age of the Kangerdlugssuaq sedimentary series is considered to be late upper Senonian to early Eocene in age.

The igneous activity began with the extrusion of vast quantities of basaltic liquid, giving the plateau basalt series which covers an area of approximately 60,000 sq. km (Wenk, 1961).

Most of the lava flows in Knud Rasmussen Land and the Blosseville coast were extruded subaerially. A sinking of the lava pile evidently took place concomitantly with its accumulation, as marine sediments of lower to middle Eocene age, the Kap Dalton sediments, formed soon after the extrusion of the highest basalts (Wager, 1934). Just south of Scoresby Sund, pillow lavas which grade upwards into subaerial lavas and hydroclastic breccias have been recorded and indicate that in this area shallow-water conditions existed at least during one stage of volcanic activity.

The lower basalts between Kangerdlugssuaq and J. C. Jacobsen Fjord are associated with evenly and well-bedded tuffs and agglomerates, and were extruded under water (Deer, in press); the estimated total thickness of the lava–tuff series exposed in J. C. Jacobsens Fjord is 6.5 km.

Recent K/Ar dating of a number of basalts from the Kap Brewster area on the southside of Scoresby Sund corroborate the paleontological dating of the volcanic activity; the values lie between 55 and 66 m.y. (Beckingsale *et al.*, 1970).

Petrographically, the plateau basalts are poorly documented. They include porphyritic as well as nonporphyritic varieties, and most of them are vesicular. Chemically, the basalts include tholeiites and olivine tholeiites; more alkaline basalts appear to be absent (Fawcett *et al.*, 1966; Brooks, 1973).

The upper basalts of the Prisen af Wales Bjerge are different, they do not form parallel flows and presumably were more viscous than the regular plateau lavas. These lavas consist of a porphyritic lava group which includes olivine, pyroxene, and plagioclase porphyritic lavas of intermediate composition. The lavas are more alkalic in character than most of the lavas of the main plateau series since they all contain analcite in the groundmass (Anwar, 1955).

Between Kap Hammer and Nansen Fjord the pile of lavas dips toward the Denmark Strait at angles between 45° and 50°; the dips of the lavas flatten inland and are generally not greater than 10° at the heads of the fjords, while in the Gunnbjørn Fjeld areas the lavas are practically horizontal. These seaward-increasing lava dips are interpreted as part of a large-scale crustal flexure (Wager and Deer, 1939).

Associated with the flexure is the more striking dike swarm, which reaches a total length of approximately 520 km. East of Kangerlugssuaq the densest part of the dike swarm lies on the headlands where the dip of the lavas is greatest, and there is a clear correlation between the intensity of the dike swarm and the direction and magnitude of the coastal flexure. It appears that dike formation occurred when tension in the upward part of the flexure reached a critical value, dependent on the hydrostatic pressure of the dike magma and the tensile strength of the rocks.

Except for a small area of the Skaergaard peninsula, which may not be representative (Vincent, 1953), little work on the petrography and chemistry of the dike swarm has been made available so far.

In contrast to the region north of Scoresby Sund where our knowledge of the plutonic centers was accumulated gradually the discovery of the Tertiary plutonic centers along the margin of the coastal mountains and on some of the offshore islands between Kap Gustav Holm and Kangerdlugssuaq was due to one man—the late Prof. L. R. Wager—in 1930 (Deer, 1967). He first showed the existence of the central complexes at Kap Gustav Holm, Kialineq, Nûgalik, Igdlitarajik, and Kangerdlugssuaq (Wager, 1934), and on a later expedition he further discovered the center in the Lilloise Mountains north of Kangerdlugssuaq. Some smaller occurrences of plutonic igneous rocks have been found by subsequent parties (Brooks, 1973).

The igneous rocks of the plutonic complexes closely resemble those described from the region north of Scoresby Sund (p. 405–406).

Apart from a preliminary report of their content of igneous rock types, the plutonic complexes listed below are still almost unknown:

Kap Gustav Holm	gabbro and syenite
Nûgalik	gabbro and microgranite
Igdlitarajik	gabbro
Nordre Aputiteq	gabbro
Lilloise, Borgtinderne	syenites, nepheline syenites, and mafic rocks

The area of Kangerdlugssuaq is the best known, mainly because of the presence there of the now classical Skaergaard gabbro intrusion (Wager and Deer, 1939; Wager and Brown, 1968). The Kap Edward Holm layered gabbro intrusion has lately been investigated by Elsdon (1970) and Abbot and Deer (1972), while the Kangerdlugssuaq quartz syenite–foyaite has been described recently (Kempe et al., 1970; Kempe and Deer, 1971), and the Kialineq center has been briefly treated by Brown (1968).

Two plutonic phases can be recognized, the first characterized by gabbroic rocks, the second mainly represented by syenites. Thus the Skaergaard layered intrusion (Deer, in press), the Kap Edward Holm layered gabbros (Deer and Abbot, 1972; Elsdon, 1969), the Kaerven basic intrusion (Ojha, 1966), and the

Basistoppen Sheet (Wager and Deer, 1939; Hughes, 1956; Douglas, 1964) belong to the first phase. To this phase also belong three gabbroic macrodikes which are closely associated with the Skaergaard intrusion. The Kap Deichman Syenite, the Kap Boswell Syenite (Beckinsale *et al.*, 1970), the Kangerdlugssuaq alkaline intrusion (Wager, 1965; Kempe *et al.*, 1970; Kempe and Deer, 1970; Deer, in press; Beckinsale *et al.*, 1970), and the Kialineq composite center (Brown, 1968) belong to the second plutonic phase.

The sequence of igneous events affecting the area south of Scoresby Sund can be briefly summarized as follows. Igneous activity between Scoresby Sund and Kangerdlugssuaq began with the extrusion of voluminous plateau basalts. This was succeeded by the intrusion of a number of basic plutonic complexes, which in terms of the volume of magma involved was a comparatively minor phase.

These events were succeeded by a major tectonic phase, during which the coastal flexure and its associated dike swarm were formed. This was apparently accompanied by major subcrustal transport of material and is considered the second major event.

The second and final plutonic phase, which like the first did not involve very large mass movement on a local scale, involved the intrusion of the more differentiated, syenitic, foyaitic, and granitic magmas.

The succession of magmatic events therefore appears to be the same as in the northern area, where, however, the absence of a flexure and dike swarm on such a scale makes relative dating less precise.

E. The Faeroe Islands

The Faeroe Islands have a surface area of 1400 sq. km and consist exclusively of basalt lavas which were poured out subaerially in the earliest Tertiary, and which form a pile 3 km thick (Rasmussen and Noe-Nygaard, 1969, 1970) (Fig. 8). The first geological observation from the Faeroe Islands goes back to Debes (1673). More comprehensive descriptions of geological phenomena were given by Born (1792, 1793, 1797), Landt (1800), Mackenzie (1814), Allan (1813), Forchhammer (1824), Helland (1880), Geikie (1880), and Walker and Davidson (1936). Rasmussen (1946) compiled a bibliography of the existing geological literature, and a supplementary list is given by Rasmussen and Noe-Nygaard (1969) along with the text to a new geological map in six sheets, scale 1:50,000.

The strata of the Faeroe Islands can be divided into five members (1–5 below (Fig. 9).

1. The lower lava series, about 900 m thick, consists of flows with marked columnar jointing and an average thickness of about 20 m. The lavas were

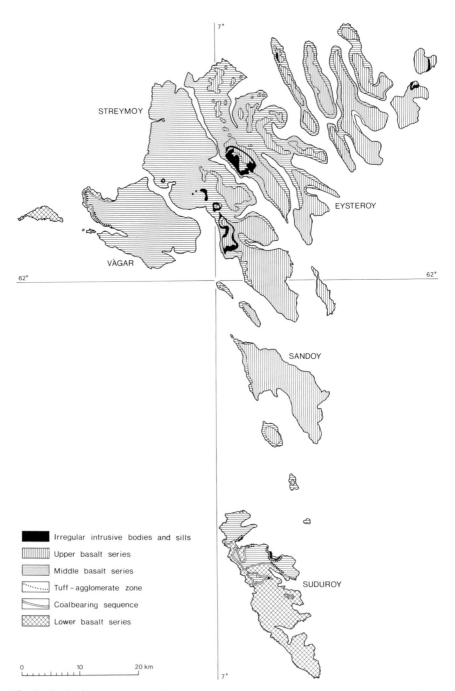

Fig. 8. Geological map of the Faeroe Islands (after Rasmussen and Noe-Nygaard, 1969).

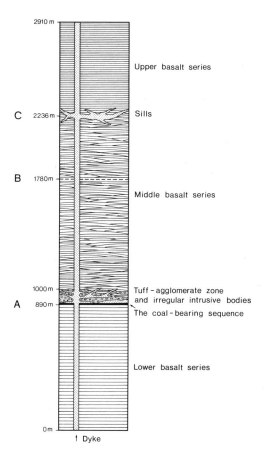

Fig. 9. Schematic section through the lava sequence of the Faeroe Islands (after Rasmussen and Noe-Nygaard, 1969).

most probably produced by fissures, and there was a noncontinuous pattern of volcanic activity. During periods of quiescence thin layers of interbasaltic tuff–clay sediments were repeatedly laid down. The lavas are very monotonous, aphyric or cryptophyric tholeiites. They have been described by Walker and Davidson (1936) and by Noe-Nygaard and Rasmussen (1968). Erosion of the lower basalt series has produced typical "trap country" with big steps and long, gently inclined slopes.

2. The coal-bearing series. The deep weathering of the lavas underlying the sedimentary strata indicates that deposition followed an erosional interval of some length. The average thickness of the coal-bearing series is 10 m; it consists of hardened clays with local intercalations of sand, thin conglomerates, and one workable coal seam. Pollen and poorly preserved plant remains

have been found, indicative of an Eocene age (Rasmussen and Koch, 1963; Laufeld, 1965).

3. The tuff–agglomerate zone. The volcanic activity following the deposition of the coal-bearing series began with a highly explosive initial phase of which scoriae and tephra (bombs, lapilli, and ash) were the main products. The deposits of this phase have been named the tuff–agglomerate zone; they contain fragments of the underlying sedimentary sequence and are often intruded by irregular bodies of basalt, which belong to the final period of minor intrusions (Rasmussen, 1952).

4. The middle lava series, about 1350 m thick, is made up of very thin, ropy lava flow units, rich in zeolites and generally with a red skin on their upper surface. The thin flow units may be welded together to form more compact masses; columnar jointing is practically absent.

The rock types which build up the middle series show great variations, but porphyritic lavas are dominant; they are all tholeiitic. Interspersed with these are sets of thin olivine porphyritic lavas, generally rich in zeolites and rare picritic flows.

The volcanic activity which led to the formation of the lavas of the middle basalt series seems to have been localized in vents of limited extent. Ten such vents have been found (Rasmussen, 1962; Rasmussen and Noe-Nygaard, 1969). Stratigraphically, they are known to occur from a little above the middle of the series to its very top. As the vents all lie at the present fjords, they must be related to a system of fractures which determined the position of these.

In the middle lava series erosion has produced vertical, "laminated" cliffs on the coasts and smooth, convex slopes inland (Fig. 10).

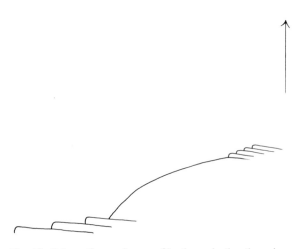

Fig. 10. Schematic erosion profile through the three lava series of the Faeroe Islands.

5. The upper lava series, about 625 m thick, consists of clearly separable lava flows similar to those of the lower lava series, but, on average, only half as thick. There is a tendency for columnar jointing to occur in the thickest flows.

The main rock type is a hard, bluish, basaltic lava with small phenocrysts of olivine and plagioclase. It often shows fluxion banding and has a platy fracture and smooth fracture surfaces. The lava is an olivine tholeiite (Noe-Nygaard and Rasmussen, 1968). Lenticular zeolite-filled amygdules are common. The groundmass texture can be intergranular, but subophitic and ophitic textures dominate. Some intercalated plagioclase porphyritic flows of the middle series type occur in the lower part.

The lavas from the upper part of the middle series and the lower part of the upper series show two different forms of eruption, either as flows from craters aligned along a fissure (Hald, Noe-Nygaard, and Waagstein, 1969) or from flat "scutulum" type shield volcanoes (Noe-Nygaard, 1968).

The extrusive phase of activity was followed by a period of tension during which dikes and sills were intruded. In some cases the dikes have been feeders of the sills. Within the coalbearing sequences intrusions are of irregular form (Rasmussen, 1952).

Petrographically the dominant igneous rock type is the tholeiitic basalt (Noe-Nygaard, 1966b). From wet analysis and x-ray fluorescence studies (Rasmussen and Noe-Nygaard, 1969) it appears that the earlier lavas were erupted at lower temperatures than the later flows and that there is a decrease in the silica content with increasing height in the lava pile (Noe-Nygaard, 1967).

The volcanic activity which led to the formation of the Faeroe Islands began in the west, presumably near the crest of the present Wyville–Thompson ridge, with the eruption of aphyric, quartz tholeiites. Subsequently the zone of activity migrated eastward to a location almost central in the present group of islands, to finally move further east away from the ridge producing then undifferentiated olivine tholeiite lavas which are transgressive from the east.

The three basalt series which form the Faeroe plateau are separated by weak unconformities. A slight easterly tilting occurred followed by dike injection after the formation of the lower lava series. The jointing pattern developed at this time may have controlled the trend of the fjords along which the explosive vents of the middle lava series are located (p. 412). Further tilting occurred prior to the extrusion of the upper lava series. A complex pattern of jointing was developed with repeated adjustments along master joints forming lamellar zones (Peacock, 1928; Noe-Nygaard, 1940; Rasmussen, 1955).

The ridge to the west continued to rise until the upper lava series required its low easterly to southeasterly dip. In the northern group of islands a clear but low-amplitude anticline with an easterly plunge is found (Noe-Nygaard, 1962, 1966*a*) which is clearly apparent in the gravity map (Saxov and Abrahamsen, 1964, 1966). The Eocene age of the Faeroes from paleontological evidence is corroborated by radiometric dates (Tarling and Gale, 1968) so that the date assigned earlier (Rasmussen and Noe-Nygaard, 1966) must be abandoned.

F. The British Isles

Concentrated along the western seaboard of Scotland and in northeast Ireland occur the plutonic and volcanic rocks of the so-called Hebridean Province. These rocks have received attention from an early period and have played an important role in the development of petrological thought. In this connection may be mentioned particularly the way in which the publication of the Ardnamurchan memoir of the Geological Survey (Ritchey *et al.*, 1930) stimulated thought on the interrelationships of basaltic magmas (see, e.g., Turner and Verhoogen, 1960, p. 225). Similarly, the close association of acid and basic magmas found throughout the province has formed a background for many studies into the origin of granitic rocks (see, e.g., Thompson, 1969, and succeeding discussions).

Recent general descriptions of the province, particularly field relations, are given for Ireland by Charlesworth (1963) and for Scotland by Stewart (1965).

The province has long been regarded as having connection with the igneous rocks of similar age in the Faeroes, East and West Greenland, Baffin Island, and possibly even the more recent ones of Iceland. The recent development of the concept of plate tectonics has given impetus to this notion by postulating that all these rocks have arisen in the process of breaking up and drifting apart of the earlier supercontinent of Laurasia to form the present North Atlantic Ocean (Brooks, 1973).

Lavas resting in a number of places, both in Scotland and Ireland, on Upper Cretaceous chalk which has been uplifted prior to volcanism, establish a maximum age for the onset of magmatic activity. Various age estimates have also been made using the flora of interbedded sediments (Hallam, 1965, p. 409; Charlesworth, 1963, p. 393) but with widely varying results. The age of most, if not all, the magmatic rocks is now regarded as Paleocene/Eocene on the basis of radiometric age determinations (Beckinsale *et al.*, 1970).

In Scotland, Tertiary magmatic rocks are concentrated around a number of centers, the most important of which are in Arran, Ardnamurchan, Mull, Rhum, Skye, and St. Kilda (Fig. 11). A further center is believed to lie far out in the Atlantic, of which the islet of Rockall is the only visible part. The most important occurrences of lavas are in Skye and Mull.

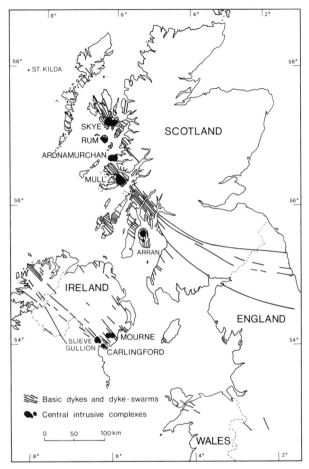

Fig. 11. Map showing distribution of Tertiary northwest dikes in relation to Tertiary plutonic districts of the British Isles (reprinted from *The Geology of Ardnamurchan*).

In Ireland, an extensive basalt plateau covers large areas of the northeast (Antrim plateau), while plutonic centers are located at Carlingford, Slieve Gullion, and the Mourne Mountains. Outlying representatives are found in a number of places (Emeleus and Preston, 1969) of which the dike swarms of the Barnesmore Granite, Donegal (Walker and Leedal, 1954), and the Porcupine bank, 200 miles west of Galway (Cole and Crook, 1910), are worthy of mention.

In England, the island of Lundy (Dollar, 1941), which is composed of granite, has been identified as Britain's most southerly Tertiary intrusive by radiometric age determinations (Miller and Fitch, 1962; Dodson and Long, 1962), while swarms of tholeiitic dikes radiating from the Scottish centers penetrate far into the north of England (Holmes and Harwood, 1929). In the south, Eocene volcanic ash has recently been identified, which is, however, believed to be derived from the Danish area (Elliot, 1971).

Four main episodes of igneous activity have been traditionally recognized:

1. Voluminous outpourings of plateau lavas
2. Establishment of central vents and agglomerate formation
3. Intrusion of plutonic and hypabyssal rocks at a number of centers
4. Intrusion of regional dike swarms

However Stewart (1965, p. 424) questions the validity of such a sequence.

Lavas are preserved in Skye, Mull, the Small Isles, Ardnamurchan, and Antrim, of which the latter is by far the biggest area (ca. 3800 sq. km). That their extent was once much greater is shown by the abrupt terminations of the lava plateau around its margins and the presence of inclusions of basalt in a number of intrusions presently lying outside the basalt areas (e.g., Slieve Gullion).

The lavas were erupted subaerially with significant time lapses, as indicated by the frequent occurrence of boles, red partings, and lateritic horizons. The paucity of pyroclastics indicates eruption was nonexplosive. Both fissure- and central-type eruptions are known.

The lavas of Antrim, which are about 900 m thick in places, are divided into three groups by two interbasaltic horizons. The lower basalts are alkali basalts (belonging to what is traditionally known as the plateau–magma series) usually carrying olivine phenocrysts and sometimes plagioclase phenocrysts in addition. The lower, or main, interbasaltic horizon is composed of lateritic and bauxitic weathering products of the basalts themselves and of rhyolites and acid pyroclastics which are found at this horizon in the Tardree area (Cameron and Sabine, 1969). Lignites also occur in some places. The middle basalts are tholeiites (nonporphyritic central magma type) which are characteristically columnar (as at the Giants Causeway) and are more restricted in their occurrence, thinning out and disappearing to the south. Above the upper interbasaltic horizon, which is very poorly exposed, are the upper basalts, alkali basalts like the lower basalts. They almost invariably carry phenocrystic olivine. Chemical analyses of the lavas have been published by Patterson (1952) and Patterson and Swaine (1955). A regional zonation of the zeolite minerals has been described by Walker (1951, 1959, 1960a, 1960b).

In Mull the basalts are made up of a ca. 900 m thick plateau group (alkali basalts), which include aphyric together with olivine- and plagioclase-phyric types and some mugearites, and the 900 m thick central group (tholeiitic basalts, largely aphyric but with some feldspar-phyric types). The central group lavas are preserved largely within a subsidence caldera, whereas the plateau group covers large areas of the island outside the caldera.

Most of the northern part of Skye is covered by lavas, mainly alkali basalt or its derivatives, mugearite and trachyte (Anderson and Dunham, 1966). They reach a thickness of at least 600 m.

Other more limited occurrences of lavas are to be found on the Small Isles, Rhum, Eigg, Muck, and Canna (Ridley, 1971) and are preserved within the central ring complex of Arran.

The principal centers of plutonic activity in Ireland are at Carlingford, Slieve Gullion, and the Mourne Mountains. In Scotland Arran, Rhum, Skye, and St. Kilda are important, to which must be added the isolated outcrop of Rockall.

The Carlingford complex is made up of a number of gabbros and acid intrusions, together with several generations of cone sheets (LeBas, 1960) and possibly ring dikes. Slieve Gullion is primarily a ring dike of porphyritic felsite and granophyre (Emeleus, 1962) together with a central complex of rocks whose origins have been much debated (Reynolds, 1950, 1951; Wager and Bailey, 1953; Bailey and McCallien, 1956). The Mourne Mountains, on the other hand, consist of five separate granite intrusions, with two in the Western Mournes associated with massive cone sheets. Relevant data have been provided by Brown (1956), Brown and Rushton (1960), and Emeleus (1955).

In Scotland the best known plutonic complexes are on Arran (Stewart, 1965) and Skye (Thompson, 1969). In Arran, Tyrrell (1928) recognized three groups of sills according to age in southern Arran. In the center of the island is a ring complex (Stewart, 1965), while a large granite massif occupies most of the northern part of the island. Plutonic rocks also occupy most of the central part of Skye, with the layered basic and ultrabasic complex of the Cuillin Hills and the granitic complexes of the Eastern and Western Red Hills.

The basic and ultrabasic complex consists of several intrusive phases emplaced in conjunction with ring faulting. The magmas, like those of Rhum and other plutonic centers, appear to have been of tholeiitic type, in contrast to the predominance of alkali olivine types among the lavas, and underwent crystallization by bottom accumulation of the separating phases to give the layered sequences (Wager and Brown, 1968). This complex is associated with a number of basic dikes, conesheets, and peridotite dikes, the latter being the subject of a number of investigations (Bowen, 1928; Drever and Johnston, 1958; Gibb, 1968). An interesting granophyre in this area is believed to have been derived by melting of arkose (Wager and Bailey, 1953).

The granites, which are later than the Cuillins complex and lie to the east, have arcuate outcrops and are believed to have ring-dike form. They are made up of a considerable number of units, some being slightly peralkaline in character, and are associated with a suite of hybrid rocks. Intensive work has been carried out on them and is still in progress, but, although the consensus favors an origin by anatexis of the basement, it does not yet appear to be possible to rule out a derivation by fractional crystallization of the abundant basaltic magmas. A more recent review of this problem has been given by Thompson (1969) in which further references can be found.

Additional features of interest in this area are the zoned skarns in the limestones adjacent to one of the granites (Tilley, 1951) and the fine composite sills of the Broadford area, which have basic margins and acid centers.

In the north of Skye and adjacent islands, a complex of thick teschenite sills with layered structures occur which are believed to be examples of flow differentiation (Simkin, 1967).

Rhum is perhaps best known for its large plutons of ultrabasic cumulates which have been emplaced by upfaulting along a ring fault. The cumulates have been subdivided into a number of rhythmic units (Brown, 1956; Wadsworth, 1961; Wager and Brown, 1968). Small granophyric and acid rock bodies occur along with lavas, pyroclastics, and a well-developed set of radial dikes.

St. Kilda is an outlying center which contains a complex array of ultrabasic, basic, intermediate, and acid rocks (Harding, 1966). Good examples of basic magma chilling against acid magma have been described. It is likely that the exposed rocks form part of a ring complex (Stewart, 1965).

Rockall was described by Sabine (1960). This very isolated rock, 190 miles west of St. Kilda, is of considerable interest in reconstructing the history of the North Atlantic Ocean (Vine, 1966). Along with the Porcupine bank and the Faeroes platform it now appears likely that it consists of continental material (Bott and Watts, 1971). The islet itself consists of aegirine granite with acmite, riebeckite, and complex zirconosilicates, while blocks of basalt, phonolitic trachyte, aegirine granite, and sandstone have been dredged from the surrounding bank.

Meighan and Preston (1971) have recently drawn attention to the arrangement of Tertiary activity along several north–south lineaments, which appear to swing more east–west than to the south. These include a westerly belt through Rockall and the Porcupine bank, one through St. Kilda, the Barnesmore Granite (Co. Donegal) into Wales and the English Midlands, and a further through the Scottish Tertiary centers. These lineaments are marked by intrusive centers, dike swarms, and magnetic anomalies.

No detailed discussion of the petrogenesis can be given here, the reader being referred to Stewart (1965). Problems still center on the classical ones, namely, the relationships between the two magma types (plateau or alkali olivine basalt type and the central or tholeiitic type) and the origin of the granitic rocks (basaltic differentiates or partial melts of basement). Both are problems central to petrology as a whole and not peculiar to this province, although work carried out here has been of considerable importance to the whole science of petrology.

Work is still very actively in progress, and our information on petro-chemistry (e.g., Ridley, 1971), isotopic compositions (e.g., Moorbath and Welke, 1969a, 1969b), and geophysics (e.g., McQuillan and Tuson, 1965) has in the last few years been considerably augmented. However, no final answer to the two central problems are yet at hand.

G. Iceland

Iceland is situated in the North Atlantic where two submarine structures meet, the mid-Atlantic ridge and the Brito-Arctic ridge. The area of Iceland is 103,106 sq. km. The northernmost point lies at lat. 66°33′ N; the southernmost mainland point lies at lat. 63°23′ N. The shortest distance from Iceland to Greenland is less than 300 km, and to Scotland about 800 km. The average height is approximately 500 m above sea level.

Volcanism seems to have been more vigorous and continuous in this part of the North Atlantic basin than elsewhere; it has been active in the production of lavas, tephras, and intrusives ever since Miocene. About 85–90% of the volume of the island consist of volcanic rocks, the rest mainly of sedimentary rocks transported over short distances. The main structure and past volcanism of Iceland is best explained if one assumes that the volcanic activity, since late Tertiary, has steadily been confined to the present active belt, and that sea-floor spreading is responsible for the present structure and the areal distribu-tion of the various volcanic zones (Bödvarsson and Walker, 1964). The rocks of Iceland may be divided into: (1) a Tertiary stage, the products of which were erupted from middle Miocene to Pliocene; (2) a Quaternary stage, which again is subdivided into the Gray Basalt Formation (Gauss and Matuyama epochs), the Móberg Formation (Brunhes epoch), and the Holocene (post-glacial) formation (Einarsson, 1971).

Olafsson and Pálsson (1772) gave good descriptions of various geological phenomena in Iceland and made a clear distinction between the "orderly basalt formation" (the Tertiary plateau basalts) and the "disorderly basalt formation" (Quaternary hyaloclastic rocks).

From the first half of the last century the works of von Nidda (1837) and von Waltershausen (1847) are still worth mentioning. It was von Walters-hausen who introduced the name "Palagonite Formation" as a collective term for the Quaternary, mainly, hyaloclastic rocks which reminded him of rock types which he knew from Sicily.

With the last two decades of the century a period was initiated in which Icelandic scientists themselves took the leading part in the geological investiga-tion of their country; the most prominent of these were Th. Thoroddsen and H. Pjeturss.

Tertiary rocks cover approximately 35–40% of the area of Iceland and

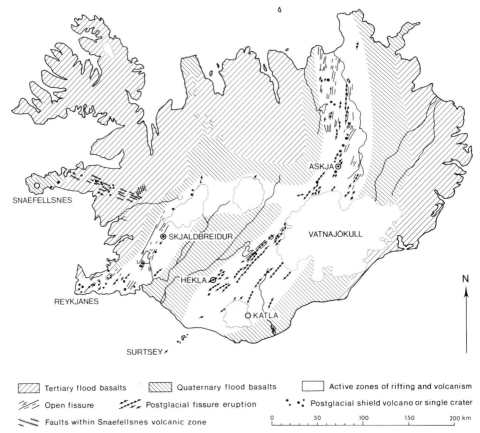

Fig. 12. Geological map of Iceland (modified after a sketch by Sveinn Jakobsson).

crop out roughly symmetrically on both sides of the Quaternary formations (Fig. 12). The bulk of the well-stratified Tertiary rocks consists of lavas which generally dip a few degrees inland, i.e., toward the present active zones. The oldest exposed rocks are thus found along the northwestern and eastern coasts. Opinions as to the age of the Tertiary rocks have varied with time. Heer (1868) dated the older part of the formation as lower Miocene based on studies of plant fossils, but later it was suggested and generally believed that the oldest rocks were of lower Tertiary age, based on paleontological datings of plant beds in Great Britain and East Greenland (Pjeturss, 1905, 1910; Hawkes, 1938). Recent K/Ar datings (Gale et al., 1966; Dagley et al., 1967; Moorbath et al., 1968) have shown that the age of the oldest exposed rocks is probably only 15–20 m.y. or middle Miocene. The upper limit of the Tertiary is determined by the oldest moraines of regional distribution.

The Tertiary areas first became known in outline through the intensive studies of Thoroddsen (1905). Since then work has been carried out from time to time on more restricted areas mainly in eastern Iceland (Hawkes, 1924; Cargill, Hawkes, and Ledeboer, 1928; Hawkes and Hawkes, 1933; Anderson, 1949; Dearnley, 1954; Jónsson, 1954; Tryggvason and White, 1955; Jux, 1960). In the 1950's, G. P. L. Walker began systematic mapping in the region around and south of Reydarfjördur. In the Reydarfjördur area, an east–west profile through 4500 m of lavas revealed (Walker, 1959) that tholeiites constitute 48% by volume, olivine-bearing basalts 23%, plagioclase-porphyritic basalts 12%, andesite lavas 3%, and rhyolitic to dacitic lavas 8%. Pyroclastics and sedimentary intercalations were estimated at 6%. The average thickness of 550 measured basalt lava flows was around 10 m. The basaltic lavas (flood basalts) are believed to have been formed mainly through fissure volcanism. Dikes, which are usually parallel to the strike of the lavas, can rarely be traced to a lava flow. Walker (1960) has, however, shown that in the Berufjördur–Breiddalur area the number of dikes diminishes upwards in the lava pile and reaches zero by extrapolation to an altitude deduced to be the original top of the lavas from studies of the distribution of zeolite zones.

Dikes occur occasionally in big swarms and are then related to central volcanoes (Fig. 13), which are situated in the middle of such swarms (Walker, 1963; Hald *et al.*, 1971). A total of perhaps 12 volcanic centers has been dis-

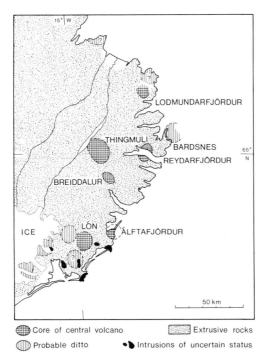

Fig. 13. Central volcanoes of East Iceland (simplified after Walker, 1963).

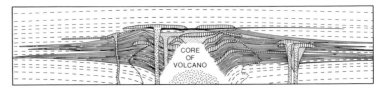

Fig. 14. Cross section of a cedar tree volcano (Breiddalur). The envelope of the volcano consists of flood basalts, the main mass of the volcano of basic and intermediate flows. Rhyolitic lavas are shown by vertical ruling, tuffs and agglomerates are dotted. Intrusions occur in the core of the volcano and in the feeders to parasitic craters. (After Walker, 1963.)

covered in eastern Iceland, notably the Breiddalur and Thingmúli volcanoes (Walker, 1963; Carmichael, 1964). Some of the centers are recognizable as central volcanoes which at times stood as flat cones (300–600 m high) above the surrounding basalt plains. Fissure eruptions and central volcanic activity ran simultaneously (Walker, 1963), resulting in the contemporary piling up of lavas from two sources. The term "cedar tree volcano" has been proposed (Fig. 14) for the engulfed central volcanoes. The products of the central volcanoes are distinguishable from the flood basalts since intermediate and acid extrusive and intrusive rocks are predominantly found in the core of the central volcanoes, where the rocks may be drastically altered. Around the volcanic aureoles, secondary minerals can often be traced (Walker, 1963). The presence of palagonitic pyroclastic rocks and pillow lavas suggests the former presence of a caldera and crater lake in some volcanoes. Well-formed conesheets and ring dikes have been described by Sigurdsson (1966, 1970) in the Setberg centers in west Iceland, indicating the existence of temporary magma batches at 2–3 km depth.

Some of the larger intrusions cannot be directly related to central volcanic activity, as, for example, the acid Sandfell laccolite (Hawkes and Hawkes, 1933) and the gabbro–granophyre intrusions on Austurhorn and Vesturhorn (Cargill et al., 1928; Walker, 1966). In the two last-named intrusions there is evidence for the coexistence of acid and basic magmas, so that an intimate connection between the two is suspected (Walker, 1966). Other Icelandic examples of this phenomenon are also known.

Walker (1960) has further been able to estimate the approximate altitude of the highest land surface at the time of zeolitization of the strata (p. 421). This altitude varies from about 700 m above the present sea level west of Lagarfljót (middle East Iceland), to 1000–1500 m in most of eastern Iceland, to 1800 m or more in parts of southeastern Iceland. Intrusions of gabbro and granophyre are generally found in places where erosion has reached the deepest levels of the lava pile.

Large areas of Tertiary rocks in northern and northwestern Iceland are

very poorly described; northern Iceland seems structurally to be more irregular than eastern and western Iceland. 'Central volcanic complexes have
been described from Vididalur (Annells, 1968). On the northwestern peninsula
the Króksfjördur central volcano has recently been described (Hald, Noe-
Nygaard, and Pedersen, 1971), and a section through 1800 m of Tertiary
plateau lavas has been sampled by Noe-Nygaard and Waagstein, but the
results are not published yet. A lava sequence at Brjánslaekur which contains
the well-known beds containing a Tertiary flora (Thoroddsen, 1899) has been
described by Friedrich (1966).

The volcanic activity on a regional scale probably continued without
interruption into the Quaternary. The onset of the glaciation has been dated
at approximately 3–3½ m.y. ago, based on isotope and paleomagnetic datings
in Jökuldalur in east Iceland, Tjörnes (Einarsson, 1971), and Snaefellsnes
(Sigurdsson, 1970). Knowledge of the age and distribution of rocks which are
now known to be indisputably Quaternary (Pleistocene) was primarily advanced through the discoveries of Pjeturss (1900, 1910) who found morainic
horizons both within the Móberg formation and the Gray Basalt Formation,
most of which was believed by former authors to be "preglacial."

The distribution of Quaternary rocks is uncertainly known in many
regions and is only shown schematically in Fig. 12. They crop out along
and adjacent to the present active volcanic zones. The presence of an icecap
meant that the outer environment for the volcanism changed drastically.
Subglacial eruptions resulted, with the formation of pillow lavas, pyroclastic
breccias, and tuffs which were subsequently palagonitized. Mountain ridges
and isolated mountains consisting of palagonitic rocks formed under the ice,
but because of subsequent glaciations they were eroded and partly leveled off,
and they are now found as irregular horizons between subaerial lava flows,
often intercalated with morainic deposits. In interglacial times extensive lavas
were formed similar to the Tertiary; shield volcanoes especially are of common
occurrence, as for instance, Lyngdalsheidi and Skálpanes (Kjartansson, 1960)
in the Reykjanes zone and Vadalda and Grjótháls in the northern zone.

The Móberg Formation, sensu strictu, was mainly formed within the last
two glaciations (Kjartansson, 1960; Saemundsson, 1967), and is primarily
made up of two volcanic forms: subglacial fissure eruptions resulting in the
piling up of ridges, and eruptions from circular vents which often resulted in
the formation of a stapi (table mountain) with a lava cap on the top when
the eruption at last managed to break through the ice cover. The elucidation
of the origin and mode of formation of these mountains was mainly due to
Pjeturss (1910), Nielsen and Noe-Nygaard (1936), Noe-Nygaard (1940), Kjartansson (1943, 1967), and Bemmelen and Rutten (1955). The petrography of
the palagonitic rocks has been treated by Peacock (1926), Noe-Nygaard
(1940), and Jakobsson (1972b).

Volcanic activity also continued in central volcanoes during this period but, because of periodic, explosive volcanic activity under ice cover, the resulting volcanoes are often of much bigger dimensions than those in the Tertiary, e.g., Eyjafjallajökull, Öraefajökull, and Snaefellsjökull. The Setberg volcanic centers in Snaefellsnes were active in the Quaternary period (Sigurdsson, 1970).

The postglacial volcanism, which follows directly upon that of the Pleistocene, is confined to zones which intersect Iceland from southwest to northeast (Figs. 12 and 15). In the southwest the volcanic belt is divided into three individual zones—the western (Snaefellsnes) zone, the middle (Reykjanes) zone, and the eastern zone which continues to the north coast of Iceland (northern zone) (Fig. 15). Within these zones, which are among the most active on earth, averaging one eruption every fifth year, a great variety of volcanic forms occur. Thanks to the studies of mainly Thoroddsen (1906, 1925) and Thorarinsson (1953, 1960, 1964, 1967a, 1967b) our knowledge of the postglacial volcanism is relatively good.

The postglacial volcanism is probably fundamentally of the same nature as in Pleistocene and Tertiary times. One group is represented by fissure and shield volcano eruptions producing homogeneous basalts, the other by centra

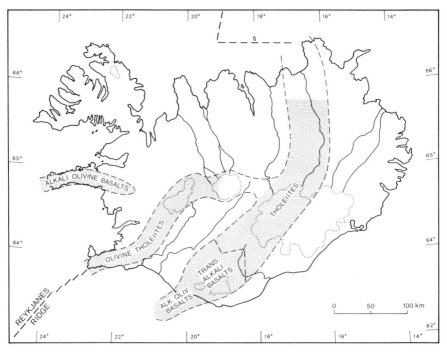

Fig. 15. Distribution of the various kindreds of the postglacial lavas of Iceland (after Jakobsson, 1973).

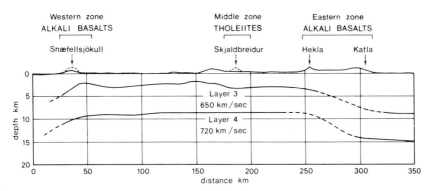

Fig. 16. East–west profile across southern Iceland (after Jakobsson, 1973).

volcanic activity producing basalts accompanied by intermediate and acid rocks.

Iceland has a wealth of various types of volcanoes, which have been classified according to shape, size, materials produced, and the number of eruptions (Thorarinsson, 1960).

Linear eruptions are most common in the area southwest of Vatnajökul, where the famous 25-km-long Laki fissure, which erupted in 1783 and caused famine in Iceland, is situated; the lava covered 565 sq. km with a volume of approximately 12 cu. km (Thorarinsson, 1969). Valagjá, near Hekla, is an example of an explosion fissure, where only a negligible quantity of lava was produced.

Of eruption sites with circular vents the shield volcano type is the most common. More than 20 postglacial shield volcanoes are known in Iceland. One of the most prominent is Skjaldbreidur in southwestern Iceland (Tryggvason, 1943). From the northern zone, Trölladyngja, Ketildyngja, and Kerlingardyngja can be mentioned.

Hverfjall, at Mývatn in the northern zone, is an example of a tephra ring volcano (Thorarinsson, 1952).

Only a few maars are known; Graenavatn in the Reykjanes zone is the most typical one.

Submarine eruptions are known to have occurred repeatedly in the offshore extensions of the volcanic zones. The submarine Surtsey eruption (1963–1967) on the shelf south of Iceland evolved and was built up in a manner greatly resembling the subglacial volcanoes on the mainland (Thorarinsson, 1964; Kjartansson, 1967).

The present glaciers of Iceland can be considered as remnants from the last ice age. Where the glaciers cover volcanic areas, subglacial eruptions similar to those which produced the Moberg Formation still take place. The best known subglacial volcanoes of today are Grimsvötn under Vatnajökull

Glacier (last eruption in 1936) and Katla under Mýrdalsjökull Glacier (last eruption in 1918). Such eruptions may be accompanied by immense glacier bursts (Icel. jökullhlaup) where the maximum discharge is recorded to exceed 1,000,000 m³/sec.

Several central volcanoes have been active in postglacial time, the most famous being Hekla with eruptions in 1845, 1947–1948, and 1970; Askja which last erupted in 1961 (Thorarinsson and Sigvaldason, 1962); Öraefajökull with eruptions in 1362 and 1727 (Thorarinsson, 1958); and Snaefellsjökull (Sigurdsson, 1970). Except for Hekla, few andesitic lavas have formed in postglacial time. Hekla shows a cyclic chemical variation, which is linear with time (Thorarinsson, 1967a). Rhyolitic eruptions are usually phreatic (Hekla, Askja), however, in the Torfajökull region acid lavas occur.

The extensive crater groups, "areal eruptions" of earlier authors, found in some big lava flows, as for instance the Landbrotshólar craters on the Eldgjá lava and the Mývatn crater groups on the younger Laxá lava, have proved to be of secondary origin (pseudocraters). They were formed when the lava spread over ground which was soaked with water, which then escaped as steam through the lava flow with the resulting formation of numerous small craters (Thorarinsson, 1953).

Tephrochronological studies by Thorarinsson (1960, 1967) have made it possible to date fairly accurately many postglacial events.

The postglacial lavas cover an area of approximately 12,000 sq. km, and calculations on the volume vary between 400 cu. km (Thorarinsson, 1967) and 480 cu. km (Jakobsson, 1972b).

Hydrothermal activity is of widespread occurrence in Iceland; it has been studied by Thorkelsson (1910), Thoroddsen (1925), Thorkelsson (1941), Einarsson (1942), Sonder (1941), and Barth (1950). Bödvarsson (1960) classified the thermal activity into two main groups.

(1) The low-temperature activity group, in which the maximum temperature at the base of water circulation does not exceed 150°C, is characterized on the surface by the presence of hot springs, producing only negligible amounts of steam; they show a low degree of thermal metamorphism. The low-temperature areas are mainly found in the western and northern Tertiary districts. There are about 250 thermal areas of this type with more than 600 major springs.

(2) The high-temperature activity group is characterized by temperatures higher than 150°C and is confined to the active volcanic zones. Some 13 thermal areas are known. In this group hot springs are almost absent, but a great number of steam holes occur, and there are large areas of hot ground; a high degree of thermal metamorphism is characteristic. The Torfajökull and Grímsvötn thermal areas are the greatest with a heat output up to $750 \times 10^{\circ}$

cal/sec. The chemical character of the thermal waters and gases in Iceland has been discussed by Sigvaldason (1966).

The crustal structure of Iceland is fairly well known through the works of Båth (1960) and Pálmason (1963, 1971). In the uppermost 20 km of the crust, four layers have been distinguished and these are interpreted to be probably basaltic in composition. Layer 4, which presumably belongs to the uppermost mantle, has a P velocity of 7.2 km/sec; it occurs at a depth between 8 and 16 km.

A certain symmetry of the crustal structure appears to exist in southwestern Iceland in the continuation of the Reykjanes ridge, but at lat. 65° N a major structural change is indicated (Pálmason, 1971), possibly connected with the right-lateral offset of the crest of the mid-Atlantic ridge in central Iceland (Fig. 12). There are indications that the chemistry of the volcanics in the active volcanic zones can be correlated with the crustal structure (Jakobsson, 1972).

The modern view of the petrology of Iceland is based on work which begins with Holmes (1918), Hawkes (1924), Peacock (1926), and Cargill et al. (1928). These authors generally agreed that two rock series were present: a pre- and interglacial calc-alkali series, and a postglacial to recent alkalic series. The petrology gradually became more known through the works of Tryggvason (1943), Noe-Nygaard (1952), Carmichael (1964, 1967), and Heier et al. (1966).

Recent research (Sigvaldason, 1966) has revealed that both tholeiitic and alkalic basalts are present in late Pleistocene and postglacial time, and Jakobsson (1972b) has shown that the postglacial basalts have a distinct distribution pattern, with tholeiites in the Reykjanes and northern zones and alkali basalts in the two flank zones (the eastern and the western zones) in South and West Iceland (Fig. 15). This variation in the chemistry of the basalts can be attributed to variations in depth and the degree of partial melting. Central volcanoes can also be divided into those with tholeiitic affinities, e.g., Thingmúli, Setberg, and Askja, and those with alkali affinities like Snaefellsjökull, Torfajökull, and Öraefajökull (cf., Sigurdsson, 1970). The distribution of differentiated rocks is thus in accordance with the distribution of various basalt kindreds.

Recent studies have also shown that there is a real chemical difference between the gross chemistry of Icelandic basalts and basalts from the mid-Atlantic ridge (Sigvaldason, 1969; Jakobsson, 1972b). The Icelandic basalts are generally richer in Fe, Ti, P, and K, but poorer in Al and Mg than those from the ridge. On the other hand, a comparison with available data from the Tertiary of Iceland, the Faeroes, and East Greenland indicates a close affinity in the chemical composition for the whole region (see Table 1 in Noe-Nygaard, 1966). In terms of sea-floor spreading, this could mean that chemical composition of primary rocks produced on the Iceland part of the mid-Atlantic ridge has been constant since the opening up of the North Atlantic Ocean.

H. Jan Mayen

The volcanic island Jan Mayen covers an area of 380 sq. km; it lies between lat. 70°50′ and 71°10′ N and between long. 7°56′ and 9°05′ W. The south-western half of the island is controlled by fissure volcanism, and here numerous cumulo domes are found (Dollar, 1966); the northeastern half of the island is dominated by the great, basaltic stratovolcano Beerenberg, which rises to 2277 m above sea level, its height above the surrounding sea bottom being approximately 5000 m. The summit crater is about 1500 m in diameter and about 300 m deep and is filled with ice which feeds the Weyprectbreen. Toward the sea the mountain is mostly bordered by steep cliffs.

In the early half of the seventeenth century the island was often visited by whalers (Fotherby, 1625). Its scientific exploration began with an Austrian expedition in the Polar Year 1882–1883 (Boldva, 1886), but scattered observations had been made earlier (e.g., Scoresby, 1820). Papers by Wordie (1926) and Tyrrell (1926) for quite a long period represented the geological knowledge we had of the island.

A number of publications by F. J. Fitch and other members of the University of London expeditions in 1959 and 1961 have given us a good picture of the Beerenberg volcano on Nord Jan (Fitch, 1964). H. Carstens (1961, 1962) described the lavas of the southwestern part of the island.

The geological history of Beerenberg can be summarized in this way. All rocks prior to the formation of the summit cone are comprised in the Kapp Muyen Group; lavas which belong to the Nordvestkapp Formation of this group make up the main mass of the Beerenberg dome. The summit consists of the Sentralkrater Formation only, which marks the final stage of the central eruptions. Later volcanic events have taken place along fissures on the lower slopes of the volcano where they have given rise to a great number of parasitic cinder cones. Lavas and pyroclastics of this phase are included in the Nordkapp Group.

Fitch (1965) dates the birth of the Beerenberg to late Pleistocene. The main activity is post-Pleistocene in age. Volcanic eruptions on Jan Mayen of which we have knowledge occurred in 1633(?), 1732, 1818, and 1970.

Fumarolic activity has been recorded in two places on the south coast (Orvin, 1960).

The older lavas of the Kapp Muyen Group are dominantly ankaramites and augite–olivine porphyritic basalts, while in the upper part of the lava succession glomerophyric aggregates of plagioclase appear. The lavas of the Sentralkrateret Formation consist of glomerophyric plagioclase basalts of trachybasaltic and trachyandesitic composition. The central eruptions of Beerenberg thus underwent a gradual change with time.

The youngest lavas of the Nordkapp Group, originating through flank fissures, comprise ankaramitic as well as glomerophyric plagioclase basalts.

Xenoliths of olivinite occur in basalts of all ages.

The Beerenberg lavas belong to the alkali-olivine basaltic association and are characterized by high potash values; interstitial alkali feldspar and analcite are often found.

In the southwestern part of the island trachybasalts which carry leucite, analcite, and sodalite in the groundmass have been found (Carstens, 1961, 1962, 1963), and similar rocks may also occur among the Beerenberg rocks. Trachytes proper have not been found among the Beerenberg volcanics.

Rocks with a close resemblance to the Icelandic hyaloclastite–palagonite breccias occur near sea level on the south coast of the island in the Havhest-berget Formation (Fitch, 1965). These rocks were probably formed beneath an ice sheet.

Irregular sills and dikes are either porphyritic or aphyric rocks of ankaramitic or basaltic composition (Hawkins and Roberts, 1965).

The petrology of the volcanic and intrusive rocks of Nord-Jan has recently been treated by Hawkins and Roberts (1972). The rock types include potash ankaramites, potash basalts, trachybasalts, and trachyandesites. The Na_2O/K_2O ratio averages 1.15. Xenocrysts are common and may make up 20–40% of the rock bulk.

The xenocrysts are thought to have crystallized within the upper mantle, whereas the brown titansalites and other phases crystallized at lower, near-surface pressures. It is possible that the initial basic melt was produced by the partial melting of mantle material at 20–30 kbar which had reached the wehrlite condition. If the basic melt precipitated olivine and chromium diopside at 20–30 kbar, this could possibly lead to the production of a residuum equivalent to the primitive Nord-Jan magma.

A major eruption took place on Jan Mayen in September–October 1970 (Gjelsvik, 1970). The lava flows were first observed on September 20, but the first fissure opening may well have occurred a couple of days earlier (Siggerud, 1972). Five crater groups were formed along a 6-km-long fissure zone (Fig. 17) on the northeast slope of Beerenberg. The fissure ran from about 1000 m above sea level down to the sea. Lava from the eruption built up a new land area in the sea of approximately 4 sq. km. The minimum volume of the effusives has been estimated to 0.5 cu. km. The lavas are potash-rich basalts. In late October the eruption of lava ceased, but tephra and gases were sporadically emitted up to the last observations in June 1971.

Bulk-rock and mineral chemistry of recent Jan Mayen basalts from the 1970 flank eruption has been produced by Weigand (1972), who presents major-

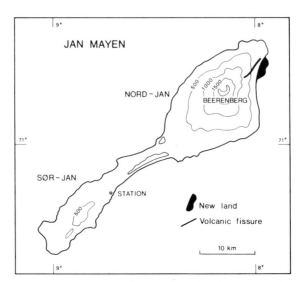

Fig. 17. Contour map of Jan Mayen showing the eruption
fissure and the addition of land during the 1970 eruption on
the flank of Beerenberg (after Siggerud, 1972).

element analyses of four alkali olivine basalts. He interprets the diopside
xenocrysts as fragments of ultramafic nodules, which are reported from the
older Jan Mayen rocks. The obvious compositional changes between the salite
phenocrysts and the titanaugite rims and groundmass point to widely differing
physical and chemical environments.

III. DISCUSSION

The formation of the present North Atlantic Ocean is likely to have begun
with a fissuring and breaking up of a huge Laurasian landmass, perhaps
already in mid-Mesozoic time. Indications of this primary fissuring may be
found in the existence of master dikes in eastern Canada and West Greenland
and in early tholeiitic basalt flows and sills in Spitzbergen.

As the separation of the individual larger landmasses, Eurasia with
Spitzbergen and the Barents Sea shelf, from Greenland and from North
America proceeded, many marginal landmasses foundered to form shelf areas;
these are generally covered with Mesozoic sedimentary strata.

As a consequence of the ongoing widening of the sea-covered areas, the
ocean floor proper was eventually laid bare, and an enormous increase in
volcanic activity began. An ocean-floor rift formed from the North Labrador
Sea northward. Off the central West Greenland region it gave rise to a voluminous production of basaltic lavas of which large remnants are still left on dry

land. Farther north where crust material is present in the Davis strait a graben structure arose in which several kilometers of Mesozoic sediments are still preserved.

East of Greenland another ocean-floor rift came into existence. It was highly volcanic and produced basalts along the east coast of that island as well as along the opposite side of the new ocean, that is, the basalts of the Faeroes and the Hebridean Province including the banks west of Scotland. The continued westward movement of Greenland may have loosened a spur of crust material which lay underneath the present ridge between East Greenland and Northwest Scotland bearing the Faeroes on its back. Volcanic activity of this episode lasted for a relatively short time, that is, 10–15 m.y. in lower Eocene.

As the North Atlantic achieved much of its present configuration, a new rifting epoch of the ocean floor began in early Miocene, probably initiated through a north-going propagation of the rifting farther south in the Atlantic, producing the northern part of the present sinuous mid-oceanic ridge. Where the new ridge cuts the scars of the old rifting, Iceland came into existence. Jan Mayen is related to the present ridge itself.

At present, no detailed discussion of the petrogenesis of the basalts can be given. The main problems are still concerned with the relationship between the two main magma types, alkali olivine basalts and tholeiitic basalts.

In East Greenland tholeiitic basalts lie nearest to the coast and lowest in the lava pile. The highest alkalinity is shown by the lavas in the Nunatak zone far inland; upward, in the coastal lava sequence, abnormal types and rocks of intermediate composition are met with.

The lowermost two-thirds of the basalt sequence in the Faroes are quartz tholeiitic; the upper third oltholeiitic.

West Greenland offers the greatest and most significant diversity of basaltic rock types. Work in progress will probably give valuable contributions to our understanding of the origin of the main basalt kindreds.

The composition of the plutons in East Greenland centers on gabbros and syenites, whereas the Hebridean Province is governed by gabbros and granites.

ACKNOWLEDGMENTS

This article is the product of a collaboration among the author and some of the other members of the "basalt group" in Copenhagen. Dr. Kent Brooks wrote on the Hebridean Province, Mr. Sveinn Jakobsson on Iceland, and Mr. Asger Ken Pedersen on West Greenland. The author has adapted the various contributions and is responsible for the article as a whole.

REFERENCES

Introduction

Baker, P. E., 1973, Islands of the South Atlantic, in: *Ocean Basins and Margins I. The South Atlantic*, (A. E. M. Nairn and F. G. Stehli, ed.), Plenum Press, New York, London.

Blackadar, R. G., 1964, Basic intrusions of the Queen Elisabeth Islands, District of Franklin: Geol. Surv. Canada, Bull. 97, 36 pp.

Böggild, O. B., 1900, The Deposits of the Sea Bottom, *The Danish Ingolf-Expedition*, Vol. I: København, p. 3.

Böggild, O. B., 1907, Sediments sous-marine receueillis dans la Mer du Grönland: Duc d'Orleans Croisière Oceanographique, Bruxelles.

Böggild, O. B., 1918, Den vulkanske Aske i Moleret: Danm. Geol. Unders. II R. No. 37.

Brooks, C. K., 1973, The Tertiary of Greenland: A volcanic and plutonic record of continental break-up: Mem. Ass. of Amer. Petroleum Geol.

Daly, R. A., 1926, *Our Mobile Earth*: New York.

Daly, R. A., 1942, *The Floor of the Ocean*: Univ. North Carolina Press.

Darwin, Ch., 1844, Geological Observation on the Volcanic Islands visited during the voyage of H. M. S. Beagle: London.

Hamberg, A., 1899, Über die Basalte des König Karl Landes. G.F.F., Vol. 21: Stockholm.

Johnson, G. L., Closuit, A. W., and Pew, J. A., 1969, Geological and geophysical observations in the Northern Labrador Sea: Arctic.

Johnson, G. L. and Eckhoff, O. B., 1966, Bathymetry of the north Greenland sea: *Deep-Sea Res.*, v. 13, London.

Johnson, G. L. and Heezen, B. C., 1967, Morphology and evolution of the Norwegian–Greenland Sea: *Deep-Sea Res.*, v. 14, Oxford.

Johnson, G. L., Vogt P. R., and Schneider, E. D., 1971, Morphology of the Northeastern Atlantic and Labrador Sea: *Deutschen Hydrogr. Zeitschr.*, v. 24, n. 1.

Nansen, F., 1904, The bathymetrical features of the North Polar Seas, in: The Norwegian North Polar Expedition Results: London.

Noe-Nygaard, A., 1965, The composition "gap" in the basalt–rhyolite association as elucidated by an example of Eocene volcanic ash in Denmark: *Medd. Dansk. Geol. Foren.*, v. 15.

Norin, R., 1934, Zur Geologie der südschwedischen Basalte: Lunds Geol. Min. Inst. Medd. No. 57.

Printzlau, J. and Larsen, O., 1973, K/Ar determinations in alkaline olivine basalts from Skaane, South Sweden: G.F.F. 794, Stockholm.

Stride, A. H., Curray, J. R., Moore, D. G., and Belderson, R. H., 1969, Marine geology of the Atlantic continental margin of Europe: *Phil. Trans. Roy. Soc.* (London), Ser. A, n. 1148, v. 264.

Tyrrell, G. W. and Sandford, K. S., 1933, Geology and petrology of the dolerites of Spitsbergen: *Roy. Soc. Edinburgh Proc.*, v. 53.

Vogt, P. R., Ostenso, N. A., Johnson, G. L., 1970, Magnetic and bathymetric data bearing on sea-floor spreading north of Iceland: *J. Geoph. Res.*, v. 75.

Watt, W. S., 1969, The Coast parallel dike swarm of southwest Greenland in relation to opening of the Labrador Sea: *Canad. J. Earth. Sci.*, v. 6.

Wegener, A., 1922, Die Entstehung der Kontinente und Ozeane (3rd edit.), Braunsweig.

Davis Strait and Baffin Island

Andrews, J. T., Guenmel, G. K., Wray, J. L., and Ives, J. D., 1972, An early Tertiary outcrop in north-central Baffin Island. N.W.T.: *Canad. Journ. Earth Sci.*, v. 9, n. 3.

Clarke, D. B., 1970, Tertiary basalts of Baffin Bay. Possible primary magma from the mantle: *Contr. Min. and Petr.*, v. 25, n. 3.

Henderson, G., 1972, The geological setting of the West Greenland basin in the Baffin Bay region: *Geol. Surv. Canada.*

Park, J., Clarke, D. B., Johnson, J., and Keen, M. J., 1971, Seward extension of the West Greenland Tertiary volcanic province: *Earth and Planetary Sci. Letters*, v. 10, n. 2.

Wilson, J. Tuzo and Clarke, D. B., 1965, Geological Expedition to Capes Dyer and Searle, Baffin Island, Canada: *Nature*, v. 205, London.

West Greenland

(*M.o.G.* = *Meddelelser om Grønland*, København)

Birkelund, T., 1965, Upper cretaceous belemnites from West Greenland: *M.o.G.*, p. 137, København.

Brooks, C. K., 1973, The Tertiary of Greenland: A volcanic and plutonic record of continental break-up: Memoir Am. Ass. Petroleum Geol.

Clarke, D. B., 1970, Tertiary basalts of Baffin Bay: possible primary magma from the mantle: *Contrib. Min. and Petrol.*, v. 25.

Clarke, D. B. and Pedersen, A. K., (in press), The Tertiary volcanic province of West Greenland, in: *The Geology of Greenland, Geol. Surv. of Greenland.* Copenhagen.

Drescher, F. K. and Krueger, H. K. E. 1928: Der Peridotit von Kaersut (Grønland) und sein Ganggefolge. Neues Jahrb. für Min. etc. 57 A. Stuttgart.

Drever, H. J., 1952, The origin of some ultramafic rocks. A preliminary survey of the evidence for and against gravitative accumulation of olivine: *Bull. Geol. Soc. Denmark*, v. 12.

Drever, H. J., 1956, The geology of Ubekendt Ejland, West Greenland. Pt. II. The picritic sheets and dykes of the east coast: *M.o.G.*, v. 137, København.

Drever, H. J., 1958, Geological results of four expeditions to Ubekendt Ejland, West Greenland: *Arctic*, v. 11.

Drever, H. J. and Game, P. M., 1948, The Geology of Ubekendt Ejland, West Greenland. Part 1. A preliminary report: *M.o.G.*, v. 134, n. 8, København.

Drever, H. J. and Johnston, R., 1966, A natural high-lime silicate liquid more basic than basalt: *Journ. Petr.*, v. 7, n. 3, Oxford.

Giesecke, K. L., 1910, Karl Ludwig Gieseckes mineralogisches Reisejournal über Grönland, 1806–1813. 2 Edit.: *M.o.G.*, v. 35, København.

Hald, N., 1971, An investigation of the igneous rocks on Hareøen and Western Nûgssuaq, West Greenland: Grønlands Geologiske Undersøgelse. Report No. 35, København.

Hald, N., 1973, Preliminary results of the mapping of the Tertiary basalts in Western Nûgssuaq. Grønlands Geologiske Undersøgelse, København.

Heim, A., 1910, Über die Petrographie und Geologie der Umgebungen von Karsuarsuk, Nordseite der Halbinsel Nugssuak, W-Grönland: *M.o.G.*, v. 47, København.

Henderson, G., 1969, Field work supplementing photogeological interpretation of Nûgssuaq: Grønlands Geologiske Undersøgelse, Report No. 19, København.

Henderson, G., 1972, The geological setting of the West Greenland basin in the Baffin Bay region: Geol. Surv. Canada. Paper 71–23.

Krueger, H. K. E., 1928, Zur Geologie von Westgrönland, besonders der Umgebung der Disko Bucht und des Umanak-Fjordes: *M.o.G.*, v. 74, København.

Munck, S., 1945, On the Geology and Petrography of the West Greenland Basalt Province. V. Two Major Doleritic Intrusions of the Nûgssuaq Peninsula: *M.o.G.*, v. 137, København.

Munck, S. and Noe-Nygaard, A., 1957, Age determination of the various stages of the Tertiary volcanism in the West Greenland basalt province: Congr. Geol. Intern. XX Sect. 1, Mexico.

Nieland, H., 1931, Beitrag zur Kenntnis der Deckenbasalten von Westgrönland: Chem. d. Erde 6, Jena.

Noe-Nygaard, A., 1942, On the Geology and Petrography of the West Greenland Basalt Province. III: *M.o.G.*, v. 137, n. 3, København.

Nordenskjöld, A. E., 1870, Redogörelse för an Expedition til Grönland år 1870, in: *Övers. af Kongl. Videusk. Akad. No. 10*: Stockholm.

Pedersen, A. K., 1970, A petrological investigation of Tertiary volcanic rocks from northern Disko (in Danish): Unpublished thesis, Copenhagen University.

Pulvertaft, C., 1968, The precambrian stratigraphy of Western Greenland: 23rd Int. Geol. Congr. Vol. 4, Praha.

Ravn, J. P. J., 1918, De Marine Kridtaflejringer i Vest Grønland og deres Fauna: *M.o.G.*, v. 56, København.

Rosenkrantz, A., et al., 1940, Den danske Nugssuak Expedition 1939: *Medd. Dansk Geol. Foren.*, v. 9, København.

Rosenkrantz, A., 1952, Oversigt over Kridt- og Tertiaerformationens stratigrafiske Forhold i Vestgrønland: *Bull. Geol. Soc. Denmark*, v. 12.

Rosenkrantz, A., 1964, Le Dano-Paléocène dans Groënland occidental: Bur. Recherches Geol. et Minières Mém. 28.

Rosenkrantz, A., 1970, Marine Upper Cretaceous and lowermost Tertiary deposits in West Greenland: *Bull. Geol. Soc. Denmark*, v. 19, n. 4.

Rosenkrantz, A., Noe-Nygaard, A., Gry, H., Munck, S., and Laursen, D., 1942, A geological reconnaissance of the southern part of the Svartenhuk peninsula: *M.o.G.*, v. 135, n. 3, København.

Rosenkrantz, A. and Pulvertaft, C., 1969, Cretaceous–Tertiary stratigraphy and tectonics in northern West Greenland: Mem. Amer. Ass. Petrol. Geol., No. 12.

Steenstrup, K. J. V., 1883, Bidrag til Kjendskab til de geognostiske og geographiske Forhold i en Del af Nord-Grønland: *M.o.G.*, v. 4, København.

Thompson, R. N. and Patrick, D. J., 1968, Folding and slumping in a layered gabbro: *Geol. Journ.*, v. 6, n. 1.

von Wolff, F., 1931, *Der Vulkanismus*, II Band, 2. Teil: Ferdinand Enke, Stuttgart.

East Greenland

Abbot, D. and Deer, W. A., 1972, Geological investigations in East Greenland. X. The gabbro cumulates of the Kap Edvard Holm lower layered series: *M.o.G.*, v. 190, n. 6.

Anwar, Y. M., 1955, Geological investigations in East Greenland. Part V. The petrology of The Prinsen af Wales Bjerge lavas: *M.o.G.*, v. 135, n. 1.

Backlund, H. G., 1937, Sur quelques roches éruptives de la série basaltique de la Côte oriental du Groenland: *Compt. Rend. Acad. des Sciences*, Tome 204, Paris.

Backlund, H. G. und Malmquist, D., 1932, Zur Geologie und Petrographie der nordost-grönländischen Basaltformation: *M.o.G.*, v. 87, n. 5.

Bearth, P., 1959, On the alkali massif of the Werner Bjerge in East Greenland: *M.o.G.*, v. 153, n. 4.

Beckingsale, R. D., Brooks, K. C., and Rex, D. C., 1970, K–Ar ages for the Tertiary of East Greenland: *Bull. Geol. Soc. Denmark*, v. 20, København.

Brooks, C. K., 1973, The Tertiary of Greenland: A volcanic and plutonic record of continental break-up: Mem. Ass. of Amer. Petroleum Geol.

Brown, P., 1968, Igneous rocks of the Kialineq area, East Greenland: *Geol. Soc. Proc.* No. 1649, London.

Bütler, H., 1957, Beobachtungen an der Hauptbruchzone der Küste von Zentral–Ostgrönland: *M.o.G.*, v. 160.

Deer, W. A., 1967, Lawrence Rickard Wager: Biogr. Mem. of the Fellows of the Roy. Soc., Vol. 13, London.

Deer, W. A., (in press), The Tertiary igneous rocks between Scoresby Sund and Kap Gustav Holm.

Deer, W. A. and Abbot, D., 1972, Geological investigations in East Greenland. IX. The gabbro cumulates of the Kap Edvard Holm lower layered series: *M.o.G.*, v. 190, n. 6.

Douglas, J. A. V., 1964, Geological investigations in East Greenland. VII. The Basistoppen Sheet, a differentiated basic intrusion into the upper part of the Skaergaard complex, East Greenland: *M.o.G.*, v. 164, n. 5.

Elsdon, R., 1969, The structure and intrusive mechanism of the Kap Edvard Holm layered gabbro complex, East Greenland: *Geol. Mag.*, v. 106, London.

Fawcett, J. J., Rucklidge, J. C., and Brooks, K. C., 1966, Geological Expedition to the Tertiary basalt region of Scoresby Sund, East Greenland: *Nature*, v. 212.

Frebold, H. und Noe-Nygaard, A., 1938, Marines Jungpaleozoikum und Mesozoikum von der Traill-Insel: *M.o.G.*, v. 119, n. 2.

Haller, J., 1956, Geologie der Nunatakker Region von Zentral-Ostgrönland zwischen 72°30′ und 74°10′ N.Br.: *M.o.G.*, v. 154.

Hamilton, E. I., 1966, The isotopic composition of lead in igneous rocks: *Earth and Planet Sci. Lett.*, v. 1, Amsterdam.

Hochstetter, F. von, 1874, Geologie Ostgrönlands zwischen dem 73° und 76° nördl. Br., in: Koldewey, K.

Hughes, C. J., 1956, Geological Investigations in East Greenland. VI. A differentiated basic sill in the Skaergaard Intrusion, East Greenland and related sills injecting the lavas: *M.o.G.*, v. 137, n. 2.

Kapp, H., 1960, Zur Petrologie der Subvulkane zwischen Mesters Vig und Antarctic Havn (Ost-Grönland): *M.o.G.*, v. 153, n. 2.

Katz, H. R., 1952, Ein Querschnitt durch die Nunatakzone Ostgrönlands (ca. 74°N.B.): *M.o.G.*, v. 144.

Koch, L., 1929, Geology of East Greenland: *M.o.G.*, v. 73.

Kempe, D. R. C., Deer, W. A., and Wager, L. R., 1970, Geological Investigations in East Greenland. VIII. The Petrology of the Kangerdlugssuaq alkaline intrusion, East Greenland: *M.o.G.*, v. 190, n. 2.

Kempe, D. R. C. and Deer, W. A., 1970, Geological Investigations in East Greenland. IX. The mineralogy of the Kangerdlugssuaq alkaline intrusion, East Greenland: *M.o.G.*, v. 190, n. 3.

Koldewey, F. von, 1874, Die zweite deutsche Nordpolarfahrt in den Jahren 1869 und 1870 unter Führung des Kapitän Karl Koldewey, 2 Vols., Leipzig.

Krokström, T., 1944, Petrological studies on some basaltic rocks from East Greenland: *M.o.G.*, v. 103, n. 6.

Lenz, O., 1874, Spezielle Darstellung der geologischen Verhältnisse Ostgrönlands, in: Koldewey, K.

Nathorst, A. G., 1901, Bidrag til nordöstra Grönlands geologi (with petrographic descriptions by H. Bäckström): *G.F.F.*, v. 23, Stockholm.

Nielsen, E., 1935, The Permian and Eotriassic vertebrate-bearing beds at Godthaab Gulf (East Greenland): *M.o.G.*, v. 98, n. 1.

Noe-Nygaard, A., 1937, Die Palaeozoischen Eruptivgesteine von Canning Land: *M.o.G.*, v. 118, n. 6.

Noe-Nygaard, A., 1940, Syenitforekomsten ved Antarctic Havn: *Medd. Dansk Geol. Foren.*, v. 9, København.

Nordenskjöld, O., 1908, On the geology and physical geography of East Greenland: *M.o.G.*, v. 28, n. 5.

Ojha, D. N., 1966, Petrology of the Kaerven layered intrusion, East Greenland: *J. Geochem. Soc. India*, v. 1.

Orvin, A. K., 1931, A fossil river bed in East Greenland: *Norsk Geol. Tidsskr.*, v. 12, Oslo.

Putallaz, J., 1961, Géologie de la partie médiane de Traill Ø (Groenland oriental): *M.o.G.*, v. 164, n. 2.

Scoresby, Wm. (Jun.), 1823, Journal of a voyage to the northern whale fishery, including researches and discoveries on the eastern coast of West Greenland, made in the summer of 1822, in the ship *Baffin* of Liverpool: Edinburgh.

Stauber, H. P., 1938, Stratigraphische Untersuchungen postdevonischer Sedimente auf den Inseln Traill und Geographical Society: *M.o.G.*, v. 114, n. 1.

Stauber, H. P., 1942, Zur Geologie der Traill Insel (Nordostgrönland): *Eclog. Helv.*, v. 35, n. 1, Zürich.

Stern, P., 1964, Zur Petrographie von Nordhoeks Bjerg und Nörlunds Alper, Hudson Land (Zentral Ostgrönland): *M.o.G.*, v. 168, n. 5.

Tyrrell, G. W., 1932, The petrography of some Kainozoic igneous rocks, and of the Cap Parry alkaline complex, East Greenland: *Geol. Mag.*, v. 69, London.

Vincent, E. A., 1953, Hornblende-lamprophyre dykes of basaltic parentage from the Skaergaard area, East Greenland: *Q.J.G.S.*, v. 109, London.

Vischer, A., 1943, Die postdevonische Tektonik von Ostgrönland zwischen 74° und 75° N. Br.: *M.o.G.*, v. 133.

Wager, L. R., 1934, Geological Investigations in East Greenland. I. General geology from Angmagssalik to Cape Dalton: *M.o.G.*, v. 105, n. 2.

Wager, L. R. and Brown, M., 1969, *Layered Igneous Rocks*: Oliver and Boyd, Edinburgh and London.

Wager, L. R. and Deer, A. W., 1938, A dyke swarm and crustal flexure in East Greenland: *Geol. Mag.*, v. 75, London.

Wager, L. R. and Deer, W. A., 1939, Geological Investigations in East Greenland. III. The petrology of the Skaergaard intrusion, Kangerdlugssuaq, East Greenland: *M.o.G.*, v. 105, n. 4.

Wager, L. R. and Hamilton, E. I., 1964, Some radiometric rock ages and the problem of the southward continuation of the East Greenland Caledonian orogeny: *Nature*, v. 204, p. 4963, London.

Wenk, E., 1961, Tertiary of Greenland, in: Raasch, G. O., *Geology of the Arctic*, I and II, Toronto.

Wordie, J. M., 1927, The Cambridge expedition to East Greenland in 1926: *The Geogr. Journ.*

The Faeroe Islands

Allan, Thomas, 1814, An account of the mineralogy of the Faroe Islands: *Trans. of the Royal Soc. of Edinb.*, v. VII, Edinburgh.

Born, C. L. U. v., Captain, 1792, Om Basalt-Bierge paa Faerøerne. Udtog af Brev fra Hr. Capitain Born til Hr. Etatsraad Rothe: *Skr. af Nat. Hist. Selsk.*, v. II, København.

Born, C. L. U. v., Captain, 1793, Fortsaettelse af Brevvexlingen imellem Hr. Capit. Born og Hr. Etatsraad Rothe, om de Faerøske Basalt-Bierge: *Skr. af Nat. Hist. Selsk.*, v. III, København.

Born, C. L. U. v., Captain, 1797, Fortsaettelse af Brevvexlingen fra Capitain Born til T. Rothe, om den Faerøiske Basalt: *Skr. af Nat. Hist. Selsk.*, v. IV, København.

Debes, L. J., 1673, *Faeroe et Faroa Reserata*, København.

Forchhammer, J. G., 1823–24, Beretning om Faerøerne: Overs. over Vid. Selsk. Forh., København.

Geikie, James, 1880, On the Geology of the Faeröe Islands: *Trans. of the Royal Soc. of Edinb.*, v. XXX, Edinburgh.

Hald, N., Noe-Nygaard, A., and Waagstein, R., 1969, On extrusion forms in Plateau Basalts, 2. The Klakksvík Flow, Faeroe Islands: *Medd. D.G.F. (Bull. Geol. Soc. Denmark)*, v. 19, n. 1, København.

Helland, Amund, 1880, Om Faerøernes Geologi: *Geografisk Tidsskrift*, v. IV, København.

Landt, Jørgen, 1800, Forsøg til en Beskrivelse over Faerøerne (4-21, 92-105, 161-175), København.

Laufeld, O., 1965, Sporomorphs in Tertiary Coal from the Faeroe Islands: *G.F.F.*, Stockholm.

Mackenzie, Sir George Steuart, 1814, An Account of some Geological Facts Observed in the Faroe Islands: *Trans. of the Royal Soc. of Edinb.*, v. VII, Edinburgh.

Noe-Nygaard, Arne, 1940, Om gjógv-Systemernes Alder på Faerøerne: *Medd. D.G.F. (Bull. Geol. Soc. Denmark)*, v. IX, København.

Noe-Nygaard, Arne, 1962, The Geology of the Faeroes: *Q.J.G.S.*, v. 118, London.

Noe-Nygaard, Arne, 1966*a*, Chemical composition of tholeiitic basalts from the Wyville-Thompson Ridge belt: *Nature*, v. 212, London.

Noe-Nygaard, Arne, 1966*b*, The invisible part of the Faeroes: *Medd. D.G.F. (Bull. Geol. Soc. Denmark)*, v. 16, København.

Noe-Nygaard, Arne, 1967, Variation in titania and alumina content through a three kilometres thick basaltic lava pile in the Faeroes: *Medd. D.G.F. (Bull. Geol. Soc. Denmark)*, v. 17, København.

Noe-Nygaard, A., 1968, On extrusion forms in Plateau Basalts. Shield volcanoes of "scutulum" type: *Scientia Islandica*, Reykjavík.

Noe-Nygaard, A. and Rasmussen, J., 1968, Petrology of a 3000 metre sequence of basaltic lavas in the Faeroe Islands: *Lithos*, v. 1, Oslo.

Peacock, M. A., 1927–28, Recent lines of fracture in the Faeroes, in relation to the theories of fiord formation in northern basaltic plateaux: *Trans. of the Geol. Soc. of Glasgow*, v. XVIII, Glasgow.

Rasmussen, Jóannes, 1946, Oversigt over den geologiske litteratur vedrørende Faerøerne: *Medd. D.G.F. (Bull. Geol. Soc. Denmark)*, v. 11, n. 1, København.

Rasmussen, Jóannes, 1951, Transgressive sillintrusioner i Faerøplateauet: *Medd. D.G.F. (Bull. Geol. Soc. Denmark)*, v. 12, København.

Rasmussen, Jóannes, 1952, Bidrag til forståelse af den faerøske lagseries opbygning: *Medd. D.G.F. (Bull. Geol. Soc. Denmark)*, v. 12, København.

Rasmussen, Jóannes, 1955, Nøkur ord um gjair i Føroyum—Upprunna teirra og alder: *Frodskaparrit*, v. 4, Tórshavn.

Rasmussen, Jóannes, 1962, Um goshálsar i Føroyum: *Frodskaparrit*, v. 11, Tórshavn.

Rasmussen, J. and Koch, E., 1963, Fossil Metasequoia from Mikines, Faroe Islands: *Frods-kaparrit*, v. 12, Tórshavn.

Rasmussen, J. and Noe-Nygaard, A., 1966, New data on the geological age of the Faeroes: *Nature*, v. 209, London.

Rasmussen, J. and Noe-Nygaard, A., 1969, Beskrivelse til Geologisk Kort over Faerøerne (1:50,000): *D.G.U.* I Row, v. 24, København.

Rasmussen, J. and Noe-Nygaard, A., 1970, Geology of the Faeroes: *D.G.U.* I Row, v. 25, Copenhagen.

Saxov, S. and Abrahamsen, N., 1964, A note on some gravity and density measurings in the Faeroe Islands: *Boll. di geofis. teor. ed appl.*, v. 4.

Saxov, S. and Abrahamsen, N., 1966, Some geophysical investigations in the Faeroe Islands: *Zeitschr. f. Geophys.*, Würtzburg.

Tarling, D. H. and Gale, N. H., 1968, Isotopic dating and paleomagnetic polarity in the Faeroe Islands: *Nature*, v. 218, London.

Walker, Frederick and Davidson, Charl. F., 1936, A contribution to the geology of the Faeroes: *Trans. of the Royal Soc. of Edinb.*, v. LVIII, Edinburgh.

British Isles

Andersen, F. W., Dunham, K. C., 1966, The geology of Northern Skye: Memoirs Geol. Surv. Scotland.

Bailey, E. B. and McCallien, W. J., 1956, Composite minor intrusions and the Slieve Gullion Complex: *Lpool. Manchr. Geol. Journ.*, v. 1.

Beckinsale, R. D., Brooks, C. K., and Rex, D. C., 1970, K–Ar ages for the Tertiary of East Greenland: *Bull. Geol. Soc. Denmark*, v. 20.

Bott, M. H. P. and Watts, A. B., 1971, Deep structure of the continental margin adjacent to the British Isles: Inst. Geol. Sci. Rept. No. 70/14.

Bowen, N. L., 1928, The evolution of the igneous rocks: Princeton and Oxford.

Brooks, C. K., 1972, The Tertiary of Greenland: A volcanic and plutonic record of continental break-up: Memoir, Am. Ass. Petrol. Geologists.

Brown, G. M., 1956, The layered ultrabasic rocks of Rhum, Inner Hebrides: *Phil. Trans. Roy. Soc.*, B, v. 240.

Brown, P. E., 1956, The Mourne Mountains granite—A further study: *Geol. Mag.*, v. 93.

Brown, P. E. and Rushton, B. J., 1960, Some chemical data on the Mourne Mountain Granite, G 2: *Geochim. Cosmochim. Acta*, v. 18.

Cameron, I. B. and Sabine, P. A., 1969, The Tertiary welded-tuff vent agglomerate and associated rocks of Sandy Braes, Co. Antrim: Inst. Geol. Sciences Rept. No. 69/6.

Charlesworth, J. K., 1963, *Historical Geology of Ireland*: Oliver and Boyd.

Cole, G. A. J. and Crook, T., 1910, On rock specimens dredged from the floor of the Atlantic, etc.: Mem. Geol. Surv. Ireland.

Dodson, M. H. and Long, L. E., 1962, Age of Lundy granite, Bristol Channel: *Nature*, v. 195.

Dollar, A. T. J., 1941, The Lundy Complex: its petrology and tectonics: *Q. J. Geol. Soc. Lond.*, v. 97.

Drever, H. I. and Johnston, R., 1958, The petrology of picritic rocks in minor intrusions—A Hebridean group: *Trans. Roy. Soc. Edinb.*, v. 63.

Elliot, G. F., 1971, Eocene volcanics in southeast England: *Nature*, v. 230, n. 9.

Emeleus, C. H., 1955, The granites of the western Mourne Mountains, County Down: *Sci. Proc. Roy. Dublin Soc.*, v. 27(NS).

Emeleus, C. H., 1962, The porphyritic felsite of the Tertiary ring complex of Slieve Gullion, Co. Armagh: *Proc. Roy. Irish Acad.*, v. 62B.

Emeleus, C. H. and Preston, J., 1969, Field excursion guide to the Tertiary volcanic rocks of Ireland: Int. Assoc. Volcanol. Chem. of Earth's Interior (Oxford).

Gibb, F. G. F., 1968, Flow differentiation in the xenolithic ultrabasic dykes of the Cuillins and the Strathaird peninsula, Isle of Skye, Scotland: *Journ. Petrol.*, v. 9.

Hallam, A., 1965, Jurassic, Cretaceous and Tertiary Sediments, in: *The Geology of Scotland* F. Y. Craig, ed.: Oliver and Boyd.

Harding, R. R., 1966, The Mulloch Sgar Complex, St. Kilda, Outer Hebrides: *Scottish. J. Geol.*, v. 2.

Holmes, A. and Harwood, H. F., 1929, The tholeiite dikes of the north of England: *Min. Mag.*, v. XXII.

LeBas, M. J., 1960, The petrology of the layered basic rocks of the Carlingford Complex. Co. Louth: *Trans. Royal Soc. Edinb.*, v. 64.

Mc Quillan, R. and Tuson, J., 1965, An interpretation from gravity measurements of the sizes of some British Tertiary granites: *Proc. Geol. Soc. London, 1621.*

Meighan, I. G. and Preston, J., 1971, Tertiary volcanism in Ireland: U. K. Contr. Upper Mantle Project. Final Rept. (Royal Society).

Miller, J. A. and Fitch, F. J., 1962, Age of the Lundy Island granites: *Nature*, v. 195.

Moorbath, S. and Welke, H., 1969a, Lead isotope studies on igneous rocks from the Isle of Skye, Northwest Scotland: *Earth Planet Sci. Letters*, v. 5.

Moorbath, S. and Welke, H., 1969b, Isotopic evidence for the continental affinity of the Rockall Bank, North Atlantic: *Earth Planet Sci. Letters*, v. 5.

Petterson, E. M., 1952, A petrochemical study of the Tertiary lavas of northeast Ireland: *Geochim. Cosmochim. Acta*, v. 2.

Patterson, E. M. and Swaine, D. J., 1955, A petrochemical study of Tertiary tholeiitic basalts: The middle lavas of the Antrim plateau: *Geochim. Cosmochim. Acta*, v. 8.

Reynolds, D. L., 1950, The transformation of Caledonian granodiorite to Tertiary granophyre on Slieve Gullion, Co. Armagh: Intern. Geol. Congr. Proc. Part III, Sect. B. London.

Reynolds, D. L., 1951, The geology of Slieve Gullion, Foughill and Carrickcarnan: *Trans. Royal Soc. Edinb.*, v. 62.

Ridley, I., 1971, The petrology of some volcanic rocks from the British Tertiary Province: the Islands of Rhum, Eigg, Canna and Muck: *Contr. Mineral Petrol.*, v. 32.

Ritchey, J. E., *et al.*, 1930, The geology of Ardnamurchan, northwest Mull and Coll: Mem. Geol. Surv. Scotland.

Sabine, P. A., 1960, The geology of Rockall, North Atlantic: *Bull. Geol. Surv. Gt. Britain*, v. 16.

Stewart, F. H., 1965, Tertiary igneous activity, in: *The Geology of Scotland*, G. Y. Craig, ed.: Oliver and Boyd.

Simkin, T., 1967, Flow differentiation in picritic sills of north Skye, in: *Ultramafic and Related Rocks*, Wyllie, P. J., ed.: New York and London.

Thompson, R. N., 1969, Tertiary granites and associated rocks of the Marsco area, Isle of Skye: *Quart. J. Geol. Soc. Lond.*, v. 124.

Tilley, C. E., 1951, The zoned contact skarns of the Broadford area, Skye: a study of boron-fluorine metasomatism in dolomites: *Min. Mag.*, v. 29.

Turner, F. J. and Verhoogen, J., 1960, Igneous and metamorphic petrology: MacGraw-Hill.

Tyrrell, G. W., 1928, The geology of Arran: Mem. Geol. Surv.

Vine, F. J., 1966, Spreading of the ocean floor: new evidence: *Science*, v. 154.

Wadsworth, W. J., 1961, The layered ultrabasic rocks of south-west Rhum, Inner Hebrides: *Phil. Trans. Roy. Soc. Lond.*, v. 244B.

Wager, L. R. and Bailey, E. B., 1953, Basic magma chilled against acid magma: *Nature*, v. 172.

Wager, L. R. and Brown, G. M., 1968, *Layered Igneous Rocks*: Edinburgh.

Wager, L. R., Weedon, D. S., Vincent, E. A., 1953, A granophyre from Coire Uaigneich, Isle of Skye, containing quartz paramorphs after tridymite: *Min. Mag.*, v. 30.

Walker, G. P. L., 1951, The amygdale minerals in the Tertiary lavas of Ireland. I: *Min. Mag.*, v. 29.

Walker, G. P. L., 1959, The amygdale minerals in the Tertiary lavas of Ireland. II. The distribution of prehnite: *Min. Mag.*, v. 32.

Walker, G. P. L., 1960a, The amygdale minerals in the Tertiary lavas of Ireland. III. Regional distribution: *Min. Mag.*, v. 32.

Walker, C. P. L., 1960b, The amygdale minerals in the Tertiary lavas of Ireland. IV. The crystal habit of calcite: *Min. Mag.*, v. 32.

Walker, G. P. L. and Leedal, G. P., 1954, The Barnesmore Complex, County Donegal: *Scient. Proc. R. Dublin Soc.*, v. 26 (NS).

Iceland

Anderson, F. W., 1949, Geological observations in south-eastern and central Iceland: *Trans. R. Soc. Edinb.*, v. 61.

Annells, R. N., 1968, A geological investigation of a Tertiary intrusive centre in the Vididalur-Vatnsdalur area, N-Iceland: Ph. D. Thesis, Univ. of St. Andrews.

Barth, T. F. W., 1950, Volcanic geology, hot springs and geysers of Iceland: Publ. Carn. Inst. 587.

Båth, M., 1960, Crustal structure of Iceland: *J. Geophys. Res.*, v. 65.

Bemmelen, R. W. van and Rutten, M. G., 1955, Table Mountains of northern Iceland: Leiden.

Bödvarsson, G., 1960, Hot springs and the exploitation of natural heat resources, in: *On the Geology and Geophysics of Iceland*: Reykjavik.

Bödvarsson, F. and Walker, G. P. L., 1964, Crustal drift in Iceland: *Geophys. Journ.*, v. 8.

Cann, G. R., 1971, Major element variations in ocean floor basalts: *Phil. Trans. Roy. Soc. London*, v. 268.

Cargill, H. K., Hawkes, L., and Ledeboer, J. A., 1928, The major intrusion of south-eastern Iceland: *Quart. Journ. Geol. Soc.*, v. 84.

Carmichael, I. S. E., 1964, The petrology of Thingmuli, a Tertiary volcano in Eastern Iceland: *Journ. Petrol.*, v. 5.

Carmichael, J. S. E., 1967, The mineralogy of Thingmuli, a Tertiary volcano in eastern Iceland: *Amer. Min.*, v. 52.

Dagley, P., Wilson, R. L., Ade-Hall, J. M., Walker, G. P. L., Haggerty, S. E., Sigurgeirsson, T., Watkins, N. D., Smith, P. J., Edwards, J., and Grasty, R. L., 1967, Geomagnetic polarity zones for Icelandic lavas: *Nature*, v. 216.

Dearnley, R., 1954, A contribution to the geology of Lodmundarfjördur: *Acta Nat. Isl.*, v. 1, n. 9.

Einarsson, Tr., 1942, Über das Wesen der heissen Quellen Islands: *Soc. Sci. Isl. Rit.*, v. 26.

Einarsson, Th., 1971, *Jardfraedi* (Geology): Heimskringla, Reykjavik, 254 p.

Friedrich, W., 1966, Zur Geologie von Brjánslaekur (Nordwest-Island) unter besonderer Berücksichtigung der Fossilen Flora (Dissert.), Köln.

Gale, N. H., Moorbath, S., Simons, G., and Walker, G. P. L., 1966, K–Ar ages of acid intrusive rocks from Iceland: *Earth & Plan. Sci. Letters*, v. 1.

Hald, N., Noe-Nygaard, A., and Ken Pedersen, A., 1971, The Króksfjördur central volcano in NW-Iceland: *Acta Nat. Isl.*, v. II, n. 10.

Hawkes, L., 1924, On an olivine-dacite in the Tertiary volcanic series of eastern Iceland: The Raudaskrida (Hamarsfjord): *Quart. J. Geol. Soc.*, v. 80.

Hawkes, L., 1938, The age of the rocks and topography of middle northern Iceland: *Geol. Mag.*, v. 75.

Hawkes, L. and Hawkes, H. K., 1933, The Sandfell laccolith and "dome of elevation": *Quart. J. Geol. Soc.*, v. 89.

Heer, O., 1868, Die Miocene Flora von Island: *Flora Fossilis Arctica*, v. 1 (III).

Heier, K. S., Chappell, B. W., Arrieus, D. A., and Morgan, J. W., 1966, The Geochemistry of four Icelandic basalts: *Norsk. Geol. Tidsskr*: Oslo.

Holmes, A., 1918, The basaltic Rocks of the Arctic Region: *Min. Mag.*, v. 18, London.

Jakobsson, S. P., 1972a, On the consolidation and palagonitization of the tephra of the Surtsey volcanic Island: Surtsey Res. Progr. Rep. VI.

Jakobsson, S. P., 1972b, Chemistry and distribution pattern of recent basaltic rocks in Iceland: *Lithos*.

Jónsson, J., 1954, Outline of the geology of the Hornafjördur region: *Geogr. Ann.*, v. 36.

Jux, U., 1960, Zur Geologie des Vopnafjördur-Gebietes in Nordost-Island: *Geologie*, v. 28.

Kjartansson, G., 1943, Árnesingasaga I: Reykjavik.

Kjartansson, G., 1960, The Móberg formation, in: *On the Geology and Geophysics of Iceland*: Reykjavik.

Kjartansson, G., 1967, Volcanic forms at the sea bottom: *Soc. Sci. Isl.*, v. 38.

Melson, W. G., Thompson, G., and Andel, T. H., 1967, Volcanism and metamorphism in the mid-Atlantic ridge, 22°N. latitude: *J. Geophys. Res.*, v. 73.

Moorbath, S., Sigurdsson, H., and Goodwin, R., 1968, K–Ar ages of the oldest exposed rocks in Iceland: *Earth & Plan. Sci. Letters*, v. 4.

Nielsen, N. and Noe-Nygaard, A., 1936, Om den islandske "Palagonitformations" Oprindelse: *Geogr. Tidsskr.*, v. 39.

Noe-Nygaard, A., 1940, Sub-glacial volcanic activity in ancient and recent times: *Fol. Geogr. Dan.*, Tom. 1, n. 2.

Noe-Nygaard, A., 1952, A Group of Liparite Occurrences in Vatnajökull, Iceland: *Fol. Geogr. Dan.*, v. I, n. 3.

Noe-Nygaard, A., 1966, Chemical composition of tholeiitic basalts from the Wyville-Thompson ridge belt: *Nature*, v. 212.

Olafsson, E. and Pálsson, B., 1772, Rejse gjennem Island: Sorø, 2 vols.

Pálmason, G., 1963, Seismic refraction investigation of the basalt lavas in northern and eastern Iceland: *Jökull*.

Pálmason, G., 1971, Crustal structure of Iceland from explosion seismology: *Soc. Sci. Isl.*, v. 40.

Peacock, M. A., 1926, The petrology of Iceland. The basic tuffs: *Trans. Roy. Soc. Edinb.*, v. 55.

Pjeturss(on), H., 1900, The glacial palagoniteformation of Iceland: *Scott. Geogr. Mag.*, v. 1900.

Pjeturss(on), H., 1905, Om Islands geologi: *Medd. D. Geol. For.*, v. 11, København.

Pjeturss(on), H., 1910, Island. Handb. d. Reg. Geologie.

Saemundsson, K., 1967, Vulkanismus und Tektonik des Hengill-Gebietes in südwest-Island: *Acta Nat. Isl.*, v. II, n. 7.

Sigurdsson, H., 1966, Geology of the Setberg area, W. Iceland: *Soc. Sci. Isl.*, Greinar IV.

Sigurdsson, H., 1970, Petrology of the Setberg volcanic region and acid rocks of Iceland: Ph. D. Thesis, Univ. of Durham.

Sigvaldason, G. E., 1966, Chemistry of thermal waters and gases in Iceland: *Bull. Volc.*, v. 29.

Sigvaldason, G. E., 1969, Chemistry of Basalts from the Icelandic rift zone: *Contr. Min. and Petrol.*, v. 20.

Sonder, R., 1941, Studien über heisse Quellen in Island: Zürich.

Thorarinsson, S., 1944, Tefrokronologiska Studier på Island: *Munksgaard*, København (et Geogr. Annaler, Stockholm).

Thorarinsson, S., 1952, Hverfjall: *Natt. Fr.*, v. 22, p. 113–139.

Thorarinsson, S., 1953, The crater groups in Iceland: *Bull. Volc.*, v. 16.

Thorarinsson, S., 1958, The Öraefajökull eruption 1362: *Acta Nat. Isl.*, v. II, n. 2.

Thorarinsson, S., 1960, The postglacial volcanism, in: *On the Geology and Geophysics of Iceland*: Reykjavik.

Thorarinsson, S., 1964, Surtsey. The new island in the North Atlantic: Reykjavik.

Thorarinsson, S., 1967a, The eruption of Hekla 1947–48. I: *Soc. Sci. Isl.*

Thorarinsson, S., 1967b, Some problems of volcanism in Iceland: *Geol. Rdsch.*, v. 57.

Thorarinsson, S., 1969, The Lakagigar eruption of 1783: *Bull. Volc.*, v. 23.

Thorarinsson, S. and Sigvaldason, G. E., 1962, The eruption in Askja 1961: *Am. J. Sci.*, v. 260.

Thorkelson, T., 1910, The hot springs of Iceland: D. kgl. danske Vidensk. Selsk. Skr. 7. R. Nat. og natv. Afdl. 8, 4.

Thorkelsson, Th., 1941, On thermal activity in Iceland: Reykjavik.

Thoroddsen, Th., 1899, Explorations in Iceland 1881–1898: *The Geogr. Journ.*, v. 13, London.

Thoroddsen, Th., 1906, Island: Grundriss der Geographie und Geologie. Peterm. Mitteil. Ergänz. Heft Nr. 152.

Thoroddsen, Th., 1925, Die Geschichte der Isländischen Vulkane: D. kgl. danske Vidensk. Selsk. Skr. Afdl. 8, R. IX.

Tryggvason, T., 1943, Das Skjaldbreidur-Gebiet auf Island: *Bull. Geol. Inst. Uppsala*, v. 30.

Tryggvason T. and White, D. E., 1955, Rhyolitic tuffs in lower Tertiary plateau basalts of eastern Iceland: *Am. Journ. Sci.*, v. 253.

von Nidda, Krug, 1837, Geognost. Darstellung der Insel Island: Karstens Archiv.

von Waltershausen, Sartorius, 1847, Physisch-geographische Skizzen von Island: Göttingen.

Walker, G. P. L., 1959, Geology of the Reydarfjördur area, Eastern Iceland: *Quart. Journ. Geol. Soc.*, v. 114.

Walker, G. P. L., 1960, Zeolite zones and dike distribution in relation to the structure of basalts of eastern Iceland: *Journ. Geol.*, v. 68.

Walker, G. P. L., 1963, The Breiddalur central volcano, eastern Iceland: *Quart. Journ. Geol. Soc.*, v. 119.

Walker, G. P. L., 1966, Acid volcanic rocks in Iceland: *Bull. Volc.*, v. 29.

Ward, P. L., Pálmason, G., and Drake, C., 1969, Microearthquake survey and the mid-Atlantic ridge in Iceland: *J. Geophys. Res.*, v. 74.

Jan Mayen

Boldva, A. B. von, 1886, Oesterreichische Polarexpedition. Jan Mayen 1882–83: Beobachtungsergebnisse, Wien.

Carstens, H., 1961, Cristobalite-trachytes of Jan Mayen: *Norsk Polarinst.*, Skrifter 121, Oslo.

Carstens, H., 1962, Lavas of the southern part of Jan Mayen. Årbok 1961: *Norsk Polarinst.*, Oslo.

Carstens, H., 1963: Leucite- and sodalite-bearing trachybasalts of Jan Mayen. Årbok 1962: *Norsk Polarinst.*, Oslo.

Dollar, A. T. J., 1966, Genetic aspects of the Jan Mayen Fissure Volcano Group on the mid-Oceanic submarine Mohns ridge, Norwegian Sea: *Bull. Volc.*, v. XXIX, Napoli.

Fitch, F. J., 1961, The University of London 1961 Beerenberg Expedition: *Nature*, v. 194, London.

Fitch, F. J., 1964, The Development of Beerenberg Volcano, Jan Mayen: *Proc. Geol. Ass. London.*

Fitch, F. J., 1965, (Discussion): *Proc. Geol. Soc. London*, v. 1622, p. 73.

Fotherby, R., 1625, Purchas His Pilgrimes etc.: London (New Edition. Glasgow 1905–07).

Gjelsvik, F., 1970, Volcano on Jan Mayen alive again: *Nature*, v. 228, London.

Orvin, A. K., 1960, The place-names of Jan Mayen: *Norsk Polarinst.*, Skr. 120, Oslo.

Roberts, B. and Hawkins, T. R. W., 1965, The geology of the area around Nordkapp, Jan Mayen. Årbok 1963: *Norsk Polarinst.*, Oslo.

Roberts, B. and Hawkins, T. R. W., 1972, The Petrology of the volcanic and intrusive rocks of Nord-Jan, Jan Mayen. Årbok 1970: *Norsk Polarinst.*, Oslo.

Scoresby, W. Jr., 1820, *Account of the Artic Regions*: Edinburgh.

Siggerud, T., 1972, The volcanic eruption of Jan Mayen 1970. Årbok 1970: *Norsk Polarinst.*, Oslo.

Tyrrell, G. W., 1926, The Petrography of Jan Mayen: *Trans. Roy. Soc. Edinburgh*, v. LIV, n. III, Edinburgh.

Weigand, P. W., 1972, Bulk-rock and mineral chemistry of recent Jan Mayen basalts. Årbok 1970: *Norsk Polarinst.*, Oslo.

Wordie, J. M., 1926, The Geology of Jan Mayen: *Trans. Roy. Soc. Edinburgh*, v. LIV, n. III, Edinburgh.

Chapter 12

THE OCEANIC ISLANDS: AZORES

W. I. Ridley*
Lunar Science Institute
Houston, Texas

N. D. Watkins
Graduate School of Oceanography
University of Rhode Island
Kingston, Rhode Island

and

D. J. MacFarlane
Shell-BP
Lagos, Nigeria

I. INTRODUCTION

An island or archipelago is a geologically anomalous feature of any ocean basin. An understanding of the origin and features of any island or archipelago is, therefore, ideally obtained by the process of recognizing the associated anomalous features, formulating models to explain the features, and testing the model so produced, preferably by use of several independent approaches.

* Present address: Lamont–Doherty Geological Observatory of Columbia University, Palisades, N. Y. 10964.

This discussion of the Azores will consist of presentation of the known geological, geophysical, and geochemical data, some of which are previously unpublished. To place these data in the proper context, we also must present similar data for the adjacent sea floor, which we arbitrarily confine to the area roughly between lat. 30–45° N and long. 20–50° W. Emphasis will be given to what we consider to be those unusual features suspected to be critically relevant to an understanding of the origin of the islands.

II. REGIONAL SETTING

The Azores (Fig. 1) consists of nine islands trending linearly but obliquely across the mid-Atlantic ridge between lat. 37° and 40° N at the western terminus of the world's major east–west tectonic element, which is called the Alpide Tectonic Zone. This major feature extends from Indonesia through southeast Asia to the Mediterranean, and finally from Gibraltar to the Azores in the form of the East Azores fracture zone (EAFZ), which, locally, possesses up to 1000 m of relief (Laughton *et al.*, 1972). It is suspected that the EAFZ may extend further westward (Fig. 2), after slight northward offset, as the linear but broad positive element between the Azores and the southern edge of the Sohm Abyssal Plain (Tolstoy, 1951; Krause, 1965). The eastern fracture is defined seismically (Menard, 1965; Barazangi and Dorman, 1969), but the western feature is inactive at this time, although Drake and Woodward (1963) infer ancient activity by suggesting that major offset in New England structural elements, as well as the genesis of the New England seamounts, may be related to movement along the structural element trending west from the Azores.

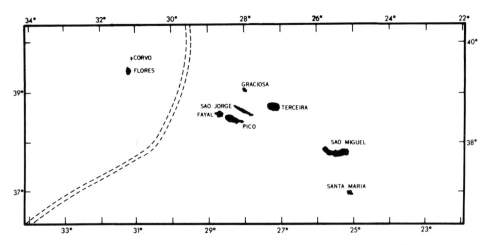

Fig. 1. Map of the Azores Islands. Approximate position of mid-Atlantic rift is shown. Location of map is included in Fig. 2. (From Krause and Watkins, 1970.)

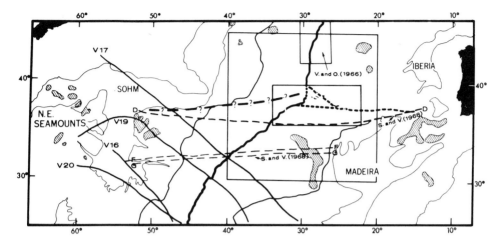

Fig. 2. Map showing location of the Azores in north-central Atlantic, and major tectonic elements, abyssal plains (Sohm, Madeira, Iberia), seamounts (shaded), survey areas, and some of the track lines that are referred to in the text. Large survey block is area covered by bathymetric map (Fig. 3); small survey block to the north is that area surveyed by Vogt and Ostenso (1966); central survey block is that area showing detailed island locations in Fig. 1. Tracks identified as S and V indicate U.S. Naval Oceanographic tracks (Schneider and Vogt, 1968). (From Krause and Watkins, 1970.)

The EAFZ (which is a unique feature in the North Atlantic) and the mid-Atlantic ridge (Fig. 2) form boundaries of major tectonic units, according to Morgan (1968), and therefore constitute a triple junction (McKenzie and Morgan, 1969) in the vicinity of the Azores, separating the eastern North Atlantic into two separate crustal plates. Whether or not the western North Atlantic can be legitimately assumed to be a complete single plate will be considered later.

The mid-Atlantic ridge extends through the island chain (Fig. 1) and changes trend markedly from northeast–southwest to north–south at lat. 38.5° N (Figs. 1 and 2). The mid-Atlantic ridge loses much of its bathymetric definition in the Azores platform, which is a regional broadening of the ridge eastward (Fig. 3) 5.8 million sq. km in area (Kaula, 1970). Krause (1966) has discussed the relationship of the ridge to the platform. Machado (1959) has described the Terceira rift, which extends at least from Graciosa to Terceira and Sao Miguel, oblique to the trend of the EAFZ. Krause and Watkins (1970) have provided additional bathymetric definition of the trough or rift (Fig. 4). This relatively short tectonic element (Terceira rift) probably forms a critical segment of the triple-point system. Some topographic features on Faial reflect the tensional nature of the rift, in addition to the volcanic activity essential to creation of the islands.

Major fractures are found across the mid-Atlantic ridge immediately north of the platform at lat. 43° N (Phillips *et al.*, 1969) and between lat. 43° and 44° N (Keen, 1970). The Atlantis fracture zone traverses the mid-Atlantic ridge at lat. 30° N. This brief description of the regional setting of the Azores clearly indicates that many unusual features (as well as the islands) exist in this part of the central Atlantic, for example, the seismic EAFZ and its aseismic westward continuation (after northward offset as the West Azores feature), the major change in trend in the mid-Atlantic ridge and its broadening to the east, and the Terceira rift. Any model of the genesis of the Azores should incorporate these regional features. Additional constraints will emerge during the following discussion of finer details of the geology, geochemistry, and geophysics of the region.

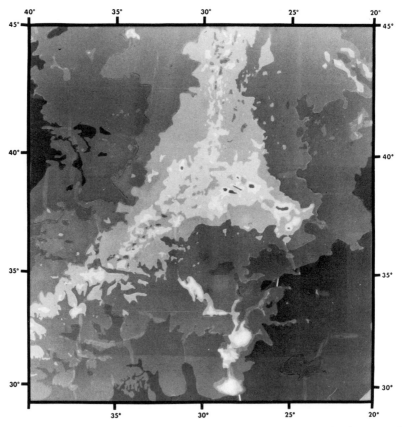

Fig. 3. Photograph of a bathymetric contour map of the north-central Atlantic survey block indicated in Fig. 2. Contour interval is 500 fathoms. The major change of grey tone used is at the 1500 fathoms level. This map is constructed from six U.S. Naval Oceanographic Office charts (Numbers BC 306, 307, 308, 406, BC 0407N, 0408N). (From Krause and Watkins, 1970.)

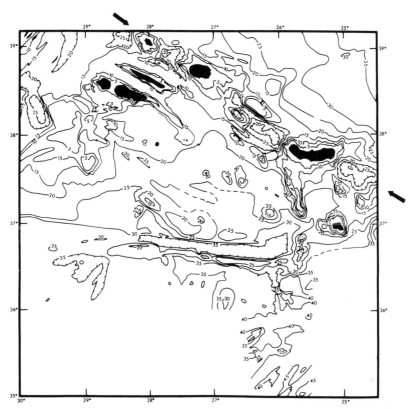

Fig. 4. Bathymetric map resulting from *R. V. Trident* survey. Contours labeled in hundreds of meters. Islands in black. The line of the Terceira rift indicated by arrows. (From Krause and Watkins, 1970.)

III. GEOLOGY

A. Surface Geology

The Azores Islands are composed principally of volcanic rocks, predominantly basaltic lavas. Many of the islands also display caldera structures that appear to be associated with extensive trachytic pumice deposits. Trachytic lavas, as well as numerous adventive cones (commonly arranged in linear trends that reflect underlying fissures), are found on some islands. Detailed geological maps of individual islands are unavailable, those presented here being a compilation of surface volcanic features by Hadwen and Walker. The more recent geological descriptions concentrate mainly on the voluminous, superficial trachytic ash deposits (Self, 1971; Walker and Croasdale, 1971).

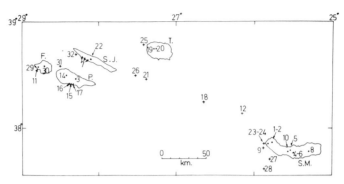

Fig. 5. Map and table showing dates and locations of recorded volcanic eruptions in the central Azores area, from Weston (1964). •, eruptions in the island; +, sea eruptions. Dates of the eruptions are as follows: 1439(1), 1460(2), 1562(3), 1563(4, 5), 1564(6), 1580(7), 1630(8), 1638(9), 1652(10), 1672(11), 1682(12), 1713(13), 1718(14, 15, 16), 1720(17, 18), 1761(19, 20), 1800(21), 1808(22), 1811(23), 1861(24), 1867(25), 1902(26), 1907(27), 1911(28), 1957(29), 1958(30), 1963(31), 1964(32).

Historic eruptions have occurred on Sao Miguel, Terceira, Sao Jorge, Pico, and Faial (Zbyszewski, 1963; Machado, 1962, 1965; Fig. 5). Hot springs and fumaroles exist on all the islands except Corvo and Santa Maria (Fig. 6).

Two periods of volcanic activity are represented in the rocks of *Santa Maria*. The oldest volcanics are thin, ankaramitic, subaerial lavas and tuffs, dated by Abdel Monem *et al.* (1968) as 8.12–6.08 m.y., and dissected by a swarm of ankaramitic dikes. The younger series, part of which has been dated at 4.3 m.y. (Abdel Monem *et al.*, 1968), forms a peripheral belt 150–200 m thick composed of uplifted sediments and submarine volcanics. The latter include boulder beds, palagonitic breccias, and basaltic pillow lavas interbedded with calcareous mud, tuffaceous sandstone, and fossiliferous (Coquina) limestone. Using the fossiliferous limestone, Zbyszewski *et al.* (1962) have dated the volcanic activity as Miocene to Quaternary.

Sao Miguel consists essentially of four volcanoes. Three major volcanoes, aligned approximately east–west, form the uplands in the eastern half of the island. Each volcano has a summit caldera, and the western volcano is associated with a southeast-trending fissure zone. Historic submarine eruptions indicate extension of this zone to the northwest. Abdel Monem *et al.* (1968) give ages from 4.01 m.y. for the oldest exposed lavas to 0.95 m.y. for the youngest flows of the andesite series. The subaerial volcanism of Sao Miguel was, thus, roughly coeval with part of the post-Coquina volcanism on Santa Maria. Two dates of 4.65 and 4.0 m.y. for basalts from the Formigas Islets suggests that these basalts are also the stratigraphical equivalent of the post-Coquina basalts.

Two large calderas in the east and central regions and a small caldera associated with a large volcano in the northwest are the main surface features on *Terceira*. All the calderas are aligned in a west-northwest direction, whereas in the northeast a graben strikes in a northwest direction. Trachytic pumice and lava are widespread. The oldest rocks in the east are highly altered ankaramites, succeeded by relatively young feldspar basalts, trachytes, and olivine basalts. An eruption of basaltic lava occurred in 1761. Hot springs exist in the center of the island.

Sao Jorge can be broadly divided into two parts separated by a north–south-trending fault zone (Machado and Forjaz, 1965). The eastern part of the island is probably the older and comprises basaltic lavas overlain by interbedded, highly altered basalts and pyroclastics. In the central region,

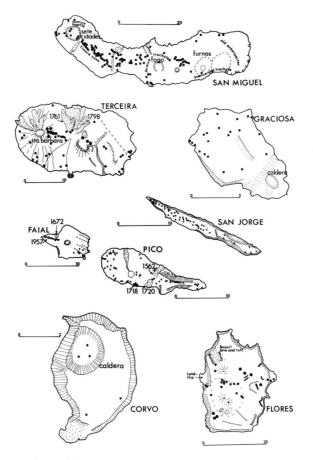

Fig. 6. Surface geology of Azores Islands, after Hadwen and Walker. Filled circles are scoria cones; filled polygons and irregular areas are trachytic plugs and dikes, respectively. Dotted lavas are trachytic; hachured lavas are basaltic.

young basalts, extruded from a narrow active zone along the island spine, have built up coastal platforms. All lavas west of the north–south fault zone have a southwesterly dip, as though they originated from vents north of the island. This finding is compatible with Hadwen and Walker's suggestion that the north coast is a fault scarp. The dips of the eastern rocks suggest an independent history. Weston (1964) lists eruptions in 1580, 1808, and 1964, and several severe earthquakes have been recorded. Machado and Forjaz (1965) have analyzed the 1964 earthquake and recognized epicenters at 4- and 7-km depths in the central region and 10- and 20-km depths near the south coast.

Two separate trend lines are visible on *Pico*—an east–west trend in the east, and a northwest trend in the west. The eastern part is long and narrow, built along an east–west fissure zone marked by abundant adventive cones, and largely covered by basaltic lavas. The only older rocks exposed are ankaramitic lavas. The western region is dominated by the huge volcano Pico de Pico that exposes young basaltic lavas and pyroclastics. The adventive cones and fissures are aligned northwest in continuation of similar features on the island of Faial.

Eruptions have been recorded in 1562, 1718, 1720, and 1963 (Weston, 1964). The western platform is subject to severe earthquakes with epicenters west and north of Pico. Machado (1954) and Machado and Forjaz (1965) suggest the presence of two interconnected magma chambers under Pico at about a 5-km depth, one under the northwest flank and a second near the center of the island.

The island of *Faial* is a symmetrical volcano with a 2-km-diameter summit caldera, extended slightly westward by a linear arrangement of overlapping adventive cones. The eastern flank of the volcano is characterized by a series of northwest-trending fault scarps, traces of which continue on Pico (Zbyszewski and da Veiga Ferreira, 1962). The rocks are largely feldspathic basalts and pyroclastics and trachytic plugs and dikes; trachytic pumice mantles the summit caldera. Weston (1964) tabulates eruptions in 1672, 1957, and 1958. Historic eruptions, decreasing in age westward, are restricted to the west portion of the island. The 1957 and 1958 eruptions of Capalinhos increased the island's length by 1 km. Slight seismic tremors have epicenters west of Faial or between Faial and Pico (Machado and Forjaz, 1965).

On *Flores*, the highest point is northwest of the center and comprises a pyroclastic cone with young basalts on the northwest slopes. To the south is a large shallow basin with a 0.5-km diameter, containing seven deep, water-filled craters and two 150-m-high cones. High cliffs surround the island; some of these cliffs may be the result of collapse. Rock types range from olivine basalt lavas and pyroclastics to trachytic plugs, lavas, and pyroclastics.

B. Petrochemistry

Analyses of Azores volcanics are reported by Esenwein (1929), Berthois (1953), Jeremine (1957), Assunçao (1961 and unpubl. data), Schmincke and Weibel (1972), and Schmincke (1973). The volcanics are all strongly alkaline (Fig. 7), and all fresh basalts are nepheline normative (Fig. 8). Ankaramites, alkali olivine basalts, and trachytes form the most abundant analyzed groups, although Schmincke and Weibel (1972) and Schmincke (1973) report many analyses of comenditic trachytes from Sao Miguel and comendites, pantellerites from Terceira. Intermediate volcanics have also been analyzed (Fig. 9), although to a lesser extent than the more basic and silicic rocks. Such a bimodal analytical distribution should be interpreted cautiously in the absence of statistically significant numbers of analyses (Cann, 1968).

Intermediate volcanics are also nepheline normative, but more evolved trachytes and rhyolites may be quartz or nepheline normative (Schmincke and Weibel, 1972). Girod and Lefeure (1972) note that the intermediate lavas from the Azores, previously referred to as andesites and latites, are clearly much more alkalic and undersaturated than calc-alkaline intermediate rocks, and are actually hawaiites and mugearites.

The presence of oversaturated, peralkaline rhyolites on some islands poses a problem since these evolved rocks are distinct from the strongly undersaturated phonolitic trachytes and phonolites that appear on many oceanic

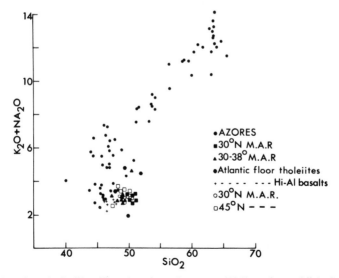

Fig. 7. Plot of total alkali's: silica, data from Assunçao (1961, and unpublished) for Azores volcanics. Other data from Muir and Tilley (1964, 1966), Nicholls (1965), Miyashiro et al. (1969), Kay et al. (1970). Note the Azores volcanics are consistently more alkaline than the surrounding volcanics from the sea floor.

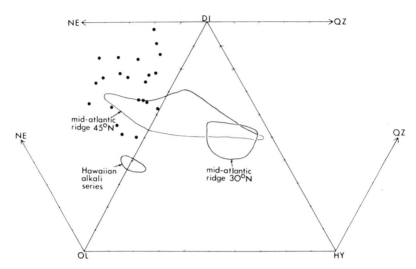

Fig. 8. Plot of the Azores basaltic volcanics in *part* of the basalt tetrahedron, after Yoder and Tilley (1962). The plagioclase component has been omitted for clarity. All the Azores basalts lie within the nepheline (NE)–olivine(OL)–diopside(DI) phase volume, reflecting their alkalic, undersaturated chemistry. Mid-Atlantic ridge basalts and Hawaiian alkalic basalts (after Kay *et al.*, 1970) largely lie in the transitional field olivine–diopside–hypersthene (HY), but none are truly tholeiitic, i.e., lie within the diopside–hypersthene–quartz(QZ) field.

islands. These latter rock types can be shown to evolve from an alkali olivine basalt parent by crystal fractionation involving principally calcic clinopyroxene, olivine, titanomagnetite, and calcic plagioclase. Bass (1971) points out that pantelleritic rocks on oceanic islands are logically related to a transitional basalt parent, and Schmincke (1973) suggests that peralkaline rhyolites on

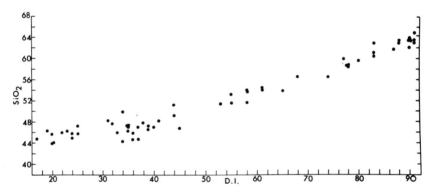

Fig. 9. Plot of silica values of Azores volcanics against the Thornton–Tuttle Differentiation Index (normative Qz + Or + Ab + Ne). The linearity results largely from the chosen coordinates, but provides a useful method of classifying a highly differentiated suite of volcanics. In the Azores, alkali olivine basalts (D.I. < 40) can be distinguished from trachybasalts and trachyandesites (40 < D.I. < 75) and trachytes (D.I. > 75).

Terceira may have a transitional basalt heritage, although the location of such basalts remains unspecified.

Schmincke and Weibel (1972) and Schmincke (1973) use the K_2O/Na_2O ratios for Azores rocks for comparisons with other Atlantic islands. Although the division of Atlantic island volcanics based on this ratio had been recognized earlier (Baker, 1969) and discarded in favor of a simpler approach, it does nonetheless underline the potassic nature of the Azores volcanics. In this respect, the Azores rocks, from basalts through hawaiites to strongly silicic differentiates, are most similar to volcanics from Gough, Tristan, and Jan Mayen but distinct from the Canaries, Cape Verde, St. Helena, Madiera, and Porto Santo. Girod and Lefeure (1972) emphasize the high K_2O/Na_2O ratios of hawaiites and mugearites as reflected in the presence of andesine phenocrysts and a significant anorthoclase component in the groundmass of these intermediate types. It is clear, however, that significant inter-island variations in K_2O/Na_2O are also present.

Although few analyses are available, it appears that there are systematic inter-island chemical variations. Terceira appears to have less alkalic, more silicic basalts than Sao Miguel, and this feature persists into the silicic differentiates (Schmincke and Weibel, 1972). Assunçao and Canilho (1970) also distinguish two alkalic groups, Flores and Corvo west of the mid-Atlantic ridge and Sao Miguel and Santa Maria east of the mid-Atlantic ridge; Pico, Faial, Terceira, San Jorge, and Graciosa form a transitional group. However, more detailed inter-island analyses may grossly alter this classification.

Mineralogical data are lacking. Boone and Fernandez (1971) cite a Syracuse University thesis by Fernandez (1969) concerning the petrology of the Nordeste complex, Sao Miguel. A detailed study of olivine from olivine nodules, basalts, ankaramites, and olivine-rich trachybasalts from Sao Miguel and Santa Maria has been made by Boone and Fernandez (1971). They note the overlap in olivine compositions between nodules, ankaramites, and alkali olivine basalts, but all are more iron-rich than typical olivine in Hawaiian lherzolites. A shallow-level origin for all these olivine-bearing samples is indicated by low Ni, high Ca, and distinct zonation. Fernandez (1973) suggests that the alkali basalt-trachyte series of the Nordeste Complex results from fractional crystallization of plagioclase $+$ titanomagnetite $\pm$ Ti–Al clinopyroxene $\pm$ olivine. The presence of mildly alkaline, sub-alkaline, and alkali basalts suggests varying degrees of partial melting within the mantle produced this spectrum of basaltic rocks.

Gast (1968a) and Oversby (1971) have measured the lead isotopic composition of several Azores volcanics. Oversby (1971) points out that the $^{206}Pb/^{204}Pb$ ratios for Capelinhos lavas (Faial) are considerably less than observed in basalts from other Azores Islands (Gast, 1968a) and suggests small-scale heterogeneities in the mantle beneath the Azores. Further analyses of Azores volcanics by Sun (pers. comm.) indicate a wider range of $^{206}Pb/^{204}Pb$ ratios than observed for Capelinhos alone, but relatively small variations in $^{207}Pb/^{204}Pb$. Data for Tristan (Oversby, 1971) and Gough (Gast, 1968a) suggest that these islands contain significantly less radiogenic lead than observed in the Azores.

Mildly alkaline volcanics have been reported from rift zones at lat. 45° N by Muir and Tilley (1964; Fig. 8), but generally the rift-zone and sea-floor basalts in this region are either olivine tholeiites or transitional mildly alkaline basalts (Muir and Tilley, 1966; Nicholls, 1965; Kay et al., 1970). The strongly alkaline character of the Azores volcanics, so close to the mid-Atlantic ridge, is thus unique and requires explanation in any evolutionary model of the Azores.

IV. GEOPHYSICS

A. Paleomagnetism

The only published paleomagnetic data are those due to Serughetti and Roche (1968) and Saucier and Roche (1964) from the volcanics of Flores, and Faial and Graciosa, respectively. The former are all of normal polarity, and are considered to be confined to the Brunhes Epoch ($t = 0$–0.7 m.y.). The collection from Faial and Graciosa are dominantly normal polarity, but two reversed bodies were sampled on Faial: these would be of pre-Brunhes age (older than 0.7 m.y.). Ellwood et al. (1973) have recently presented a paleomagnetic study of Brunhes Epoch geomagnetic secular variation using specimens from 33 separate lavas on Terceira. All lavas sampled on the island were Brunhes in age.

B. Marine Magnetic Surveys

Despite the very rapid extensions of Vine's (1966) confirmation of crustal spreading, in the form of publication of similar analyses of many marine magnetic profiles of mid-oceanic regions (see particularly those papers in *Journal of Geophysical Research*, v. 73, n. 6, p. 2069–2136, 1968), it is only recently that analysis of data from the central North Atlantic has been published (Pitman et al., 1971; Pitman and Talwani, 1972). Both these studies stress that the anomaly pattern near the Azores is poorly defined, but nonetheless provide data to show that spreading rates immediately north of the Azores have varied between 0.4 and 2.5 cm/yr, and further suggest that between $t = 9$ and 38 m.y., spreading was 0.4 cm/yr. A figure of 2.0 cm/yr is suggested for the period prior to $t = 38$ m.y., and they propose 1.3 cm/yr since $t = 9$ m.y. South of the Azores, they propose that the period from $t = 9$–38 m.y. was also marked by a slow spreading rate of 1.0 cm/yr, which is, however, much faster than for the same period to the north of the Azores. This difference in spreading rates north and south of the Azores occurs both east and west of the mid-Atlantic ridge, and has resulted in offset in the anomaly patterns, providing substantial proof of the tectonic nature of the westward extension of the EAFZ, which can therefore be reasonably called the West Azores fracture zone (WAFZ). Williams and McKenzie (1971) disagree with the Pitman *et al.* (1971) identification of anomaly 13 ($t = 38$ m.y.) in profiles of the northeast Atlantic, resulting in a preference for no marked slowing of spreading at that time. Why the zone is at present aseismic is not known. The present crustal plates, which are defined by the seismicity of only the last decade (Morgan, 1968), would not necessarily be in the same general configurations in the North Atlantic for analyses of crustal motion during the middle Tertiary and earlier. Pitman and Talwani (1972) believe that spreading north of the Azores has involved two plates only since $t = 47$ m.y.

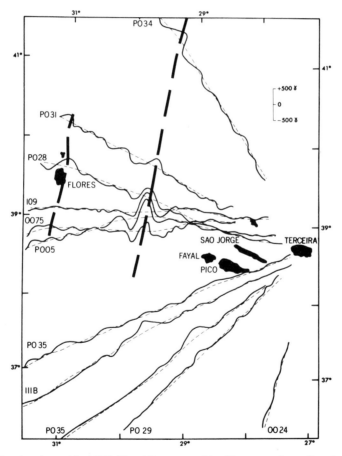

Fig. 10. Map showing residual U.S. Naval Oceanographic office magnetic profiles plotted along their track positions over the mid-Atlantic ridge in the vicinity of the Azores. On profiles black is positive; clear is negative. Note the subdued anomaly amplitude except over the ridge.

Krause and Watkins (1970) have shown that the available U.S. Naval
Oceanographic magnetic profiles between long. 21° and 35° W and lat. 35°
and 43° N also feature low-amplitude anomalies. Some of the details of the
anomalies close to the Azores are shown in Figs. 10 and 11. Only the central
anomaly, over the median rift of the mid-Atlantic ridge, is well defined in
Fig. 10. This is quite significant, however, in view of the recently proposed
model of Machado *et al.* (1972) for the Azores region. They envisage the
mid-Atlantic ridge to be locally offset to the east by a series of eight major
east–west transform faults, so that the islands (Fig. 1) occupy axial positions
over the offset ridge segments. One requirement of this model is that the
mid-Atlantic ridge is absent between Flores and Graciosa and Faial (Fig. 1).
As Fig. 10 clearly shows, the ridge axis is present. There is no evidence to support the preferred ridge configuration of Machado *et al.* (1972), whose model
is therefore not tenable. It follows that the recent tectonic analyses of Burke
et al. (1973), which employs a "ridge-jump" concept largely based on the

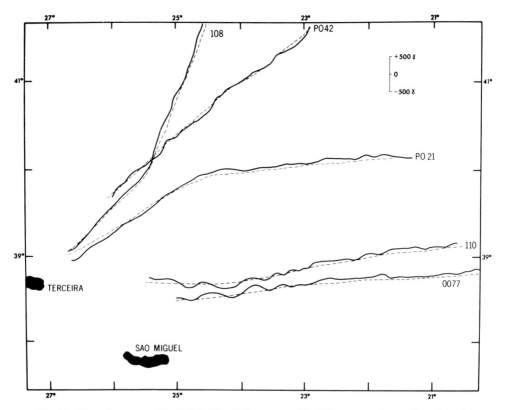

Fig. 11. Map showing residual U.S. Naval Oceanographic Office magnetic profiles plotted
along their track positions, as in Fig. 10. These data show the very subdued magnetic signa-
tures immediately east of the Azores.

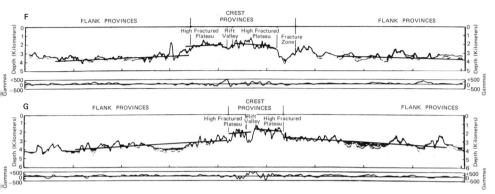

Fig. 12. Magnetic and bathymetric profiles F and G at lat. 36°N from Schneider and Vogt (1968). For track location, see Fig. 2. (From Krause and Watkins, 1970.)

similarity of spreading ridge configurations on Iceland and the model of Machado *et al.* (1972) for the Azores, is similarly untenable. An incidental part of the presentation of Machado *et al.* (1972) is assignment of parts of Faial, San Miguel, and Terceira to the Brunhes, Matuyama, or Gauss geomagnetic epochs simply on the basis of the anomaly signs on a surface magnetic map. As Watkins and Richardson (1971) have demonstrated, however, the marine magnetic anomaly pattern Machado *et al.* (1972) have in mind cannot be derived from a magnetic survey on the surface of intrusive bodies. Similarly, Schneider and Vogt (1968) have shown profiles (Figs. 2 and 12) at lat. 36° N that feature none of the classical anomalies ideally reflecting the known polarity history. Such a pattern is defined at lat. 40° N (Fig. 13), however, as shown by Krause and Watkins (1970). This profile provides a calculated spreading rate of 1.33 cm/yr (Fig. 13). Keen (1970) believes that such a spreading rate may exist up to lat. 43°05′ N near the Chaucer rise, where Vogt and Ostenso (1966) made a detailed magnetic and dredged-rock survey.

That the positive element of the Azores platform does not feature the classical linear anomaly pattern is not surprising (Fig. 14). The Azores platform is a massive anomalous buildup of upper mantle material, which is, in our opinion, inherently unlikely to facilitate the regular dike injection (Watkins and Richardson, 1971) and symmetry of the less anomalous segments of the mid-Atlantic ridge to the north and south. Although Phillips *et al.* (1969) and Shand (1949) report much serpentinite (which is often only weakly magnetic) in dredge hauls near the Azores, we suspect that the explanation of the low-amplitude magnetic anomalies is unlikely to be a strong function of the local crustal petrological type. As Watkins (1968) has summarized, slow crustal spreading, irregular dike injection, and extrusive buildup can also reduce anomaly amplitudes by reducing anomaly wavelengths to undetectable limits at sea level and by extrusive overlap that effectively cancels anomalies arising from bodies of opposite polarity. We consider it probable that the Azores platform must comprise a greater fraction of extrusive material than normal ridge areas. Igneous accumulation of material with only a single polarity, also, would not be expected to create large-amplitude magnetic anomalies, unless very large relief existed.

A survey of the mid-Atlantic ridge at its junction with the Azores (Krause and Watkins, 1970), while revealing low-amplitude anomalies, nevertheless used tracks sufficiently close for detailed correlation of anomalies (Fig. 14). The curvature of the mid-Atlantic ridge and the magnetic expression of the continuation of the EAFZ into the ridge are revealed. The central positive anomaly (over the ridge) is identified, as is anomaly 5 (Heirtzler *et al.*, 1968) which is of normal polarity, with an age between 8.79 and 9.94 m.y. The central positive anomaly extends under Flores and Corvo, providing maximum ages for the initiation of these islands and the possibility of estimates for the local rate of igneous productivity. Analysis of three profiles across the Terceira rift will be presented later, during discussion of a model of the genesis of the islands.

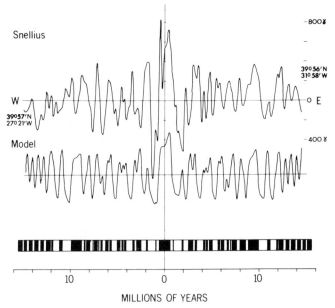

Fig. 13. Computation of the crustal spreading rate at lat. 40°N, using a magnetic profile obtained by the *R. V. Snellius*. The model profile is computed for sea level using an intensity of magnetization of 5×10^{-3} emu · cm^{-3}, no induced component, an axial dipole field, upper and lower block surfaces at 1.6 and 2.5 km below sea level, respectively, the known and predicted geomagnetic polarity time scale of the Heirtzler *et al.* (1968); and a constant spreading rate of 1.33 cm/yr. The polarities of the blocks are normal when black, and reversed when clear. Exact locations of the ends of the profiles are added. (From Krause and Watkins, 1970.)

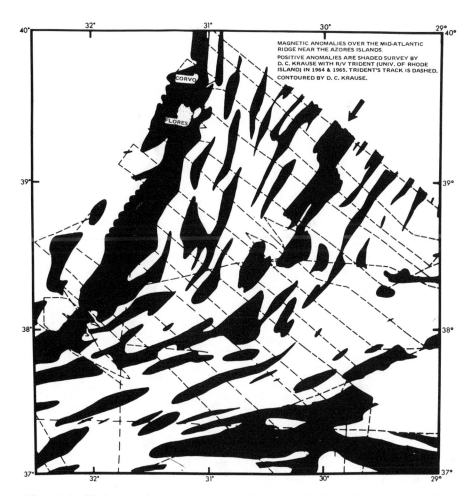

Fig. 14. Residual magnetic anomaly map resulting from *R. V. Trident* survey. Positive anomalies are black; negative anomalies are clear. Edge of coverage indicated by scalloped anomaly contour. Tracks indicated by dashed lines. Axis of mid-Atlantic ridge indicated by an arrow. (From Krause and Watkins, 1970.)

C. Gravity

Improved determination of the earth's-gravity field from satellite orbit data (Gaposchkin and Lambeck, 1970) has led to a more accurate definition of regional free-air and isostatic gravity anomalies (Kaula, 1970). The compressive nature of the Alpide Tectonic Zone is confirmed by elongate, positive, isostatic anomalies and adjacent large negative anomalies associated with the Mediterranean. The large positive and negative anomalies are similar to those in trench areas where oceanic lithosphere is being consumed. Kaula

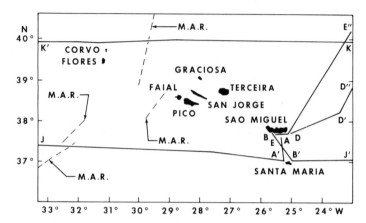

Fig. 15. Location of marine gravity tracks across the Azores plateau, after MacFarlane (1968). Profiles DD′ and EE′ were made by *R. S. S. Discovery* (University of Cambridge) in 1965; the remainder were made by *H. N. M. Snellius* (Delft Technological University) in 1965. Dashed line is the mid-Atlantic ridge and its offset portion.

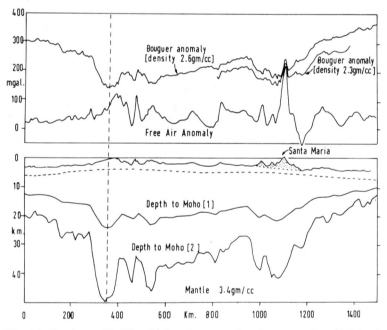

Fig. 16. Gravity profile JJ′ and inferred sturctural section across the mid-Atlantic ridge and Santa Maria (MacFarlane, 1968). For location of profile see Fig. 15. The two Bouguer anomaly profiles assume densities of 2.6 and 2.3 gm/cm³, respectively.

(1970) shows that the Azores platform has the largest mean positive free-air anomaly (16 mgal, referred to a fifth-degree spherical harmonic figure) in the Atlantic. The causative mass excess is most simply explained as the result of the pileup of material to form the platform. Indirect (geological) observations of ancient sea levels would assist in resolution of the ambiguities inherent in predicting the prospects of isostatic equilibrium being reached in this region.

There are several continuous gravity tracks in the Azores region (Fig. 15) that aid a more detailed evaluation of the structure of the oceanic lithosphere near the Azores plateau. Bouguer and free-air anomalies computed for these profiles are shown in Figs. 16, 17, and 18.

The free-air anomaly is strongly positive over the plateau to the west of Santa Maria, the mean value for 100 km being $+43$ mgal, in contrast to an easterly negative anomaly decreasing to -55 mgal. The Bouguer anomaly curve, showing an almost linear gradient of 0.19 mgal/km decreasing toward the crest of the mid-Atlantic ridge, is similar to that obtained over the ridge at lat. $32°$ N (Talwani *et al.*, 1965).

This smooth gravity curve is disrupted by a 350-km-wide negative anomaly with a -100-mgal amplitude centered on Santa Maria. The low gravity field definitely extends northward to Sao Miguel (Fig. 17), reflecting a body with

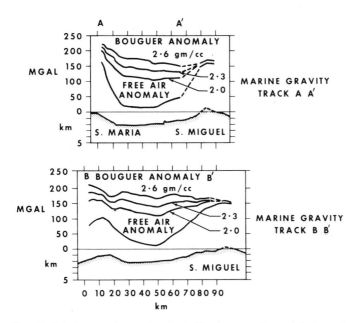

Fig. 17. Marine gravity tracks AA′, BB′ between Santa Maria and Sao Miguel. For locations see Fig. 15. Bouguer anomaly curves are computed for three possible densities (2.6, 2.3, 2.0 gm/cm³) of the upper layers of the oceanic crust. (Data from MacFarlane, 1968.)

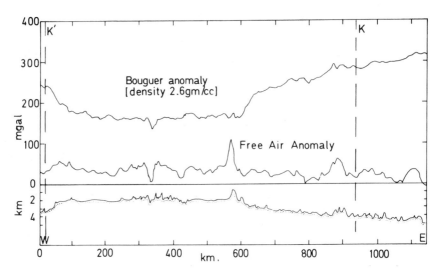

Fig. 18. Continuous gravity and topographic profile KK′ across the mid-Atlantic ridge (MacFarlane, 1968). See Fig. 15 for profile location.

an anomalously low density. The maximum computed depth to the upper surface of the body is 67 km (Bott and Smith, 1958; Skeels, 1963).

All marine gravity tracks that cross from oceanic crust onto the Azores plateau display a sudden and marked decrease in free-air anomaly (Fig. 17), and the western part of track JJ′ (Fig. 16) also demonstrates a similar anomaly decrease over the mid-Atlantic ridge. The resemblance between the patterns over the Azores plateau and the ridge strongly suggests that the lithosphere structure beneath these two structures is very similar.

Regional data from the area of the closely associated islands of Faial, Pico, and San Jorge also indicate a pronounced east–west gradient that reduces the gravity field from 124 to 86 mgal west of Faial. A regional eastward gradient appears to be present, becoming steeper toward the axis of the mid-Atlantic ridge (Figs. 18 and 19).

Superimposed upon the regional gravity pattern are the local Bouguer anomaly patterns for the individual volcanoes. These are discussed below inasmuch as they indicate the crustal structure beneath individual volcanoes or volcanic island.

On *Santa Maria* (Fig. 20), the center of the high Bouguer anomaly coincides with the uplands of the ancient core of the islands. A broad region of high Bouguer anomaly extends westward to another center that appears to be located offshore to the northwest of the island. The offshore anomaly is possibly the source of the submarine volcanics found on the island. A gravity nose extends southeastward from the main anomaly center (Fig. 20) and presumably indicates the rift zone that produced the northwest–southeast alignment of the topography and dike swarms in that area.

Two centers of high Bouguer values are aligned in an approximately east–west direction on *São Miguel* (Fig. 20). The larger center is located near the geologically oldest volcano of the island (Zbyszewski and da Veiga Ferreira, 1962) and has a maximum value of 203 mgal. The amplitude of the anomaly is approximately 60 mgal. In the west, an anomaly is centered to the southeast of the well-defined western caldera. The absence of major anomalies associated with the two more westerly volcanoes of the eastern group suggests that the volcanoes are lateral offshoots from the oldest and principal center and do not extend to any great depth. A region of low Bouguer anomaly across the west center of the island may represent a northwest–southeast-trending fracture zone. Such a possibility is in excellent agreement with the topographic and fault trends and with numerous adventive cones in this area.

On *Terceira*, the Bouguer anomaly (Fig. 20) reflects the WNW–ESE trend of the island topography and bathymetry and confirms the existence of an active belt that follows this trend across the center of the island. The gravity relief is small; the observed anomaly ranges from 118 to 140 mgal, with the higher values in a WNW–ESE belt. The distribution of gravity stations around the eastern volcano is poor, but the volcano appears to represent a gravity high. The western volcano may be a younger, smaller version of the eastern caldera, where a region of high gravity values encloses a smaller area of lower Bouguer values. These low gravity values and the large quantity of ignimbrite

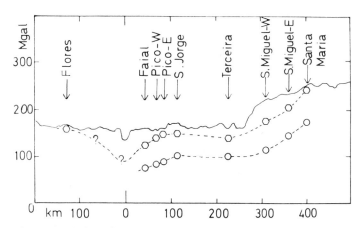

Fig. 19. Variation of gravity field across the Azores plateau related to distance from the mid-Atlantic ridge (MacFarlane, 1968). Polygons are the maximum Bouguer anomalies observed on each island (see Figs. 20–22); circles are estimated regional Bouguer anomalies. The continuous line is Bouguer anomaly curve based on marine gravity data discussed in text.

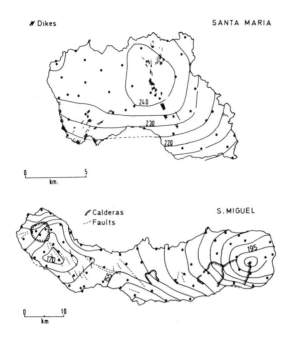

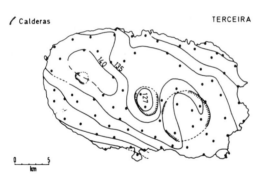

Fig. 20. Bouguer anomaly maps for the islands of Santa Maria, Sao Miguel, and Terceira. Contour interval is 5 mgal; filled circles are individual gravity stations. For methods of computations, errors, etc., see MacFarlane and Ridley (1968). Surface geological features, possibly relating to the gravity patterns, are from Hadwen and Walker.

deposits surrounding the calderas suggest that these calderas belong to the same category as those on Sao Miguel.

The highest Bouguer anomaly on *Sao Jorge* (Fig. 21) is 149 mgal, the lowest 108 mgal, with a possible east–west gradient along the island. The main feature is a gravity high in the center of the eastern, and geologically older, part of the island. To the northwest, the gravity contours are elongated along the topographic trend of the island.

Figure 21 reflects the east–west topographic trend on *Pico* and confirms the rift-zone origin of at least the eastern part of the island. A gravity nose from the eastern center extends towards the west-central anomaly. The possibility exists that an offshoot from a magma chamber under the eastern part

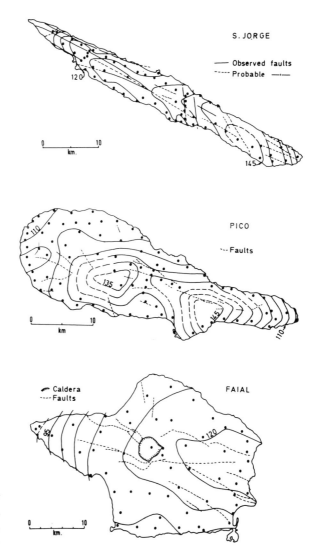

Fig. 21. Bouguer anomaly maps for Sao Jorge, Pico, and Faial (MacFarlane, 1968). Details as in Fig. 20.

of the island extends northwestward along a line of weakness associated with the tectonic trends of the western half of the island. This possibility offers a satisfactory explanation for the faults, fissures, adventive cones, and historic eruptions that characterize this part of the island.

The removal of the regional gradient for *Faial* (Fig. 21) shows a center of high Bouguer anomaly over the summit caldera. The anomaly extends eastward along a zone of faults and grabens and westward along the rift zone that has produced the western peninsula. The high gravity anomaly observed at the caldera bottom is unique among the Azores calderas, although the

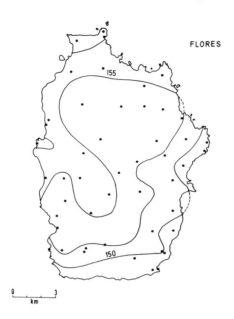

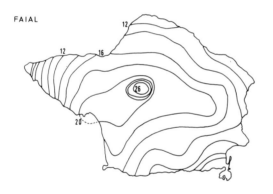

Fig. 22. Bouguer anomaly map for Flores, and residual gravity anomaly over Faial after removal of a regional field (MacFarlane, 1968). Details as in Fig. 20.

pumice mantle is similar to that over other calderas. The anomaly requires a high-density body close to the surface of the caldera.

The Bouguer anomaly for *Flores* (Fig. 22) is remarkable for its low relief, the lowest observed value being 146 mgal and the highest 161 mgal. No major gravity center is visible on the island, although the maximum anomaly in the southwest coincides with a region of fumarole activity.

D. Seismology

Mention has already been made of the seismicity of the EAFZ and the aseismic nature of the westward extension. These characteristics are demonstrated by Barazangi and Dorman's (1969) epicenter map (Fig. 23). Superimposed on Fig. 23 are the solutions by Banghar and Sykes (1969) of the two earthquakes, following the initial study by Isacks *et al.* (1968). The part tensional, part strike-slip solution at the west end of the EAFZ (Fig. 23) contrasts with the compressional event at the east end, which Roberts (1970) believes to be consistent with the overthrusting required to explain gravity, bathymetric, and seismic profiler results in the Gulf of Cadiz. In contrast, Ritsema (1969) has presented two earthquake solutions that apparently are consistent with the EAFZ being a right-lateral transcurrent fault, as assumed by Morgan (1968) in his global tectonic analysis, and as preferred by Lopez Arroyo and Udias (1972) and Udias and Lopez Arroyo (1972) to explain their analyses of very recent earthquakes on the EAFZ.

We know of no local seismic refraction surveys. Lee and Taylor (1966) refer to Ewing and Ewing (1959) as showing in a series of three refraction lines that the Moho depth west of the Azores (long. 39–35° W) decreases from 10.2 to 7.0 km. At the same time, the upper mantle velocity decreases from 8.1 to 7.9 km/sec. A line to the east (long. 26° W) detects the Moho at a 9.4-km depth. Fairly symmetrical shallowing of the mantle toward the ridge is indicated. Barrett and Aumento (1970) have summarized the results of refraction surveys of the mid-Atlantic ridge at lat. 45° N. Five discrete layers are found with velocities ranging from 2.2 to 6.8 km/sec and total thickness from 3.5 to 4.0 km overlying material of mantle velocity (7.8–8.0 km/sec). The Moho depth beneath the Azores plateau has not been studied in detail, although Nafe and Drake (1969) show an increase of Moho depth from 6 (on the ridge at lat. 33° N) to 11 km under the Azores.

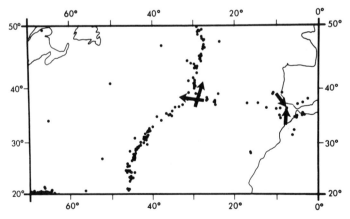

Fig. 23. Map showing locations of earthquake epicenters in the north-central Atlantic for the period 1960–1968, from Barazangi and Dorman (1969). The location and solutions of earthquakes studied by Banghar and Sykes (1969, Fig. 2) are shown. Both slip vectors are shown, since, as Banghar and Sykes state, "it is difficult to distinguish the fault plane and auxiliary plane." (From Krause and Watkins, 1970.)

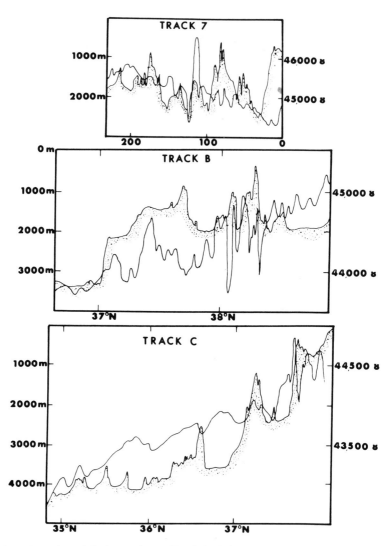

Fig. 24. Magnetic and bathymetric profiles (stippled) along tracks 7, B, and A of *R. V. Trident* surveys. Track 7 is northwest–southeast across the mid-Atlantic ridge, just south of Flores; track B is (left half) north–south across the East Azores fracture zone at long. 30°W and (right half) parallel to track 7; track C is south of Sao Miguel across the East Azores fracture zone. (From Krause and Watkins, 1970.)

The published reflection surveys around the Azores appear to be restricted to crossings of the mid-Atlantic ridge at various latitudes south and north of the Azores plateau. Ewing and Ewing (1967) (Fig. 2) show that the sediment thickness at lat. 40° N increases abruptly at a distance from the mid-Atlantic ridge equivalent to a crustal age of 10 m.y. and suggest that this thickening might be caused by a halt in crustal spreading that resumed at 10 m.y. In-conclusive results from deep-sea drilling in the North Atlantic (Peterson *et al.*, 1970) cannot confirm the suggested halt in spreading for the restricted area under discussion. Using magnetic anomaly data, however, Pitman *et al.* (1971) suggest that evidence exists to show a cease in crustal spreading 38 m.y. ago and a resumption 10 m.y. ago.

The seismic profiler data published by Schneider and Vogt (1968), for lat. 36° N (Fig. 12), illustrate the great relief across the mid-Atlantic ridge and flank provinces. Figure 24 shows similar data close to the Azores, including sections across the EAFZ.

E. Heat Flow

No heat-flow measurements have been made on the islands. Phillips *et al.* (1969) have defined areas of high and low heat flow north and south, respec-tively, of a fracture zone at lat. 43° N. Serpentinized peridotite predominates in dredge hauls to the north, and basalts predominate to the south of the zone. No heat-flow data are published regarding the immediate vicinity of the Azores, according to the compilation of Lee and Uyeda (1965).

V. ORIGIN OF THE AZORES

A. The Model and Some Testing

We now insert the previously outlined observations into a regional syn-thesis. A testable model for the origin of the features so far defined is presented.

The major enigma of the Azores area, the contrasting seismicity of the EAFZ and WAFZ, is first explained by using a mirror image of the Ball and Harrison (1969) model, which was applied to equatorial Atlantic struc-tures. The model (Fig. 25) consists of a ridge migrating eastward at its own crustal spreading rate, which is greater to the north than to the south of a transverse fracture system. As Fig. 25 shows, seismic east and aseismic west fracture zones result, with maximum seismicity in the ridge–ridge transform fault region. Left-lateral motion in the offset region contrasts with right-lateral motion along the east fracture. This model is in turn utilized as the first stage of a crustal development model for the Azores (Fig. 26), as envisaged by Krause and Watkins (1970). A change in the direction of crustal spreading (Fig. 26) (stage 2) leads to the development of a leaky transform fault (stage 3) with the resulting growth of a secondary spreading center. Northward offset of the aseismic west fracture increases, relative to the seismically active east fracture zone. Four different crustal spreading rates are suggested. Inserting

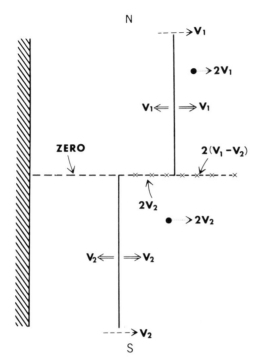

Fig. 25. Diagram illustrating the seismic–aseismic transverse fracture model with spreading rate greater to the north than to the south, and southern ridge offset to the west. Stable reference block hachured column on west side. The ridge to north of the east–west fracture system has a spreading rate V_1. The ridge to the south of the east–west fracture system has a spreading rate V_2, and is migrating to the east at a velocity of V_2 relative to the stable western reference point; other indicated velocities (including "zero") represent relative velocities across the transverse fracture system. Crosses indicate zones of potential seismic activity. Spreading rate V_1 is greater than V_2. (From Krause and Watkins, 1970.)

known lengths and angular relationships into the final (present) stage shown in Fig. 26 gives Fig. 27. In this simple geometrical model, three unknown crustal spreading rates can be calculated, given one known spreading rate. The magnetic profile at lat. $40°$ N (Fig. 13) is, perhaps atypically, highly amenable to conventional spreading analysis. A spreading rate of 1.33 cm/yr is suggested and used as V_1 in Fig. 27. Resulting calculations (Fig. 27) provide spreading rates of 0.2 cm/yr for the secondary spreading center and rates of 0.8 and 1.1 cm/yr for other segments of the mid-Atlantic ridge. Thus, an estimate can be made of the time required to create the secondary spreading center and, therefore, the time since the proposed change in crustal spreading direction. Assuming that no major changes in crustal spreading rate have occurred, a period of 45 m.y. would be needed. Phillips and Luyendyk (1970) use the regular, slightly arcuate configuration of the 1000-km-long Atlantic fracture zone west from the mid-Atlantic ridge at lat. $30°$ N to infer no change in the regional crustal spreading direction during the past 40 m.y. Given that this tenuous method of defining crustal motion directions is valid, it would be interesting to know the fracture configuration in the older, western parts of the fracture, where the oceanic crust reaches pre-Eocene ages. We stress our doubt that the simple linear magnetic anomaly configuration of Fig. 26 would be maintained in reality, particularly within the secondary spreading center; zones of rotation, offset, and effusive buildup could be expected.

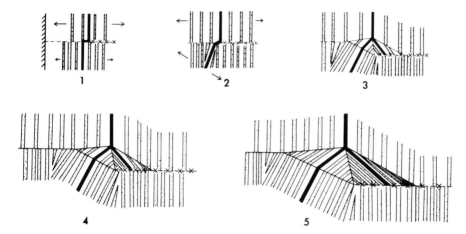

Fig. 26. Simplified model showing proposed development of the Azores crust. Initial condition is stage 1; present condition is stage 5. The difference in relative spreading is illustrated by use of dotted stripes, which are intended to represent arbitrary polarity changes at regularly spaced time intervals: the faster spreading is indicated by a wider spacing between stripes. The mid-Atlantic ridge is indicated by a heavy line. *For detailed description and the many limitations, particularly within the secondary spreading centers, see text.* Note continuation of arbitrary anomaly patterns between different spreading segments, except on the south edge of the secondary spreading center. Small crosses indicate the zones of potential seismic activity outside the main and secondary ridges. (From Krause and Watkins, 1970.)

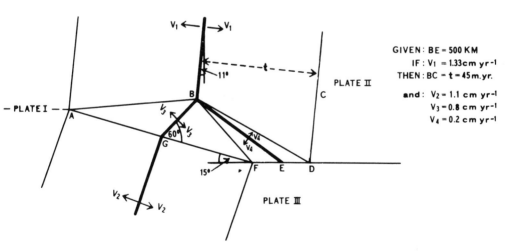

Fig. 27. Diagram showing basic geometrical framework of the suggested present stage in the development of the Azores oceanic crust (Fig. 26, Stage 5), based on the available physiographic data. Also shown is a table of resulting predicted crustal spreading rates together with the time of initiation of change in crustal spreading direction. (From Krause and Watkins, 1970.)

McKenzie (1972) has criticized the Krause–Watkins model (Fig. 25) stating that it is unnecessarily complicated, preferring instead a much simpler triple-junction system. The simpler model results, however, from ignoring both the existence and northward offset of the WAFZ. Not surprisingly this simpler model does not require a change in regional crustal spreading to create the secondary spreading center. We consider it essential that any model provides a triggering mechanism to bring the secondary spreading center into existence. The Krause–Watkins model provides this mechanism in the form of a change of spreading direction which also provides the offset of the WAFZ. McKenzie's model provides no such mechanism, requiring what appears to be spontaneous growth of the center.

The model we use results mainly from bathymetric, seismic, and marine magnetic data. We now examine details of these data and some of the other observations already presented, as tests of the possible validity of this model.

Bathymetry. The Terceira rift (Fig. 4) acts as the axis of the secondary spreading center. The Azores plateau (Fig. 3) represents the buildup caused by the secondary spreading center, which is the predominant local igneous source and, therefore, can be expected to be manifested as an eastward broadening of the ridge, as observed.

Seismicity. The observed great activity along the ridge, a less active EAFZ, and an aseismic WAFZ are consistent with the model. In addition, the solution of Bhangar and Sykes (1969) (Fig. 22) can be interpreted to be movement along and normal to the secondary spreading center. This movement would be an inherent property in the growth of such a feature.

Marine Magnetic Data. While we stress our doubt that classical linear anomalies will be detected over the platform (for reasons previously discussed), the predicted spreading rates (Fig. 27) are nevertheless testable, particularly in terms of the width of the central anomaly (0–0.7 m.y.), which may be expected to be the best defined of any anomalies present. Reasons for our optimism include the greater expected width of this anomaly because the time period involved is twice as great as the average time of constant polarity during the Tertiary (Heirtzler *et al.*, 1968) and the minimal effect of spontaneous decay processes, which decrease the intensity of magnetization of submarine basalts (Watkins and Paster, 1971; Banerjee, 1971) and the corresponding anomaly amplitude. The effects of spontaneous decay on reduction of magnetic anomaly amplitude would, by definition, be minimal in the central anomaly region.

Figure 28 shows an attempt at conventional crustal spreading analyses of three profiles that include the secondary spreading center (the Terceira rift). Although imperfect matching to the polarity scale (Fig. 28, lower) results, particularly for the older parts of the scale, the central anomaly is, nevertheless,

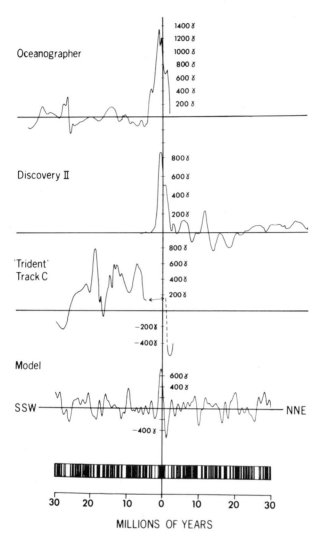

Fig. 28. Three magnetic profiles across the newly discovered Terceira rift, projected normal to the rift, and their comparison with a crustal spreading model. For exact track locations see Krause and Watkins (1970). The sea-level profile of the model is calculated using a spreading rate of 0.25 cm/yr. For the other parameters, see Fig. 13. Note the gap in the profile from *Trident* track C, which corresponds to the position of the island of Sao Miguel. A large negative anomaly exists immediately northeast of Sao Miguel. (From Krause and Watkins, 1970.)

relatively well defined and conceivably consistent with the low predicted spreading rate of 0.2 cm/yr. The central anomalies are certainly highly consistent with a central younger zone within the spreading zone. The magnetic data over the mid-Atlantic ridge shown in Fig. 14 may be interpreted to result in part from small offsets in the ridge, which might be expected to exist in such a relatively complex tectonic regime.

Gravity. Kaula's (1970) regional isostatic anomaly map is insufficiently defined for detailed interpretation; nevertheless, some inferences can be made on the marine gravity and local Bouguer anomaly data. The conclusion that the lithosphere structure beneath the Azores and mid-Atlantic ridge are substantially similar suggests that these regions have evolved by essentially the same process, i.e., sea-floor spreading. Both have low free-air anomalies, comparable in value to those observed over Iceland, and probably reflecting "anomalous" mantle material (Einnarrson, 1954; Bath, 1960; Talwani *et al.*, 1965). The fact that Iceland is itself a surface expression of the spreading ridge again suggests that the Azores region is also a spreading center.

Profile JJ' continues east of Santa Maria, close to the line of the "Azores–Gibralter" ridge, but there is no continuation of the low free-air anomaly that might indicate a mass deficiency similar to that over the Azores plateau. Thus, although characterized by considerable seismicity (Heezen *et al.*, 1959), this ridge is probably structurally unlike either the mid-Atlantic ridge or EAFZ and is unlikely to be associated with anomalous mantle material. This part of the ridge does not show the extensive volcanic buildup associated with the EAFZ, compatible with a model involving secondary spreading and attendant volcanism from that part of the fracture zone most closely related to the ridge.

Petrogenesis. The mechanism proposed here for the development of the EAFZ (differential spreading in the north and south leading to transcurrent faulting and formation of a secondary spreading center) has recently been invoked by Sigurdsson (1970) for western Iceland. In the latter case secondary spreading was accompanied by eruption of alkalic and transitional basalts, in contrast with the extensive tholeiitic plateau volcanism associated with the major spreading axis of Iceland.

A similar situation is present in the Azores region, where the subaerial alkalic volcanism contrasts with the tholeiitic volcanism at the mid-Atlantic ridge. McBirney and Gass (1967) have correlated regions of high heat flow with tholeiitic volcanism at the crest of mid-ocean ridges, whereas alkalic volcanism is more closely related to regions of lower heat flow away from the ridges. This view of relating magma type with heat flow remains an attractive proposition, and Bass (1971) has attempted to introduce slow spreading rate as a factor in producing more alkalic volcanism. Thus it appears feasible to relate the subaerial alkalic volcanism on the Azores islands to low heat flow and slow spreading of the EAFZ relative to the mid-Atlantic ridge.

Morgan (1971) indicates the general regions of triple junctions (Galapagos, Reunion, Azores) are the sites of mantle plumes (hot spots) and proposes a mechanism whereby initial tholeiitic volcanism is directly related to initiation of a hot spot, whereas alkalic magma is generated within the litho-sphere following displacement from the hot spot during plate motion. Although this process assumes *a priori* that alkalic volcanics on oceanic islands simply cap a more tholeiitic core, an assumption yet to be proven in the Azores, it does have the advantage of explaining the provinciality of volcanism in the region of the EAFZ.

Green and Ringwood (1967), O'Hara (1968), and Green (1971) have shown experimentally that basalt chemistry is strongly dependent on depth of magma generation. Alkalic basalts are generally considered to develop at greater depth and with smaller degrees of partial melting than tholeiites. Applying this to the Azores region, the voluminous tholeiitic volcanism at the mid-Atlantic ridge requires very rapid depression of the geotherms outside the immediate zone of magma eruption to produce alkalic volcanics on the Azores islands. Rapid perturbation of the geotherms has been demonstrated by Oxburgh and Turcotte, and appears to be reflected in rapid depression of the low-velocity zone (Talwani *et al.*, 1965; Archambeau *et al.*, 1969) away from ocean ridges.

Formation of the EAFZ and subsequent tensional tectonism and slow spreading provided a suitable structural environment for release of basalt magma to the surface, from a part of the lithosphere essentially independent of the thermochemical system of the mid-Atlantic ridge. More detailed analysis of the evolution of the mantle beneath the Azores requires significantly more detail concerning vertical and horizontal chemical variations within individual islands and across the Azores plateau as a regional entity. Nonetheless, it is evident that the extensive alkalic volcanism in the Azores would be compatible with the present evolutionary model.

Isotopic and Paleontological Ages. The proposed secondary spreading center cannot exceed 45 m.y. in age. The K/Ar and paleontological ages for the islands previously presented are much younger. The distribution of ages within the spreading center does not conflict with the ages that can be inferred for different parts of the spreading center, if the proposed spreading rates are used.

Heat Flow. Although no heat-flow data exist for the immediate vicinity of the Azores, a secondary spreading center ideally may be distinguished by heat-flow measurements, for several reasons. These reasons include the fact that small secondary spreading centers are unlikely to feature convection cells of mantle depth, which are favored (but not proved) for mid-ocean spreading processes, and slow spreading may result in less surface thermal release. Hanks (1971) has proposed a model relating surface heat flow and vertical velocities of mass transport beneath oceanic rises; this model may be ap-propriately tested near the Azores.

B. Speculations on the Cause of Spreading Direction Changes

A critical aspect of the proposed genesis of the Azores region is the change in direction of crustal spreading, essential for the development (on the convex side of the ridge) of the secondary spreading center, which in turn has created most of the Azores plateau. The cause of this proposed change in crustal spreading approximately 45 m.y. ago or earlier is an appropriate subject for speculation.

A change in mantle thermal conditions would be required to facilitate a change in crustal spreading directions. This change could be accomplished by a change in the vertical mantle thermal gradient, which, if greatly increased, could drastically affect the configuration of any mantle convection cells (Torrance and Turcotte, 1971). The extreme effect could be bifurcation of cells, which, if separated across a horizontal plane in mid-mantle regions, would greatly enhance prospects of rearrangement of shallower convection cells. These cells would be those more directly involved in crustal motions. The motion of adjacent cells would conceivably involve less frictional drag at the cell interfaces than at interfaces between cells and the core–mantle boundary. Because of reduced viscous drag in at least the upper half of the mantle, adjacent continental masses then could more readily influence the configuration of convection cells under the oceanic crust. Continental mass impingement could lead to a change in spreading direction, if continental edges were appropriately irregular. In summary, a process increasing the vertical mantle temperature gradient (the flanking, drifting continents impinging on features capable of slowing the drift and spreading rate?) could lead to re-arrangement of convection cell configuration, which in turn could be expressed as a change in spreading direction. This sequence of events would imply that for initiation of changes in crustal spreading directions (and later, secondary spreading center development), slowing or even halting of crustal spreading is required. Some evidence exists for a halt in spreading in the North Atlantic (Pitman *et al.*, 1971); however, the halt was between 38 and 10 m.y. ago, which is after our inferred spreading direction change.

The foregoing highly speculative series of arguments is therefore testable in its final stages, but even if a relationship between spreading rates and direction changes is confirmed, the earlier stage arguments would not, at this time, necessarily be supported.

ACKNOWLEDGMENTS

Parts of this work were carried out while D. J. MacFarlane held a Royal Dutch Shell Fellowship at Imperial College, London. W. I. Ridley was supported by a National Research Council Postdoctoral Fellowship at Johnson

Space Center, Houston, and by the Lunar Science Institute, which is operated by the Universities Space Research Association under NASA contract No. NSR 09-051-001. We acknowledge the help of Dr. B. M. Gunn (University of Montreal) in compiling the short but elusive list of chemical analyses of Azores rocks, and the University of Cambridge and Delft Technological University for access to marine geophysical data. This paper constitutes the Lunar Science Institute Contribution No. 154.

REFERENCES

Abdel-Monem, A., Fernandez, L. A., and Boon, G. A., 1968, Pliocene–Pleistocene minimum K–Ar ages of older eruptive centers, Eastern Azores (abstract): *Trans. Am. Geophys. Union*, v. 49, n. 1, p. 363.

Archambeau, C. B., Flinn, A. E., and Lambert, D. G., 1969, Fine structure of the upper mantle: *Jour. Geophys. Res.*, v. 74, p. 5825–5866.

Assunçao, C. F. Torre de, 1961, Estudo petrographico da ilha de S. Miguel (Acores): *Commun. Serv. Geol. Port.*, v. 45, p. 81–176.

Assunçao, C. F. Torre de and Canilho, M. H., 1970. Notas sobre petrografia comparada das ilhas Atlanticas: *Lisboa Univ. Fac. Ciênc. Mus. e Lab. Mineral. e Geol. Bol.*, v. 11, p. 305–342.

Baker, I., 1969, Petrology of the volcanic rocks of St. Helena Island, South Atlantic: *Bull. Geol. Soc. Am.*, v. 80, p. 1283–1310.

Ball, M. M. and Harrison, C. G. H., 1969, Origin of the Gulf and Caribbean and implications regarding ocean ridge extension, migration, and shear: *Gulf Coast Assoc. Geol. Soc. Trans. 19th Annual Meeting*, p. 287–294.

Banerjee, S. K., 1971, Decay of marine magnetic anomalies by ferrous ion diffusion: *Nature*, v. 229, p. 181–183.

Bangher, A. R. and Sykes, L. R., 1969, Focal mechanisms of earthquakes in the Indian Ocean and adjacent regions: *J. Geophys. Res.*, v. 74, n. 2, p. 632–649.

Barazangi, M. and Dorman, J., 1969, World seismicity maps compiled from ESSA, Coast and Geodetic Survey, epicenter data, 1961–1967: *Bull. Seism. Soc. Am.*, v. 59, n. 1, p. 369–381.

Barrett, D. L. and Aumento, F., 1970, The Mid-Atlantic Ridge 45°N, XI. Seismic velocity, density, and layering of the crust: *Can. Jour. Earth Science*, v. 7, p. 1117–1124.

Bass, M. N., 1971, Variable abyssal basalt populations and their relation to sea-floor spreading rates: *Earth Plan. Sci. Letters*.

Bath, M., 1960, Crustal structure of Iceland: *J. Geophys. Res.*, v. 65, p. 1793–1807.

Berthois, L., 1953, Contribution à l'étude lithologique de archipel des Azores: *Commun. Serv. Geol. Port.*, v. 34.

Boone, G. M. and Fernandez, L. A., 1971, Phenocrystic olivines from the eastern Azores: *Mineral. Mag.*, v. 38, p. 165–178.

Bott, M. H. P. and Smith, R. A., 1958, The estimation of the limiting depth of gravitating bodies: *Geophys. Pros.*, v. 6, p. 1–10.

Burke, K., Kidd, W. S. F., and Wilson, J. T., 1973, Plumes and concentric plume traces of the Eurasian Plate: *Nature Physical Sciences*, v. 241, p. 128–129.

Cann. J. R., 1968, Bimodal distribution of rocks from volcanic islands: *Earth Plan. Sci. Letters*, v. 4, p. 479–480.

de Veiga, Ferreira, O., 1961, Afloramentos de calcario miocenico da Ilha de Santa Maria, Azores: *Commun. Serv. Geol. Port.*, v. 45, p. 493–501.

Drake, C. L. and Woodward, H. O., 1963, Appalachian curvature, wrench faulting and offshore structures: *Trans. New York Acad. Sci.*, v. 26, n. 2, p. 48–63.

Drew, W., 1965, Paleomagnetic measurements in Quaternary lavas from the Azores Islands: Unpublished, Ph. D. Thesis, University of Newcastle-upon-Tyne.

Einarrson, T., 1954, A survey of gravity in Iceland: *Soc. Sci. Isl. Reykjavic*, 22 p.

Ellwood, B. B., Watkins, N. D., Amerigian, C., and Self, S., 1973, Brunhes epoch geomagnetic secular variation in the Azores: *EOS (Trans. American Geophysical Union)*, v. 54, p. 254 (abstract).

Esenwein, P., 1929, Zur Petrographic der Azoren: *Zeitschr. für Vulkanol.*, v. 12, p. 108–227.

Ewing, J. and Ewing, M., 1959, Seismic refraction measurements in the Atlantic Ocean basins, in the Mediterranean Sea, on the Mid-Atlantic Ridge, and in the Norwegian Sea: *Geol. Soc. Am. Bull.*, v. 70, p. 291–318.

Ewing, J., and Ewing, M., 1967, Sediment distribution on the mid-oceanic ridges with respect to spreading of the sea-floor: *Science*, v. 156, p. 1590–1591.

Fernandez, L. A., 1973, Petrology of the Nordeste Volcanic Complex, São Miguel Island, Azores: *Geol. Soc. Amer.* (abstracts with Programs) p. 618.

Gaposchkin, E. M. and Lambeck, K., 1970, 1969 Smithsonian Standard Earth II: *Smithsonian Astrophys. Obs. Spec. Rep. 315*, 93 pp.

Gast, P. W., 1968a, Upper mantle chemistry and evolution of the earth's crust, in: *The History of the Earth's Crust*, R. A. Phinney ed.: Princeton University Press, p. 15–27.

Gast, P. W., 1968b, Trace element fractionation and the origin of tholeiitic and alkaline magma types: *Geochim. Cosmochim. Acta*, v. 32, p. 1057–1086.

Girdler, R. W., 1963, Geophysical evidence on the nature of magmas and intrusions associated with rift valleys: *Bull. Volcan.*, v. 26, p. 34–47.

Girod, M. and Lefeure, C., 1972, The so-called andesites of the Azores: *Contrib. Mineral. and Petrol.*, v. 35, p. 159–167.

Green, D. H., 1971, Composition of basaltic magmas as indicators of conditions of origin: Application to oceanic volcanism: *Phil. Trans. Roy. Soc. London*, v. 268A, p. 707–725.

Green, D. H. and Ringwood, A. E., 1967, The genesis of basaltic magmas: *Contrib. Mineral. and Petrol.*, v. 15, p. 103–190.

Hadwen, P. and Walker, G. P. L., A geological review of the Azores, in press.

Hanks, T. C., 1971, Model relating heat-flow values near, and vertical velocities of mass transport beneath the Oceanic Rise: *J. Geophys. Res.*, v. 76, p. 537–544.

Heezen, B. C., Tharp, M., and Ewing, M., 1959, The floors of the oceans, 1. The North Atlantic: *Geol. Soc. Am. Spec. Paper*, v. 65.

Heirtzler, J. F., Dickson, G. O., Herron, E. M., Pitman, W. C., and LePichon, X., 1968, Marine magnetic anomalies, geomagnetic field reversals and motions of the ocean floor and continents: *J. Geophys. Res.*, v. 73, p. 2119–2136.

Isacks, B., Oliver, Jr., and Sykes, L., 1968, Seismology and the new global tectonics: *J. Geophys. Res.*, v. 73, p. 5855–5899.

Jeremine, E., 1957, Etude microscopique des roches de la region de Furmas (S. Miguel, Acores): *Commun. Serv. Geol. Port.*, v. 38, p. 65–90.

Kaula, W., 1970, Earth's gravity field: relation to global tectonics: *Science*, v. 169, p. 982–985.

Kay, R., Hubbard, N. J., Gast, P. W., 1970, Chemical characteristics and origin of Oceanic Ridge Volcanic rocks: *J. Geophys. Res.*, v. 75, p. 1585–1613.

Keen, M. J., 1970, Fracture zones on the Mid-Atlantic Ridge between 43°N and 44°N: *Canadian Jour. Earth Sci.*, v. 7, p. 1352–1355.

Krause, D. C., 1965, East and West Azores fracture zones in the North Atlantic: *Submarine Geology and Geophysics* (17th Colston Symposium, Bristol, 1965), p. 163–173.

Krause, D. C., 1966, Relation of the Mid-Atlantic Rift to the Azores Platform: *Second Int. Ocean. Cong.* (Moscow), p. 207 (abstract).

Krause, D. C. and Watkins, N. D., 1970, North Atlantic crustal genesis in the vicinity of the Azores: *Geophys. Journ. Roy. Astron. Soc.*, v. 19, p. 261–283.

Laughton, A. S., Whitmarsh, R. B., Rusby, J. S. M., Somers, M. L., Revie, J., McCartney, B. S., and Nafe, J. C., 1972, A continuous east-west fault on the Azores-Gibralter ridge: *Nature*, v. 237, p. 217–220.

Lee, W. H. K., and Taylor, P. T., 1966, Global analysis of seismic refraction measurement: *Geophys. Journ. Roy. Astron. Soc.*, v. 11, p. 389–413.

Lee, W. H. K., and Uyeda, S., 1965, Review of heat flow data, in: *Terrestrial Heat Flow* (Geophysical Monograph, American Geophysical Union), v. 8, p. 87–190.

Le Pichon, X., Houtz, R., Drake, C. L., and Nafe, J. E., 1965, Crustal structure of mid-oceanic ridges, 1 Seismic refraction measurements: *J. Geophys. Res.*, v. 70, p. 319–339.

Loncarevic, B. D., Mason, C. S., and Mathews, D. H., 1966, Mid-Atlantic Ridge near 45°N, 1. The median valley: *Canadian Journ. Earth Sci.*, v. 3, p. 327–349.

Lopez Arroyo, A., and Udias, A., 1972, After shock sequence and focal parameters of the February 28, 1964 earthquake of the Azores-Gibralter Fracture Zone: *Bull. Seis. Soc. America*, v. 62, p. 699–720.

Lui-Kai, M., 1966, Unpublished M. Sc. Thesis (University of Newcastle-upon-Tyne), Department of Physics.

MacFarlane, D. J., 1968, The structure of some Atlantic Islands as deduced from gravity and other geophysical data: Ph. D. Thesis, University of London.

MacFarlane, D. J. and Ridley, W. I., 1968, An interpretation of gravity data for Lanzarote, Canary Islands: *Earth Plan. Sci. Letters*, v. 6, p. 431–436.

Machado, F., 1954, Earthquake intensity, anomalies and magma chambers of Azorean Volcanoes: *Trans. Am. Geophys. Union*, v. 35, p. 833–837.

Machado, F., 1959, Submarine pits of the Azores plateau: *Bull. Volcan.*, v. 11, n. 21, p. 109–116.

Machado, F., 1962, Sobre o mechismo du erupcao dos Capelinhos: *Geol. Serv. Port. Memoir*, n. 9, 19 p.

Machado, F., 1965, Volcanismo das Ilhas de Cabo Verde e das Outras Ilhas Atlantidas: *Junta Invest. do Ultramar* (Lisbon), v. 117, 83 p.

Machado, F. and Forjaz, V. H., 1965, Seismic swarm in the Azores Feb. 1964. Preliminary Report: *Bol. Soc. Geol. Port.*, v. 15, p. 201–206.

Machado, F., Quintino, J., and Monteiro, J. H., 1972, Geology of the Azores and the Mid-Atlantic Rift: *Proc. XXIV Int. Geol. Cong.*, Sec. 3, p. 134–142.

McBirney, A. R. and Gass, I. G., 1967, Relations of oceanic volcanic rocks of mid-ocean rises and heat flow: *Earth Planet Sci. Letters*, v. 2, p. 265–276.

McKenzie, D., 1972, Active tectonics of the Mediterranean region: *Geophys. Jour. Roy. Astron. Soc.*, v. 30, p. 109–186.

McKenzie, D. P. and Morgan, W. J., 1969, Evolution of triple junctions: *Nature*, v. 224, p. 125–133.

Menard, H. W., 1965, Sea floor relief and mantle convection: *Phys. Chem. Earth*, v. 6, p. 315–364.

Miyashiro, A., Shido, F., and Ewing, M., 1969, Diversity and origin of abyssal tholeiite from the Mid-Atlantic Ridge near 24° and 30° North latitude: *Contrib. Mineral. and Petrol.*, v. 23, p. 38–52.

Morgan, W. J., 1968, Rises, trenches, great faluts, and crustal blocks: *J. Geophys. Res.*, v. 73, p. 1959–1982.

Morgan, W. J., 1971, Convection plumes in the lower mantle: *Nature*, v. 230, p. 42–43.

Muir, I. D. and Tilley, C. E., 1964, Basalts from the northern part of the rift zone of the Mid-Atlantic Ridge: *J. Petrol.*, v. 5, p. 409–434.

Muir, I. D. and Tilley, C. E., 1966, Basalts from the northern part of the Mid-Atlantic Ridge: 11. The Atlantis collections from 30°N: *J. Petrol.*, v. 7, p. 193–201.

Nafe, J. E. and Drake, C. L., 1969, Floor of the Atlantic—Summary of Geophysical Data, in: *North Atlantic—Geology and Continental Drift* (Memoir 12, Amer. Assoc. Petrol. Geol.), p. 59–87.

Nicholls, G. D., 1965, Basalts from the deep ocean floor: *Mineral Mag.*, v. 34, p. 373–388.

O'Hara, M. J., 1968, The bearing of phase equilibria studies in synthetic and natural systems on the origin and evolution of basic and ultrabasic rocks: *Earth Sci. Rev.*, v. 4, p. 69–133.

Oversby, V. M., 1971, Lead in oceanic islands: Faial, Azores and Trindade: *Earth Planet. Sci., Letters*, v. 11, p. 401–406.

Peterson, M. N. A., Edgar, N. T., Cita, M., Gartner, S., Goll, R., Nigrini, C., and von der Borch, C., 1970, Initial reports on the deep-sea drilling project (U. S. Government printing office, Washington, D. C.), v. 2, p. 424–426.

Phillips, J. D. and Luyendyk, B. P., 1970, Central North Atlantic plate motions over the last 40 million years: *Science*, v. 170, p. 727–729.

Phillips, J. D., Thompson, G., Von Herzen, R. P., and Bowen, V. T., 1969, Mid-Atlantic Ridge near 43°N latitude: *J. Geophys. Res.*, v. 74, p. 3069–3081.

Pitman, W. C. and Talwani, M., 1972, Sea floor spreading in the North Atlantic: *Geol. Soc. Amer. Bull.*, v. 83, p. 619–646.

Pitman, W. C., Talwani, M., Heirtzler, J. R., 1971, Ages of the Atlantic Ocean from magnetic anomalies: *Earth Planet. Sci. Letters*, v. 11, p. 195–200.

Ridley, W. I., 1970, The petrology of the Las Canadas volcanoes, Tenerife, Canary Islands: *Contrib. Mineral. and Petrol.*, v. 26, p. 124–160.

Ritsema, A. R., 1969, Seismic data of the West Mediterranean and the problem of oceanization: *Verhand. Kon. Ned. Geol. Mijn Gen.*, v. 26, p. 105–120.

Roberts, D. C., 1970, The Rif-Betic orogen in the Gulf of Cadiz: *Marine Geology*, v. 9, p. M31–M37.

Saucier, H. and Roche, A., 1964, Etude paleomagnetique de laves de Faial et de Graciosa: *Commun. Geol. Serv. Port.*, v. 48, p. 5–13.

Schmincke, H. U., 1973, Magmatic evolution and tectonic regime in the Canary Madeira and Azores Island Groups: *Geol. Soc. Amer. Bull.*, v. 84, p. 633–648.

Schmincke, H. U. and Weibel, M., 1972, Chemical study of rocks from Madeira, Porto Santo, and Sao Miguel, Terceira (Azores): *Neuss. Jb. Miner. Abh.*, v. 117, p. 253–281.

Schneider, E. D., and Vogt, P. R., 1968, Discontinuities in the history of sea-floor spreading: *Nature*, v. 217, p. 1212–1222.

Self, S., 1971, The Lajes ignimbrite, Iiha Terceira, Azores: *Comm. Geol. Serv. Portugal*, v. 55, p. 165–180.

Serughetti, J. and Roche, A., 1968, Etude paleomagnetique de laves de l'Ile de Flores: *Computes Rendus Acad. Sci. Paris*, v. 267, p. 1185–1188.

Shand, S. J., 1949, Serpentinite from oceanic rocks near the Azores: *J. Geology*, v. 57, p. 89–92.

Sigurdsson, H., 1970, Structural origin and Plate tectonics of the Snaefellsnes volcanic zone, Western Iceland: *Earth Planet. Sci. Letters*, v. 10, p. 129–135.

Skeels, D. C., 1963, An approximate solution of the problem of maximum depth in gravity interpretation: *Geophysics*, v. 28, p. 724–735.

Talwani, M., Le Pichon X., and Ewing, M., 1965, Crustal structure of the Mid-Ocean Ridges. 2) Computed model from gravity and seismic refraction data: *J. Geophys. Res.*, v. 70, p. 341–352.

Tolstoy, I., 1951, Submarine topography in the North Atlantic: *Bull. Geol. Soc. Am.*, v. 62, p. 441–450.

Torrance, K. E. and D. L. Turcotte, 1971, Structure of convection cells in mantle: *J. Geophys. Res.*, v. 76, p. 1154–1161.

Udias, A., and Lopez Arroyo, A. 1972, Plate tectonics and the Azores-Gibralter Ridge: *Nature*, v. 237, p. 67–69.

Vine, F., 1966, Spreading of the ocean floor: new evidence: *Science*, v. 154, p. 1405–1415.

Vogt, P. R. and Ostenso, N. A., 1966, Magnetic survey over the Mid-Atlantic Ridge between 42° and 46°N: *J. Geophys. Res.*, v. 71, p. 4389–4412.

Walker, G. P. H. and Croasdale, R., 1971, Two Plinian-type eruptions in the Azores: *J. Geol. Soc. London*, v. 127, p. 17–55.

Watkins, N. D., 1968, Comments on the interpretation of linear magnetic anomalies: *Pure and Applied Geophys.*, v. 69, p. 179–192.

Watkins, N. D. and Paster, T., 1971, Magnetic properties of igneous rocks from the ocean floor: *Phil. Trans. Roy. Soc. Lond.*, v. A-268, p. 507–550.

Watkins, N. D. and Richardson, A., 1971, Intrusives, extrusives, and linear magnetic anomalies: *Geophys. Jour. Roy. Astron. Soc.*, v. 23, p. 1–13.

Weston, F. S., 1964, List of recorded volcanic eruptions in the Azores with brief reports: *Boletim. do Museu e Lab. Mineral. e. Geol. Lisboa*, v. 10, p. 3–18.

Williams, C. A. and McKenzie, D., 1971, The volution of the north-east Atlantic: *Nature*, v. 232, p. 168–173.

Yoder, H. S. and Tilley, C. E., 1962, Origin of basalt magmas: An experimental study of natural and synthetic rock systems: *J. Petrol.*, v. 3, p. 342–529.

Zbyszewski, G., 1963, Les phenomenes volcaniques modernes dans l'Archipel des Azores: *Comm. Geol. Serv. Portugal*, v. 47, 277 p.

Zbyszewski, G. and da Veiga Ferreira, O., 1962, La Faune Miocene de l'ile de Santa Maria: *Communic. de Servicos Geologicos de Portugal*, v. 46, p. 247–290.

Chapter 13

TECTONIC AND RADIOMETRIC AGE COMPARISONS

Frank John Fitch

Birkbeck College
University of London, U.K.

John Arthur Miller and Diana Mildred Warrell*

Department of Geodesy and Geophysics
University of Cambridge, U.K.

and

Susan Carole Williams

Birkbeck College
University of London, U.K.

I. INTRODUCTION

The present-day structure of the North Atlantic Ocean is dominated by the mid-Atlantic ridge and the three major crustal plates which meet at (and probably originate at) this feature. These major crustal plates are the Eastern North Atlantic or Eurasian, the Western Atlantic or American, and the African plates. The oldest dated ocean-floor rocks from the area are of Upper Jurassic (Oxfordian) age and are thought to indicate that the opening of the North Atlantic began about 180 m.y. ago (Early Jurassic, Le Pichon and Fox, 1971).

* Present address: Institute of Geological Sciences, London, U. K.

The mid-Atlantic ridge is the site of present-day volcanism: eruptions have occurred within the last five years in the Azores, Iceland, and Jan Mayen. Modern dating research has generally confirmed the predictions made by Wilson (1963) that the ocean floor and oldest volcanic rocks on the Atlantic oceanic islands would be found to increase in age away from the ridge (Funnel and Smith, 1968; JOIDES results; Moorbath *et al.*, 1968; Aumento *et al.*, 1968; Fleisher *et al.*, 1968; and many other authors).

Much further work on age patterns, paleomagnetism, and structure of the ocean floor is required before a final synthesis of oceanic development can be made for the North Atlantic. Seven distinct continental areas appear to be carried on the major Atlantic crustal plates at present; North America, Greenland, Rockall, Eurasia, Iberia, Africa, and South America. This chapter will be concerned only with the coastal areas of North America, Greenland, Eurasia, and Iberia directly abutting the North Atlantic, although such areas as the Mauritanides of northwest Africa and the Innuitian belt of Arctic Canada also have an important bearing on the problem.

The figures in this chapter have been drawn using the reconstruction of the North Atlantic by Bullard *et al.* (1965). This reconstruction was made by assembling the continents in a position of best fit and then refining this model by computation of the best least squares fit. It should be pointed out that several recent reconstructions show important differences; Flinn (1971) has demonstrated that in the case of Greenland and northwest Europe there are two possible "best-fit" positions which are about 5° (560 km) apart. Both fits are so good that it is impossible to make a choice between them on geometric grounds, but geological evidence favors the alternative reconstruction to that used by Bullard *et al.* (1965). Recent work in northwest Shetland (Miller and Flinn, 1966; Pringle, 1970) has shown that the Caledonian front passes through these islands, which would require "kinking" of the front, on previous reconstructions, between Greenland and Europe. Bott and Watts (1971) have also proposed an alternative fit for the continents incorporating the whole of the Faeroe rise as a continental fragment. Le Pichon and Fox (1971) have shown that a better fit of the major oceanic fracture zones is obtained if Africa is moved southward in relation to North America.

A major problem in geochronological correlation is the lack of consistency in decay constants used in the calculation of apparent ages. Several different time scales have been proposed using different decay constants within each isotopic system. From the geological viewpoint compatibility between the time scales used is more important than the *absolute* time scale.

Unless otherwise stated, all Rb/Sr data in this paper has been calculated using the decay constant $\lambda_{Rb} = 1.39 \times 10^{-11}$ yr^{-1}.

II. STRUCTURAL AND RADIOISOTOPIC DATA FROM THE NORTH ATLANTIC CONTINENTAL AREAS

Discussion of Precambrian data is complicated by the multiplicity of classifications in use. Table I compares the terminology of various authors. Bridgwater and Windley (1971a) and Salop (1968) present cogent reasons for not making a division between Katarchaean and Archaean time at 3000 m.y., and numerous further criticisms can be made of the presently accepted nomenclature. For the purpose of this review data from the Precambrian are discussed in three sections under the headings early, middle, and late Precambrian complexes.

Early Precambrian complexes include all rocks in which the major orogenic events, metamorphism, and intrusive activity are older than 3000 m.y. The early Precambrian complexes constitute the "low-level, high-grade Archean terrains" of Windley and Bridgwater (1971). They are composed, in the main, of quartzo-felspathic gneisses with intercalated basic igneous and anorthositic layered complexes and supracrustal belts all at the same very high grade of metamorphism. Occasional interleaving of lower grade supracrustal belts with migmatitic marginal zones and the observation of basement/dike swarm/cover rock relationships (e.g., McGregor, 1969) make it likely that most early Precambrian complexes have a multicycle origin. Postorogenic granites belonging to a number of these cycles are common. Due to partial or complete overprinting, most of the radiometric ages obtained from early Precambrian complex rocks are discrepantly low. It is likely, however, that the major orogenic events in these complexes were completed by 3500 m.y. and that orogeny between 3000 and 3500 m.y. was more localized, for during this latter period supracrustal rocks of a distinctive character began to accumulate in a number of linear basins throughout the world. These post-3500-m.y. basins are typically filled with a sequence of extrusive ultramafites, intrusive and extrusive basic igneous rocks, cherts, banded ironstones, and flysch-type deposits. Usually they have been involved in major orogenic events between 2400 and 2950 m.y. and are now seen as the typical low-grade greenstone–schist belts to be found within the "high-level Archaean crust" of Windley and Bridgwater (1971).

Middle Precambrian complexes include those rocks involved in 2400–2950 m.y. orogeny and/or reworking but not affected by post-1950-m.y. events. They constitute, therefore, the "high-level, low-grade Archaean terrains" of Windley and Bridgwater (1971).

In the late Precambrian, related families of mobile belts can be more easily distinguished. Two main groups are recognized: the Svecofennid–Churchill group with major orogenic activity within the period 1950–1550 m.y., and the Grenville group with major orogenic activity within the period 1200–850 m.y.

A. Early Precambrian Complexes

Major orogeny before 3500 m.y.; apparent ages mostly 2950 m.y.; Fig. 1.

1. *Europe*

In the British Isles the oldest apparent ages obtained from the Lewisian complex are all less than 3000 m.y. in age (middle Precambrian or younger), but geological evidence suggests that older rocks are preserved within the Scourian; Dearnley and Dunning (1968) propose a pre-Scourian period during which grey gneisses and microcline pegmatites were formed. The evidence for this comes from the early pegmatites which apparently underwent high-grade metamorphism and yield an apparent age of 2560 ± 80 m.y. (Rb/Sr age

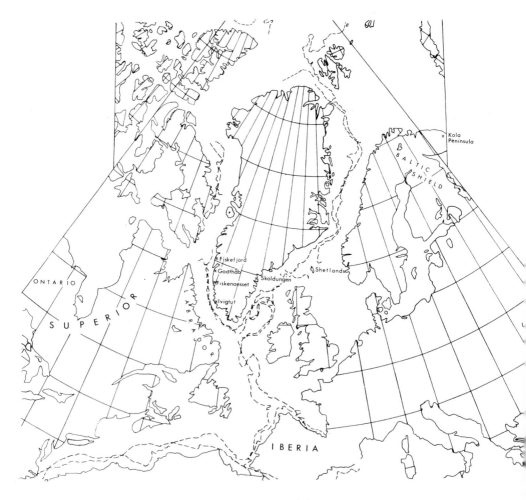

Fig. 1. Early Precambrian areas.

TABLE I

Comparative Table of the Division of Precambrian Time

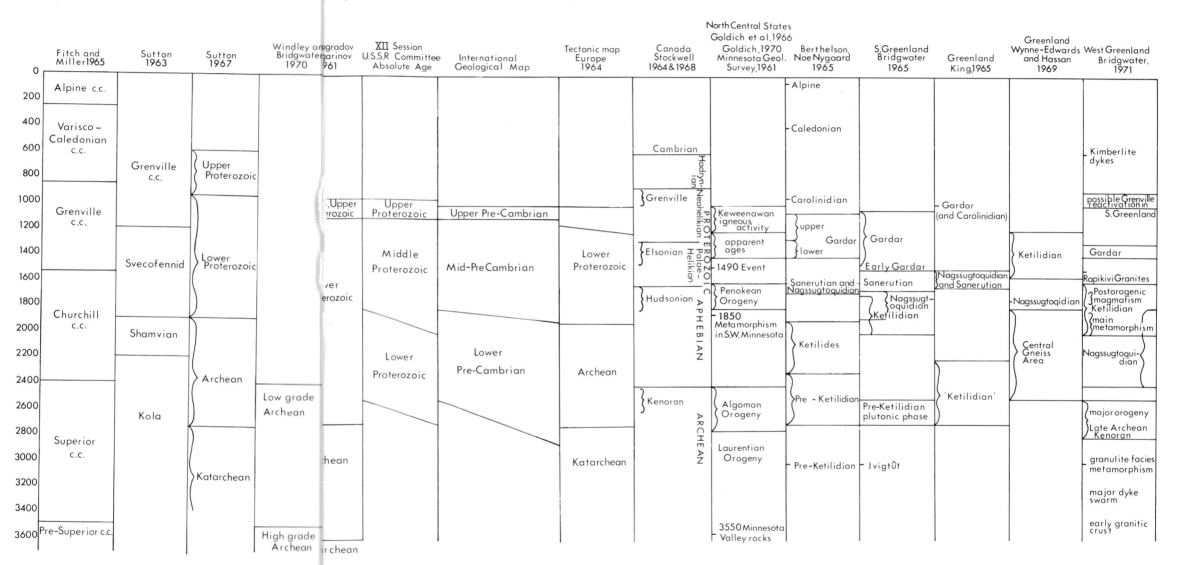

calculated using $\lambda_{Rb} = 1.4292 \times 10^{-11}$ yr^{-1}) (Lambert, Evans, and Dearnley, 1970). Lead isotope evidence suggests that some part of early Scourian pyroxene–granulite facies metamorphism may have occured before 2900–3000 m.y. ago showing that most of the Lewisian complex was in existence at this time (Moorbath *et al.*, 1969). Windley and Bridgwater (1971) recognize areas in Scotland which are typical of their early Precambrian low-level, high-grade metamorphic terrains; among these are the high-grade supracrustal rocks described by Coward *et al.* (1969) and the calcic anorthositic rocks on Lewis (Watson, 1969).

The oldest rocks in continental Europe have apparent ages in the range 3600–3300 m.y. (see Table I). They occur in the northeast of the Baltic shield, the Kola peninsula, and the Ukraine. In both areas the old rocks occur as discrete masses within apparently younger migmatites and granite gneisses, which are thought to represent, at least in part, reworked early Precambrian material. Although rocks of the Kola peninsula (termed the Saami protoplatform by Semenenko, 1970) have undergone high-grade metamorphism, they can be seen to include psammatic and pelitic metasediments. The oldest formations give apparent ages of 2700–2000 m.y. (the Polmos Series) and 3300–3400 m.y. (the Monchegorsk peridotite). A few older dates around 3600 m.y. have been reported from this area (Sutton, 1967).

Kratz *et al.* (1968) believe that most of the Baltic shield developed fairly continuously during the interval from 3500–1650 m.y. but find that the pattern of age zonation does not uphold the continental accretion theory (Hurley, 1970; Hurley and Rand, 1969; and others). They suggest that the oldest rocks occur in the Saami–Karelian zone in the east of the Kola peninsula where basement rocks have produced numerous apparent ages over 2600 m.y. by K/Ar, Rb/Sr, and U/Th/Pb methods.

The Ukranian shield forms the southern and oldest part of the eastern European platform; Semenenko *et al.* (1968) group rocks of apparent age 3500–2600 m.y. together as the Dneiper Konski protoplatform. Two orogenic cycles have been recognized within this group with apparent ages of 3500–3100 m.y. (the Konkian) and 3100–2700 m.y. (the Aulian). Thus the Aulian is typical middle Precambrian in age. Most of the age determinations of these rocks are U/Th/Pb on accessory minerals or K/Ar on amphiboles.

2. *North America and Greenland*

Although there are no large areas of unaltered early Precambrian complex rocks in North America, there is abundant evidence of an ancient crust over 3000 m.y. old which has been almost completely reworked by later events. Apparent ages of 3000 m.y. and older have been found in Ontario (Wells, in Gross and Ferguson, 1965; Catanzaro, 1963; Slawson *et al.*, 1963; Kanasewich and Farquhar, 1965), Manitoba (Ozard and Russell, 1971), and south-west Minnesota (Catanzaro, 1963). The oldest apparent rock ages obtained in North America are around 3500 m.y. (Goldich et al., 1970, 1966, 1961). Distinctive high-grade basement gneisses of pre-3000-m.y. age are found in the Superior and Great Slave provinces, in Labrador (Bridgwater et al., 1971), in the Godthåb Fjord region of West Greenland (Black et al., 1971), and East Greenland (Bridgwater and Gormsen, 1968, 1969). Calc-anorthosite rocks formed before 3000 m.y. occur in the West Greenland coastal strip from Fiskefjord–Ivigtut (Windley, 1969a), East Greenland (Bridgwater and Gormsen, 1968, 1969), and the Fiskenaesset complex of West Greenland (Ghisler and Windley, 1967). Rb/Sr dating in South Greenland has detected no evidence so far for pre-Ketilidian (middle Precambrian) basement in this area (Van Breeman et al., 1971). Geological evidence from West Greenland shows that most of the "Archaean" rocks of this area were formed before 3000 m.y. and were re-worked during the Kenoran orogeny. Lead isotopes from galenas of the Ivigtut region show initial ages of over 3500–3000 m.y. Hornblende from the granulite facies complex of the Fiskenaesfjorden region of southern West Greenland was given a minimum age of 3210 m.y. by Lambert and Simons in 1969. Amphibolite grade felspathic gneisses from Godthåb have given the oldest recorded Rb/Sr isochron of 3980 ± 170 m.y. and Pb/Pb whole-rock isochron of 3620 ± 100 m.y. (Black et al., 1971).

A late igneous stage in the metamorphism of these high-grade areas has been detected in the norite–charnockite complex at Skjoldungen in southeast Greenland (Bridgwater and Gormsen, 1969).

B. Middle Precambrian Complexes

Major orogeny between 2400 and 2950 m.y.; Fig. 2.

1. Europe

The largest block of middle Precambrian rock in Europe extends from the White Sea southward for 900 km through Karelia to Lake Onega and Lake Ladoga, northeast of Leningrad. Smaller areas of middle Precambrian are also found in northwest Scotland, Sweden, Finland, and the Lofoten Islands.

The early Precambrian basement, already described, shows evidence in many areas of partial or complete overprinting between 2500 and 1700 m.y. with a pattern of discrepantly low conventional K/Ar age determinations from biotites, muscovites, and amphiboles. Cutting across this zone is the east–west Belmorian (or White Sea) subzone which exhibits a mixed pattern of radiometric ages, the oldest apparent ages averaging around 2800 m.y. The great majority of individual dates, however, fall in the range 1600–1900

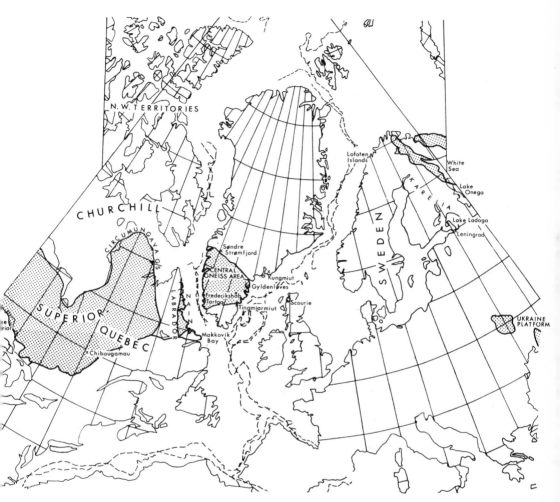

Fig. 2. Middle Precambrian areas.

m.y. (amphiboles, biotites, and muscovites) and are discrepantly low ages. U/Th/Pb dates from pegmatites show a peak at 1960 m.y., which probably represents the final magmatic event in the development of the Belemorides.

Further south, the Voronezh and Volga-Ural blocks within the Russian platform have yielded apparent ages in the range 2400–2000 m.y. (Semenenko, 1970).

In the Ukraine platform, drilling has proved the presence of large areas of middle Precambrian rocks below the present cover of young rocks. Extensive granite and migmatite complexes were formed during the interval 3100–2700 m.y. (the Aulian orogeny). In the north Ladoga region, Gorokhov *et al.* (1970) have described basement gneisses which give an apparent Rb/Sr age of 2320 ± 100 m.y., but most of this area has been overprinted during late Precambrian time.

To the west of the Baltic shield are the younger belts of the Svecofennides and the Caledonides, but the old Precambrian dates are encountered to the northwest of these belts in northwest Scotland and Norway. The earliest apparent ages found in the Lewisian complex are from Scourian rocks (Evans, 1965). The following major episodes have been recognized in the Scourian by many authors (Sutton and Watson, 1951, 1959; Giletti et al., 1961; Dearnley and Dunning, 1968; Evans, 1965).

1. "Pre-Scourian" metamorphism, 2900 m.y. and older.
2. Granulite facies metamorphism, 2600 m.y. and older.
3. Pegmatite intrusion and further plutonism, 2600–2200 m.y.
4. Amphibolite facies metamorphism around 2200 m.y., and intrusion of Scourie dikes.

Pidgeon (in Dunning, 1972) has reported apparent ages of 2700–2800 m.y. (U/Pb) from zircons in granulite facies rocks of Scourian age. An important feature in the Scourian is the intrusion of the end-Scourian dike swarm, which is over 250 km wide and which has given apparent ages of 2200–1900 m.y. (Evans and Tarney, 1964). Dearnley has shown from geological evidence that the dikes were intruded into cold crust and suffered later reheating between 1900 and 1700 m.y. Park (1970), in a review of Lewisian chronology, suggests that reliable correlation within the Lewisian is very difficult due to the widespread occurrence of multiple remetamorphism within the complex, and proposes a threefold division of the Lewisian into Badcallian (early Scourian), Inverian (late Scourian), and Laxfordian episodes. The early granulite facies metamorphic episodes give apparent ages of 2600 m.y. or older (Evans, 1965), while later pegmatites have given apparent ages of 2560 ± 80 m.y. (Lambert, Evans, and Dearnley, 1970) and 2540 m.y. in the Scourie area (Giletti, 1961).

Metamorphic events at 2800 and 1800 m.y. in the Lofoten Islands of northern Norway have been dated by K/Ar determinations on biotites while Re/Os dating has yielded apparent ages of 2080 and 2290 m.y. (Herr and Merz, 1958). However, most of the rocks in this area appear to have been completely overprinted during late Precambrian times (Heier and Compston, 1969). To the north of the Lofoten Islands, Heier and Griffin (1971, unpublished) have obtained an apparent age of 2800 m.y. (Rb/Sr) from a granulite facies gneiss on Vesterålen.

2. North America and Greenland

On a correlation across the Atlantic it would appear that the western part of Greenland, Canada, and America were stabilized areas with a few basic dike intrusions while the Belmoride orogeny was still in progress in Europe (see Table I). It seems unlikely that the middle Precambrian gneisses were all formed as a result of a single phase of folding and metamorphism, and in western and southern Greenland there is evidence of several chronological divisions (Bridgwater and Windley, 1971a). In Canada there are several profound unconformities within the Precambrian rocks, but the dates of most of these have been masked by the later Kenoran orogeny. It seems that during the middle Precambrian time metamorphic belts were superimposed so that few rocks escaped complete reworking by later metamorphic phases. A brief plutonic event dated at 2740–2700 m.y. in northwest Quebec can be traced for wide areas.

The first main orogenic event of the middle Precambrian has been dated by Burwash (1969) at 2500–2700 m.y. The younger of these dates were obtained by U/Pb dating of accessory minerals, while the older dates were obtained by K/Ar determinations on micas.

In the Lake Superior region the Laurentian granite intrudes Keewatin greenstones and is overlain by middle Precambrian rocks showing geological evidence of early plutonism. There are, however, no dates to confirm this event because of subsequent overprinting. In Minnesota there was a period of metamorphism and granitic intrusion and this is dated at 2400–2750 m.y. (the Algoman orogeny) and may compare with the Kenoran orogeny of the Canadian shield. King (1969) regards the Canadian provinces as individual fold belts. Thus the middle Precambrian consisted of the Superior, Slave, and Nain fold belts. The Slave Province once may have formed a single block with the Superior Province but is thought by Gibb (1971) to have split away at about 2390 m.y. ago (the end of the Kenoran orogeny). Some evidence suggests that the whole of the Canadian shield area is underlain by early Precambrian rocks making a splitting hypothesis unlikely (Gibb, 1968). Isolated Kenoran dates have been obtained within the Churchill Province, which could indicate that small remnants of ancient complexes of middle Precambrian are reworked in the younger Churchill mobile belt. Gibb (1971) suggests continental accretion around the Slave and Superior cratons. The Nain belt, defined by Taylor (1971), forms a narrow strip 65 miles wide along the east coast of Labrador and is closely linked with the central gneiss area of West Greenland (Bridgwater and Windley, 1971a) (see Fig. 2).

During the middle Precambrian (Kenoran) event most of the early Precambrian rocks of the Superior Province were reworked and suffered reheating, migmatization, and metamorphism. In the west there are no plutonic rocks with apparent ages less than 2500 m.y. old. Fold belts of this age converge toward the Arctic, and in Labrador the Kenoran orogeny is characterized by northerly trending structures. Dearnley (1966) claimed that middle Precambrian fold belts are not confined within forelands of unfolded rocks and that few areas have avoided repeated reworking. In Canada and Greenland clastic sediments and calc-alkaline volcanics formed in complex synclinal belts between intrusive and metamorphic rocks. These are low-grade metamorphic greenstone belts which can be traced for large distances from the Northwest Territories of Canada through the Canadian shield to West Greenland. Dating of the rocks is masked by the 2500 m.y. Kenoran reworking (Wanless *et al.*, 1967; Roscoe, 1971; Goodwin and Ridler, 1970). It is claimed that the greenstone belts formed a closed system to loss of radiogenic elements earlier than the underlying high-grade metamorphics and that in places the radiometric dates can give a false impression that the underlying basement rocks were younger than the surface supracrustal rocks. In the Canadian shield volcanic rocks of the middle Precambrian, termed the Keewatin volcanics, include dacites, rhyodacites, and rhyolites with numerous basalts and some andesites. In Greenland there are fewer volcanic rocks, but these are similar to their Canadian counterparts (Windley *et al.*, 1966). A preorogenic series of sediments and volcanic rocks were deposited in southeast Greenland. In West Greenland the central gneiss area reaches from Frederikshåb to Søndre Strømfjord and is dated at 1650–2000 m.y. by Pulvertaft (1968), and in southeast Greenland from Gyldenløves Fjord to Tingmiarmiut Fjord where inhomogeneous gneiss and biotite gneisses are dated at 2190 m.y. by Larsen (1969) and Bridgwater and Gormsen (1969) (Fig. 2).

Metamorphic grade increases northward in both the Superior Province and Greenland, the highest grade being granulite facies which belongs to the early Precambrian. The Hopedale gneiss of the Makkovick Bay area of Labrador is dated at 2655 m.y. by Wanless *et al.* (1965) and 2430 m.y. by Leech *et al.* (1963). A transition from low- to high-grade metamorphic rocks has been distinguished by Windley *et al.* (1966) in West Greenland. These rocks were later migmatized to hornblende and biotite gneisses.

The late igneous intrusions in the Kenoran orogeny compare with those of the Phanerozoic fold belts in form. Early plutonism occurred with migmatite fronts advancing on early Precambrian crystalline rocks, while older middle Precambrian volcanics and sediments give a peak of apparent ages at 2700 m.y. In West Greenland the earliest plutonic rocks were dated at 2500–2700 m.y. by Allaart *et al.* (1969). Later intrusions included pegmatites, granites, gabbros, and anorthosites at 2400 m.y. In Labrador amphibolite metamorphism and acid plutonic activity and extensive migmatization characterizes the Kenoran event (Bridgwater and Windley, 1971*a*). Late tectonic granite invaded the gneisses in southeast Greenland.

A 2000–2600 m.y. postorogenic phase of dike intrusion occurred throughout much of Greenland and the Nain Province of Labrador (Bridgwater and Windley, 1971*a*). In southeast Greenland swarms of basic dikes cut the gneisses, and the hypersthene dolerite dikes of the Kungmiut area are comparable with similar dikes of West Greenland (Berthelsen and Bridgwater, 1960). This late-stage dike intrusion was associated with middle Precambrian faulting. It is at this period of the orogeny that Gibb (1971) suggests the major rifting about the Circum-Ungava geosyncline with the movement of the Slave Province craton away to the west took place. Areas of granulite facies metamorphosed rocks of middle to early Precambrian age are found in the Agto district with overprinted apparent ages of 1710–1650 m.y. (Larsen and Møller, 1968a,b).

C. Late Precambrian Complexes

Precambrian mobile belts younger than 2400 m.y.; Fig. 3.

The regime of mobile belt activity that was initiated by the greenstone belts of the middle Precambrian became more clearly defined during the late Precambrian. Two main groups of mobile belts can be recognized; the Svecofennid–Churchill group (see Fig. 3) and the later Grenville group (see Fig. 4). Orogenic activity occurred at 1950–1550 m.y. (Svecofennid–Churchill) and at 1200–850 m.y. (Grenville). Between these two major periods of activity there were some minor events which will be considered separately under the heading of Elsonian–Gothian events.

1. The Svecofennid–Churchill Group

The Svecofennid Group of Europe. In western Europe sediments of Svecofennid age rest unconformably on folded early and middle Precambrian rocks older than 2500 m.y.; in eastern Europe it is difficult to distinguish between the continuing effects of middle Precambrian events and those of the late Precambrian. Sediments overlie basement rocks dated at 2100–2000 m.y. in Krivoi Rog and Kursk and at 2500–2300 m.y. in Finland and Sweden.

The European Svecofennid belt occupies the central and western part of the Baltic shield and continues south into the Ukraine where it forms the Ingulo–Krivoi–Rog belt. Further east 2000 m.y. apparent ages come from deformed Precambrian rocks in the median massifs of the lower Paleozoic Urals belt. Apparent ages within the Svecofennides indicate initial plutonism between 1800 and 1600 m.y., with a major peak at 1750 m.y. in Scandinavia, Karelia, and the Ukraine. The main metamorphic event in this belt is thought to have occurred between 1775 and 1900 m.y., and an apparent age of 1885±20 m.y. has been obtained from the Ladoga region of Karelia (Gorokhov *et al.*, 1970). Further west, basement granite gneisses in the northwest Nord-Trondelag area of Norway have given apparent ages in the range 1700–1800 m.y. (Z. W. O., 1968). In the northwest part of the Ukraine, Semenenko *et al.* (1968) date an orogenic episode between 1700 and 1500 m.y. Postorogenic granites in the north Ladoga region have given an apparent age of 1815 m.y.

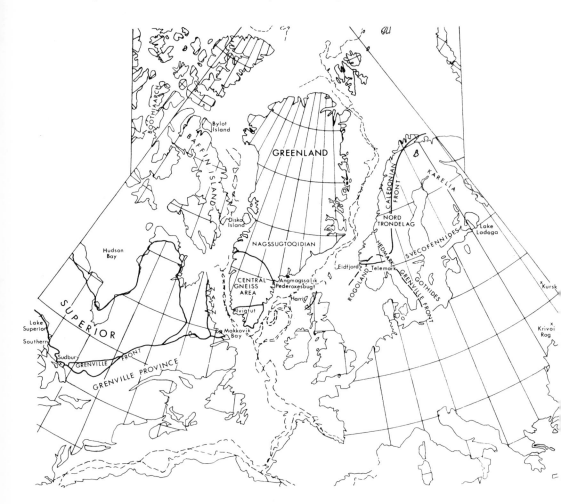

Fig. 3. Late Precambrian–Svecofennid/Hudsonian areas.

Svecofennian of the British Isles. The Svecofennian belt is represented in Scotland by a period of metamorphism and granite intrusion between 1700 and 1900 m.y. (the Laxfordian). Prolonged activity caused the lowering of many ages in this range to 1500–1400 m.y., and there is also evidence of a further lowering of ages at around 1200 m.y., possibly associated with extensive faulting.

A Rb/Sr whole-rock isochron of 1850 ± 50 m.y. has been obtained from Laxfordian gneisses by Lambert and Holland (1971), who suggest that this event gives a minimum age for the climax of Laxfordian metamorphism. The highest apparent K/Ar mineral age from the type area is 1750 m.y. Younger apparent ages from minerals in the same area and close to it range down to

1575 ± 50 m.y., with a further peak at 1400 m.y.; these have been attributed to late metamorphic effects, pegmatite formation, and the slow cooling of the system as a whole. Lambert and Holland (1971) describe this disturbed belt of low ages as a region of largely structural (as opposed to thermal) disturbance along shear zones. Apparent ages of 1500, 1565, and 1720 m.y. have been reported by Lambert, Evans, and Dearnley (1970) from the Outer Hebrides where they are related to the age of regional metamorphism and pegmatite intrusion.

Van Breemen *et al.* (1971) report an apparent age of 1750 ± 34 m.y. (Rb/Sr whole-rock isochron) from the Harris Granite. U/Pb determinations on zircons from the granite give a minimum age of about 1715 m.y.

Bridgwater and Windley (in discussion of Lambert and Holland, 1971) demonstrated that the Laxfordian forms part of the thermally active belt of the Svecofennides, which was later affected by the anorthosite event that will be discussed below. They suggest that this area represents a zone of thermal activity which remained in existence for 600–800 m.y. after the main orogenic episode and which was largely responsible for the scatter of apparent ages between 1200 and 1700 m.y. Park (1970) proposes alternatively that the pattern of isotopic ages may be due to three separate events: a metamorphic episode at, or before, 1850 m.y., a widespread migmatitic or pegmatite-forming episode between 1500 and 1400 m.y. (with a peak of apparent ages at 1400 m.y.), and, finally, a minor reheating event at about 1100 m.y.

Churchill Group in North America. Sediments and volcanics were already being deposited in a shifting system of basins before the end of the middle Precambrian, for instance, the Great Slave Group and the Huronian Group which was derived from the rising Superior craton to the north of the Sudbury region and dated at 2170–2100 m.y. Sedimentation in these basins spans a wide time range, and some has been effected by the Hudsonian folding (Roscoe, 1971).

Rocks of age 2000–1600 m.y. are found in several areas in the West Atlantic region—the Canadian Churchill Province, the Southern Province, the Western Nain Province of Labrador [now redefined as part of the Churchill Province by Taylor (1971)], and the north central and southern strips of Greenland (Fig. 3). The dates give evidence for a very widespread metamorphic and orogenic event in the late Precambrian (Hudsonian, Ketilidian, and Nags-sugtoqidian) which effected post-Kenoran sediments and volcanics within the new system of cross-cutting mobile belts as well as older basement material. The metamorphic event masks the distinction between middle Precambrian and supracrustal rocks in much of the area although the unconformity can be seen in Labrador and South Greenland (Higgins, 1970; Higgins and Bondesen, 1966). In the Labrador trough and the Circum-Ungava geosyncline detail

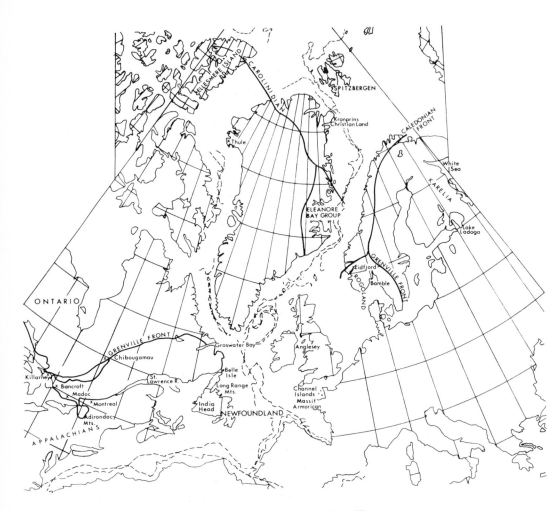

Fig. 4. Late Precambrian–Grenville areas.

of sedimentation (Dimroth, 1970) and stratigraphy has not been completely obliterated, and these sedimentary troughs prove an interesting study as the earliest basins confined between middle Precambrian cratonic forelands which are comparable with Phanerozoic sedimentary basins. Churchill rocks bounding the Labrador geosyncline are dated at 1580 m.y. by Gibb and Walcott (1971). In the Wollaston fold belt of Saskatchewan the main Hudsonian orogeny and folding has been dated at 1750 m.y., although the surrounding basement —giving an apparent age of 2200 m.y. by Rb/Sr determination—was not effected by the Hudsonian metamorphism. (Money, 1968; Beall et al., 1963). This has lead Fahrig and Eade (1968) to suggest that the Hudsonian orogeny was merely an overprinting event. In the Churchill Province in northern

Manitoba and Saskatchewan the basal Precambrian was involved in two tectonic epochs, the latest of which is dated at 1700 m.y. by Milligan (1970). The northwest shelf also shows two main periods of deformation (Gibb, 1971).

Metamorphism increases away from the Superior craton, and in Labrador rocks of low greenschist facies abut the craton and pass into a zone of metasediments and migmatites in the center of the region with high-grade migmatites and supracrustal rocks of granulite facies on the eastern margin, furthest from the Superior craton. Apparent ages of 1640–2380 m.y. can be traced south of the recognized middle Precambrian margin to the Lac St. Jean area of Quebec where Rb/Sr dates of 1750 m.y. have been recorded (Frith and Doig, 1971).

In the Southern Province of northern Michigan and Minnesota two periods of metamorphism are traced, the younger of which is dated at 1700–1900 m.y. by Goldich *et al.* (1961), correlating with the Hudsonian orogeny. The earliest plutonism in northwest Canada began in 1800–1600 m.y.; in the Makkovik Bay area of Labrador metamorphism and intrusion is dated at 1600 m.y. (Gandhi *et al.*, 1969). Appinitic and calc-alkaline igneous activity is dated at 1600–1700 m.y. and corresponds closely with that in Greenland. In the area west of Hudson Bay, plutonism had ended by 1600 m.y.

Much of Baffin Island is underlain by metasedimentary rocks of probable late Precambrian age which include sequences of basic and ultrabasic sills and dikes associated with basalt flows and very high-grade iron deposits (Jackson, 1966, 1969, 1971). Blackadar (1967*a*) has dated the Churchill Province basement gneiss of granulite–amphibolite facies from the Boothia Arch as 1630–1670 m.y., which must be the final closing of the system for K/Ar elements. These gneisses are typical of deep-level deformation in mobile belts and show three phases of deformation (Brown *et al.*, 1969). A granulite facies belt of metamorphism stretches across Baffin Island from Bylot Island to the North Barnes Ice Cap and covers most of southern Baffin Island.

Churchill Province in Greenland (Ketilidian, Nagssugtoqidian, and Sanerutian orogenies). In southern Greenland the Ketilidian mobile belt is characterized by widespread syntectonic granite and basic magmatism dated at 1750–1950 m.y. (Windley and Bridgwater, 1971). To the south the rocks are increasingly severely deformed, metamorphosed, and intruded. The same three zones of metamorphic grade are traced here as in Labrador (Bridgwater *et al.*, 1971) with a marginal zone of greenchist facies, a central zone of gneisses and granites including the Julianehåb Granite, and the high-grade granulite facies rocks which outcrop in the far south of Greenland.

In the northwest of Greenland the 1650–1750 m.y. (Nagssugtoqidian) orogeny is correlated with the Ketilidian and Hudsonian events in southern

Greenland and America (Larsen and Møller, 1968). In this belt the metamorphic grade increases toward the center where a regional thermal event with refolding has occurred. Granite intrusion on Disko Island is dated at 1850 m.y., and plutonism ended by 1700 m.y. (Bridgwater et al., 1971). To the east the Nagssugtoqidian is overprinted by the later 900 m.y. Carolinidian event.

In southern Greenland a 1500–1640 m.y. late tectonic episode with rapakivi granite intrusion called the Sanerutian orogeny is comparable with posttectonic granite intrusion in Scandinavia. In the Ivigtut region Sanerutian granites are dated at 1600 m.y. On the east coast this overprint is found as far north as Angmagssalik (King, 1969). After 1600 m.y. plutonism was restricted to smaller areas in the mobile belts of eastern Canada and southern Greenland.

2. Elsonian–Gothian Events Across the North Atlantic

In the postorogenic period after 1650 m.y. Bridgwater and Windley (1971a) note variation in the geological evolution of different parts of the shield. In some areas, minerals yield ages which are 50–100 m.y. younger than Rb/Sr age determinations of whole-rock samples, and this has been attributed to normal postorogenic uplift and cooling effects; but in other areas a much more complicated geochronological pattern emerges. Apparent age determinations on metamorphosed country rocks are of up to 600 m.y. younger than those for the intrusions causing regional metamorphism determined by more sophisticated methods. This discrepancy is too large to be attributable to slow cooling. Widespread thermal activity (dated in the range 1200–1700 m.y.) is not comparable in scale or character with major Precambrian mobile belts (Bridgwater and Windley, 1971a).

In Labrador this period of metamorphic activity is known as the Elsonian orogeny but it represents such a wide spread of ages that it is not a single orogenic event (Bridgwater et al., 1971). Wynne-Edwards and Hassan (1970) have recognized no specific peak of ages for this event. Emslie (1970) regards the minimum age of anorthosite intrusion in this area as 1500 m.y. There is no obvious fold belt associated with the Elsonian age group of rocks in Labrador, and it appears to be a minor overprinting event associated with batholith intrusion and possibly downwarping of the crust (Taylor, 1971).

By 1200 m.y. there was a great restriction in the size of the mobile belts, and in southern Greenland plutonic activity was reduced to a few isolated post-tectonic granites and intrusions associated with faulting. The Gardar sequence of Southern Greenland included sandstone deposition associated with alkaline intrusion. The earliest Gardar intrusions are dated at 1300–1400 m.y.

A phase of dike intrusion at 1200 m.y. was followed by the major alkaline complex intrusions of southwest Greenland dated at 1255–1020 m.y. by Bridgwater (1965).

In Canada and Baffin Island a few post-tectonic dikes were intruded at this time, and in southwest Greenland olivine tholeiitic dikes were intruded parallel to the coast. The Duluth and Muskox intrusions of central Canada were also intruded at this time (Robertson, 1969).

In Europe this episode is generally known as the Gothian, and is characterized by apparent ages in the range 1500–1300 m.y. (Z.W.O., 1968) mainly obtained from the "Gothides" of southern Sweden. The Dala porphyries in the Dalrna region of Sweden have been dated at 1590 ± 65 m.y., while the sub-Jotnian basement volcanics in east Hedmark give an age of 1541 ± 69 m.y. (Rb/Sr whole-rock isochron). Veersteve (1970) has reported ages in the range 1550–1600 m.y. for the Telemark supracrustal rocks. In Rogoland to the southwest of Telemark an apparent age of 1478 ± 78 m.y. has been reported from basement migmatites which were largely overprinted at about 1100 m.y. A zone of hyperite intrusion in southern Sweden has given an apparent K/Ar age of 1550 ± 100 m.y. (Z.W.O., 1968), and the same age has been reported from the basement gneisses of the Eidfjord area, Norway. In the Ukraine region Semenenko et al. (1968) correlate the east–west Ovruch–Volnian mobile belt giving apparent ages of 1500–1200 m.y. with the Gothides to the west.

The area of high thermal activity described by Bridgwater and Windley (1971a) corresponds to the zone of massive anorthosites and rapakivi granites which form part of the "anorthosite event" described by several authors (Windley, 1970; Hernes, 1966; Payne et al., 1965; Doig, 1970). Geochronological evidence suggests that this postorogenic zone may young in a westerly direction; in the Ukraine, the rapakivi granites and anorthosites were probably emplaced between 1700 and 1800 m.y. ago while in North America the majority give apparent ages in the range 1400–1000 m.y. The emplacement of the "postorogenic" igneous suites was commonly controlled by tectonic lineaments and grabenlike structures which later controlled the deposition of the continental sandstones. The time limits of this event are between 1750 and 1100 m.y. (Herz, 1969) with a concentration of dates at 1300 ± 200 m.y. Herz (1969) suggests that the anorthosite belt does not appear to be related to any known tectonic trends in the Precambrian, but he points out that many of the intrusions tend to lie in the same belt as the younger Grenville Province. Bridgwater and Windley (1971a) conclude that the whole Svecofennian belt and the control of subsequent metamorphic events during the initiation of the Grenville orogeny was controlled by a single major feature in the crust and upper mantle; the type of metamorphism and the particular magmatic rocks produced point to the fact that this belt may have been the site of a major upwelling of mantle material between 1000 and 2000 m.y. The concentration of these rocks in a relatively limited part of the earth's history is ascribed to a unique combination of high heat flow and stable crust.

3. *The Grenville Period* (1200–850 m.y., Fig. 4)

Introduction. Toward the end of the Proterozoic great thicknesses of sediments were deposited in a number of long mobile belts bounding stable areas consolidated during Svecofennid–Churchill times. One such belt runs from western North America through Canada, North Greenland, Spitzbergen, and into Norway where it splits into two branches; one branch passes east along the north side of the Baltic shield to Novaya Zemlya, the Timan peninsula, and then south through the Urals, while the other runs along the southeast coast of the United States. An eastward extension of the Scandinavian–British belt is thought to have run westward through central Europe, but here later Variscan and Alpine orogenic activity has overprinted earlier events.

Dating of the Grenville is confused. In the southern branch, major activity appears to be confined to the interval 1200–950 m.y., while elsewhere dates of 650 m.y. or less are assigned to the Grenville. Recent reinterpretation (mainly on the Canadian Grenville Province) indicates that the Grenville period shows a pattern of repeated tectonic overprinting (complete references in Wynne-Edwards and Hassan, 1970). For example, low K/Ar dates of rocks effected by the Carolinidian orogeny in Greenland may be due to overprinting by Paleozoic thermal events. It was obviously a long and complex period with activity in different parts of the mobile belt at different times.

The Grenville Province in North America—The Southern Branch of the Grenville Cycle. The Grenville front (Fig. 4) forms a boundary between rocks of distinct deformation and metamorphism, and in many places it is a line of northwest thrusting. It forms the northern margin of the Grenville Province in which rocks of 800–1100 m.y. predominate although relict age pattern first developed in the Hudsonian and Kenoran orogenies can be traced southward through the province (Frith and Doig, 1971; Stockwell, 1964). The front truncates at high angles all structures of the earlier provinces and can be traced south of the Benedict Mountains and through Pottles Bay in Labrador (Stevenson, 1970) or through the Smokey Archipelago (Grasty *et al.*, 1969) or through Groswater Bay on paleomagnetic data (Fahrig, in Taylor, 1971). Further east the front passes south of Greenland and only minor faulting during the Grenville period seems to have affected the rocks of southern Greenland. In Ontario dating along the Grenville front in the Kilarney district by Krogh (1971) indicates that structural features typical of the Grenville front were established at least 1460 m.y. ago and the possibility that some of the lower apparent ages obtained within the Grenville province reflect the age of final uplift or late overprinting.

In Canada the metamorphism and plutonism of the Grenville cycle seems to have begun before much sediment had accumulated, and it is thought to be chiefly a reworking of the middle Precambrian and earlier basement. Many rocks still retain older ages south of the Grenville front, and Stockwell (1964) has traced early Precambrian and middle Precambrian rocks retaining relict ages through the Grenville fold belt to Chibougamau.

Much detailed work has been carried out on the dating of Grenville rocks. Wynne-Edwards and Hassan (1970) find a sharp peak of K/Ar apparent ages at about 950 m.y. showing very little subsequent reworking. Rb/Sr apparent ages are generally much older and probably reflect middle Precambrian and late Precambrian ages of rocks involved in Grenville reworking and overprinting (Frith and Doig, 1971; Wanless et al., 1970). In the Grenville area it appears that intense metamorphism was followed by plutonism, but Baer (1971) considers that all deformation and metamorphism was associated with anorthosite intrusion in the fold belt from Nain through the Grenville Province. The anorthosites in the Grenville area show no deformation subsequent to intrusion and probably were intruded in the 1400–1000 m.y. anorthosite phase (see p. 499). The small amount of granite intruded, lack of structural grain parallel to the boundaries, and lack of a single strong regional metamorphic event make the Grenville crustal disturbance in this area differ from what is usually referred to as an orogenic belt. Wynne-Edwards and Hassan (1970) believe that in the Grenville we see only the basement of a major orogenic sequence, and overlying metasediments and granitic intrusions have already been eroded away. Bridgwater and Windley (1971a) consider that the anorthosites were emplaced prior to the last tectonic events in the Grenville orogeny, thus contradicting Baer (1971).

It is likely that the North American Grenville Province once stretched far southeast of its present limits and has since been mostly reworked by the Appalachian mountain building episodes. North of Montreal the chief penetrative deformation is dated at 1175–1110 m.y. Carbonatite intrusions from the north of Montreal are dated at 822–1005 m.y. by Doig and Barton (1968). It is thought that the first appearance of the carbonatite sequence is associated with the St. Lawrence rift system. Detailed work on the Madoc and Bancroft areas north of Lake Ontario by MacIntyre et al. (1967) has shown a very complex history for the Grenville Province with several stages of intrusion and metamorphism. They consider that the chief igneous and metamorphic events took place 1000–1300 m.y. ago in several different stages. Different types of dating give different ages, and K/Ar apparent ages are consistently young at about 950 m.y. Apparent ages of 1700–1500 m.y. from Ontario by Dence et al. (1971) are thought to date initial crystallization in this area which was later affected by slight thermal metamorphism.

Numerous inliers of Grenville aged rocks occur along the axis of the Appalachian belt, in the Long Range Mountains of Newfoundland (dated by Williams, 1969; Lowdon *et al.*, 1963; Pringle *et al.*, 1971; Silver and Lumbers, 1965; and Wanless *et al.*, 1969), and in New England (dated by Brace, 1953; Williams, 1969; Long and Kulp, 1962; and Silver, 1968). Apparent ages by K/Ar analysis for these inliers are generally low, from 800 to 945 m.y., but Rb/Sr dating reveals age for intrusion at 1160 m.y. (Pringle *et al.*, 1971). Williams (1969) believes the anorthosite intrusions in the Long Range Mountains may be Elsonian in age and later overprinted by the Grenville. In southern Quebec a phase of alkali igneous activity is dated at 900 m.y. (Doig and Barton, 1968). Recent work by Fahrig *et al.* (1971) has distinguished the very widespread basic igneous intrusions and extrusive activity, including dike swarm intrusion, in the Coppermine and McKenzie districts of Canada.

The Northern Branch of the Grenville Belt in North America and Greenland. A Grenville belt passes up the west coast of North America through Arctic Canada, across northern Baffin Island, North Greenland (Thule area), and through eastern Greenland (the Carolinidian). Evidence of a geosynclinal trough along the northeast edge of the Canadian shield which formed before 1000 m.y. ago has been masked by superimposed Phanerozoic orogenies and mobile belts. In Arctic Canada deposition in this late Precambrian belt (the Franklinian geosyncline) was on a very large scale (Trettin, 1969a,b, 1971). The Eleonore Bay Group and the Thule Group of sediments (Berthelsen and Noe-Nygaard, 1965), although not closely dated, are thought to have been deposited largely in pre-Carolinidian times. In Kronprins Christian Land arenaceous sediments were deposited before 988 m.y. and were invaded by basic intrusions (Henriksen and Hepson, 1970). The sediments are comparable with the Hecla Hoek of Spitzbergen (Harland, 1969). In the Carolinidian orogeny the major episode of folding and metamorphism is dated at about 1000–800 m.y., and the apparent ages found in the range 788–988 m.y. are probably discrepantly low due to subsequent overprinting.

Intrusion of a postorogenic dike swarm followed metamorphism and can be traced over a wider area than the orogenic phase. Postorogenic tholeiitic dikes found in Baffin Island are dated at 915 m.y. (Blackader, 1970). Dikes off the coast of Labrador are dated at 990 m.y. by Gandhi *et al.* (1969) and in Newfoundland are dated at about 950 m.y. (Pringle *et al.*, 1971).

A later phase of igneous activity producing dikes and lavas is dated at about 655–700 m.y. by Fahrig *et al.* (1971), and this can be traced across huge distances from Baffin Island, where they are related to faulting, to Coronation Gulf in the Northwest Territories of Arctic Canada. Paleomagnetic evidence from the two groups indicated a change in polarity between the two periods of intrusion.

The Grenville Period in Europe. The Grenville front in Europe is thought to pass across the southern part of the Lewisian outcrop and beneath the Moine thrust. In Scandinavia there is a major interface between age zones of 950 and 1750 m.y.

The main occurrence of Grenville ages in Europe is in the Sveco-Norwegian zone of southern Norway and Sweden. Between 1600 and 1300 m.y. the Telemark and Dalslandian sedimentary successions were laid down in this zone, and at the same time the Jotnian continental succession was deposited to the northwest. The extrusion of widespread Jotnian basalts and the deposition of the Jotnian sandstones is thought to have been largely controlled by regional grabenlike structures with east–west trends. Other sediments laid down during this interval include the Salmi Series of Lake Ladoga, which cuts transgressively across the rapakivi granites of Karelia, and the sandstones of the Tersky coast on the White Sea (Kratz *et al.*, 1968).

The Dalslandian sedimentation was followed by a period of peralkaline and alkaline magmatism in Sweden and Finland. The main Dalslandian (Sveco-Norwegian) orogenic episode is thought to have taken place between 1150 and 900 m.y. (Z.W.O., 1968); this was a widespread metamorphic event causing granulite facies metamorphism in the center of the mobile belt. A Rb/Sr isochron age of 1160 ± 100 m.y. has been reported from rocks of the Telemark suite which have undergone metamorphism to greenschist facies, while Michot and Pasteels (1968) have obtained a minimum age of 1050 ±100 m.y. for the main metamorphic event from Rb/Sr isochron studies on gneisses in southern Rogoland. Further east in Norway (in the Bamble area) O'nions *et al.* (1969) suggest a minimum age of 1100 m.y. for the main metamorphism, based on K/Ar age determinations on amphiboles. There is widespread evidence for late activity at about 950 m.y.; late pegmatites give apparent ages of 1000 m.y. (O'nions *et al.*, 1969) and were followed by the emplacement of granites, which give apparent ages of 921 m.y. in the east Hedmark area and 965 ± 30 m.y. in the Eidfjord area. Much detailed geochronological work has been reported in the yearly progress reports of the Z.W.O. Laboratorium voor Isotopen-Geologie, Amsterdam.

Apparent ages in the "Grenville" range have been reported from several areas in Scotland. Rb/Sr ages for micas from the Knoydart pegmatite in the northwest Highlands were reported by Giletti *et al.* (1961) and Long and Lambert (1963) within the broad range 445 ± 55 to 740 ± 15 m.y. Conventional K/Ar muscovite ages of 446–459 m.y. have been determined from the marginal parts of a large muscovite book (Miller and Brown, 1965; Dodson and Snelling, in Giletti *et al.*, 1961), the core of which gives an age of 744 ± 10 m.y. by ^{40}Ar/^{39}Ar age spectrum analysis (Fitch *et al.*, 1969) and is thought to represent the age of crystallization of the mica. Long and Lambert (1963) have also reported apparent ages of 747, 720, and 662 m.y. (Rb/Sr) for three pegmatites from Carn Gorm. Pidgeon (*in* Dunning, 1972) has made a detailed study of muscovites from one of these pegmatites and has obtained an overall Rb/Sr isochron age of 780 m.y. U/Pb results on zircons from the nearby Carn Chuinneag Granite suggest a very old 1100–1200 m.y. apparent age, with major lead loss occurring at 510 m.y. Rb/Sr ratio plots (Pidgeon *in* Dunning, 1972) do not confirm this high age. Rb/Sr isochron determinations suggest an apparent age of 562 ± 30 m.y. for the granite (Long, 1964). In other parts of Britain it has been suggested that rocks of Grenville age can be found in the Mona complex of Anglesey, the Malvernian gneisses, and the Precambrian rocks of Charnwood Forest. These are comparable with the Carolinidian of northeast Greenland (Dalziel, 1969). The major part of the central area of the British Isles was involved in the cross-cutting Caledonian and Variscan mobile belts during Phanerozoic times, and reworking and metamorphism has overprinted the ages from the Precambrian basement.

Precambrian basement rocks appear again in the Armorican massif of northwest France and in the Channel Islands. Some of the high-grade metamorphic rocks of the Channel Isles have been compared on lithological evidence to those of the Lizard and Start Point areas in southwest England (Sutton and Watson, 1957) and to the Pentevrien succession in Britanny and Normandy. The Pentevrien has also been correlated with the Scottish Lewisian on structural, lithological, and stratigraphic evidence (Rast and Crimes, 1969).

In northwest France, the following sequence of events can be seen in the Armorican massif.

(1) Folding of the Pentevrien basement, with apparent ages in the range 1200–900 m.y. (K/Ar and Rb/Sr determinations), showing a peak at 1000 ± 100 m.y. (Leutwein, 1968).

(2) Unconformable deposition of the late Precambrian Brioverien, the oldest sediments of which are dated at 750–780 m.y.

Spitzbergen lies within the Caledonian metamorphic zone, but a few questionable older ages in the range 541 ± 24 m.y. to 636 ± 20 m.y. (Gayer *et al.*, 1966) may represent the remnants of a Precambrian foreland. Harland (1969) has compared the succession found in the Caledonian (the Hecla Hoek succession) with that of East Greenland, northern Norway, and the North Atlantic region—all of which are characterized by late Precambrian to early Paleozoic sedimentation.

D. Phanerozoic

Following the major "Grenville" orogenic events that occurred between 1200 and 950 m.y. a new system of mobile belts developed during the uppermost Precambrian and on into the Phanerozoic. These were frequently superimposed upon, and cross-cut, the Grenville belts. Individual Phanerozoic orogenies often appear to be relatively simple, with deformation and metamorphic episodes closely related in a single cycle restricted in both space and time. Nevertheless, the late Precambrian and Paleozoic fold belts represent a coherent post-Grenville mobile belt system. The plate motions initiated after the end of Variscan orogenesis became a clearly recognizable pattern at about 200 m.y. ago. There may be a world-wide rhythm of orogenic activity within the period 850–250 m.y., with maxima perhaps around the end of the Precambrian, late Cambrian–early Ordovician, middle–late Ordovician, late Silurian–early Devonian, middle and late Carboniferous, and mid-Permo-Triassic, but the precision of the geochronological evidence currently available is not sufficient for detailed correlation to be made over large distances. The difficulty does not lie in the volume of data but in its very variable quality. Unseen discrepancies are present in many published age determinations, and the average value of a group of variably discrepant determinations has little or no significance, and may be less accurate than the indication given by the least discrepant determinations of the group. Until the major Phanerozoic events throughout the world have been redated, or have had their currently accepted dates confirmed using techniques which can eliminate possible discrepancy, any long-distance correlations must be suspect.

For this reason the history of the uppermost Precambrian and Paleozoic mobile belts is discussed in a number of separate areas. At present it is not possible to say that the sequence of events in each area follows a similar time table, with only the intensity of activity varying along the length of the mobile belt, or whether there is a significant variation in both sequence and time scale along the length of the belt.

The North American Appalachian belt is superimposed upon older rocks along the southeast coast of America, and continues eastward through Newfoundland to Ireland and Britanny, where it divides, one branch passing through southern Britain and northern France to Germany, the other through northwest Britain, Norway, and East Greenland to Spitzbergen. In East Greenland this branch meets a further belt which runs from the Canadian Arctic through Ellesmere Island and northern Greenland, where it is known as the Innuitian fold belt (see Fig. 5).

In Upper Palaeozoic times the Northwest European (Caledonian) branch of this system became inactive and in southern Britain the later Hercynian (Variscan) belt is transgressive across the older trends (Fig. 5).

It has been suggested that during the Caledonian/Appalachian cycle the present North Sea was the site of a triple junction between three major crustal plates (Dewey *et al.*, 1970). These were the Laurentian craton, together with the orthotectonic belt of the Appalachian/Caledonian chain, the Afro-European plate, including the paratectonic belt and the stable foreland to the south,

Fig. 5. Phanerozoic areas.

and the Baltic plate. This hypothesis was proposed on the basis of the known provincialism of the lower Paleozoic faunas; it has been known for some time that the present close juxtaposition of the different faunas might be explained by some form of postdepositional movements, and it can be seen that the concept of gradual contraction of an earlier North Atlantic Ocean during lower Paleozoic times could explain the changing degree of provincialism exhibited by the "Pacific," "Atlantic," and "Baltic" faunas with time.

1. *Appalachian Belt*

A recent summary of dating in the Appalachian belt has been presented by Lyons and Faul (1968). The first major tectonic event to affect the whole of the Appalachian belt occurred in the Ordovician period, although several minor localized events can be recognized within the late Precambrian and early Paleozoic.

The Avalonian Orogeny. The Avalon peninsula of Newfoundland is thought to have been a volcanic island arc during late Precambrian times which became consolidated during the Avalonian orogeny (Hughes and Brückner, 1971). Geochronological evidence of the Avalonian orogenic episode has been found in the Maritime Provinces and New England and on the continental shelf. Volcanics with an apparent minimum age of 500 m.y. occur on Cape Breton Island and in eastern Newfoundland. Petrochemical analysis of the Harbour Maine volcanics of Newfoundland by Papezic (1970) has revealed a striking similarity to the Uriconian volcanics of Britain (Greig *et al.*, 1968). Papezic (1970) suggests that a belt of postorogenic faulting and volcanism extended from Newfoundland across to the British Midlands with activity similar to the Basin and Range Province of America. A series of granites was intruded at this time in southeast Newfoundland, and the Holyrood Granite in Conception Bay has been dated at 575 m.y. (McCartney *et al.*, 1966). Other granites in New England give apparent ages in the range 570–580 m.y. (Fairbairn *et al.*, 1960). In Nova Scotia granites give apparent ages of 500 m.y. (Fairbairn *et al.*, 1960).

Neponset Disturbance. A set of ages obtained from rocks along the northeastern edge of the Appalachians may show a thermal event dated in New England at 650–540 m.y. (Fairbairn *et al.*, 1967).

The St. Lawrence Graben System. Further dates close to 600 m.y. are associated with the rift system which Kumarapeli and Saull (1966) trace from the St. Lawrence graben across the Atlantic to Sweden. Intrusive carbonatites along the rift are dated at about 565 m.y., and Doig and Barton (1968) have traced the 565-m.y. event over wide areas. King and Maclean (1970) and Keen *et al.* (1970), however, find no evidence to support the extension of the rift along the Laurentian Channel, since seismic reflection profiling shows structural trends cutting across the channel. Bird *et al.* (1971) consider that the 565-m.y. event dates the initial rifting of the North Atlantic continents. In Labrador the Aillik Bay carbonatites give an apparent age of 570 m.y. (Kranck, 1953), as do lamprophyre dikes of the Makkovik Bay area (Leech *et al.*, 1963) which may be comparable with intrusions restricted to the rift system.

Pre-Taconic/Post-Avalonian Events. The Penobscot disturbance in Maine (Hall, 1966; Neuman and Rankin, 1966; Pavlides *et al.*, 1968) gives apparent ages in the range 510–475 m.y. (Fairbairn, 1971). This localized Late Cambrian disturbance was noted along the northwest side of the Appalachian belt, and has been dated at 495 m.y. in the Quebec Granite (Lyons and Faul, 1968), and at 490 m.y. in Massachussetts (Zartman, 1969). The Penobscot disturbance in Maine and similar events in Newfoundland, Quebec, and Gaspé are all early Paleozoic events and have been compared with the Arenig deformation in Ireland by Hall (1969).

The Taconic Orogeny. John Rodgers (1971) has recently reviewed the Taconic orogeny in North America and regards all the orogenic events from the Avalonian to the Allegheny as minor pulses within one complex orogeny which formed the Appalachian belt.

The chief period of activity, termed the Taconic orogeny, occurs from Early Ordovician to Silurian and has broad age limits. The orogeny produced a large landmass in a similar area to the Avalonian orogeny. The culmination of very intense volcanic activity along the northwest flank of the belt during the Silurian has been compared to an island arc formation separated from the eastern flank by very deep water which may be the proto-Atlantic (Bird *et al.*, 1971).

The Northern boundary of the Appalachian belt nearly impinges on the Precambrian rocks in Canada, but is separated by a narrow Paleozoic foreland of undeformed rocks. In Quebec the front is a low-angled thrust fault called Logan's Line which is continued beneath the St. Lawrence estuary. Geophysical work by Sheridan and Drake (1968) traces rocks worked by the Taconic orogeny across the continental shelf east of Newfoundland, and they consider that this may have continued across to Europe prior to continental drift. There is definite geological evidence for a period of folding and magmatism in the Ordovician period, but the isotopic dating of these events seems generally unreliable. Later periods of folding and metamorphism have caused overprinting, and some rocks dated at 430–450 m.y. may have been formed earlier.

A profound unconformity between mid-Ordovician and mid-Silurian rocks can be found in the Appalachian region and is evidence of a widespread orogenic event associated with metamorphism, plutonism, and volcanism. The unconformity is greatest in the northwestern border of Quebec but absent in Maine and Gaspé. In the Taconic mountains the orogeny culminated in extensive thrusting; overthrusting also occurred in northwest Newfoundland where two large Taconic klippen were formed. Much further south, de Cserna (1971) has described similar klippen of Taconic age resting on Grenville gneisses in eastern Mexico.

The Taconic orogeny is regarded as a generally mild continental disturbance by Zen (1967). There was widespread metamorphism which is detected in Vermont, New York, and the Blue Ridge Mountains. Intrusive granites and gabbros have been dated in New Brunswick at 462 m.y. (Wanless *et al.*, 1965), at 432 m.y. in Nova Scotia (Wanless *et al.*, 1967), and at 440 m.y.

in Newfoundland (Leech *et al.*, 1963). Igneous activity occurred in the Central Appalachians from 410 to 430 m.y. and in Nova Scotia around 430 m.y. (Fairbairn, 1971).

The Salinic Disturbance. This name was proposed by Boucot (1962) and represents the break between Upper Silurian and Lower Devonian rocks, although there is no geological evidence of plutonism at this time. Volcanic rocks in Nova Scotia have given apparent ages of 384 m.y. and in Massachussetts of 335 m.y. Magmatism from 390 to 400 m.y. in the Appalachians is clear but not continuous. In northern Maine the Salinic unconformity at the top of the Silurian dated at 410–430 m.y. is overshadowed by later Devonian movements (Fairbairn, 1971).

The Acadian Orogeny. There are many conflicting views on the correlation of the Acadian orogeny of the Appalachian belt with Europe. Wynne-Edwards and Hassan (1970) link the Acadian with the Hercynian (Variscan) of Europe in one time zone of rocks with apparent ages of about 300 m.y. and believe the Acadian represents a diverging branch of the European Hercynian orogenic belt. However, in the Appalachians there seems to be no distinct break between the Taconic and Acadian and it is more likely that they represent different spasms of activity along the same mobile belt. Fairbairn (1971) compares both the Taconic and Acadian with the late Caledonian of Britain, as both are terminated by continental Upper and Middle Devonian deposits. The strong 360-m.y. Acadian overprint is not found in the Caledonides, although some discrepantly low apparent ages do show evidence of weak overprinting at this time. Sheridan and Drake (1968) find geophysical evidence that rocks affected by the Acadian orogeny die out half way across the continental shelf east of Newfoundland, and this event probably did not extend across to Europe as a major orogeny. Some authors interpret this as a sign of major rifting of continents between the Taconic and Acadian events, but it may be interpreted as a fading out of intensity in Acadian activity along the belt.

Recent Rb/Sr dates for the Acadian granites of Vermont by Naylor (1971) are older than previous dates for the Acadian, and on this evidence the Silurian/ Devonian boundary is placed at 410 m.y. and late-stage granite emplacement at about 380 m.y. Considering that early Devonian sediments are included in the orogeny and are deformed and metamorphosed to the same degree as the oldest rocks involved, it seems that the orogeny must have been a very brief event, at most 30 m.y. long. The Rb/Sr ages are probably accurate for the time of emplacement of the granites, and the much lower K/Ar figures obtained are probably due to later reworking or overprinting. This implies that in the Appalachian region there must have been a minor event which overprinted the rocks during the middle to late Devonian. Naylor (1971) believes that the Acadian orogeny occurred during the closing of a former ocean basin, accounting for the recumbent folding and sliding in southern New England, the basin remaining open further north leaving highly contorted sediments between the two plates. This movement had ended by Devonian time. King (1969) considers that there is no orogeny at the Silurian/Devonian boundary comparable with the classic end-Silurian phase seen in Britain and Scandinavia, but in some areas continental Upper and Middle Devonian deposits overlie Acadian rocks (Fairbairn, 1971) placing an upper limit to the Acadian orogeny. Spooner and Fairbairn (1970) consider the 400-m.y. event of granite plutonism near Calais in Maine as pre-Acadian and post-Taconic in an intermediate event between the two main orogenies, while Williams (1969) has dated the earliest batholiths in northeast Newfoundland not affected by metamorphism to be of early Acadian age in the range 450–400 m.y. Boucot (1962) has suggested that the Acadian activity started in northern Nova Scotia and Maine and moved southward, accounting for the older dates for the orogeny found toward the north. Thus in New Brunswick, northern Maine, and Nova Scotia K/Ar apparent ages of 360–400 m.y. are obtained, but ages decrease southward in Maine and Nova Scotia (Lyons and Faul, 1968; many authors). Rodgers (1971) considers apparent ages in Maine and the maritime provinces to be consistent with two phases of plutonism in the Devonian period with perhaps different volcanic centers. Cormier (1969) has dated the metamorphism in Nova Scotia by the Rb/Sr method at 391 m.y., and volcanics in Cape Breton Island are dated at about 396 m.y. (McDougall et al., 1966; Cormier and Kelly, 1964). Metamorphism associated with the orogeny has an apparent age of 340 m.y. in Nova Scotia (Fairbairn et al., 1960).

During the Acadian orogeny there was a median fold belt from central Newfoundland through the Maritime Provinces to eastern Connecticut and western Rhode Island. The core of the belt lies slightly to the east of the core of the Taconic belt. From recent work it seems that little value can be given to the comparison of individual ages of rocks in the Appalachian region until complete uniformity of dating techniques can be attained. The strong geological and paleontological control in this region reveals the inaccuracy of the dates where apparent ages differing by 20 to 30 m.y. may be found for a narrowly defined paleontological unit.

Allegheny Disturbance. In southeastern New Brunswick, Nova Scotia, and New England an intra-Carboniferous unconformity is evidence for Carboniferous–Permian movements (Webb, 1963; Rodgers, 1967). In eastern Massachussetts and Rhode Island where Carboniferous rocks are severely metamorphosed deformation postdates the mid-Pennsylvanian. This Carboniferous–Permian event is the major orogeny in the Appalachian plateau and ridge provinces. Movement was accompanied by granitic intrusion. Dating of this event has recently been carried out by Zartman and Marvin (1971) and has yielded apparent ages around 260 m.y. in New England and Nova Scotia (Fairbairn *et al.*, 1960). The dates of this disturbance are comparable with Variscan events in southern Britain, but they still lie within the Appalachian belt and there is no distinct break in the cycle. Severe faulting occurred during the Carboniferous; for example, the Cabot fault described by Wilson (1962) cuts through Newfoundland, the Maritime Provinces, and New England.

Palisade Disturbance. In the Late Triassic between 190 and 200 m.y. basaltic flows and sills occur, such as the New York Palisades and Newark Group of flood basalts. These volcanic outpourings are associated with doming and rifting in the Atlantic, and Burek (1970) suggests 200 m.y. as a date for initial rifting.

The White Mountain plutonic series was intruded at 180 m.y. (lead-alpha ages of Lyons *et al.*, 1957) and the Conway Granite was intruded at about the same time, although Foland *et al.* (1971) find a range of 183–110 m.y. for the granite. The nearby Monteregian Hills intrusions show an initial crystallization at about 125 m.y., according to Rb/Sr whole-rock determinations of Fairbairn *et al.* (1963), but biotite separated from the rocks gives a very low figure of 99 m.y. showing subsequent reworking. Doig and Barton (1968) consider that the Monteregian Hills intrusions may represent the latest alkali intrusive phase of the St. Lawrence graben system. Widespread alkali intrusions occur along the west side of the Appalachians, and samples from Vermont, New York, and Virginia have been dated at 120–150 m.y. by Zartman *et al.* (1965). A 70-m.y. spread of apparent ages has been shown by Foland *et al.* (1971) for Jurassic and Cretaceous magmatic events in New York, New Hampshire, Maine, and South Quebec.

Lamprophyre dikes were intruded during the Jurassic and Cretaceous in Newfoundland and are comparable with dikes intruded in Greenland (Wanless *et al.*, 1965; Wanless *et al.*, 1967).

2. The Caledonian Belt in Great Britain

In view of the extremely large amount of published geological and geochronological data on the Caledonian belt in Britain, it is only possible in this paper to give an outline of the major events which have been dated. For descriptive purposes, several authors have found it convenient to divide the belt into two parts: an orthotectonic belt to the north and a paratectonic belt to the southeast, separated by the northeast–southwest-trending Highland Boundary fault. The rocks of the orthotectonic belt show a complex structural and metamorphic history, with the main period of deformation occurring in Late Cambrian–Early Ordovician times, while in the paratectonic belt the rocks are characterized by simpler deformation styles and low-grade metamorphism, with the climax of metamorphism occurring in Late Silurian–Early Devonian times. The two-stage development of the mobile belt has been described in terms of plate tectonics by Dewey (1971a). He envisages the development of the orthotectonic belt in Early Ordovician times as the result of the descent of a lithospheric plate beneath the Midland Valley of Scotland and the Highlands, while the Late Silurian growth of the paratectonic belt is attributed to the approach and subsequent collision of two continental plate areas. Roberts (1971) has questioned the strict labeling of ortho- and para-tectonic belts, particularly when extended to Scandinavia, where the highly deformed and metamorphosed lower Paleozoic rocks of the "orthotectonic zone" can be traced into the British paratectonic belt on the basis of the age of deformation—Silurian in both areas. He points out that the intensity of the Silurian phase varied appreciably along the Caledonian–Appalachian belt, being strongest in Norway, East Greenland, and New England, where it would be classified as orthotectonic, and weakest in Britain and Ireland, where it is classified as paratectonic. It therefore seems preferable to base any division on the actual timing of the main metamorphic event in any one area, and to recognize that overlapping events will occur in some parts of the belt.

One major problem in the Scottish Highlands has already been described under the heading of the Grenville episode—this is the strong radiometric evidence for a Precambrian event within the metamorphic Moinian rocks, whereas the geological evidence all points to an important event in Ordovician times. However, there is now a growing awareness that metamorphic episodes previously correlated across large areas within the Moinian may in fact be of purely local significance, and Powell (1971) has presented strong evidence of two quite separate metamorphic events within the Moinian, widely spaced in time.

Rast and Crimes (1969) suggested the following terminology for a general sequence of events in the Caledonian belt, comprising four episodes of orogenic activity; each episode involves one particular part of the belt, represents a major interruption of the Caledonian cycle of sedimentation, and produces an interpenetrative deformation and metamorphism with associated granite intrusion.

The Cadomian Orogenic Episode. This is represented by the sub-Cambrian unconformity in northern France, the English Midlands and Anglesey, and involves both tectonic and metamorphic activity and the development of post-tectonic granites. A large number of dates from this region fall in the range 570–600 m.y.; these dates are mainly indications of overprinting found in the inliers of crystalline basement within the paratectonic belt. Uriconian rocks have yielded ages of 677 and 632 m.y., Malvernian rocks give ages in the range 590–710 m.y., and ages of 680 m.y. have been found in the Precambrian rocks of Charnwood Forest. The post-Cadomian emergence of land areas is suggested by the presence of terrestrial ignimbrites in Anglesey, dated at 580 m.y. (Fitch *et al.*, 1969) and also by significant stratigraphical breaks at this horizon.

The Grampian Orogenic Episode. This is the major phase of deformation in the orthotectonic belt of the British Caledonides. It is defined by Rast and Crimes (1969) as the post-Lower Cambrian, pre-Arenig episode of deformation and metamorphism affecting the Grampian Highlands of Scotland and the Dalradian of western Ireland; in the paratectonic belt it is represented only by small stratigraphic breaks in the sedimentary sequence, while in the shelf sequence to the north of the mobile belt there appears to have been continuous sedimentation during this period. Much recent research on the stratigraphic age of this episode seems to point to a diachronous deformation sequence in the orthotectonic belt as a whole, although at any one locality it was probably completed within a very short interval. Dewey *et al.* (1970), in a study of the stratigraphic evidence from Britain and Norway, conclude that the stratigraphic age of the orogenic episode is within the lower Arenig. The evidence seems to suggest that in western Ireland deformation was completed before Arenig times (Dewey, 1971*b*, unpublished) but spread diachronously northward, occurring before Llanvirn times in Norway.

Phillips and Kennedy (1971, unpublished) think that the geological and geochronological evidence from Ireland suggests that deformation and recrystallization in the mobile zone had already started at depth while upper Dalradian turbidites were still being deposited at the surface (along with associated basic volcanics). Additional palynological evidence comes from the Dalradian succession on the northeast coast of Scotland where Downie (1971) has described microfossils of probable Middle Ordovician age, suggesting that sedimentation continued later in this area.

Radioisotopic evidence for the age of the Grampian episode comes from several sources. Pankhurst (1970) has obtained a Rb/Sr isochron age of 501 $\pm$17 m.y. for the synmetamorphic basic intrusions of northeast Scotland, and Pidgeon (*in* Dunning, 1972) has shown that a major overprinting event caused extensive lead loss from zircons in granite gneisses at around 510 m.y.

In Ireland Leggo and Pidgeon (1970) date the main tectonic phase as occurring between 520 and 500 m.y.; synmetamorphic gabbros have been dated by Pidgeon (1969) by U/Pb analysis at 515 $\pm$ 15 m.y., while the post-tectonic Donegal granite has given an apparent age of 498 $\pm$ 5 m.y. (whole-rock Rb/Sr isochron). Apparent ages obtained from metamorphic minerals and rocks in the Moines range from 420 to 430 m.y., increasing westwards to 500 m.y., while those in the Dalradian fall in the range 425–440 m.y., increasing to 475 m.y. in Connemara.

Two hypotheses have been advanced to explain this spread of mineral ages; these are the "slow cooling" and "overprinting" hypotheses. The slow cooling hypothesis attributes the distribution of ages to gradual cooling of the Moine and Dalradian rocks on uplift during Caledonian times, the higher level rocks such as the Dalradian cooling more rapidly to explain the peak of older ages in these rocks. The overprinting hypothesis postulates that the spread of apparent ages is caused by the overprinting of early ages by an end-Caledonian event at 420 m.y. Overprinting, with the production of "mixed" ages in the deeper levels of the Moinian, would obviously be more complete than at the higher levels of the Dalradian, explaining the older peak in the Dalradian ages. ^{40}Ar/^{39}Ar age studies (Fitch *et al.*, 1969) have demonstrated that virtually all the apparent K/Ar mineral ages from the Moine and Dalradian rocks are mixed or overprinted ages. Moine micas from Morar, for example, have at least two separate components in a mixed age of 542 ± 24 m.y. (Miller and Brown, 1965); the rock in fact was affected by one major event at about 600 m.y. and a later event between 430 and 420 m.y. Similarly, results of ^{40}Ar/^{39}Ar determinations on some Dalradian rocks show a major event at 470 m.y. followed by a later event at 425 m.y.

The Lakelandian Orogenic Episode. This is a far less widespread event than the Grampian episode, and is known mainly from breaks in the sedimentary sequence—the mid-Ordovician movements recorded in Wales, for example— and by granite intrusion, e.g., the Tanygriseau Granite in North Wales, dated at 477 ± 13 m.y. (Fitch *et al.*, 1969).

The Cymrian (Major End-Caledonian) Orogenic Episode. This is the major event affecting rocks in the paratectonic belt, where it can be seen to affect rocks ranging in age up to Upper Silurian. It caused low-grade regional metamorphism, and was accompanied by widespread granite intrusion—the so-called newer granites, found in both the orthotectonic and paratectonic zones, and giving apparent ages in the range 390–410 m.y.

The Variscan Orogeny in Britain. The European Hercynian mobile belt trends ESE–WNW across southwest Ireland, southwest England, northern France, and north Germany. The northern margin of the belt is marked locally by a strong thrust in Ireland and also in continental Europe, although here it is largely obscured by later cover rocks. The nature of the Variscan orogeny has been disputed by a number of authors: Simpson (1962) postulates a single phase of orogenesis near the end of the Carboniferous, while Stille (1924) described a series of events beginning in the Devonian and extending into the Permian (based on evidence from continental Europe). Dodson and Rex (1969, 1971) have obtained ages from slates and phyllites in southwest England which suggest that a definite sequence of events can be recognized, as follows.

1. Apparent ages in the range 365–345 m.y. from South Cornwall, suggesting an early Hercynian (Taconic?) phase of folding with a minimum stratigraphic age of Upper Devonian–Lower Carboniferous.

2. Apparent ages in the range 340–320 m.y. from the east–west Devonian slate belt, suggesting an approximate age of folding in mid-Carboniferous time equivalent to Stille's "Sudetic" phase (Stille, 1924).

3. Scattered apparent ages in the range 270–290 m.y. marking the well-known late Carboniferous events.

The early dates correlate with work on the Lizard area of southwest England (Miller and Green, 1961) which gave a minimum age of 390 m.y. for the Lizard gneisses, although ages of up to 492 m.y. were reported. The Lizard area is therefore thought to represent a much older basement which suffered later metamorphism during nappe emplacement in an early (360–390 m.y.) phase of tectonism. Dodson and Rex correlate this phase with the "Bretonic" phase described by Stille from Brittany (where there is a marked Devonian–Carboniferous unconformity).

There appears to be good structural evidence in this area for the over-printing of Caledonian trends (northeast–southwest) by later Variscan (east–west) structures. Dearman et al. (1969) have correlated the regional pattern of K/Ar ages with a series of tectonic events; the zone of 365–345 m.y. apparent ages in South Cornwall corresponds to a continuous Lizard–Start–Dodman structural zone which has a well-marked Caledonoid trend and which they suggest represents an actively rising cordillera in Devonian times, bringing up pre-Devonian basement and culminating in an end-Devonian metamorphic event. This zone is truncated to the north by a zone of east–west structures (the main Devonian slate belt) which corresponds to the 340–320-m.y. apparent age zone of Dodson and Rex. Remnants of earlier northeast–southwest structures can be found in this zone and may account for the few anomalous older ages reported for this area.

The postorogenic granite suite of southwest England has yielded apparent ages ranging from 250 to 306 m.y., while K/Ar ages from the slates give a minimum age of 280 m.y. for the main end-Variscan folding; however, as Permian lavas in this area have given an apparent age of 280 ± 10 m.y., the true age of emplacement of the granites is thought to be 295 ± 5 m.y. (Fitch and Miller, 1964).

3. Phanerozoic Orogenesis in Europe

France. The Precambrian rocks of northwest France are divided into a highly metamorphosed basement complex (the Pentevrien) and an unconformably overlying less-metamorphosed sequence (the Brioverien).

The main orogenic episode in the Brioverien occurred between 650 and 670 m.y. (Leutwein, 1968) and is named the Cadomian. In a late phase of the orogeny a series of granitic intrusions were emplaced, e.g., the Athis Granite; all the granites give apparent minimum ages around 560 m.y. (Leutwein, 1968). To the east in the Vosges ages correlating with those of the mid-Brioverien schists are found (in the Schistes de Villers); these rocks show a very similar lithology to that of the Brioverien rocks (which have a minimum age of 615 ± 20 m.y.) and are thought to be analogous. Adams (1968) and Michot and Deutsch (1970) have reported dates from the Gneiss de Brest, which predates the main regional Cadomian metamorphism. The gneiss gives a minimum age of 690 ± 40 m.y. (Rb/Sr isochron) while a postorogenic granite gives an Rb/Sr isochron àge of 565 ± 40 m.y.; this therefore gives a rough limit to the length of the Cadomian episode, from 565 to 690 m.y.

There is a considerable structural break between the deformed and metamorphosed upper Brioverien and the overlying basal Cambrian conglomerates.

Dates in the range 460–470 m.y. have been obtained from granites in Brittany and from associated mineralization phenomena (Leutwein, 1968); although these are described as "late Cadomian" by Leutwein it seems possible that they might be connected with events occurring in the Lakelandian episode in Britain.

In northern Brittany, K/Ar mineral ages from pre-Ordovician rocks fall in the range 290–345 m.y., suggesting overprinting of earlier events during Hercynian times. Ages of 330 m.y. are reported from Brioverien metamorphic rocks (Michot and Deutsch, 1970). Adams (1968) has described an older group of granites, which give a whole-rock Rb/Sr isochron age of 340 ± 5 m.y., and a younger group which give an Rb/Sr isochron age of 300 ± 3 m.y.

In western France, uranium mineralization has been dated by the lead isotope method and gives apparent ages in the range 250–260 m.y., which Kosztolanyi and Coppins (1970) place within the Saalian phase of the Hercynian.

The Channel Islands. K/Ar mineral ages from the metamorphosed Brioverien rocks all fall in the range 530–600 m.y. with a spread of young ages in the range 530–570 m.y., broadly similar to ages reported from diorites on Jersey (530–550 m.y.) (Adams, 1967). Adams suggests that the main Cadomian metamorphism occurred at or before 590 m.y. and that the slightly later intrusion of the diorites at 550 m.y. caused a lowering of the older ages. Rb/Sr ages (whole rock) from early Cadomian igneous rocks in the Channel Isles give an upper time limit to the metamorphism at about 650 m.y.

Spain. A similar relationship can be seen in northwest Spain where basal Cambrian quartzites and carbonates lie unconformably on a late Precambrian tillite sequence (Leutwein, 1968). In this area Priem *et al.* (1970) have described a pre-Hercynian basement retaining evidence of older orogenic and igneous events. Preliminary Rb/Sr investigations on granite gneisses indicate an apparent age in the range 460–430 m.y. for the age of emplacement. These rocks in turn intrude metasediments recording an earlier regional metamorphism. Priem *et al.* correlate this 460–430 m.y. event with a peak of mineral ages found in the British Caledonian belt. There is general agreement that Caledonian orogenic effects were absent or very weak in the Iberian peninsula; the only effect seen is a local unconformity between the Cambrian and the base of the Ordovician. In southern Spain there appears to have been a continuous sequence from Precambrian to Cambrian.

The Hercynian "Hesperic massif" covers northwest, west, and central Spain and the greater part of Portugal, and consists of Precambrian and Paleozoic rocks which were folded, metamorphosed, and invaded by granitic magmas during the Hercynian episode. On the basis of geological and geochronological evidence the granites are divided into a younger and an older group; the Rb/Sr isochron age of the younger suite is 280 ± 11 m.y. while that of the older suite is 298 ± 10 m.y. (Priem *et al.*, 1970). Whole-rock Rb/Sr isochron studies on four samples of gneissic, highly deformed granites suggest an age of 349 ± 10 m.y., and Priem *et al.* correlate this with a major folding phase of mid-Westphalian age described in Portugal (Schermerhorn, 1956). Capdevila and Vialette (1970) date "younger" granites in northwest Spain at 304 ± 10 and 303 ± 6 m.y.

Central Europe. In continental Europe there is widespread evidence of Paleozoic orogenic events, as this is the main area of the Hercynian mobile belt. In many places there is clear evidence of overprinting of earlier events —for example, in the Tatra Mountains in Poland, Burchart (1969) has described two periods of metamorphic activity, one producing apparent ages in the range 410–430 m.y. in the regional gneisses, and a later, Hercynian phase of granite intrusion between 290 and 310 m.y.

In the southern Schwarzland Wendt *et al.* (1970) have obtained Rb/Sr apparent ages in the range 315–276 m.y. from Hercynian granites, while mineral ages from the same granites have been determined by K/Ar and Rb/Sr methods, and give apparent ages in the range 320–327 m.y. Faul and Jäger (1963) have dated samples from different Hercynian ranges in central Europe and have reported ages in the range 280–330 m.y.

Scandinavia. The first major event in the Norwegian Caledonides was of Late Cambrian–Early Ordovician age, equivalent to the Grampian episode in Britain, and this event has been described in West Finmark and western Norway. It may be represented further to the east by a period of extensive volcanism in eastern and central Norway (Roberts, 1971). On Söröy island in northern Norway this metamorphic event was accompanied by the intrusion of large volumes of mantle-derived mafic and ultramafic magmas, which Sturt (1971) correlates with the major orogenic metamorphism of the Scottish Dalradian. Dates are sparse but suggest a minimum age of about 500 m.y. for this event. Pringle (1971) has obtained ages of 530 ± 35 m.y. (Rb/Sr isochron) on rocks from Söröy and 525 ± 45 m.y. from shales interbedded with tillite just below the Lower Cambrian (this represents the age of metamorphism of the shales). An apparent age of 660 ± 40 m.y. was obtained from un-metamorphosed shales interbedded with the tillite; this is thought to represent the age of diagenesis and would appear to correlate with the range of apparent ages of 750–650 m.y. obtained from similar late Precambrian tillites in many parts of the world. Shales from a horizon below the tillite give a Rb/Sr isochron of 805 ± 90 m.y. Rb/Sr apparent ages from synorogenic intrusions suggest that although the overall age of the major Caledonian metamorphism is around 500 m.y., it was diachronous and ranged from 400 to 600 m.y. (Pringle, 1971).

A Middle–Late Ordovician metamorphic event is claimed by Kvale (1960) in the Bergen area, although there are very few dates in this area. Conglomerates overlie an unconformity of late Llanvirn, early Llandovery age.

A major end-Silurian, pre-Devonian event is recorded in many parts of the Norwegian Caledonian belt with ages giving a peak at 430 m.y.

Spitzbergen. The whole of Spitzbergen lies within the northern extension of the British–Scandinavian Caledonian belt, and apart from a small number of old dates which may indicate remnants of a Precambrian basement, all the age determinations from the central area of the island give Caledonian ages. The metasedimentary and associated igneous rocks of the Hecla Hoek succession form a Precambrian to mid-Ordovician sequence which shows broad similarities with the late Precambrian–early Paleozoic successions of northern Norway, East Greenland, and the Appalachians (Fig. 5).

Harland (1969) describes three small groups of "Precambrian" dates which occur outside the central mobile zone in lower Hecla Hoek rocks; in the northwest, the age range is 529–541 ± 15-24 m.y., in the northeast it is 537–636 ± 30-20 m.y., and in the south it is 584–586 ± 26 m.y. The age of the main metamorphic event in Spitzbergen is between 430 and 420 m.y., with a range of later dates down to 380 m.y. Extensive late plutonic activity gave rise to a series of granite plutons which give apparent ages of 400 ± 20 m.y.

4. *The Phanerozoic of Greenland and the Arctic Islands*

In Greenland the nonorogenic Paleozoic development consisted of a post-Carolinidian basic dike swarm and various continental deposits. During later deformation supracrustal rocks were overthrust to the west, and further south basement rocks were involved in folding and migmatization. A 600-m.y. thermal event has been identified in the Scoresby Sund area by Larsen (1969).

The East Greenland Fold Belt. This belt of activity is exposed between lat. 70° and 82° N along the eastern coast of Greenland and underwent a continuous period of tectonic activity from the Silurian to the Carboniferous. Haller (1970) recognizes three phases in the orogeny, with a main phase from 420 to 400 m.y., late spasms from 400 to 350 m.y., and a minor episode from 350 to 270 m.y. The chief phase of movement was in Late Silurian and Early Devonian time; regional metamorphism was strongest in the central area of East Greenland, and late orogenic activity was also confined to this region. Intrusive activity was continuous from 450 to 390 m.y., and a major peak occurs at 400 m.y. with a small rise at 430 m.y. as seen in Norway. Fairbairn (1971) suggests the granite pegmatites from north Greenland, dated at 395–400 m.y. (Haller and Kulp, 1962), may be comparable with the Salinic disturbance intrusions of the Appalachians. Sediments laid down during Devonian time and the underlying folded rocks were reworked during the "younger Caledonian" orogeny. This mid- to upper Devonian phase of folding occurs at the same time as the main Innuitian folding in the North Greenland fold belt and is also the same age as the Acadian orogeny of the Appalachians. All are intruded by postorogenic granites dated at about 395 m.y. (King, 1969).

Some Mesozoic igneous activity occurred in southwestern Greenland, and intrusions of this period were dated by the Rb/Sr method at 225 m.y. (Pringle, in Andrews and Emeleus, 1971). A series of coast-parallel dikes are linked with rifting between Greenland and Labrador by Andrews and Emeleus (1971).

North Greenland Fold Belt. A recent paper by Dawes (1971) reviews current work on the North Greenland Fold Belt. It is situated at the extreme north of Greenland and trends in an east–west direction. Diastrophism occurred here from late Silurian to late Devonian and possibly into the Carboniferous. The folding history is long and complex with at least three phases of folding in northern Peary Land (Dawes and Soper, 1971). The main folding is pre-mid-Pennsylvanian (Carboniferous) and postdates upper Silurian sediments. Some Caenozoic movements are thought to be an eastern extension of the Innuitian fold belt activity seen in Peary Land. The Laramidian thermal episode of regional metamorphism is restricted to eastern Peary Land. In the Arctic Islands three orogenic phases occur within the Silurian Period. In Axel Heiberg Island the main orogeny is equivalent to the 420–400-m.y. event of Greenland, and the final orogenic event is in mid- to late Devonian time (Trettin, 1971; Tozer and Thorsteinsson, 1964).

5. Tertiary Activity in Britain

Volcanism in the British Tertiary Province is thought to have begun at, or shortly before, 66 m.y. in the earliest Paleocene time (Evans *et al.*, 1973). The apparent age of the major plutonic centers is between 60 and 58 m.y., but the plutonic activity continued at various localities until at least 50 m.y. in the Eocene. Extensive areas of Tertiary volcanic rocks occur in northwest Britain, where they are exposed on many islands in the Inner Hebrides and on the adjacent mainland. They also occur in the Antrim plateau of northern Ireland, on the island of Lundy in the Bristol Channel, and on the outlying islands of St. Kilda and Rockall.

Tarling and Gale (1968) have shown that the plateau lavas of the Faeroe Islands were extruded in the same period (i.e., Paleocene–Eocene), with apparent ages ranging from 49 to 62 m.y. Moorbath *et al.* (1968) have found the oldest lavas on Iceland to be 16 m.y. old, i.e., middle Miocene. Jan Mayen island is still volcanically active, and the oldest lavas have been dated at 0.5 m.y. (Fitch *et al.*, 1965) (Fig. 6).

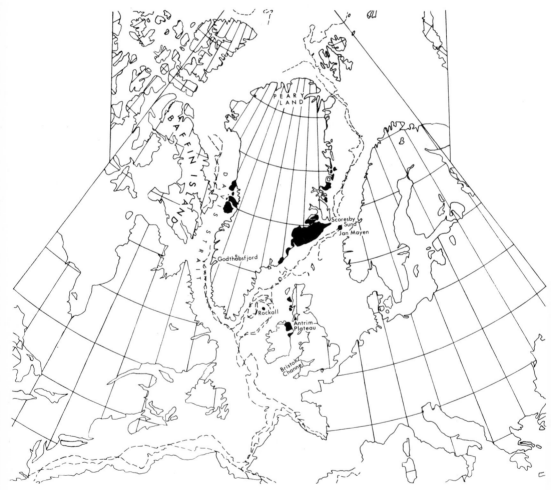

Fig. 6. Tertiary basalt provinces.

6. Tertiary Intrusives of Greenland

In East Greenland the largest area of plateau basalts was extruded from 55 to 60 m.y. in Scoresby Sund (Fig. 6), during a brief event dated by Beckinsale *et al.* (1970). These basalts are the same age as those on the Faeroes. Haller (1970) describes a coastal flexure in East Greenland at this time following intrusion of syenite and alkaline granite with associated post-mid-Eocene dike intrusion in the areas of strongest flexure. The latest event in East Greenland was dated at 50 m.y. by Beckinsale *et al.* (1970).

A minor dike swarm in the Godthåbsfjord area (Fig. 6) has been dated at 57.3 m.y., and basalts on either side of the Davis Strait were dated at 50–60 m.y. (Le Pichon *et al.*, 1971). Keen *et al.* (1970) suggest that the Baffin Island lavas are younger than the Greenland samples. In general, the British and West Greenland intrusives seem to give higher ages than the East Greenland basalts (Beckinsale *et al.*, 1970).

In northern Greenland, Dawes and Soper (1971) have shown that dating of supposed Paleozoic metasediments and volcanics in North Peary Land all give Cretaceous and Tertiary dates, giving evidence of regional metamorphism in the Cretaceous and a possible thermal event in the Tertiary. Dawes (1971) detects important Tertiary tectonism with folding and faulting and thrusting in North Peary Land, but this is not seen in East Peary Land. In South Peary Land dikes cutting Cambrian sediments are dated at 72.2 m.y. (Bridgwater, 1970).

III. BREAKUP OF THE CONTINENTS

Many papers have been written on the breakup of the continents bordering the North Atlantic—Le Pichon *et al.* (1971), Le Pichon and Fox (1971), Laughton (1971), Vogt *et al.* (1971), Vogt (1970), McElhinney and Burek (1971), Keen (1970), Smith (1971), and Freeland and Dietz (1971) to name but a few.

The geological evidence suggests that the North Atlantic continents were a single structural unit until at least some time later than the youngest structures which can be traced across the Atlantic; these are of Hercynian age, probably early Permian (Smith, 1971). Results from the JOIDES drilling program have shown the presence of Upper Jurassic sediments on the continental rise off Cape Hatteras, and this is thought to indicate a time of opening of about 175–180 m.y.

From other marine sediments, a 200–165 m.y. age of rifting seems probable (Funnell and Smith, 1968; Talwani *et al.*, 1969), and the older age of 200 m.y. has been proposed by several authors, including McElhinney and Burek (1971) and Burek (1970). The age of rifting has been interpreted as Triassic by Vogt *et al.* (1971), Bott and Watts (1971), and Williams and McKenzie (1971); Permian by Emery *et al.* (1970) and Heirtzler and Hayes (1967); or much older (Drake and Nafe, 1968). It seems certain that the opening of at least part of the North Atlantic preceded that of the South Atlantic; the latter is thought to be no more than 130 m.y. old.

Magnetic anomaly patterns are well established in the central zone of the Atlantic, and recent spreading rates have been estimated from these rocks. However, there is a "quiet zone" beyond anomaly 32 (dated at 79 m.y.) on either side of the Atlantic which contains no clear magnetic anomaly patterns and which can only be dated by extrapolation outward from the younger central zone. This quiet zone is thought to represent a period of approximately 50 m.y. in which no magnetic reversals took place (Helsley and Steiner, 1969). In the Bay of Biscay a series of anomalies have been mapped which appear to be truncated by the quiet zone east of anomaly 32, suggesting that here the sea floor must be at least 50 m.y. older than anomaly 32, i.e., older than 125 m.y. (Lower Cretaceous). A similar area of older magnetic anomalies has been identified off southwest Spain. Williams and McKenzie (1971) suggest that both these sets of anomalies were formed during the period in which the Bay of Biscay was formed by the counterclockwise rotation of the Iberian peninsula, in Upper Jurassic or Lower Cretaceous times. Hallam (1971) has suggested that the downwarping of the southern borders of the North Atlantic with associated faulting and volcanism in Upper Triassic times may mark an initial period of rifting some 50 m.y. before the actual splitting of the continents. This is supported by May (1971) who has found similar evidence of an initial tensional phase; Late Triassic dolerite dike swarms in eastern North America, West Africa, and northeast South America fall into a regular radiating pattern on the reconstruction of Dietz *et al.* (1970) converging on the Blake plateau, the Bahama platform, and the western Senegal basin. As the dikes are of typical oceanic composition, this would appear to be strong evidence of an early tensional phase in the crust, as the result of a concentration of stress in the upper mantle. At a later date the continental crust split, and May shows that the initial rifting followed the stress trajectories which can be measured in the dikes.

A generalized sequence of events in the opening of the North Atlantic can be summarized as follows.

1. Initial splitting of the continents, beginning perhaps about 200 m.y. ago in the southern part of the North Atlantic, and with the spreading ridge extending northwards into the Bay of Biscay in pre-Upper Cretaceous times.

2. Rotation of Spain and formation of the Bay of Biscay in Upper Jurassic or Lower Cretaceous times.

3. Before anomaly 32 (79–80 m.y.) there was an opening between Rockall and Europe (Le Pichon *et al.*, 1971).

4. By the time of anomaly 32, the spreading ridge extended northwest, with the motion at the northern end probably being taken up by the Gibbs fracture zone. Spreading ceased in the Bay of Biscay.

5. By the time of anomaly 31, the spreading ridge extended into the Labrador Sea. Le Pichon *et al.* (1971) regard the opening of the Labrador Sea as a two-phase event, with the first phase occurring between anomaly 32 and 24 (76 and 60 m.y.) and involving the east–west splitting of Labrador and Greenland along the Davis Strait.

6. The second phase of opening occurred between anomalies 24 and 20 (60 and 49 m.y.) and was heralded by strong compression of the Franklinian geosyncline and the Sverdrup basin in the Canadian Arctic and by compression and rotation of Ellesmere Island. The spreading ridge extended northeast at the time of anomaly 24 so that a triple junction was initiated to the south of Greenland. During this phase Greenland moved northward relative to North America. The triple-junction opening of the Labrador Sea, Reykjanes ridge, and Norwegian Sea was contemporaneous with the Eocene volcanism of Greenland, the Faeroe Islands, Rockall, Britain, and the Baffin Island and Disko Island basalts.

7. Spreading on the Labrador Sea ridge appears to have practically ceased after 49 m.y., while at the same time the axis of spreading in the Norwegian Sea shifted westward, forming Mohn's ridge, and splitting the Jan Mayen ridge from the Greenland continental shelf. The original axis is now marked by a partially buried chain of seamounts in the Norwegian basin (Johnson *et al.*, 1971).

IV. CONCLUSIONS

In this chapter we have attempted to review the literature on tectonic and radioisotopic age correlation across the North Atlantic available to December 1971. Despite the large volume of published data, conclusive evidence for detailed correlation of any particular minor orogenic or other event across the North Atlantic is still virtually unobtainable. On a larger scale, however, there are numerous close similarities between the continental areas bordering this ocean. The same major cyclical history is seen throughout, with major orogenic–radioisotopic maxima occurring at >3500 m.y., 2950–2400 m.y., 1950–1550 m.y., 1150–900 m.y., and 550–250 m.y., and there is a major structural pattern of stabilized mobile belts or chelozones (Fitch, 1965) with a unique anorthosite belt that can only be resolved into a coherent and

KEY

Early Precambrian
structural zones   x

Middle Precambrian
structural zones

Late Precambriam – Sveco
Fennid /Hudsonian
structural zones

Late Precambrian – Grenville
structural zones

Phanerozoic structural zones

Alpine structural zones

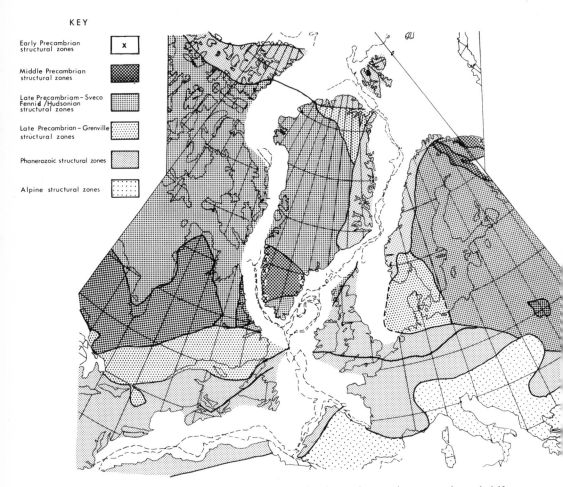

Fig. 7. Reconstruction of the North Atlantic continent prior to continental drift.

logical whole by assuming the predrift unity of the disrupted North Atlantic fragments (Fig. 7).

Early Precambrian complex rocks, or rocks that appear to be their partially or completely overprinted equivalents, are known from many localities across North America, Greenland, Britain, the Kola peninsula, and the Ukraine. There appears to be growing evidence of distinct developmental episodes (chelogenic cycles), each of which is concluded by a series of major orogenic–metamorphic–metasomatic–magmatic–radioisotopic events, as follows.

(6)   <250–200 m.y.   Alpine cycle

(5)   >200–250 m.y. ⎫
      < ~ 850 m.y. ⎭   Caledonian–Variscan cycle

(4) $> \sim$ 850 m.y. $\Big\}$ Grenville cycle
$< \sim$ 1550 m.y.

(3) $> \sim$ 1550 m.y. $\Big\}$ Svecofennid–Churchill cycle
$< \sim$ 2400 m.y.

(2) $> \sim$ 2400 m.y. $\Big\}$ Middle Precambrian cycle
$< \sim$ 3500 m.y.

(1) $> \sim$ 3500 m.y.   Early Precambrian cycle(s)

REFERENCES

Adams, C. J. D., 1967, K–Ar ages from the basement complex of the Channel Islands (United Kingdom) and the adjacent French mainland: *Earth Planet. Sci. Lett.*, v. 2, p. 52–56.

Adams, C. J. D., 1968, in: Stratigraphy and structure of part of west Finistère, France, Bishop, A. C., Bradshaw, J. D., Renouf, J. D. and Taylor, R. T.: *Proc. Geol. Soc. Lond.*, v. 1649, p. 121–127.

Allaart, J. H., Bridgwater, D., and Henriksen, N., 1969, Prequaternary geology of southwest Greenland and its bearing on North Atlantic correlation problems: *Amer. Ass. Petrol. Geol. Mem.*, v. 12, p. 859–882.

Andrews, J. R. and Emeleus, C. H., 1971, Preliminary account of Kimberlite intrusion from Frederikshab district of south west Greenland: *Grøn. Geol. Unders. Rapp.*, v. 31, 26 p.

Aumento, F., Wanless, R. K., and Stevens, R. D., 1968, K–Ar ages and spreading rates on the mid-Atlantic ridge at 45° north: *Science*, v. 161, p. 1338–1339.

Baer, A. J., 1971, Reappraisal of the Grenvillian "orogeny": *Geol. Assoc. Can. Ann. Meeting*, Sudbury, Ontario, (abstracts): p. 30.

Beall, G. H., Hurley, P. M., Fairbairn, H. W., and Pinson, W. H., 1963, Comparison of K–Ar and whole rock Rb–Sr dating in New Quebec and Labrador: *Am. J. Sci.*, v. 261, p. 571–580.

Beckinsale, R. D., Brooks, C. K., and Rex, D. C., 1970, K–Ar ages for the Tertiary of East Greenland: *Bull. Geol. Soc. Denmark*, v. 20, p. 27–37.

Berthelsen, A. and Bridgwater, D., 1960, On the field occurrence of some basic dykes of supposed Pre-Cambrian age from the southern Sukkertoppen district western Greenland. *Meddr. Grønland*, v. 123, No. 3.

Berthelsen, A. and Noe-Nygaard, A., 1965, The Precambrain of Greenland, in: *The Precambrian*, Rankama, K., ed.: New York, Interscience, Vol. II, p. 113–262.

Bird, J. M., Dewey, J. F., and Kidd, W. S. F., 1971, Proto-Atlantic oceanic crust and mantle Appalachian/Caledonian Ophiolites: *Nature Phys. Sci.*, London, v. 231, p. 28–31.

Black, L. P., Gale, N. H., Moorbath, S., Pankhurst, R. J. and McGregor, V. R., 1971, Isotopic dating of very early Precambrian amphibolite facies gneisses from the Godthaab district, West Greenland: *Earth Planet. Sci. Letters*, v. 12, p. 245–259.

Blackadar, R. G., 1967a, Precambrian geology of Boothia Peninsula, Somerset Island and Prince of Wales Island district, Franklin: *Geol. Surv. Can. Bull.*, v. 151, 62 p.

Blackadar, R. G., 1967b, Geological reconnaissance southern Baffin Island, District of Franklin: *Geol. Surv. Can. Paper*, 66–47, 32 p.

Blackadar, R. G., 1970, Precambrian geology of northwestern Baffin Island, District of Franklin: *Geol. Surv. Can. Bull.*, v. 191, 89 p.

Bott, M. H. P. and Watts, A. B., 1971, Deep structure of the continental margin adjacent to the British Isles: Institute of Geological Sciences Report 70/14, The Geology of the East Atlantic Continental Margin, 2, Europe, p. 89–110.

Boucot, A. J., 1962, Appalachian Siluro-Devonian, in: *Some aspects of the Variscan fold belt*, Coe, K., ed.: 9th. Inter University Geological Congress, Manchester University Press, p. 155–163.

Brace, W. F., 1953, The Geology of the Rutland area: *Vermont Geol. Surv. Bull.*, v. 6, 124 p.

Bridgwater, D., 1965, Isotopic age determinations from Southern Greenland and their geological setting: *Medd. Grønland*, v. 179, n. 4, 56 p.

Bridgwater, D., 1970, A compilation of K–Ar age determinations of rocks from Greenland carried out in 1969: *Grøn. Geol. Unders. Rapp.*, v. 28, p. 47–55.

Bridgwater, D., Escher, A., Jackson, G. D., Taylor, F. C., and Windley, B. F., 1971, The development of the Precambrian shield in western Greenland, Labrador and Baffin Island.

Bridgwater, D. and Gormsen, K., 1968, Precambrian rocks of the Angmagssalik area, East Greenland: *Grøn. Geol. Unders. Rapp.*, v. 15, p. 61–72.

Bridgwater, D. and Gormsen, K., 1969, Geological reconnaissance of the Precambrian rocks of southeast Greenland: *Grøn. Geol. Unders. Rapp.*, v. 19, p. 43–50.

Bridgwater, D. and Windley, B. F., 1971*a*, Anorthosites, postorogenic granites, acid volcanic rocks and crustal development in the North Atlantic shield during the mid-Proterozoic, Special Publication: *Geol. Soc. South Africa*, Granite 71.

Bridgwater, D. and Windley, B. F., 1971*b*, in: A geochronological study of the Lewisian of the Laxford area, northwest Scotland, Lambert, R. St. J. and Holland, J. C.: *Proc. Geol. Soc. Lond.*, v. 1664, p. 251–255.

Brown, R. L., Dalziel, I. W. D., and Rust, B. R., 1969, The structure, metamorphism and development of the Boothia Arch, Arctic Canada: *Can. J. Earth Sci.*, v. 6, p. 525–543.

Bullard, E. C., Everett, J. E., and Smith, A. G., 1965, The fit of the continents around the Atlantic, in: A symposium on continental drift: *Phil. Trans. Roy. Soc. Lond.*, Ser. A, v. 258, p. 41–51.

Burchart, J., 1969, Rubidium–strontium isochron age of the crystalline core of the Tatra Mountains, Poland: *Am. J. Sci.*, v. 266, p. 895–907.

Burek, P. J., 1970, Magnetic reversals: their application to stratigraphic problems: *Amer. Assoc. Petrol. Geol. Bull.*, v. 54, p. 1120–1139.

Burwash, R. A., 1969, Comparative Precambrian geochronology of North American, European and Siberian shields: *Can. J. Earth Sci.*, v. 6, p. 357–365.

Capdevila, R. and Vialette, Y., 1970, Estimation radiométrique de l'âge de la deuxième phase tectonique hercynienne en Galice moyenne (Nord-Ouest de l'Espagne): *C. R. Acad. Sci. Paris*, v. 270, p. 2527–2530.

Catanzaro, E. J., 1963, Zircon ages in southwest Minnesota: *J. Geophys. Res.*, v. 68, p. 2045–2048.

Cormier, R. F., 1969, Radiometric dating of the Coldbrook Group of southern New Brunswick, Canada: *Can. J. Earth Sci.*, v. 6, p. 393–398.

Cormier, R. F. and Kelly, A. M., 1964, Absolute age of the Fisset Brook formation and Devonian-Missippian boundary, Cape Breton Island, Nova Scotia: *Can. J. Earth Sci.*, v. 1, p. 159–166.

Coward, M. P., Francis, P. W., Graham, R. H., Myers, J. S., and Watson, J. V., 1969, Remnants of an early meta-sedimentary assemblage in the Lewisian complex of the Outer Hebrides: *Proc. Geol. Assoc. Lond.*, v. 80, p. 387–408.

Dalziel, I. W. D., 1969, Pre-Permian history of the British Isles: *Amer. Assoc. Petrol. Geol. Mem.*, v. 12, p. 5–31.

Dawes, P. R., 1971, The North Greenland fold belt and environs: *Bull. Geol. Soc. Denmark*, v. 20, p. 197–239.

Dawes, P. R. and Soper, N. J., 1971, Significance of K–Ar age determinations from northern Peary Land: *Grøn. Geol. Unders. Rapp.*, v. 35, p. 60–62.

Dearman, W. R., Leveridge, B. E., and Turner, R. G., 1969, Structural sequences and the ages of slates and phyllites from south west England: *Proc. Geol. Soc. Lond.*, v. 1654, p. 41–46.

Dearnley, R., 1966, Orogenic fold-belts and a hypothesis of earth evolution: *Physics Chem. Earth London*, v. 7, p. 1–114.

Dearnley, R. and Dunning, F. W., 1968, Metamorphosed and deformed pegmatites and basic dykes in the Lewisian complex of the Outer Hebrides and their geological significance: *Quart. J. Geol. Soc. Lond.*, v. 123, p. 335–378.

de Cserna, Z., 1971, Taconian (early Caledonian) deformation in the Huasteca structural belt of eastern Mexico: *Am. J. Sci.*, v. 271, p. 544–550.

Dence, M. R., Hartung, J. B., and Sutter, J. F., 1971, Old K–Ar mineral ages from the Grenville Province, Ontario: *Can. J. Earth Sci.*, v. 8, p. 1495–1498.

Dewey, J. F., 1971*a*, A model for the lower Palaeozoic evolution of the southern margin of the early Caledonides of Scotland and Ireland: *Scot. J. Geol.*, v. 7, p. 119–240.

Dewey, J. F., 1971*b*, Plate tectonics: a probable model for the evolution of the Appalachian/ Caledonian orogenic belt, Lecture at symposium "Dating events in the metamorphic Caledonides," Edinburgh, September, 1971.

Dewey, J. F., Rickards, R. B., and Skevington, D., 1970, New light on the age of the Dalradian deformation and metamorphism in western Ireland: *Norsk. Geol. Tidsskr.*, v. 50, p. 19–44.

Dietz, R. S., Holden, J. C., and Sproull, W. P., 1970, Geotectonic evolution and subsidence of the Bahama Platform: *Bull. Geol. Soc. Am.*, v. 81, p. 1915–1927.

Dimroth, E., 1970, Evolution of the Labrador geosyncline: *Bull. Geol. Soc. Am.*, v. 81, p. 2717–2742.

Dodson, M. H. and Rex, D. C., 1969, Potassium–argon ages of slates and phyllites from southwest England: *Proc. Geol. Soc. Lond.*, v. 1652, p. 239–240.

Dodson, M. H. and Rex, D. C., 1971, Potassium–argon ages of slates and phyllites from south-west England: *Quart. J. Geol. Soc. London*, v. 126, p. 465–500.

Doig, R., 1970, An alkaline rock province linking Europe and North America: *Can. J. Earth Sci.*, v. 7, p. 22–28.

Doig, R. and Barton, J. M. Jr., 1968, Ages of carbonatites and other alkaline rocks in Quebec: *Can. J. Earth Sci.*, v. 5, p. 1401–1407.

Downie, C., 1971, Dalradian palynology: *Jour. Geol. Soc. Lond.*, v. 127, p. 296–297.

Drake, C. L. and Nafe, J. E., Transition from ocean to continent from seismic refraction data, in: *The crust and upper mantle of the Pacific area*: Geophys. Monograph 12, Knopoff, L., Drake, C. L., and Hart, P. J., eds., p. 174–186.

Dunning, F. W., 1972, Dating events in the Metamorphic Caledonides: impressions of the symposium held in Edinburgh, September 1971. *Scot. J. Geol.*, v. 8, p. 179–192.

Emery, K. O., Uchupi, E., Phillips, J. D., Bowin, C. O., Bunce, E. T., and Knott, S. T., 1970, Continental Rise off eastern North America: *Bull. Amer. Assoc. Petrol. Geol.*, v. 54, p. 44–108.

Emslie, R. F., 1964, Potassium–argon age of the Michikamau anorthosite intrusion, Labrador: *Nature, Lond.*, v. 202, p. 172–173.

Emslie, R. F., 1970, The geology of the Michikamau intrusion, Labrador: *Geol. Surv. Can. Paper*, No. 68–57, 83 p.

Evans, A. Ll., Fitch, F. J., and Miller, J. A., 1973, Potassium–argon age determinations on some British Tertiary volcanic rocks: *Journ. Geol. Soc. Lond.*, v. 129, p. 419–443.

Evans, C. R., 1965, Geochronology of the Lewisian basement near Lochinver Sutherland: *Nature, Lond.*, v. 207, p. 54–55.

Evans, C. R., and Tarney, J., 1964, Isotopic ages of Assynt Dykes: *Nature, Lond.*, v. 204, p. 638–641.

Fahrig, W. F., and Eade, K. E., 1968, The chemical evolution of the Canadian Shield: *Can. J. Earth Sci.*, v. 5, p. 1247–1252.

Fahrig, W. F., Irving, E., and Jackson, G. D., 1971, Palaeomagnetism of the Franklin diabases: *Can. J. Earth Sci.*, v. 8, p. 455–467.

Fairbairn, H. W., 1971, Radiometric age of mid-Palaeozoic intrusives in the Appalachian Caledonides mobile belt: *Am. J. Sci.*, v. 270, p. 203–217.

Fairbairn, H. W., Faure, G., Pinson, W. H., Hurley, P. M., and Powell, J. L., 1963, Initial ratio of strontium 87 to strontium 86, whole-rock age and discordant biotite in the Monteregian Hills igneous province, Quebec: *J. Geophys. Res.*, v. 68, p. 6515–6522.

Fairbairn, H. W., Hurley, H. W., Pinson, W. H., and Cormier, R. F., 1960, Age of the granitic rocks of Nova Scotia: *Bull. Geol. Soc. Am.*, v. 71, p. 399–414.

Fairbairn, H. W., Moorbath, S., Ramo, A. O., Pinson, W. H., and Hurley, P. M., 1967, Rb/Sr age of granitic rocks of southeast Massachussetts and the age of the lower Cambrian at Hoppin Hill: *Earth Planet. Sci. Lett.*, v. 2, p. 321–328.

Faul, H. and Jäger, E., 1963, Ages of some granitic rocks in the Schwarzwald and the Massif Central: *J. Geophys. Res.*, v. 68, p. 3293–3300.

Fitch, F. J., 1965, Appendix: The structural unity of the reconstructed North Atlantic Continent: *Phil. Trans. Roy. Soc.*, v. 258, p. 191–193.

Fitch, F. J., Grasty, R. L., and Miller, J. A., 1965, K–Ar ages of rocks from Jan Mayen and an outline of its volcanic history: *Nature, Lond.*, v. 207, p. 1349–1351.

Fitch, F. J. and Miller, J. A., 1964, The age of the paroxysmal Variscan orogeny in England: *Quart. J. Geol. Soc. London*, v. 120s, p. 169–170.

Fitch, F. J. and Miller, J. A., 1965, Major cycles in the history of the Earth: *Nature, Lond.*, v. 206, p. 1023–1027.

Fitch, F. J., Miller, J. A., and Mitchell, J. G., 1969, A new approach to radio-isotopic dating in orogenic belts, in: *Time and Place in Orogeny*, Kent, P. E., *et al.*, eds.: Geological Society, London, p. 157–195.

Fleischer, R. L., Viertl, J. R. M., Price, P. B., and Aumento, F., 1968, Mid-Atlantic Ridge, age and spreading rates: *Science, N.Y.*, v. 161, p. 1339–1342.

Flinn, D., 1971, On the Fit of Greenland and North-West Europe before Continental Drifting: *Proc. Geol. Assoc.*, v. 82, p. 469–472.

Foland, K. A., Quinn, Q. W., and Giletti, B. J., 1971, K–Ar and Rb–Sr Jurassic and Cretaceous ages for intrusives of the White Mountain magma series, northern New England: *Am. J. Sci.*, v. 270, p. 321–330.

Freeland, G. L. and Dietz, R. S., 1971, Plate tectonics evolution of the Caribbean - Gulf of Mexico region: *Nature, Lond.*, v. 232, p. 20–23.

Frith, R. A. and Doig, R., 1971, Rb–Sr isotopic studies of the Grenville structural province in the Chibougamau and Lac St. Jean area: *Geol. Assoc. Canada Ann. Meeting*, Sudbury, Ontario (abstracts), p. 25.

Funell, B. M. and Smith, G. A., 1968, Opening of the Atlantic Ocean: *Nature, Lond.*, v. 219, p. 1328–1333.

Gandhi, S. S., Grasty, R. L., and Grieve, R. A. F., 1969, The geology and geochronology of the Makkovik Bay area, Labrador: *Can. J. Earth Sci.*, v. 6, p. 1019–1035.

Gayer, R. A., Gee, D. G., Harland, W. B., Miller, J. A., Spall, H. R., Wallis, R. H., and Wisness, T. S., 1966, Radiometric age determinations on rocks from Spitzbergen: *Norsk. Polarinst. Skr.*, n. 137, Oslo.

Ghisler, M. and Windley, B. F., 1967, The chromite deposits of the Fiskenaesset region, west Greenland: *Grøn. geol. Unders. Rapp.*, v. 12.

Gibb, R. A., 1968, A geological interpretation of the Bouger anomalies adjacent to the Churchill-Superior boundary in Northern Manitoba: *Can. J. Earth Sci.*, v. 5, p. 439–453.

Gibb, R. A., 1971, Origin of the Great Arc of Eastern Hudson Bay: a Precambrian drift reconstruction: *Earth Planet. Sci. Lett.*, v. 10, p. 365–371.

Gibb, R. A. and Walcott, R. L., 1971, A Precambrian suture in the Canadian Shield: *Earth Planet. Sci. Lett.*, v. 10, p. 417–422.

Giletti, B. J., Moorbath, S., and Lambert, R. St. J., 1961, A geochronological study of the metamorphic complexes of the Scottish Highlands: *Quart. J. Geol. Soc. Lond.*, v. 117, p. 233–272.

Goldich, S. S., Hedge, C. E., and Stern, T. W., 1970, Age of the Morton and Montevideo gneisses and related rocks, south west Minnesota: *Bull. Geol. Soc. Am.*, v. 81, p. 3671–3696.

Goldich, S. S., Lidiak, E. G., Hedge, C. E., and Walthall, F. G., 1966, Geochronology of the midcontinent region, United States, 2, Northern area: *J. Geophys. Res.*, v. 71, p. 5389–5408.

Goldich, S. S., Nier, A. O., Baardsgaard, H., Hoffman, J. H., and Kreuger, H. W., 1961, The Precambrian geology and geochronology of Minnesota: *Minn. Geol. Surv. Bull.*, v. 41, 193 p.

Goodwin, A. M. and Ridler, R. H., 1970, The Abitibi orogenic belt: *Geol. Surv. Can. Paper*, n. 70-40, 30 p.

Gorokhov, I. M., Varshavskaya, E. S., Kutyarin, E. P., and Lobach-Zhuchenko, S. B., 1970, Preliminary Rb-Sr geochronology of the North Ladoga Region, Soviet Karelia: *Eclogae Geol. Helv.*, v. 63, p. 95–104.

Grasty, R. L., Rucklidge, J. C., and Elders, W. A., 1969, New K–Ar age determinations on rocks from east coast of Labrador: *Can. J. Earth Sci.*, v. 6, p. 340–343.

Greig, D. C., Wright, J. E., Haines, B. A., and Mitchell, G. H., 1968, Geology of the country around Church Stretton, Craven Arms, Wenlock Edge, and Brown Clee: *Geol. Survey Gt. Britain, Mem.*, n. 166, 379 p.

Gross, G. A. and Ferguson, S. A., 1965, The anatomy of an Archaean greenstone belt: *Trans. Can. Inst. Mining Met.*, v. 58, p. 940–946.

Hall, B. A., 1966, Outline of the geology and mineralisation of the south end of the Munsungen anticlinorium, Piscataquis County, Maine: *Maine Geol. Surv. Bull.*, v. 18, p. 5–9.

Hall, B. A., 1969, Pre-Middle Ordovician unconformity in northern New England and Quebec: *Amer. Assoc. Petrol. Geol. Mem.*, v. 12, p. 467–476.

Hallam, A., 1971, Mesozoic Geology and the opening of the North Atlantic: *J. Geol.*, v. 79, p. 129–157.

Haller, J., 1970, Tectonic map of east Greenland (1:500,000). An account of tectonism, plutonism, and volcanism in east Greenland: *Medd. Grønland*, v. 171, n. 5, 286 p.

Haller, J. and Kulp, J. L., 1962, Absolute age determinations in east Greenland: *Medd. Gronland*, v. 171, n. 1, 77 p.

Harland, W. B., 1969, Contribution of Spitzbergen to understanding of tectonic evolution of the North Atlantic region: *Amer. Assoc. Petrol. Geol. Mem.*, v. 12, p. 817–852.

Heier, K. S. and Compston, W., 1969, Interpretation of Rb–Sr age patterns in high-grade metamorphic rocks, North Norway: *Norsk. Geol. Tidsskr.*, v. 49, p. 257–283.

Heier, K. S. and Griffin, C. K., 1971, Lecture given at Symposium on Lewisian rocks, University of Keele, March, 1971.

Heirtzler, J. R. and Hayes, D. E., 1967, Magnetic boundaries in the North Atlantic Ocean: *Science*, v. 157, p. 185–187.

Helsley, C. E. and Steiner, M. B., 1969, Evidence for long intervals of normal polarity during the Cretaceous Period: *Earth Planet. Sci. Lett.*, v. 5, p. 325–332.

Henriksen, N. and Hepson, H. F., 1970, K–Ar ages determinations on dolerites from southern Peary Land: *Grøn. Geol. Unders. Rapp.*, v. 28, p. 55–58.

Hernes, I., 1966, Age of the anorthosites in the Norwegian Caledonian chain: *Nature, Lond.*, v. 209, p. 191.

Herr, W. and Merz, E., 1958, On the determination of the half life of ^{187}Re. Further datings by the Rhenium–Osmium method: *Z. Naturforschung*, v. 13a, n. 3, p. 231.

Herz, N., 1969, Anorthosite belts, continental drift and the anorthosite event: *Science, N.Y.*, v. 164, p. 944–947.

Higgins, A. K., 1970, The stratigraphy and structure of the Ketilidian rocks of Midternaes, south-west Greenland: *Grøn. Geol. Unders. Bull.*, v. 87, p. 96 p.

Higgins, A. K., and Bondesen, N. E., 1966, Supracrustals of pre-Ketilidian age (the Tartoq group) and their relationships with Ketilidian supracrustals in the Ivigtut region, southwest Greenland: *Grøn. Geol. Unders. Rapp.*, v. 8, 21 p.

Hughes, C. J. and Brückner, W. D., 1971, Late Precambrian rocks of eastern Avalon peninsula Newfoundland- a volcanic island complex: *Can. J. Earth Sci.*, v. 8, p. 899–915.

Hurley, P. M., 1970, Distribution of age provinces in Laurasia: *Earth Planet. Sci. Lett.*, v. 8, p. 189–196.

Hurley, P. M. and Rand, J. R., 1969, Pre-drift continental nuclei: *Science, N.Y.*, v. 164, p. 1229–1242.

Jackson, G. D., 1966, Geology and mineral possibilities of the Mary river region northern Baffin Island: *Can. Mining J.*, v. 87, p. 57–61.

Jackson, G. D., 1969, Reconnaissance of north central Baffin Island: *Geol. Surv. Can. Paper* 69-1, pt. A, p. 171–176.

Jackson, G. D., 1971, Operation Penny Highlands, south central Baffin Island: *Geol. Surv. Can. Paper* 71-1, pt. A, p. 138–140.

Johnson, G. L., Vogt, P. R., and Avery, O. E., 1971, Evolution of the Norwegian Basin: Institute of Geological Sciences Report 70/14, The Geology of the East Atlantic continental margin, 2, Europe, p. 53–66.

Kanasewich, E. R. and Farquhar, R. M., 1965, Lead isotope ratio from the Cobalt–Noranda area, Canada: *Can. J. Earth Sci.*, v. 2, p. 361–384.

Keen, M. J., 1970, A possible diapir in the Laurentain Channel: *Can. J. Earth Sci.*, v. 7, p. 1561–1564.

Keen, M. J., Ewing, G. M., and Loncarevic, B. D., 1970, The continental margin of eastern Canada: Georges Bank to Kane Badin, in: *The Sea*, Vol. 4, Maxwell, A. E., ed.: Interscience, New York.

King, L. H. and Maclean, B., 1970, Origin of the outer part of the Laurentian channel: *Can. J. Earth Sci.*, v. 7, p. 1470–1484.

King, P. B., 1969a, The tectonics of North America: *U.S. Geol. Surv. Profess. Paper*, 628, U.S. Department of the Interior.

King, P. B. (compiler), 1969b, Tectonic map of North America: *U.S. Geol. Surv. map* G 67154 (1969), U.S. Department of the Interior.

Kosztolanyi, C. and Coppins, R., 1970, Etudes gèochronologique de la minèralisation uranisère de la Mine du Chardon (vondée France): *Eclogae Geol. Helv.*, v. 63, p. 185–196.

Kranck, E. H., 1953, Bedrock geology of the seaboard of Labrador between Domino Run and Hopedale, Newfoundland: *Geol. Surv. Can. Bull.*, v. 26, 45 p.

Kratz, K. O., Gerling, E. K., and Lobach-Zhuchenko, S. B., 1968, The isotope geology of the Precambrian of the Baltic Shield: *Can. J. Earth Sci.*, v. 5, p. 657–660.

Krogh, T. E., 1971, Isotopic ages along the Grenville Front in Ontario: *Geol. Assoc. Canada Ann. Meeting*, Sudbury, Ontario, (abstracts), p. 34.

Kumarapeli, P. S. and Saull, V. A., 1966, The St. Lawrence valley system: A north American equivalent of the East African rift system: *Can. J. Earth Sci.*, v. 3, p. 639–658.

Kvale, A., 1960, The Nappe area of the Caledonides in western Norway: *Norges Geol. Unders.*, v. 212 (e), p. 1.

Lambert, R. St. J., Evans, C. R., and Dearnley, R., 1970, Isotopic ages of dykes and pegmatitic gneiss from the southern islands of the Outer Hebrides: *Scot. J. Geol.*, v. 6, p. 208–213.

Lambert, R. St. J. and Holland, J. G., 1971, A geochronological study of the Lewisian of the Laxford area: *Proc. Geol. Soc. Lond.*, v. 1664, p. 251–255.

Lambert, R. St. J., and Simons, J. G., 1969, New K–Ar determinations from southern west Greenland, *Grøn. Geol. Unders. Rapp.*, v. 19, p. 68–71.

Larsen, O., 1969, K–Ar determinations: *Grøn. Geol. Unders. Rapp.*, v. 19, p. 62.

Larsen, O. and Møller, J., 1968a, K–Ar age determinations from west Greenland, I reconnaissance programme: *Grøn. Geol. Unders. Rapp.*, v. 15, p. 82–86.

Larsen, O. and Møller, J., 1968b, K–Ar studies in west Greenland: *Can. J. Earth Sci.*, v. 5, p. 683–692.

Laughton, A. S., 1971, South Labrador Sea and the evolution of the North Atlantic: *Nature, Lond.*, v. 232, p. 612–617.

Leech, G. B., Lowdon, J. A., Stockwell, C. H., and Wanless, R. K., 1963, Age determinations and geological studies: *Geol. Surv. Can. Paper* 63-17, p. 114–117.

Leggo, P. J. and Pidgeon, R. T., 1970, Geochronological investigations of Caledonian history in western Ireland: *Eclogae Geol. Helv.*, v. 63, p. 207–212.

Le Pichon, X. and Fox, P. J., 1971, Marginal offsets, fracture zones and the early opening of the North Atlantic: *J. Geophys. Res.*, v. 76, p. 6294–6308.

Le Pichon, X., Hyndman, R. D., and Pautot, G., 1971, Geophysical study of the opening of the Labrador Sea: *J. Geophys. Res.*, v. 76, p. 4724–4743.

Leutwein, F., 1968, Contribution à la connaissance du précambrien récent en Europe Occidentale et dévelopment géochronologique du Briovérien en Bretagne (France): *Can. J. Earth Sci.*, v. 5, p. 673–682.

Long, L. E., 1964, Rb–Sr chronology of the Carn Chuinneag intrusion, Rosshire, Scotland: *J. Geophys. Res.*, v. 69, p. 1589–1597.

Long, L. E. and Kulp, J. L., 1962, Isotopic age study of the metamorphic history of the Manhatten and Reading Prongs: *Bull. Geol. Soc. Am.*, v. 73, p. 969–995.

Long, L. E. and Lambert, R. St. J., 1963, Rb–Sr isotopic ages from the Moine Series, in: *The British Caledonides*, Johnson, M. R. W. and Stewart, F. H., eds.: Oliver and Boyd, Edinburgh, p. 217–247.

Lowdon, J. A., Stockwell, C. H., Tipper, H. W., and Wanless, R. K., 1963, Age determinations and geological studies: *Geol. Surv. Can. Paper* 62-17.

Lyons, J. B. and Faul, H., 1968, Isotope geochronology of the northern Appalachians, in: *Studies in Appalachian Geology*, Zen, E-An, White, W. S., Hadley, J. B., and Thompson, J. B. Jr., eds.: Interscience, New York, p. 305–318.

Lyons, J. B., Jaffe, H. W., Gottfried, D., and Waring, C. L., 1957, Lead alpha ages of some New Hampshire granites: *Am. J. Sci.*, v. 255, p. 527–546.

MacIntyre, R. M., York, D., and Moorhouse, W. W., 1967, K–Ar determinations in the Madoc-Bancroft area in the Grenville province of the Canadian Shield: *Can. J. Earth Sci.*, v. 4, p. 815–828.

May, P. R., 1971, Triassic–Jurassic diabase dykes around the North Atlantic: *Bull. Geol. Soc. Am.*, v. 82, p. 1285–1292.

McCartney, W. D., Poole, W. H., Wanless, R. K., Williams, H., and Loveridge, W. D., 1966, Rb–Sr age and geological setting of the Hollyrood granite, southeast Newfoundland: *Can. J. Earth Sci.*, v. 3, p. 947–957.

McDougall, I., Compton, W., and Bofinger, V. M., 1966, Isotopic age determinations on Upper Devonian rocks from Victoria, Australia, a revised estimate of the Devonian Carboniferous boundary: *Bull. Geol. Soc. Am.*, v. 77, p. 1075–1088.

McElhinney, M. W., and Burek, P. J., 1971, Mesozoic palaeomagnetic stratigraphy: *Nature, Lond.*, v. 232, p. 98–102.

McGregor, V. R., 1969, Early Precambrian geology of the Godthåb area: *Grøn. Geol. Unders. Rapp.*, v. 19, p. 28–30.

Michot, J. and Deutsch, S., 1970, U–Pb zircon ages and polycyclism of the gneiss de Brest and the adjacent formations (Brittany): *Eclogae Geol. Helv.*, v. 63, p. 215–227.

Michot, J. and Pasteels, P., 1968, Etude géochronologique du domaine métamorphique du sud-ouest de la Norvège (Note preliminaire): *Ann. Soc. Geol. Belg.*, v. 91, p. 93–110.

Miller, J. A. and Brown, P. E., 1965, K–Ar age studies in Scotland: *Geol. Mag.*, v. 102, p. 106–134.

Miller, J. A. and Flinn, D., 1966, A survey of the age relations of Shetland Rocks: *Geol. Journ.*, v. 5, p. 95–116.

Miller, J. A. and Green, D. H., 1961, Age determinations of rocks in the Lizard (Cornwall) area: *Nature, Lond.*, v. 192, p. 1175–1176.

Milligan, G. C., 1970, Geology of the George River Series, Cape Breton: *Nova Scotia Dep. Mines Mem.*, n. 7, 111 p.

Money, P. L., 1968, The Wollaston Lake fold belt system, Sasketchewan–Manitoba: *Can. J. Earth Sci.*, v. 5, p. 1489–1504.

Moorbath, S., Sigurdsson, H., and Goodwin, R., 1968, K–Ar ages of the oldest exposed rocks in Iceland: *Earth Planet. Sci. Lett.*, v. 4, p. 197–205.

Moorbath, S., Welke, H., and Gale, N. H., 1969, The significance of lead isotope studies in ancient, high-grade metamorphic basement complexes, as exemplified by the Lewisian rocks of northwest Scotland: *Earth Planet. Sci. Lett.*, v. 6, p. 245–256.

Naylor, R. S., 1971, Acadian orogeny, an abrupt and brief event: *Science, N.Y.*, v. 172, p. 558–560.

Neuman, R. B. and Rankin, D. W., 1966, Bedrock geology of the Slim Pond region, in: Guidebook, 1966, New England Intercollegiate geological conference, 58th. Ann. Meeting, Katajdin, Miami, p. 8–17.

O'nions, R. K., Morton, R. D., and Baadsgaard, H., 1969, Potassium–argon ages from the Bamble sector of the Fennoscandian shield in south Norway: *Norsk. Geol. Tidsskr.*, v. 49, p. 171–190.

Ozard, J. M. and Russell, R. D., 1971, Lead isotope studies of rock samples from the Superior geological province: *Can. J. Earth Sci.*, v. 8, p. 444–454.

Pankhurst, R. J., 1970, The Geochronology of the basic igneous complexes, in: The "younger" basic igneous complexes of north-east Scotland, and their metamorphic envelope: *Scot. J. Geol.*, v. 6, p. 83–107.

Papezic, V. S., 1970, Petrochemistry of volcanic rocks of the Harbour Main group, Avalon Peninsula, Newfoundland: *Can. J. Earth Sci.*, v. 7, p. 1485–1498.

Park, R. G., 1970, Observations on Lewisian chronology: *Scot. J. Geol.*, v. 6, p. 379–399.

Pavlides, L., Boucot, A. J., and Skidmore, W. B., 1968, Stratigraphic evidence for the taconic orogeny in the northern Appalachians, in: *Studies in Appalachian Geology*, Zen, E-An., White, W. S., Hadley, J. B., and Thompson, J. B. Jr., eds.: Interscience, New York, p. 61–81.

Payne, A. V., Baadsgaard, H., Burwash, R. A., Cumming, G. L., Evans, C. R., and Folinsbee, R. E., 1965, A line of evidence supporting continental drift: The Upper Mantle Symposium, New Delhi, 1964, Copenhagen (I.U.G.S.), p. 83–93.

Phillips, W. E. A. and Kennedy, M. J., 1971, The dating of events of the marginal metamorphic zone of the Caledonian/Appalachian belt in Ireland and Newfoundland: Lecture presented at Symposium *Dating events in the metamorphic Caledonides*, Edinburgh, Sept. 1971.

Pidgeon, R. T., 1969, Zircon U–Pb ages from the Galway granite and the Dalradian, Connemara, Western Ireland: *Scott. J. Geol.*

Pidgeon, R. T., 1971b, Lecture given at Symposium on Lewissian rocks, University of Keele, March, 1971.

Powell, D., 1971, Geological evidence for Precambrian orogenesis in the Moinian: Lecture presented at Symposium *Dating events in the matamorphic Caledonides*, Edinburgh, Sept. 1971.

Priem, H. N. A., Boelrijk, N.A.I.M., Verschure, R. H., Hebeda, E. H., and Verdurmen, E. A. Th., 1970, Dating events of acid plutonism through the Palaeozoic of the Western Iberian Peninsula: *Eclogae Geol. Helv.*, v. 63, p. 255–274.

Pringle, I. R., 1970, The structural geology of the North Roe area, Shetland. *Geol. J.* v. 7, p. 147–170.

Pringle, I. R., 1971, Radiometric dates from the Scandinavian Caledonides: Paper presented at the Symposium *Dating events in the Metamorphic Caledonides*, Edinburgh, Sept. 1971.

Pringle, I. R., Miller, J. A., and Warrell, D. M., 1971, Radiometric age Determinations from the Long Range Mountains, Newfoundland: *Can. J. Earth Sci.*, v. 8, p. 1325–1330.

Pulvertaft, T. C. R., 1968, The Precambrian stratigraphy of western Greenland: 23rd. Int. Geol. Congress, Czechoslovakia, Rep. Sect. 4, p. 89–104.

Rast, N. and Crimes, T. P., 1969, Caledonian orogenic episodes in the British Isles and Northwestern France and their tectonic and chronological interpretation: *Tectonophysics*, v. 7, p. 277–307.

Roberts, D., 1971, Timing of Caledonian evorogenic activity in the Scandinavian Caledonides: *Nature Phys. Sci. Lond.*, v. 232, p. 22–23.

Robertson, W. A., 1969, Magnetisation directions in the Muskox intrusion and associated dykes and lavas: *Geol. Surv. Can. Bull.*, v. 167, 51 p.

Rodgers, J., 1967, Chronology of tectonic movements in the Appalachian region of eastern North America: *Am. J. Sci.*, v. 265, p. 408–427.

Rodgers, J., 1971, The Taconic Orogeny: *Bull. Geol. Soc. Am.*, v. 82, p. 1141–1178.

Roscoe, S. M., 1971, Huronian and other early Aphebian rocks: *Geol. Assoc. Canada Annual Meeting*, Sudbury, Ontario, (abstracts), p. 60.

Salop, L. I., 1968, Precambrian of the U.S.S.R.: 23rd. Int. Geol. Congress, Czechoslovakia, Rep. Sect. 14, p. 61–74.

Schermerhorn, L. J. G., 1956, The granites of Trancoso (Portugal): a study in microclinisation: *Am. J. Sci.*, v. 254, p. 329–348.

Semenenko, N. P., 1970, Geochronological aspects of stabilisation of continental Precambrian platforms: *Eclogae Geol. Helv.*, v. 63, p. 301–310.

Semenenko, N. P., Tkachnk, L. G., and Afanas'yeva, I. M., 1968, Deep crustal structure in Ukranian crystalline shield: *Int. Geol. Rev.*, v. 9, p. 49–58.

Sheridan, R. E. and Drake, C. L., 1968, Seaward extension of the Canadian Appalachians: *Can. J. Earth Sci.*, v. 5, p. 337–373.

Silver, L. T., 1968, A geochronological investigation of the anorthosite complex Adirondack mountains, New York: *N.Y.S., Mus. and Sci. Service, Mem.*, n. 18, 233 p.

Silver, L. T. and Lumbers, S. T., 1965, Geochronological studies in the Bancroft–Madoc area of the Grenville province, Ontario, Canada: *Geol. Soc. Am. Abstracts Progr. 1965*, Annual Meeting, p. 153.

Simpson, S., 1962, Variscan orogenic phases, in: *Some Aspects of the Variscan Fold Belt*, Coe, K., ed.: Manchester University Press.

Slawson, W. F., Kanaseowich, E. R., Ostie, R. G., and Farquhar, R. M., 1963, Age of the North American crust: *Nature, Lond.*, v. 200, p. 413–414.

Smith, A. G., 1971, Alpine deformation and the Oceanic areas of the Tethys, Mediterranean, and Atlantic: *Bull. Geol. Soc. Am.*, v. 82, p. 2039–2070.

Spooner, C. M. and Fairbairn, H. W., 1970, Relation of radiometric age of granitic Rocks near Calais, Maine to the time of the Acadian orogeny: *Bull. Geol. Soc. Am.*, v. 81, p. 3663–3670.

Stevenson, L. M., 1970, Rigolet and Groswater Bay map-areas, Newfoundland (Labrador): *Geol. Surv. Can. Paper* 69-48, 24 p.

Stille, H., 1924, *Grundfragen der vergleichenden Tektonik*: Borntraeger, Berlin.

Stockwell, C. H., 1964, Age determinations and geological studies, Part 2, Geological studies: *Geol. Surv. Can. Paper* 64-17, v. 2, p. 1–21.

Sturt, B. A., 1971, The Timing of orogenic metamorphism in the Norwegian Caledonides: Paper presented at the Symposium *Dating events in the metamorphic Caledonides*, Edinburgh, Sept. 1971.

Sutton, J., 1967, The extension of the Geological record into the Precambrian: *Proc. Geol. Assoc. Lond.*, v. 78, p. 493–534.

Sutton, J., 1968, Development of the continental framework of the Atlantic: *Proc. Geol. Assoc. Lond.*, v. 79, p. 275–303.

Sutton, J., and Watson, J. V., 1951, The pre-Torridonian history of the Loch Torridon and Scourie areas in the Northwest Highlands and its bearing on the chronological classification of the Lewisian: *Quart. J. Geol. Soc. Lond.*, v. 106, p. 241–307.

Sutton, J., and Watson, J. V., 1957, The structure of Sark, Channel Islands: *Proc. Geol. Assoc. Lond.*, v. 68, p. 179–203.

Sutton, J. and Watson, J. V., 1959, Structures in the Caledonides between Loch Duich and Glenelg. north-west Highlands: *J. Geol. Soc. Lond.*, v. 114, p. 231–257.

Talwani, M., Pitman, W. C., and Heirtzler, J. R., 1969, Magnetic anomalies in the North Atlantic: *Trans. Am. Geophys. Un.*, v. 50, p. 189.

Tarling, D. H. and Gale, N. H., 1968, Isotopic dating and palaeomagnetic polarity in the Faeroe Islands: *Nature, Lond.*, v. 218, p. 1043–1044.

Taylor, F. C., 1971, A revision of Precambrian structural provinces in northeast Quebec and northern Labrador: *Can. J. Earth Sci.*, v. 8, p. 579–584.

Tozer, E. T. and Thorsteinsson, R., 1964, Western Queen Elizabeth Islands, Arctic Archipelago: *Geol. Surv. Can. Mem.*, n. 332, 342 p.

Trettin, H. P., 1969a, Lower Palaeozoic sediments in northwestern Baffin Island District of Franklin: *Geol. Surv. Can. Bull.*, v. 157, 70 p.

Trettin, H. P., 1969b, A palaeozoic Tertiary fold belt in northernmost Ellesmere Island aligned with the Lomonosov Ridge: *Bull. Geol. Soc. Am.*, v. 80, p. 143–148.

Trettin, H. P., 1971, Geology of lower Palaeozoic formations Hazen Plateau and Southern Grant Land mountains, Ellesmere Island, and Arctic Archipelago: *Geol. Surv. Can. Bull.*, v. 203, 134 p.

Van Breeman, O., Aftalion, M., and Pidgeon, R. T., 1971, The age of the granitic injection complex of Harris, Outer Hebrides: *Scot. J. Geol.*, v. 7, p. 139–152.

Versteeve, A. J., 1970, Whole-rock Rb–Sr isochron study of the charnockite–granitic migmatites in Rogoland, southwestern Norway: Report of Z. W. O. Laboratorium voor isotopen-geologie, Amsterdam, 1970.

Vogt, P. R., 1970, Magnetic basement outcrops of the south east Greenland continental shelf: *Nature, Lond.*, v. 226, p. 743–744.

Vogt, P. R., Anderson, C. N., and Bracey, D. R., 1971, Mesozoic magnetic anomalies, sea floor spreading, and geomagnetic reversals in south western North Atlantic: *J. Geophys. Res.*, v. 76, p. 4794–4821.

Wanless, R. K., Stevens, R. D., Lachance, G. R., and Rimsaite, R. Y. A., 1965, Age determinations and geological studies: *Geol. Surv. Can. Paper* 64-17, 119 p.

Wanless, R. K., Stevens, R. D., Lachance, G. R., and Edmonds, C. M., 1968, Age determinations and geological studies, Rep. 8: *Geol. Surv. Can. Paper* 67-2, pt. A, 141 p.

Wanless, R. K., Stevens, R. D., and Loveridge, W. D., 1970, Anomalous parent–daughter isotopic relationships in rocks adjacent to the Grenville front near Chibougamau Quebec; *Eclogae Geol. Helv.*, v. 63, n. 1, p. 345–364.

Watson, J. V., 1969, The Precambrian gneiss complex of Ness, Lewis in relation to the effects of Laxfordian regeneration: *Scot. J. Geol.*, v. 5, p. 269–285.

Webb, G. W., 1963, Occurrence and exploration significance of strike-slip faults in southern New Brunswick, Canada: *Amer. Assoc. Petrol. Geol. Bull.*, v. 47, p. 1904–1927.

Wendt, I., Lenz, H., Harre, W., and Schoell, M., 1970, Total rock mineral ages of granites from southern Schwarzwald, Germany: *Eclogae Geol. Helv.*, v. 63, p. 365–370.

Williams, C. A. and McKenzie, D., 1971, The evolution of the North-East Atlantic: *Nature, Lond.*, v. 232, p. 168–173.

Williams, H., 1969, Pre-carboniferous development of Newfoundland Appalachians: *Amer. Assoc. Petrol. Geol. Mem.*, v. 12, p. 32–58.

Williams, H. and Stevens, R. H., 1969, Geology of Belle Isle—Northern extremity of the deformed Appalachian miogeosynclinal belt: *Can. J. Earth Sci.*, v. 6, p. 1145–1157.

Wilson, J. T., 1962, Cabot Fault, an Appalachian equivalent of the San Andreas and Great Glen Faults and some implications for continental displacement: *Nature, Lond.*, v. 159, p. 135–138.

Wilson, J. T., 1963, Evidence from islands on the spreading of ocean floors: *Nature, Lond.*, v. 197, p. 536–538.

Windley, B. F., 1969, Anorthosites of southern west Greenland: *Amer. Assoc. Petrol. Geol. Mem.*, v. 12, p. 899–915.

Windley, B. F., 1970, Anorthosites in the early crust of the earth and on the moon: *Nature*, v. 226, p. 333–335.

Windley, B. F. and Bridgwater, D., 1971, The evolution of Archaean low and high grade terrains: *Geol. Soc. Australia Spec. Publ.*

Windley, B. F., Henriksen, N., Higgins, A. K., Bondesen, E., and Jensen, S. B. I., 1966, Some border relations between supra crustal and infra crustal rocks in south west Greenland: *Grøn. Geol. Unders. Rapp.*, v. 9, 43 p.

Wynne-Edwards, H. R. and Hassan, Z. U., 1970, Intersecting orogenic belts across the north Atlantic: *Am. J. Sci.*, v. 268, p. 289–308.

Z. W. O., Laboratorium voor Isotopen Geologie, Second Progress Report of the Isotopic dating project in Norway, Amsterdam, 1968.

Zartman, R. E., 1969, Early Palaeozoic plutonism near Boston, Massachusetts: *Geol. Soc. Am. Abstracts*, with Programs for 1969, pt. 1, v. 1, n. 1, 66 p.

Zartman, R. E., Bronck, M. R., Heyl, A. V., and Thomas, H. H., 1965, K–Ar and Rb–Sr ages of some alkali intrusive rocks from central and eastern United States: *Geol. Soc. Am. Spec. Paper* 87, p. 187–188.

Zartman, R. E. and Marvin, R. F., 1971, Radiometric age (late Ordovician) of the Quincy, Cape Anne, and Peabody granites from eastern Massachusetts: *Bull. Geol. Soc. Am.*, v. 82, p. 937–957.

Zen, E. A., 1967, Time and space relationships of the Taconic allochthon and Autochthon: *Geol. Soc. Am. Spec. Paper* 97, 107 p.

Chapter 14

THE GEOPHYSICS OF THE NORTH ATLANTIC BASIN

H. C. Noltimier*

Department of Geology
University of Houston
Houston, Texas

I. INTRODUCTION

The areal extent of the North Atlantic basin has been studied in some detail with regard to its geology, physiography, and bottom structure. Since the early 1960's, additional emphasis has been placed on geophysical studies such as heat flow, magnetic anomalies, and seismicity. These results have tended to support the earlier ideas of continental drift as they applied to the development of the Atlantic Ocean from an initial rift between the present continental masses of North America, South America, Europe, and Africa. Paleomagnetic studies in the Americas and Europe have held that the present Atlantic Ocean is a Late Triassic to post-Triassic feature and has evolved slowly over a period of perhaps 200 m.y. The more recent studies of both the northern and southern Atlantic basins have refined these ideas to the extent that the hypothesis of sea-floor spreading in the Atlantic is reasonably consistent with most geological observations in the basin itself and around the margins. In a fairly direct way these relationships apply constraints on geological and geophysical inferences

* Now at the Department of Geology and Mineralogy, Ohio State University, Columbus, Ohio.

regarding the history of the North Atlantic rather more rigorous than at first imagined, and in this sense encourage a more unified approach in the geological sciences.

II. PHYSICAL GEOGRAPHY OF THE NORTH ATLANTIC BASIN AND ITS PRINCIPAL GEOLOGICAL FEATURES

Referring to a recent physical globe, the approximate bilateral symmetry of the North Atlantic basin about its mid-ocean ridge system is a striking feature. This symmetry in terms of basin half-width about the ridge axis persists from near Iceland (lat. 60° N) south to the region around the Demerara Abyssal Plain and Sierra Leone rise, where it is noticeably interrupted by the equatorial zone of fracture systems which offset the ridge axis more than 30° of longitude eastward before it continues into the South Atlantic. Between Iceland and the equatorial fracture systems there are several important breaks in the ridge trend, such as the Gibbs, Oceanographer, Kane, and Doldrum fracture zones, but the symmetry of the basin about the ridge axis persists in the sections lying between the interruptions in trend. Bathymetric profiles (Vogt et al., 1969) running across the ocean basin perpendicular to the ridge axis illustrate this approximate symmetry as well, the more obvious departures being represented by the widths of the abyssal plains which lie between the ridge physiographic high and the surrounding continental margins. In particular, the Hatteras Abyssal Plain is rather larger in area than the Cape Verde Abyssal Plain off the western coast of North Africa, while the Cape Verde Abyssal Plain is somewhat deeper and the North American and North African continental escarpments are of different width.

In general, the physiography of the North Atlantic Ocean floor can be summarized by noting there are three distinct physiographic provinces arranged in five subparallel belts, distinguished by water depth (Fig. 1). The continental margin province includes the continental shelf, slope, and rise, each of which are more efficiently characterized by their general topographic gradients than by water depths, which in each case may vary considerably. The shelf in general has a gradient close to 1:1000. The slope has a much steeper gradient, 1:40 and greater. The shelf is usually divided sharply from the slope at an edge or rapid change in gradient (shelf break). At the foot of the slope is the continental rise where the gradient is usually around 1:300. The rise merges gradually into the next province, the abyssal plains of the ocean basin proper. In the North Atlantic, the continental margins are usually several hundred kilometers wide or more, with the deeps of the abyssal plains about 1000 km offshore. In other ocean basins, particularly the Eastern Pacific, the width of the continental margin is considerably less.

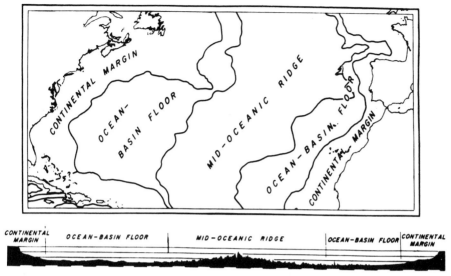

Fig. 1. Major physiographic subdivisions of the North Atlantic Ocean between the latitudes of Newfoundland and the northern limit of the Lesser Antilles. The profile is representative of the bathymetry between New England and the Sahara coast (after Heezen and Menard, 1963).

The ocean basin province lies over the abyssal plains which have very small topographic gradients, typically less than 1:1000, although these deeps are not without some local topographic features of low relief such as the abyssal hills (Heezen and Menard, 1963). In addition to the low abyssal hills, the abyssal plains are in places traversed by chains of seamounts and submarine volcanoes. A well-developed example of this is the New England seamount chain extending from the continental rise just north of Hudson Canyon southeast into the North Atlantic and dividing the Hatteras Abyssal Plain from the Sohm Abyssal Plain. This seamount chain appears to parallel a fracture zone which spans the entire North Atlantic basin, and at its eastern extremity lies another seamount–volcanic chain, the Canary Islands (Drake *et al.*, 1968). In a similar manner, the Bermuda Islands lie approximately opposite the Cape Verde Islands along the Kane fracture zone. Menard (1969) has noted that these chains of seamounts and volcanic islands show a trend of increasing age with water depth and distance from the ridge axis, with the largest and oldest volcanic islands and seamounts furthest from the ridge system.

The mid-ocean ridge province very nearly bisects the North Atlantic basin from north to south. It is of variable width, being less than 4° wide just south of Iceland and over 15° wide between the Barracuda and Kane fracture zones. The ridge province has rugged topography associated with numerous submarine volcanoes and parallel fracture zones. Water depths vary from about

1 km at the ridge crest up to about 4 km or more at the abyssal plains. Iceland is the only major feature situated at the ridge crest, while the Azores are the only other volcanic islands within 600 km of the central axis. It is notable that the vast majority of submarine volcanoes lie on the flanks of the mid-Atlantic ridge and that the islands nearest the ridge axis are much younger than those near the continental margins. A representative cross section of the North Atlantic basin is shown in Fig. 2, showing physiographic provinces and sediment types (Vogt *et al.*, 1969). In terms of overall topography and distribution of the physical provinces, the most noticeable departure from bilateral symmetry of the North Atlantic about the mid-ocean ridge axis occurs between the Oceanographer and Barracuda fracture zones and their extensions to the margins of North America and North Africa.

The latitude separation between these two fracture zones is nearly 30°. Along the western border of the basin, this separation extends from the Grand Banks south to the Blake plateau including most of the Atlantic continental margin of the United States. Along the eastern border of the basin, the 30° interval extends from Gibraltar south to the Sierra Leone rise. A comparison of the continental margins and abyssal plains of the eastern United States with Northwest Africa shows that both the continental margin province and the ocean basin province of the eastern United States is somewhat wider than that adjoining North Africa. This will be discussed in more detail in the later section on magnetic anomalies.

While as many as 100 fracture zones may exist between Iceland and St. Paul's Rocks near the present equator, a few have special prominence because of their linear extent, topographic relief, and offset of the marine magnetic anomalies. Referring to the latitude of the fault trace at its intersection with the mid-ocean ridge, the major fault zones in the North Atlantic are the Gibbs and Charlie (lat. 50° N), Oceanographer (lat. 35° N), Atlantis (lat. 30° N), Kane (lat. 25° N), and Barracuda–Guinea (lat. 10° N) fracture zones. The Atlantis fracture zone appears to trend along a small circle connecting the Canary Islands with the New England seamount chain and may extend into the North American continent between New York and Philadelphia (Maher, 1971). These fracture zones trend nearly perpendicular to the ridge axis and also nearly perpendicular to the linear marine magnetic anomalies, both east and west of the ridge province. The Terceira rift in the Azores Islands is an exception to this, however, the rift running at about 30° to the ridge axis and being confined only to the east side of the mid-ocean ridge. Doubtless, other local asymmetries will be noted in the future. In general, the only seismically active portions of these major fracture zones as well as the numerous minor ones lies between the offset sections of the ocean ridge axis.

III. SEISMIC REFRACTION STUDIES OF THE OCEANIC CRUST

The seismic refraction results for the North Atlantic, like much of its geology and geophysics, may be conveniently summarized in terms of the three physiographic provinces described in Section I, since each province has a sufficiently distinctive crustal structure to warrant discussion of its geology and geophysics separately. A clear expression of this is shown in Fig. 3, the velocity structure across the equatorial Atlantic by Leyden *et al.* (1972) along a line connecting Recife, Brazil, with Freetown, Sierra Leone. This location was chosen to avoid complications associated with trans-Atlantic fracture zones in an area where the ocean width is a minimum.

A. Ocean Basin

In the ocean basin, seismic refraction data profiles show an ocean crust with three principal layers defined by distinctive velocity intervals. The upper layer (layer I) is composed of sediments, largely unconsolidated, up to several kilometers thick but usually 0.5 km or less. Compressional wave velocities increase with depth in this layer (Ewing and Nafe, 1963; Hill, 1957), ranging from 1.5 km/sec or less in the surface sediments to 2 km/sec at depths of 0.5 km. The low velocities in the surface sediments and the positive velocity gradient with depth has meant that it has not always been possible to measure the thickness of layer I accurately, particularly in the deep-ocean basins. (In most deep-ocean areas, no refracted arrivals along the sea floor are observed, indicating that the seismic velocity in the upper layers must be equal to or less than that of the bottom water. Reflections from the sediment surface are observed, however, which indicate that the acoustic impedance of the sediment surface differs from that of the bottom water). The layer below the sediments has been identified as the oceanic basement (layer II). The compressional wave velocity in this layer is about 5 km/sec, and it has a typical thickness in the ocean basins of from 1.5 to 2 km. The composition of layer II has been the subject of some debate, partly because refractions in the layer may be masked by earlier arrivals associated with the velocity gradient in layer I. The layer was commonly missed in earlier marine refraction work (Nafe and Drake, 1969) and its widespread occurrence in the Atlantic was not recognized until adequate reflection equipment was available (Ewing and Tirey, 1961). The earlier estimates of compressional velocity were in the range of 4–6 km/sec (Keen, 1968), and there are a number of rock types which could provide velocities in that interval, including granite, consolidated sediment, and basalt. The upper surface of layer II is known to be relatively rough, from reflection measurements. Hamilton (1959) suggested that layer II may be consolidated sediment, but if so, the rough upper surface would indicate postdepositional disturbances. There is sufficient evidence that either compacted sediment or weathered basalt could form layer II, and indeed the transition from layer I to layer II may involve interbedded basalt and sediment. The results from the recent drilling program of the DSDP has shown that basalt is encountered at the base of the upper layer, although penetration well into layer II has not been possible until quite recently with the development of borehole reentry techniques in deep water. Examination of the refraction profile by Leyden *et al.* (1967) shown in Fig. 3 shows that the compressional wave velocity in the basalt exposed at the surface in the mid-ocean ridge province carries through layer II laterally away from the ridge and under layer I to the continental margins. When this evidence is considered along with the magnetic anomaly data (discussed later in the chapter) and DSDP data, it appears reasonable to identify layer II with basalt throughout the North Atlantic basin. The DSDP legs which have so far drilled in or near the deep basins of the North Atlantic are legs 2, 11, 12, and 14. The approximate locations of the sites drilled in legs 11 and 14 are shown in Fig. 4.

Beneath layer II is layer III, defined as before by a distinctive increase in compressional wave velocity from about 5 km/sec to velocities in the range of 6.4–7.1 km/sec. This layer is identified as the oceanic crust (Nafe and Drake, 1969), and its thickness varies from 4 to 5 km. Layer III has all along been identified with rocks of basaltic or gabbroic composition and is considered continuous with the "intermediate" layer between the Conrad and Mohorivicic discontinuities in sections of the continental crust, based upon the similarity

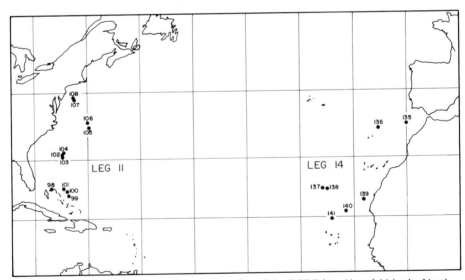

Fig. 4. Approximate locations of the drilling sites of the DSDP legs 11 and 14 in the North Atlantic.

in compressional wave velocities. Direct sampling of this layer is still questionable, but the dunites from St. Pauls Rocks (Tilley, 1947), basalts from the Iberia Abyssal Plain (Matthews, 1961), and serpentinized peridotite from the north wall of the Puerto Rico trench (Hersey, 1962) and gabbro all have compressional velocities within the range of velocities common to the layer. It should also be noted that in places metamorphic rocks have been found which may also comprise layer III locally (Matthews *et al.*, 1965) in the Indian Ocean if not in the North Atlantic. The recent models of mantle composition by Ringwood (1969) and Ringwood and Green (1969) consider the oceanic crust as a basaltic derivative of the upper 150 km of the mantle, resulting from the upward convection and partial melting of a primary pyrolite (basalt–dunite) mantle from the region of the low-velocity layer at the base of the oceanic lithosphere. If this model is generally valid, layer III may contain some peridotite which may be partially serpentinized below 500°C. The existence of serpentinites from the walls of the mid-Atlantic fracture zones is well known. Less well known is the discovery by Phillips *et al.* (1969), mentioned by Vine and Hess (1970), of serpentinites from the crest of the mid-Atlantic ridge between lat. 43° and 43.5° N in a complex fracture zone where the linear magnetic anomalies are interrupted. Dredging over an area of 1500 sq. km recovered serpentinized peridotite and no basalts. It would appear that here the basaltic layer II is absent and layer III is directly exposed at the ocean floor, giving further evidence about the general mineralogical composition in general agreement with current supposition and seismic velocities. Whatever the general composition is assumed to be, a compressional velocity in the range 6.4–7.1 km/sec is too great for granite as a major constituent.

Directly below layer III, at a depth of 4–8 km beneath the sea floor in the ocean basins, the compressional wave velocity increases to 7.8–8.5 km/sec. This velocity discontinuity marks the upper boundary of the oceanic mantle which may or may not be the same petrologically as the upper mantle beneath the continents. While the petrologic models for the mantle proposed by Ringwood (1969), Ringwood and Green (1969), and Clark and Ringwood (1964) may satisfy the majority of geological constraints (source of basalt, density, radioisotope content, compressional and shear wave velocities, and composition consistent with whole-earth chemical abundances), they must also be consistent with the stability field of basalt at the temperatures and pressures of the crust–mantle discontinuity in the ocean basins. Cohen, *et al.*, (1967) believe that the crust-mantle boundary is a basalt–eclogite transition, such that eclogite would be the stable phase at the Moho beneath the continents, while in regions of high heat flow such as ocean ridges, garnet granulite would be the stable phase between the basaltic lower crust and eclogite upper mantle. Ringwood and Green (1966) have detailed results on this transition and doubt it could be stable under the oceanic crust at the shallow depth of the oceanic Moho. No garnet granulite has been dredged up yet from the ocean crust, and Yoder and Tilley (1962) show that at depths 10–14 km beneath sea level in the ocean basins, basalt and not eclogite is the stable phase. The upper mantle may be pyroxene pyrolite in the ocean basins, while at the same time being more like garnet pyrolite at the mantle discontinuity beneath the continents. Any comparison of upper mantle compressional velocities for oceanic and transcontinental profiles shows that there are lateral inhomogeneities which may be due to several sources, compositional or phase variations being only two of the possibilities. Some better understanding of these uncertainties may result when better data are available on the sharpness of the layer III–mantle transition, under both continents and ocean basins.

B. Mid-Ocean Ridge Province

The three distinct layers above the oceanic mantle in the ocean basins become two distinct layers in the mid-ocean ridge province. Layer I is essentially absent, sediment thickness being several hundred meters thick in the carbonate zone, and virtually absent in the axial zone along the ridge crest (Nafe and Drake, 1969). The upper surface of layer II is thus the ocean floor in the ridge province and it is rough and irregular as it is beneath layer I in the basins.

There is also a change in the compressional velocities within each remaining layer. Layer II exhibits velocities in the range 4.0–8.5 km/sec, with observed values 7.2–7.7 km/sec which are not usually observed in the ocean basins (Drake and Nafe, 1968). The crust–mantle boundary rises by 4 km or more, and upper mantle compressional velocities in the range 7.2–7.7 km/sec generally replace velocities of 8 km/sec or greater compared to mantle velocities beneath the basins. These results are consistent with the upper mantle in the ridge province rising higher due to lower density and higher temperature and feeding the ridge axis volcanic crest with basaltic magmas due to its partially molten state. At the base of layer III, there is no clear-cut evidence from refraction data that the crust–mantle boundary is distinct, as it is in the basins and continental margins. This may be due to the fraction of partially melted rock in both layers at the ridge axis, or it may be due to a transition zone in which the seismic velocity changes gradually from mantle to crustal velocities and no discontinuity is apparent. The family of models of crust and mantle velocities proposed by Press (1972) are of interest in this regard, since they take into account various petrological constraints in terms of pressure and temperature at various depths beneath the ridge structure. [Various recent views about the composition of the mantle were discussed in the Birch Symposium held at Harvard in 1970 (Robertson, 1972). Theoretical fluid dynamic models for the oceanic ridge considered as a rising plume of hot mantle material have been discussed by Oxburgh and Turcotte (1968), Elsasser (1971), Torrance and Turcotte (1971), and Schubert and Turcotte (1972).]

While the overall crustal thickness in the ocean basins is of the order of 10–11 km, with the major part of the crust having a compressional velocity of 6.7 km/sec, the ridge province has a crustal thickness of 4–6 km with compressional velocity in the range of 4.5–5.5 km/sec, amenable with an oceanic crust composed essentially of basalt (Keen, 1968). This oceanic crust velocity interval could be due to an upward extension of normal mantle material with velocity in the region of 8.1 km/sec with velocity decreased by partial melting due to lowering of pressure, or through mixing of layer III material with higher velocity mantle rock (Ewing and Landisman, 1961). Also, the higher temperature gradient in the ridge province inferred from higher than normal ridge crest heat-flow measurements (Von Herzen and Lee, 1969) should diminish compressional velocities in the upper mantle in comparison with equivalent depths in the ocean basins. The decrease in velocity with higher than normal temperatures will be accompanied by a decrease in density and gravity observations, and models tend to support this conclusion (Talwani et al., 1965).

C. Continental Margins

The continental margins province is a bit difficult to define in general terms since the topography and water depths of the shelf and slope which comprise the margin province vary somewhat from the definition of these features for international reference (Wiseman and Ovey, 1955). The continental shelf is defined as the gently sloping platform extending offshore to depths of 200 m, and the continental slope as extending from 200 to 2000 m or greater depths, where it merges into the continental rise which in turn extends into the abyssal plains in the ocean basins. Along the eastern coastline of North America and the western coastline of North Africa where the most detailed geological and geophysical data is available, the general definition of the margin holds well enough not to require further qualification. As mentioned earlier, the shelf, slope, and rise are more accurately generalized in terms of typical gradients rather than bathymetry.

The continental margins are thought to be the transition zones between predominantly continental and oceanic crust. The transoceanic refraction profile of Leyden et al. (1967) illustrates on a regional basis the gradual increase in depth to mantle from the ridge flanks to the foot of the continental slopes, the gradual increase in thickness of layer III in the same interval, the more or less consistent thickness of layer II from ridge crest to abyssal plains, and the gradual increase in thickness of the sediments (consolidated and unconsolidated) away from the ridge axis with compressional velocities generally less than 3.0 km/sec. At the rise and base of the slope, there is a marked increase in sediment thickness which includes Cretaceous and younger sediments with velocities less than 3.0 km/sec, and older Mesozoic and pre-Mesozoic sediments, metamorphics, and possibly some volcanics with velocities in the range 5.2–5.8 km/sec. Beneath these Paleozoic and younger rocks, the continental crust with velocity in the range 6.1–6.5 km/sec increases rapidly in thickness to depths greater than 25 km under the outer edge of the continental slope on both the American and African coasts.

The eastern continental margin of North America has been most studied of the Atlantic margins, and several interpretive studies have been published combining seismic refraction surveys, borehole data, gravity, and magnetics. Drake and others (1959) discuss the margin north of Cape Hatteras, Drake and Nafe (1968) have interpreted velocities in terms of probable rock types, and Maher (1971) presents a comprehensive survey of geology and geophysical data for the continental shelf from Newfoundland south to Florida and into the eastern Gulf of Mexico. Fewer similar studies are as yet available from the European and African margin, but the structure of the margin of Sierra Leone and Senegal has been reported by Sheridan et al. (1969). Leg 14 of the DSDP drilled holes at several sites along the Cape Verde continental margin between October and December, 1971. The oldest sediments penetrated were limestone/ marl of Middle Cretaceous age cored at site 136 about 900 km south of Gibraltar and 160 km north of Madeira. The locations of sites 135 and 141 were chosen to obtain results for comparison with those of leg 11, in particular with the cores obtained at sites 105 and 106. The thickness of Pleistocene and Neogene sediments at site 139 near Cap Blanc is comparable to that found at site 106 off Cape Hatteras.

At site 105, Oxfordian limestone rests upon basalt, suggesting a paleontological age of 155 m.y. At sites 99, 101, 105, and 106, a hiatus in sedimentation was located at which sediments change their age from Oligocene–Miocene to Lower Cretaceous in a depth interval of less than 50 m. This hiatus represents approximately 70 m.y. of geologic time and is associated with a persistent reflector horizon (apparently in the indurated quartzose Oligocene mudstone), horizon A. The Lower Cretaceous section also contains a reflector horizon apparently associated with limestone of Lower Cretaceous–Upper Jurassic age marking the transition from terrestrial to marine deposition at horizon β. The sediment–basalt interface has been identified as horizon B.

At site 136, the drill penetrated basement at a depth of 308 m, corresponding well to horizon B. A hiatus in sedimentation between mid-Tertiary chalk ooze and Upper Cretaceous marine silty clay with ash beds, representing about 50 m.y. break in deposition, correlates well with the hiatus found in the western Atlantic, and horizon A. The paleontologic age of the oldest sediments at this site, however, is about 100–110 m.y., approximately 50 m.y. younger than that inferred for the sediments resting upon the basalt off the American east coast. Horizon β was not observed. This discrepancy between the age of the oldest sediments at the base of the continental slopes in the western and eastern Atlantic may be explainable in terms of an early proto-Atlantic ocean of pre-Cretaceous age with crust now largely covered by the continental margin sediments of North America. This will be discussed with the other interesting observed asymmetries in a later section.

The seismic structure of the North American Atlantic continental shelf is summarized by Maher (1971) and has benefitted from a few score of exploratory boreholes for petroleum drilled in water depths generally less than 100 m. Some of these boreholes have penetrated through the sedimentary section to Precambrian continental basement. The total thickness of the Paleozoic and post-Paleozoic section ranges from $\frac{1}{2}$ to 6 km of unconsolidated and consolidated sediments across the shelf from sea level to shelf break, from Newfoundland south to the Bahamas, with variation in depth of the mantle ranging from 25 to 35 km. Similar figures apply to the North African shelf, with the constraint that the North African shelf is an average of 100 km less than the eastern North American one. In addition, the structural contour map by Drake and Nafe (1968), giving the depth to velocities greater than 5.6 km for the North Atlantic, and the cross section of equal sediment thickness of the North American east coast from Halifax to Georges Bank (Drake *et al.*, 1959), show the presence of an apparent basement high running more or less continuously from Newfoundland south to the Bahama Banks under the shelf break. These maps are based upon interpretation of the depth to the bottom major reflector horizon, now identified as horizon B. If this is the case, there is a possibility that the basement high is an older island arc of Paleozoic age. Associated with the basement high is an abrupt change in magnetic gradient which also follows the shelf break as far south as the Blake plateau. As yet, there does not seem to be a similar feature identified along the edge of the North African shelf, but this may be due to lack of data at present available for the eastern Atlantic. It is important for future studies that the presence of such a structure is positively confirmed or denied for both margin provinces. The contour map also illustrates the presence of several elongate sedimentary basins on both sides of the basement high running parallel to the shelf break.

The shelf basin, modern equivalent of a miogeosyncline, appears to be absent from under the North African shelf. Like the western Atlantic rise, the North African continental rise exhibits a thickening of sediments, a modern eugeosyncline. The sediments of the American shelf are, however, somewhat thicker and appear to span a greater overall time interval.

D. Latitudinal Variations

The preceding summary of some of the refraction data for the North Atlantic has been in terms of the variations of the structure of the ocean crust from east to west, between eastern and western shorelines. There are also some variations in a north–south sense. The North Atlantic narrows considerably north from Newfoundland, and decreases in depth. The average relief of the mid-Atlantic ridge diminishes, and the thicknesses of layers II and III may also vary.

While the velocities within the sedimentary layer in the narrowing basin belt appear to change little, the velocity interval for layers II and III becomes 5.0–5.8 km/sec rather than 4.0–7.1 km/sec, while the mantle beneath it has velocity interval of 7.2–7.7 km/sec, as in more southerly latitudes. Talwani et al. (1968) observed a 4.5-km/sec velocity interval in a 1.5–3.5-km-thick layer overlying a 6.5-km/sec layer in a survey of the Reykjanes ridge parallel to the ridge axis. The 5.0–5.8-km/sec interval may therefore represent an average of these two layers (layers II and III) with compressional velocities along the Reykjanes ridge similar to the Iceland–Faeroe ridge further north. A possible reason for the change in velocity interval without invoking changes in mantle composition and its differentiates is that layer II is becoming progressively thinner toward the Arctic basin, and the velocity contrast between layer II and III is not great enough to allow resolution of the two layers, at least in the older refraction profiles. It is also possible that there are differences in the ocean crust in the region of Iceland that have so far been undetected (Gilluly, 1971).

IV. SEISMIC REFLECTION STUDIES OF THE OCEANIC CRUST

Hersey (1963) and Ewing and Ewing (1970) review the principles and techniques which have been rapidly developed to permit continuous seismic profiling at sea. By using sound sources which operate in a broad frequency bandwidth, and band pass filters at the recording devices for selectivity, the higher frequency (12 kHz) bandwidth resolves the ocean bottom depth, a 3.5-kHz band penetrates to and resolves structure and reflectors in the upper 100–200 m of sediment, and a 5–200-Hz band allows penetration into several kilometers of sediments in the abyssal plains and continental slopes. The technique as applied to penetration of sediments is best suited to the deeper ocean basins where multiple reflections are not troublesome. The profiler has been of utmost importance in confirming that the sediment cover in the oceans is thin and that the oldest oceanic sediments in the North Atlantic are Lower Cretaceous, through the location of the sediment–basalt interface, horizon B, at exposures along fault scarps (Ewing and Ewing, 1970). It has also been useful in establishing the continuity of the major reflector horizons A, β, and B through the North and South Atlantic basins. There are various subbottom reflectors besides A, β, and B through the North and South Atlantic basins, but these three appear to have geological significance in the development of the Atlantic Ocean as a whole. Ewing and Tirey (1961) introduced the continuous profiling technique, and Le Pichon et al. (1968) have extended the technique by recording wide-angle reflections and refractions while the vessel is under way. One of the most useful additional contributions from profiling has been its application to the DSDP as a guide to coring, which has permitted establishment of the ages of the several major reflector horizons.

Horizon A was first traced over the western half of the North Atlantic basin and Bermuda rise. It has tentatively been identified in the eastern Atlantic basin by the coring of leg 14 along the eastern margin of the Cape Verde Abyssal Plain. Horizon A is a prominent, nearly horizontal marker in the North American basin, and M. Ewing et al. (1964) suggested that it might be a fossil abyssal plain. J. Ewing et al. (1966) followed the horizon to an outcropping northeast of the Bahamas and the piston core samples indicated that it was associated with Late Cretaceous turbidities. Subsequent results from leg 11 indicate that the horizon is of upper to middle Tertiary age, and it appears to be associated with an increase in silica content and consolidation of middle Tertiary sediments. Horizon A has already been found on both sides of the ridge in the South Atlantic and may have a Pacific counterpart, horizon A′ (M. Ewing et al., 1966), and a Caribbean counterpart, horizon A″ (J. Ewing et al., 1967). Horizon β, the intermediate reflector, appears to be associated with a hiatus in sedimentation on both sides of the North Atlantic. It represents a 70-m.y. discontinuity between mid-Tertiary and Lower Cretaceous sediments in the western Atlantic, and a 50-m.y. discontinuity between mid-Tertiary and Upper Cretaceous sediments in the eastern Atlantic. Drilling has illuminated several important differences in the sediments above and below this hiatus reflector on either side of the North Atlantic basin.

The oldest sediments drilled at the edge of the magnetic quiet zone in the western Atlantic (leg 11) are about 155 m.y. or older, while at the edge of the quiet zone in the eastern Atlantic, the oldest sediments are 100–110 m.y. (leg 14). At the leg 11 sites, horizon β appears to be related to the upward transition from Upper Jurassic–Lower Cretaceous terrigenous sediments to marine carbonates, while at the leg 14 sites, it appears to be related to an upward transition from Upper Cretaceous sediments which are more siliceous to mid-Tertiary carbonates. The western hiatus is 20 m.y. longer and lies over oceanic crust which is about 50 m.y. older than the eastern hiatus, introducing strong evidence that the North Atlantic basin has not developed in a symmetrical way during the entire period since its apparent first beginning in the

Late Triassic. The current near-bilateral symmetry in sedimentation may be a relatively recent (post-mid-Tertiary) feature of the basin.

Reflection profiles clearly show the upper surface of layer II is rough, both in the ridge province where the profiles indicate that the sediment layer is almost totally absent except for pockets in the basalt on the ridge flanks, and in the ocean basin province where layer II is covered by sediments of layer I. Horizon B exhibits this roughness at the sediment–basalt contact on both sides of the Atlantic, except when the profiles cross the continental rises. There, the top of layer II appears to smooth out, possibly representing the first appearance of a very thin layer of well-consolidated, pre-Cretaceous sediments (M. Ewing *et al.*, 1966). Horizon β apparently represents a hiatus in sedimentation in the North Atlantic, the western one being somewhat older and longer in duration than in the east. Neither the eastern or western component of horizon β extends much further toward the central ridge province than the midline of the abyssal plains which parallel the ridge axis. On both sides of the North Atlantic, horizon β ceases before the continental slope (Heirtzler and Hayes, 1967). Horizon A appears to represent younger sediments of comparable age on both sides of the mid-Atlantic ridge, and extends from the continental slope to the ridge flanks, approximately half-way between the slope and the ridge crest. Horizon B represents a surface which ranges in age from Recent at the ridge crest to Upper Jurassic at the base of the continental slope. Horizon β represents two discontinuities in basin sedimentation, over longitudinal widths much narrower than the basin width at present. Horizon A appears to represent the upper surface of compacted and more siliceous sediment, overlain by mid-Tertiary and younger carbonates and unconsolidated turbidites. It will be interesting to see how the equivalent horizons in the South Atlantic compare in terms of sedimentary history.

V. DISTRIBUTION AND NATURE OF EARTHQUAKES IN THE NORTH ATLANTIC BASIN

Heezen and Ewing (1963) speculated that the epicentral belt of shallow tectonic earthquakes in the North Atlantic probably coincided with the rift valley zone following the ridge axis. At that time, the accuracy of epicentral determinations was within $\frac{1}{2}°$ and $1\frac{1}{2}°$ and sufficiently well defined to show clearly that the epicentral belt was about 160 km wide following the ridge axis, while the ridge province itself is more than 1600–2000 km in width and seismically quiet. The global continuity of the ridge province was not in question, but the continuity of rifted ridges was, due to (then) insufficient data. Heezen and Ewing (1963) remarked that the rift valley and associated epicentral belt might be either a continuous feature along the ridge crest, or a series of en echelon belts similar to the East African rift zone which resembles the North Atlantic ridge quite closely. At the time of their writing, the details of the numerous fracture zones cutting the North Atlantic ridge perpendicular to its strike were not as clear as now, and the accuracy of epicentral determinations has been improved even more by the completion of the World Wide Standard Seismograph Network (WWSSN). In addition, greatly improved bathymetric coverage of the ocean basins has charted many more topographic breaks in the ridge province marking the existence of the transverse fracture zones as common occurrences along the ridge system. Since 1963, the relationship of earthquakes to the tectonics of the ridge system has been significantly clarified, to a great extent through the recalculation of epicenters from older data and the recent acquisition of new data by the WWSSN. Sykes (1965, 1967, 1969) relocated epicenters along the North Atlantic ridge using earthquake data for the period 1955–1965. Barazangi and Dorman (1969) prepared a map of global seismicity for the period 1961–1967 using data from 29,000 seismic events located by ESSA. This study confirms that earthquakes are confined largely to belts associated with either oceanic trenches or the ocean ridge system. The map also indicates that the earth is divided into a relatively small number of aseismic blocks or plates with seismic zones at the boundaries. Earth models based upon this evidence were provided in quick succession by Oliver and Isacks (1967), Elsasser (1967), Morgan (1968), McKenzie (1967), and Le Pichon (1968). A comprehensive study of the global relationships was provided by Isacks et al. (1968) in terms of global seismicity and inferred relative motions at the boundaries of the seismically quiet blocks or plates. This introduced the concept of global tectonics based upon global seismicity, and the North Atlantic basin plays an important role in the global picture. The ocean ridge earthquakes are usually shallow and of magnitudes less than magnitude 5.5. Their distribution and nature are intimately related to the central ridge and transverse fault zones, between segments of the offset ridge axis, and not just to the ridge axis as previously assumed. In the North Atlantic there is only one known active trench at the Lesser Antilles Arc. The earthquakes in this region are different in character from the ridge seismicity and have been discussed by Molnar and Sykes (1969).

A. The Seismicity of the North Atlantic Ridge System

Generally speaking, shallow earthquakes associated with normal faulting are confined to the central graben of the North Atlantic ridge, the fracture surfaces paralleling the ridge axis (Sykes, 1967). A few earthquakes with normal fault plane solutions were not found to be associated with any known fracture zones. These normal fault earthquakes are therefore confined to the seismically active belt about 160 km in width along the ridge crest, and are associated with the narrow zone of higher than normal heat flux and rift volcanism. It is interesting to note the comparison here between the central seismically active zone of the North Atlantic ridge with the central seismically active zone of the East African rift zone through Lake Nyasa and Lake Tanganyika (Heezen and Menard, 1963). Heezen and Ewing (1963) remark upon the unusual similarity between the topography and seismicity of these two tectonically active rift systems and their apparent similarity in origin. Current interpretations of the gravity anomalies along and across both rifts have come to favor an explanation in terms of extension (Heiskanen and Vening Meinesz, 1958; Talwani and Le Pichon, 1969). If correct, the simple extension model for the central rift zone in both the Atlantic and East African structures identifies the central seismic zone with normal faulting resulting from tensional graben development.

Nearly perpendicular to the ridge axis, the transverse fracture zones cut the ridge axis and apparently displace it by distances up to several hundred kilometers. Sykes (1967) shows that the mechanism of earthquakes located along these fracture zones are predominately of a strike-slip nature. Furthermore, the majority of observed and accurately determined epicenters for these earthquakes lie between the offset segments of the ridge axis, based on the map of ridge axis and transverse fracture zones by Heezen and Tharp (1965). The fault plane solutions for these fracture zone earthquakes require active displacement of the oceanic crust in a sense opposite to the apparent displacement of the ridge axis. This data appears to confirm the suggestion by Wilson (1965) that the transverse fault zones represent a new class of fault, the transform fault, with active displacements opposite to the sense of transcurrent faults already familiar in the continental crust. This concept has been refined and Wilson (1970) presents a detailed comparison of the properties of transform and transcurrent faults based largely upon geophysical observations of the oceanic ridge systems.

However, the concept of the transform fault as defined by Wilson (1965), has been specifically applied to the apparent displacement of the ridge crest by the intervening active fracture zone. The implicit assumption in Wilson's analysis is that the offset ridge axis has unique merit as a reference for crustal displacement. However, the sense of relative displacement of the linear magnetic anomalies across the fracture zone between ridge axes is in the same sense as the fault plane solutions, and hence using them as evidence of crustal displacement, the transverse fracture zones resemble transcurrent faults. If the ridge offsets are original features of the rift, and no further displacement has occurred since the inception of the present active ridge system, the relative displacement of the ocean crust bordering the transverse fracture zone is confined to the region between the ridge axes and is rather similar to Wilson's definition of the class of transcurrent faults. In fact, considering the magnetic anomalies as the major geological features offset by faulting, the only distinction remaining between transcurrent and transverse faults given by Wilson (1970) is that the relative motion along the trace of a transform fault is confined to the offset part of the trace, while relative motion continues all along the trace of a transcurrent fault. This difference can also be removed, if one considers the portions of the transverse fracture zones beyond the offset ridge axes as merely topographic expressions of the original offset of the prism-shaped vertical cross section of the ridge system. Since plates behave as rigid shells, the active fault zone must be confined to the offset region between the ridge axes where tensional stress permits injection of new basalts into the ocean crust, unless the fracture zone is also a plate boundary, and the adjacent plates are moving with different velocities. This comment upon transform faults is offered by the author principally because the record of magnetic reversals in the ocean crust appears to be more of a permanent feature than the topographic expression of an active ridge system, while the relative displacement of linear magnetic anomalies associated with crustal spreading at a ridge axis can be considered over large areas and considerable periods of geologic time. With regard to the present shape of the mid-Atlantic ridge, closely paralleling both coastlines of the northern and southern basins, it would appear that the ridge axis curvatures are indeed substantially inherited features from the initial rifting of the Atlantic basin, and that few major active offsets of the pure transform type need have actually occurred. This would suggest that most of the ridge fracture zones may be regarded as transcurrent faults, and that transform faults, in the strict sense, may be rare.

B. The Seismicity of the Lesser Antilles Arc

The only active trench north of the equator in the North Atlantic basin lies just east of the Lesser Antilles Arc, at the eastern boundary of the Caribbean Sea. This trench structure exhibits a marked gravity low and is seismically active. Sykes and Ewing (1965) relocated the epicenters of approximately 500 earthquakes (from the period 1950–1964). Molnar and Sykes (1969) relocated additional earthquakes for the period 1954–1962 and discussed the tectonics of the Caribbean and Central America using both sets of relocated data. The seismic activity associated with the trench occurs along a surface or Benioff zone dipping at about 50° from the horizontal from east to west under the Lesser Antilles Island Arc. The fault plane solutions suggest a downthrusting or compression of the southwestern Atlantic crust under the eastern margin of the Caribbean. Earthquake foci are distributed along this surface from a few kilometers to depths of about 250 km. The focal mechanisms vary somewhat from the trench proper down to depths of several hundred kilometers under and behind the arc. At the trench axis the oceanic lithosphere is most sharply flexed and the shallow earthquakes have mechanisms of the double-couple type representing horizontal tension normal to the trench axis (Stauder, 1962). Further landward from the trench, the mechanisms indicate under-thrusting of the arc lithosphere by the oceanic lithosphere, the axis of com-pression is parallel to the Benioff zone surface, and the axis of tension is normal to the Benioff zone surface. There is an equivalent solution in which the axis of compression is normal to the Benioff surface and the tension axis parallels it. It is well to keep both possibilities in mind until we are sure whether the litho-sphere is pulled or pushed down into the mantle beneath and behind the arc.

C. The Oceanic Crust and Mantle from Earthquake Seismology

Analysis of earthquake waves which have traversed the North Atlantic basin has contributed to the understanding of the stratified oceanic crust as well as providing some insight into the properties of the upper mantle beneath the ocean basins and beneath the ocean ridge system. While the resolution of the various crustal layers is not as sharp as with refraction or profiler techniques, using higher frequency waves, earthquake seismology has permitted exploration in the upper mantle for regional variations in compressional and shear velocity, as well as providing evidence for the low-velocity layer at about 100 km depth in the upper mantle. Both explosion and earthquake seismology have indicated that a thick layer with velocity intermediate between layer II and the upper mantle underlies the central belt of the mid-Atlantic ridge (Ewing and Ewing, 1959; Tryggvason, 1962). Press (1972) has reported on various petrological models for the core of the ridge system which also satisfy the observations of gravity anomalies over the ridge province (Talwani et al., 1965).

It is instructive to review the structure of the ocean basins provided by the analysis of earthquakes (Brune, 1969). A typical ocean basin consists of 5 km of sea water overlying 0.1 km of unconsolidated sediments with shear wave velocity in the range 0.5–1.0 km/sec. Beneath the sediment lies 5 km of rock with compressional velocity about 6.4 km/sec. The mantle begins at a depth of 11–12 km with an abrupt increase in compressional velocity to 8.1 km/sec and shear wave velocity to 4.7 km/sec. At a depth of 100 km, the low-velocity layer begins with shear wave velocity in the range 4.4–4.5 km/sec. This layer has been delineated by studies of attenuation as a function of wavelength, giving values for shear anelasticity as a function of depth (Anderson, 1965). The anelasticity of the low-velocity layer (Q approximately 100) appears to be a maximum for the mantle. Application of surface waves to the study of regional crustal structure has developed rapidly in the past 20 years. Due to the fact that layering causes dispersion in longitudinally–vertically polarized Rayleigh waves and transversely polarized Love waves (velocity dependent upon frequency), surface waves can be used in conjunction with refraction data to help interpret crustal structure. These studies have resulted in a classification of continental and oceanic crust into seven types (Brune, 1969). The parameters most important in arriving at the various classes of crust are crustal thickness, upper mantle (P_n) velocity, tectonic characteristics such as heat flow and Bouguer gravity values (and hence density variations with depth), sediment thickness, and water depth. More refined analysis techniques with surface waves may permit the definition of low-velocity layers within the continental crust where velocity may be affected by regional variations in surface geology and crustal heat flow.

VI. HEAT-FLOW STUDIES IN THE NORTH ATLANTIC

The measurement of thermal gradients in recent sediments on the ocean floor and the calculation of thermal heat flow through the sediment from the product of thermal gradient and thermal conductivity measurements was first successfully carried out by Bullard (1954) and Revelle and Maxwell (1952) in the northeastern Pacific. Since then, probably about 10^4 heat-flow determinations have been made and have been reported by Bullard (1963), Lee and Uyeda (1965), Birch (1966), Von Herzen and Langseth (1965), and Von Herzen and Lee (1969), among others. The importance of heat flow to tectonics and geology in general is now recognized since the earth has its own internal heat source in the decay of the long-lived radioisotopes of uranium, thorium, and potassium, which have been significant sources of heat for the past 4.5×10^9 years, and will continue to supply internal heat for at least that long again in the future. Isotope geochemistry in the crust and mantle has been reviewed by Hart (1969), and the general studies of the thermal history of the earth have been reviewed by Lubimova (1969).

The present value for the mode of both oceanic and continental heat-flow data is 1.3 μcal/cm^2 sec, calculated and averaged only for regions where data is available. (1.0 μcal/cm^2 sec is one heat flow unit, or hfu.) If representative values are assumed for the area where data is missing, a value of 1.39 hfu is obtained. Assuming that both of the above are representative values for the earth's heat flux, over 6×10^{12} hfu flows through the earth's surface per second, or over 2×10^{20} cal/yr. Over geologic time scales, the amount of internal heat available from radioisotope decay as an energy source for upper mantle and crustal tectonics appears to be adequate. The total estimated heat flux at the earth's surface due to volcanism is barely 1% of the heat flux due to conduction (Horai and Uyeda, 1969). While volcanic activity in local regions may be extremely significant as a carrier of internal heat, it appears to be virtually insignificant on a global scale. This is an important point since it suggests that the adiabatic gradient and global surface heat flux are not much affected by regional volcanism, nor would regional volcanism have much affected the cooling rate and hence the thermal history of the earth.

All oceanic thermal gradient measurements have been made with either the Bullard probe (Bullard, 1954) or the Ewing thermograd (Gerard et al., 1962). Thermal conductivities have been determined on shallow sediments taken by piston coring using a static method on disk-shaped specimens (Ratcliffe, 1960) and more recently by the transient method (Von Herzen and Maxwell, 1959).*

A comprehensive table of thermal conductivities for common terrestrial rocks as well as minerals and oceanic sediments is given in the Handbook of Physical Constants (Clark, 1966). A short table of thermal properties of ocean sediments is given by Bullard (1963), Lee and Uyeda (1965), and Langseth and Von Herzen (1970). Langseth and Von Herzen (1970) discuss the variability of thermal conductivity determinations and the general agreement between determinations by the static and transient methods, as well as by the transient method measurement of thermal conductivity made in situ in the sea floor during piston coring (Corry et al., 1968). The variations in thermal conductivity determinations on similar sediments appear to be in the range 5–10% and are therefore unlikely to contribute significantly to the variability of calculated heat-flow values.

The principal feature of oceanic heat-flow determinations is the great variability, particularly between the values calculated for the ocean ridge provinces as compared to the deep basins and trenches. The ridges are belts of higher than normal heat flow, the basins appear to have near normal heat flow, and the trenches have heat flow usually below normal. However, values calculated for each of these three thermal regions often show a great scatter in themselves so that while great differences appear to exist, the standard deviations of the results themselves are relatively large. There are several physical reasons why heat-flow calculations from ocean measurements could exhibit large scatter, and in practice most of the important reasons involve the measurement of the thermal gradient. While instrumentation is now quite reliable, various geological effects may have occurred in the recent past, or

---

* In the latter method, which is less time consuming, a small-diameter hollow needle with an internal heating element is inserted in the sediment core at discrete points along its length and change in temperature with time is plotted while a known amount of heat is dissipated at a constant rate in the needle. If Q is the rate of joule heating per unit length of the needle, and t is the time, then the temperature T as a function of time is given by the equation

$$T = (Q/4\pi k) \ln(t) + \text{const}$$

(Langseth and Von Herzen, 1970). k, the thermal conductivity of the sediment, can be easily computed from a graph of the temperature increase vs. the heating time.

occur at present, to disturb the basic assumption of thermal equilibrium in a flat, solid slab of uniform thermal conductivity. It is necessary to assume in the calculation of heat flow that the heat flux is equal to the product of thermal conductivity and thermal gradient with the system in thermal equilibrium. Considering the poor thermal conductivity of normal rocks and sediments, the time required for thermal equilibrium to be reached is of the order of 10^6 yr for a layer 1 km thick near the earth's surface. Therefore, the assumption of thermal equilibrium requires this equilibrium on time scales of millions of years or longer for problems of geological interest. Due to the possibilities of submarine slumping, recent and rapid additions of additional layers by turbidity currents, local roughness in topography which can affect the local thermal gradient, and recent or current percolation of ground water into or out of a sedimentary layer, it is not surprising that heat-flow calculations show variability above the errors of measurement.

A. Heat Flow in the Mid-Atlantic Ridge Province

The global average heat flow for the ocean ridge axes is 2.5 hfu. The global average value for the ridge flanks is 0.8 hfu (Langseth and Von Herzen, 1970). Oxburgh and Turcotte (1969) have estimated the heat flux which is contributed to the axial zone by the cooling of molten basalt injected at the ridge axis at a rate given by the average ridge extension rate and the thickness of the oceanic crust. Most of the initial cooling of basaltic dikes would occur by heat transfer to deep-ocean water in about 60,000 yr. The amount of heat lost would amount to about 6×10^{18} cal/yr for the entire 70,000-km-long ridge system. This would contribute about 0.4 hfu to the ridge axis heat flow, 0.04 hfu to the net oceanic heat flow, and may help account for the uncertainty in the average flux value. The average heat flow for the North Atlantic ridge is 1.72 hfu, with a standard deviation of 1.61 calculated from individual measurements, Reykjanes ridge not included. Paralleling the relatively narrow axial zone of high heat flow, there are two bands of low heat flow having average values about 0.7 hfu or half that in the adjacent abyssal plains province calculated on a $5° \times 5°$ grid basis. The definition of these low-heat-flow bands is difficult from lack of data. In the equatorial mid-Atlantic fracture zone, there are no observed low values, and regional heat-flow averages by $5° \times 5°$ grid basis show uniformly higher than normal heat-flow values over the entire width of the fracture zone.

Although the maximum values for the North Atlantic ridge are not statistically very significant, it is interesting to note their magnitude. Less than 3% of the observed results give heat flow in excess of 6 hfu, although more than 30% of the observed results give values in excess of 2.0 hfu. A plausible reason for the nature of these results is that in the central ridge axial zone, sediment thickness is slight, less than 100 m at the deepest in local topographic lows. The thin sediment layer restricts average penetration of the heat probe, tending to minimize the measured gradient which may reduce the thermal gradient by 10%. In addition, the ridge province has the roughest topography, increasing the surface area, and tending to diminish the surface temperature gradient by perhaps 10% (Birch, 1967). Directly cooled submarine basalts near the ridge axis are not sampled by the sediment probes, and this might increase the axial heat flow by 0.5 units. But perhaps most significant, Le Pichon and Langseth (1967) have shown that rough topography increases the standard deviation by about 0.37 of the net heat flow for any given region. Von Herzen and Uyeda (1963) have observed that lower values of heat flow were determined in small flat basins of sediment on the flanks of the ridge system, whereas much higher values were found on nearby hilly topography. Thus, it may be that the number of really high heat-flow values is smaller than realistic, and that as the total coverage of the world's ocean basins increases, the regional heat-flow averages may eventually increase in the future over their values thought most reliable at present. This has been occurring over the past decade or so, as can be seen by the values quoted in Bullard (1963). It is also of interest that the highest values of ridge heat flow for the North Atlantic have been found north of lat. 46° N by Bullard and Day (1961) and Reitzel (1961), and, in general, the coverage of the far North Atlantic is less complete than the regions further south. This situation can be clearly seen in the regional maps of heat flow and data points in the review by Langseth and Von Herzen (1970). But, it is equally true that the width of the higher heat-flow zone in the north is rather narrower than in the equatorial zone where the heat flow is not so extreme at the crest but distributed over a much broader area.

While Langseth and others (1966) conclude from the narrow distribution of high heat flow in the North Atlantic that continuous continental rifting is not possible and that the current zone of high heat flow suggests a renewed period of lateral extension during the past 10 m.y. or less, their interpretation ultimately assumes that the source of heat lies at the base of the lithosphere, approximately 50 km below the ocean floor, with the ridge heat-flow anomaly being driven and maintained by the flow of basalt magma to the surface from that depth. The cooling of a new incremental strip of lithosphere 50 km thick would be slow enough to remain quite warm for distances of several hundred kilometers from the ridge axis, depending upon the rate of lateral extension. However, if the ridge heat-flow anomaly were due to partial fusion of layer III basalt at depths of only 10 km or so, the width of the central hot zone would be much narrower, since the rate of cooling would be much higher. While the rate of lateral extension in the North Atlantic may well be episodic, the idealized models of MacKenzie (1967) and MacKenzie and Sclater (1969) contain many ill-known parameters. The behavior of the basalts, the agent of heat transfer, with changes in temperature and pressure, must be kept in mind for possibilities like the stagnation zone of Bott (1967) and the laterally extended molten zone of McBirney (1963). Current models for the ridge cross section down to the low-velocity zone (Press, 1972) seem to involve some aspects of these latter models.

B. Heat Flow in the Ocean Basin and Margin Provinces

The global average heat flow in the ocean basins is 1.3 hfu with a standard deviation of 0.4. The average heat flow through the floors of the North Atlantic basins is 1.13 hfu with a standard deviation of 0.36. The basins in the equatorial Atlantic have somewhat higher values of 1.32 hfu, with a standard deviation of 0.46. The latter value is near the mode for all basins and continents, while the former is below it. The average heat flow in all of the Atlantic basins is 1.21 hfu, with a standard deviation of 0.36 (Langseth and Von Herzen, 1970).

Within the North Atlantic basin, there are some significant variations from the basin average. The northwestern Atlantic and eastern North America continental shelf are regions of below-basin-average heat flow, while the Northern Appalachians are a local heat-flow high near the boundary between the ocean basin and continental shield. This is seen on heat-flow profiles plotted from the west coast of North America to the west coast of North Africa, and from the Gulf of Mexico to Gibraltar. Heat-flow values do not change across the continental slopes of ocean basins with tectonically stable margins. The Atlantic and most of the Indian Ocean have stable margins and meet this criterion. The major exception in the North Atlantic region is the Paleozoic Appalachian fold belt, a heat-flow high which has been long lived. Enrichment of radioactive elements in the upper lithosphere within island arcs or mountain arcs associated with active trenches will inhibit heat flow from depth and produce excess heat flow persistent over long intervals. Paleozoic fold belts in eastern Australia also have high heat flow in the range 1.6–2.3 hfu. The heat flow in Mesozoic and Cenozoic orogenic belts is estimated at 2–5 hfu (Lee and Uyeda, 1965). Heat flow appears to be important in the interpretation of past orogeny along continental margins.

Generally speaking, ocean basin heat flow varies inversely with the thickness of sediment and depth to oceanic crust, and with distance from the ocean ridge. Since the sediment thickness in the basins is rarely more than 0.5 km, the major effect of the sediment cover is to reduce heat flow from the ocean crust by insulating it with a very poor thermal conductor. (Thick layers of sediments in continental basins can generate their own internal heat, depending upon the content of radioactive minerals.) Since basin sedimentation increases with age, the older basins should exhibit lower heat flow than younger ones. The northeastern Atlantic has higher heat flow than the northwestern

Atlantic, and a thinner sediment cover. The ages of the two major reflector horizons in the northeastern Atlantic appear to be younger as well, and thus heat-flow results appear to confirm that the northwestern Atlantic is somewhat older. This is important for the understanding of the development of the entire North Atlantic basin.

VII. THE NATURE AND DISTRIBUTION OF THE MAGNETIC ANOMALIES IN THE NORTH ATLANTIC AND ALONG ITS MARGINS

Bullard and Mason (1963) gave a comprehensive review of the field of marine geomagnetism, including the then recent discovery of the linear magnetic anomalies in the northeastern Pacific by Mason (1958), Mason and Raff (1961), and Raff and Mason (1961). At the time of publication of the Bullard–Mason paper, Hess (1962) and Dietz (1961) had suggested their theories of ocean-floor spreading, and Drake and others (1963) had found magnetic lineation off the east coast of North America. Following the Bullard–Mason paper, Vine and Matthews (1963) provided a means of explaining these magnetic lineations, without invoking unrealistic susceptibility contrasts in the oceanic basement, through a simple mechanism of magnetic polarization reversals of the earth's field as the oceanic crust is cooled in the axial zone of the ridge systems. This hypothesis, strongly supported by a subsequent wealth of data from the world rift system, has linked the paleomagnetic observations of geomagnetic polarity reversals to the process of crust renewal at mid-ocean ridges and is now being usefully employed to chart in some detail the history of the major ocean basins, in particular the North Atlantic (Pitman and Talwani, 1972; Phillips and Forsyth, 1972). Because the magnetic lineations can be related to geologic ages through the history of geomagnetic reversals known from other rocks (Cox, 1969; Opdyke, 1968), marine geomagnetism has assumed great importance in the study of the earth's crust. Furthermore, since the magnetic lineations do not appear to have been deformed since their formation any more than the continental margins in comparable times, their existence and relationship to sources of new crust at the ocean ridge systems contributed to the theory of plate tectonics which would seem to be the first global theory of the earth sciences (Morgan, 1968; Heirtzler et al., 1968). A comprehensive up-to-date review of the state of the art is given by Heirtzler (1970).

A. The Mid-Atlantic Ridge

The axial anomaly along the mid-Atlantic ridge had been observed by Ewing et al. (1957). The first detailed aeromagnetic survey of the axial region of a known mid-ocean ridge was completed in the Reykjanes ridge in 1963 (Baron *et al.*, 1965), after it was known that there were striking magnetic lineations off the Pacific and Atlantic coasts of North America. This survey showed that the magnetic lineations were parallel to the ridge axis and symmetrically disposed about it. Keen (1963) had already found that there were many similar magnetic anomalies along the ridge axis between lat. 60° N and lat. 42° S. Heirtzler and Le Pichon (1965) reported on 58 profiles widely spaced along the ridge axis and accounted for the central anomaly alone by a block of ocean crust 10 km wide and with high magnetic susceptibility. The association of a central magnetic anomaly with all ridges was presented by Vine and Matthews (1963) and Matthews (1967). The Reykjanes survey showed a distribution of magnetic anomalies that corroborated the ocean-floor spreading ideas of Hess and Dietz, as well as the hypothesis of Vine and Matthews on the formation of linear and symmetrical magnetic anomalies about ridge axes. Pitman and Heirtzler (1966) and Vine (1966) determined a lateral extension rate of 1 cm/yr for the Reykjanes ridge for the past 10 m.y., based upon the geomagnetic reversal time scale worked out by Cox *et al.* (1963, 1964) and Doell and Dalrymple (1966). Subsequently, the estimated lateral extension rate has been modified for various locations along the ridge and for various periods in the geologic past (Phillips, 1967; Van Andel and Bowin, 1968). A detailed analysis of the magnetic anomalies in the North Atlantic between lat. 15° and 63° N (Pitman and Talwani, 1972) indicates there is considerable fine structure in the history of the magnetic anomalies and the development of the North Atlantic basin. The magnetic anomaly studies of Keen (1963), Heirtzler and Le Pichon (1965), and Matthews (1967) indicate a long-wavelength (several hundred kilometers) axial anomaly upon which shorter wavelength anomalies representing past geomagnetic reversals are superimposed. For definition of the longer wavelength anomalies, the choice of the regional field is important. Some of the studies have removed the regional field graphically, while others use the analytic expressions of the regional field provided by Cain *et al.* (1964, 1967) and the International Geomagnetic Reference Field (IGRF), giving the coefficients of the spherical harmonic expansion of the earth's main field. As interpretations become more detailed, the use of a standard reference for the earth's main field becomes necessary so that the main field may be removed from the observed data in a systematic manner, preserving the anomalies with wavelengths of several hundred kilometers for comparison over large geographical distances. At present, the long-wavelength variation of total magnetic intensity across ridge axes is not always observed (Heirtzler, 1969). The main field is not so crucial in the long-distance correlation of the short-wavelength anomalies, but there are difficulties being noted in the study of these features by higher resolution deep-tow magnetometers (Spiess and Mudie, 1970; Luyendyk, 1969). Much of the detail in the short-period anomalies appears to be of a local statistical sort, not regionally consistent. Spiess and Mudie (1970) remark that the long-distance correlation of the ridge magnetic anomalies would possibly have gone unnoticed had the initial marine magnetic surveys been done near the ocean floor rather than near the sea surface, a large distance from the magnetic regions in the oceanic crust.

The linearity and symmetry of the magnetic anomalies in the North Atlantic are seen at their best in the results of the survey of the Reykjanes ridge (Heirtzler et al., 1966), shown in Fig. 5. The anomalies shown extend outward from the ridge axis to the neighborhood of anomaly 5, which is correlated with radiometric ages and geomagnetic reversals about 10 m.y. before the present. The time sequence of magnetic reversals worked out by Cox et al. (1963) was applied to the Reykjanes profiles by Pitman and Heirtzler giving a spreading rate of about 1 cm/yr. Phillips (1967) found a spreading rate of 1.25 cm/yr for the ridge between lat. 26° and 29° N. The spreading rate in the equatorial Atlantic between lat. 0° and 10° S has been estimated to be about 2.0 cm/yr by Dickson et al. (1968). The variation in spreading rate with latitude will be discussed later. Talwani and others (1971) have done a detailed survey of the Reykjanes ridge both across and along the trend of the ridge axis. Across the trend, the magnetic anomalies show little correlation with bottom topography. Along the trend, the anomalies are less pronounced and show good correlation with bottom roughness. The two sets of calculations for models of ocean crust magnetic intensity suggest that a relatively thin (500 m) surface layer of magnetic polarization intensity (0.015 c.g.s.) is responsible for gross character of the surface anomalies, the highs corresponding to addition of the crustal magnetization to the ambient field, lows to subtraction because of reversed polarity. This thickness and intensity of the magnetic oceanic crust is somewhat different from similar estimates made by earlier authors who assumed the magnetization resided in a layer 2 km thick (Pitman and Heirtzler, 1966; Vine, 1966). Some differences in the estimates of the intensity of oceanic crust magnetization are bound to arise as deep-tow magnetometer studies increase in coverage and more detail is confronted. Because seismic profiling has resolved no discrete bodies within the ocean crust in the ridge province, or elsewhere, which can be identified with the magnetic lineations, it is assumed that the anomalies originate within the upper part of layer II, which is basalt, and is capable of providing the magnetic intensity required in the anomaly models. Furthermore, a recent study by Marshall and Cox (1971) on the magnetic properties of oceanic lavas and pillow basalts confirms that the natural remanent magnetization (NMR) of submarine lava is of the right magnitude for the Vine and Matthews model and raises the possibility that the linear magnetic anomalies are due to pillow basalts as much as to dikes, as in the original hypothesis. The current research with deep-tow mag-

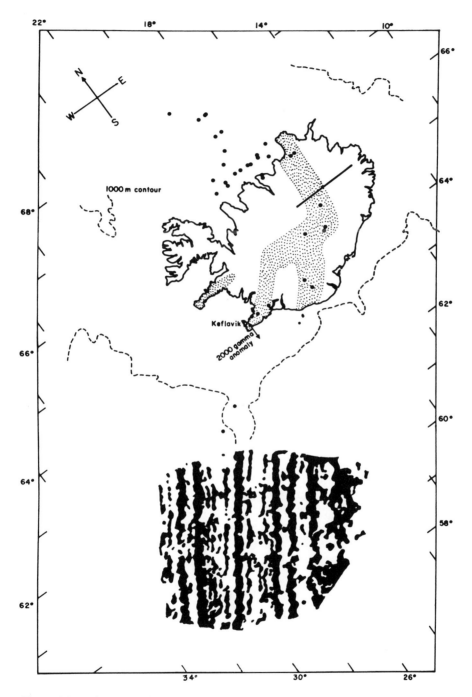

Fig. 5. Magnetic anomalies along the Reykjanes ridge south of Iceland. The black striped areas indicate the positive anomalies (after Heirtzler *et al.*, 1966).

netometers may shed light on this question. Already, there are interesting lines of evidence available from the studies of the sheeted dike structures and superjacent pillow basalts in the Troodos complex of Cyprus (Gass, 1967, 1968; Vine and Moores, 1969), which may be an exposed segment of a ridge spreading center active in the Mediterranean during Cretaceous time. Unfortunately, no reversal in the magnetization of these basalts has been found, either in dikes or pillows. If, however, the magnetization of the Troodos complex were acquired during the long period of normal polarity during the Cretaceous (Helsley and Steiner, 1969), this absence of reversals is correct. The association of magnetic lineations with active ridge spreading centers appears to be established as a general result of crustal formation concurrent with magnetic reversal. If strong magnetic lineations are to be absent from a portion of oceanic crust derived from a ridge spreading center, either there were no reversals for an extended period in the geologic past, or else the upper magnetic layer of the ocean crust which preserves the reversal history is missing. The former appears to be the case in the Troodos complex, the latter in a limited region about the North Atlantic ridge between lat. 43° and 43½° N (Phillips et al., 1969).

B. Magnetic Anomalies of the Continental Margins and Abyssal Plains

There is a distinctive boundary observed in the oceanic magnetic profiles seaward of the continental slope. The rough, lineated total field profiles typical of the ridge and ocean basin provinces smooth out to generally much smaller amplitude and longer wavelength anomalies which extend to the continental shelf. Heirtzler and Hayes (1967) noted this result along both the North American and North African margins. The lineations are lost in the "quiet" zone and are replaced shoreward on the shelf by anomalies related to coastal structures. There is yet some speculation about the reason for the existence of these two magnetically smooth belts which are about as wide as the continental margin provinces (and occur approximately in the same locations) on both sides of the main North Atlantic basin. Were the Atlantic to have formed at an average of 1.2 cm/yr rate, inferred from the well-defined magnetic anomalies observed in the ridge province, the oceanic crust which may underlie the quiet zone would be 270–220 million years old or Permian–Lower Triassic. Since the number of known reversals of the geomagnetic field are relatively few during this interval (the Kiaman magnetic interval), there would be few anomalies to be expected in layer II of this age. However, the preliminary report of the results from leg XI of the DSDP (1970) indicate that the oceanic basement at the seaward margin of the western quiet zone is about 160 m.y. old. This would require that the quiet zone be in the interval of 190–165 m.y. ago or Late Triassic in age when again few reversals occurred (DeBoer, 1968;

Burek, 1970; McElhinny and Burek, 1971). Phillips and Forsyth (1972) also assume this age interval for the quiet zone.

Heirtzler and Hayes (1967), Anderson *et al.* (1969), and Vogt *et al.* (1970) have identified a sequence of anomalies known as the Keithley sequence which occur east and parallel to the seaward edge of the quiet zone between lat. 25° and 35° N. An equivalent sequence occurs along the quiet zone in the eastern Atlantic, reported by Vogt *et al.* (1969) and Rona *et al.* (1970). Emery *et al.* (1970) discuss the various theories which have been offered to explain the presence of the quiet zone, but in the light of the recent DSDP results on the age of the hiatus in sedimentation marked by horizon β, and the difference in the time interval represented by horizon β on both sides of the Atlantic (see Section IV), the possibility of an older Paleozoic ocean (Drake and Nafe, 1968) or the possibility of a sudden shift of the ridge spreading axis between 160 and 200 m.y. ago seems to merit consideration. Pitman and Talwani (1972) remark that if the quiet zone does possibly represent a narrow Permian ocean basin, the seaward edge of the quiet zone on both sides of the North Atlantic marks the boundary along which rifting reoccurred possibly in the Jurassic. If this were so, they expect a large discontinuity in sediment thickness. This has not been observed, but a marked hiatus in sediment age above and below horizon β could be equivalent evidence, were the age of the hiatus appropriate.

Pautot *et al.* (1970) have observed that diapiric structures in the eastern North Atlantic occur along or near and parallel to the quiet zone boundary. The diapirs are related by Pautot *et al.* (1970) and Schneider (1970) to salt formed shortly after rifting of the North American and African plates. The ages assumed for the Keithley sequence seem important here. The Keithley anomalies are parallel to the quiet zone boundary and Pitman and Talwani (1972) assign ages from 155 to 130 m.y. for the sequence. Vogt *et al.* (1970) assigned an interval of from 190 to 160 m.y. to the sequence. North of Bermuda, Emery *et al.* (1970) have correlated a suite of lineations parallel to the quiet zone boundary which appear to bracket the age of the Keithley sequence north of the Bahamas. Emery *et al.* (1970) suggested an age interval of from 200 to 120 m.y. for the northern sequence, assuming that the quiet zone was generated during the long Permian Kiaman interval of predominately normal polarity. Pitman and Talwani (1972) suggest from 140 to 110 m.y. for this northern anomaly sequence, permitting the initial rifting of the North Atlantic to be Triassic–Jurassic. Equivalent data for the eastern Atlantic quiet zone will be of great interest, and may help remove some of the present uncertainties. At present, the magnetic age for both sides of the Atlantic will appear to be about 155 m.y. at the quiet zone boundary. But, anomalies perhaps 180 m.y. old are thought to exist very near the western shelf (Pitman and Talwani, 1972). Older ocean crust may underlie the western Atlantic basin.

VIII. GRAVITY RESULTS

Marine gravity survey results for the North Atlantic have recently been summarized by Nafe and Drake (1969), Talwani (1970), and Talwani and Le Pichon (1969). Gravity profiles for the continental shelves have been summarized by Maher (1971) for the North American east coast and earlier by Drake and others (1959) for the east coast of North America north of Cape Hatteras. The North African continental margin has been studied in detail more recently and is reported by Talwani and Le Pichon (1969) and Talwani (1970), with some of the North Atlantic pendulum measurements made by Vening Meinesz (1948), Worzel (1965), and Worzel and Schurbet (1955) giving preliminary results for regional trends.

Kaula (1967) has discussed the broad variations of the earth's gravity field determined from satellite observations which are capable of resolving the harmonics of the gravity field up to spherical harmonics of degree and order 12. The regional gravity field obtained from surface ship and submarine pendulum measurements averaged over a $5° \times 5°$ grid represent much higher harmonics than obtained from satellite data and shows correlation with sea-floor topography. For purposes of comparison with satellite data, the North Atlantic surface and submarine data are averaged over a $20° \times 20°$ grid resulting in good agreement. Several gross features of the gravity field in the North Atlantic are apparent on the $5° \times 5°$ grid average. The prominent feature of the surface gravity map is the minimum in the western basin of more than -60 mgal in the region of the Puerto Rico trench, with a consistent gravity minimum running along the entire western margin from Newfoundland south to the Falkland plateau. This marginal minimum is typically -20 mgal between lat. 50° N and lat. 50° S with the major departures from this value at the Puerto Rico trench and the Rio Grande rise. The exact shape of the marginal -20-mgal contour is probably better known in the north than in the south, due to better control. The eastern basin in the North Atlantic has gravity values lying between 0 and -20 mgal and is nearer isostatic compensation. The region of the mid-Atlantic ridge north of lat. 20° N has typically positive gravity values, greater than $+40$ mgal in some crestal locations. South of lat. 20° N, the ridge gravity values are much closer to 0 and indicate a ridge nearer isostatic compensation. This is also the beginning of the region of the equatorial Atlantic fracture system (Heezen and Tharp, 1961). Figure 6 shows the Atlantic 5° grid average surface free-air anomaly maps contoured at 20 mgal intervals (Talwani and Le Pichon, 1969). In the North Atlantic there appears to be a consistent correspondence between regional free-air gravity anomalies and the topography, in the sense that the $+20$ mgal over the mid-Atlantic ridge, the -20 and -40 mgal contours in the western part of the basin, and the 0 mgal contour in the eastern part of the basin are closely

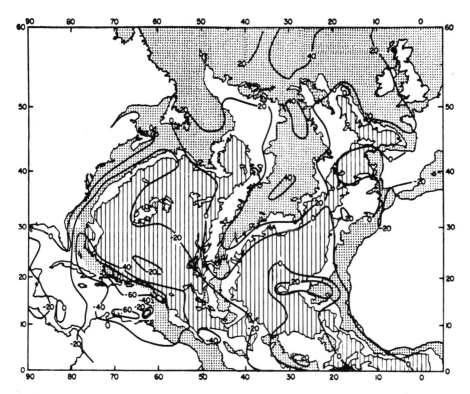

Fig. 6. North Atlantic 5° grid average surface free-air gravity anomaly map. The 5° averages are thought to be good to within 10 mgal. Generalized bathymetry is from the U.S. Navy Hydrographic Office World Chart. Areas with depths less than 2000 fathoms are stippled; areas with depths greater than 2500 fathoms are indicated by vertical lines. Note that the western basin has greater areal extent than the eastern basin (after Talwani and Le Pichon, 1969).

parallel to topographic contours. Were isostatic compensation achieved locally and at shallow depth, the free-air anomalies averaged over a 5° grid would be approximately zero and show no persistent correlation with topography. Since this is not the case, compensation must occur over greater areas and possibly at greater depths than 30 km in the region of the mid-Atlantic ridge. Kivioja (1963) used 5° grid average topography values over the globe to compute the combined effects of topography and compensation, using the Airy–Heiskanen hypothesis ($T = 30$), equivalent to a regional topographic isostatic correction for the average elevation of each 5° grid area. This reduced the magnitude of the average North Atlantic anomaly from 22 to 15 mgal, but leaves large-scale residual anomalies having little correlation with crustal features in the northern mid-Atlantic suggesting that the residual gravity field over an area in excess of 10^6 sq. km must be due in part to inhomogeneities or dynamics in the suboceanic mantle (Nafe and Drake, 1969), particularly in the region of the northern mid-Atlantic ridge. The persistent difference between the eastern and western margins of the basin, the western margin being more negative than the eastern margin, seems consistent with the greater width of the western continental margin and the thicker sequence of sediments along the North American continental shelf, as shown by seismic data and discussed earlier in this chapter.

A. Gravity Model of the Northern Mid-Atlantic Ridge

Talwani and others (1965) have provided three models for the mid-Atlantic ridge consistent with gravity and seismic data near lat. 32° N (Fig. 7). There were some difficulties in devising a model which was consistent with both the seismic and gravity data for the ridge flanks and axial zone, since normal seismic velocities are observed beneath layer III in the flank province while under the axial zone compressional velocities are lower than those normally associated with mantle material. In addition, the total thickness of layer II and III as determined by seismic data appear to remain early constant from the axial zone to the flanks of the ridge and beyond, the thickness of layer III increasing and that of layer II decreasing as one moves away from the ridge crest. This means that one set of assumptions about thicknesses and densities will not satisfy the gravity data over both regions with the corresponding changes in water depth (topography). A laterally inhomogeneous mantle was invoked to resolve the fact that while water depth decreases across the flanks from basins to ridge crest, the total crustal thickness does not increase as required by Airy isostacy, rather it remains constant or slightly decreases. This would imply an increasing mass excess and an increasing free-air anomaly from the basins to the ridge axis, of the order of 100 mgal/km change of water depth. No such increase in anomaly is observed. The three models provided by Talwani et al. (1965) all involve a density reversal in the mantle beneath the ridge crest province with some fine structure as indicated in the figure. The dimensions and densities allow agreement with both the observed gravity values and the steep slope of the Bouguer anomaly over the ridge flank, indicating that a large part of the compensation of the ridge structure is achieved at depths of 40 km or less.

The gravity minimum over the western Atlantic basin, which persists over very large regions of different crustal structure, suggests that this regional anomaly is due principally to mantle inhomogeneities and not isostatic effects or thickness of Mesozoic and younger sediments on the continental shelf. The lateral and vertical mantle inhomogeneity at the continental margin may be of a larger scale than any at the mid-ocean ridge.

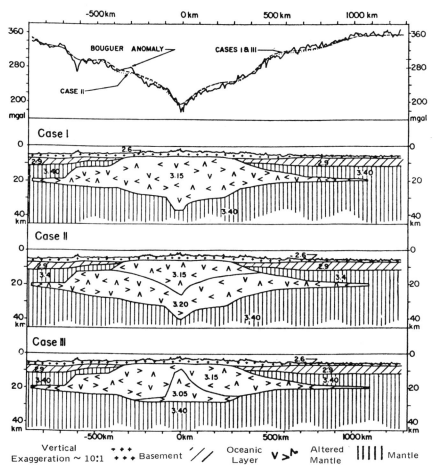

Fig. 7. Three possible models for the oceanic crust across the mid-Atlantic ridge which satisfy the observed gravity anomalies and are consistent with available seismic refraction data. In all three models the anomalous mantle found seismically under the ridge crest is assumed to lie beneath the normal mantle under the ridge flanks. In case I, the anomalous mantle is assigned a uniform density; in case II, its density is assumed to increase with depth; in case III, the anomalous mantle is assumed to be lighter near the axis of the ridge (after Talwani, Le Pichon, and Ewing, 1965).

B. Gravity Results for the Puerto Rico Trench

At the time of the writing of this review article, there would seem to be reluctance in the minds of some authors (Talwani, 1970; Nafe and Drake, 1969) to apply the gravity data at trenches to detailed models of crust–mantle structure. The most prominent surface gravity anomaly of the 5° grid average map of the North Atlantic is the −60-mgal anomaly in the region of the Puerto Rico trench, but it is difficult to separate the anomaly due to sediments associated with the trench from the larger effect of topography in free-air anomalies, and for isostatic anomalies the crustal structure must be deduced before the mass anomaly profile may be computed. Nevertheless, Talwani *et al.* (1959) have studied the Puerto Rico trench in some detail and provide a detailed mass distribution cross section of the trench with good seismic velocity data, coverage and densities chosen according to the Nafe–Drake velocity–depth curve (Worzel and Harrison, 1963). The Talwani *et al.* (1959) interpretation has been supported recently by gravity observations made on the floor of the trench in the Bathyscaphe Archimedes reduced with calculations by Talwani based upon structure inferred from his earlier surface measurements (Nafe and Drake, 1969). It is interesting that the largest negative free-air anomaly recorded, −346 mgal, occurs here, where the crust is intermediate in thickness between oceanic and continental crust, and the mass deficiency is mainly represented by the low-density sediments held within the down-bulge of the Moho. However, there may be an asymmetry to the mass deficiency in the trench, and the resulting free-air anomaly. Talwani (1970) shows that for many deep-sea trenches, the "5-km-anomaly" minimum is displaced from the axis of the trench toward the landward wall, and results from a mass deficiency over the landward wall, a situation particularly applicable to the Puerto Rico trench (Bunce *et al.*, 1970). The "5-km anomaly" computed by Talwani (1970) is obtained by reducing the free-air anomaly to a constant water depth of 5 km. All water depths greater than 5 km are "filled in" with material of density 2.60 g/cm^3, and all depths less than 5 km are "replaced" by sea water. It is this anomaly which shows clearly the asymmetry of negative values with respect to the trench axis. The asymmetry is not as clearly seen in the Bouguer or free-air anomaly profiles. This asymmetry would appear to be consistent with an underthrusting mechanism operating at the ocean trenches (Bunce *et al.*, 1970) with the asymmetry being related to the under-thrust side, the landward side in the case of the Puerto Rico trench. The under-thrust side would be expected to have piled up along it a greater thickness of sediments "scraped off" the downgoing oceanic floor. This is compatible with the seismic structure sections described by Officer *et al.* (1959) for both the Puerto Rico trench and the Barbados ridge, and the hypothesis that the Atlantic basin is widening by spreading from the mid-Atlantic ridge.

IX. ASYMMETRIES IN GEOLOGY AND GEOPHYSICS ABOUT THE MID-OCEAN RIDGE IN THE NORTH ATLANTIC BASIN

In the previous sections of this chapter numerous differences between the eastern and western basins of the North Atlantic have been cited. Either the initial rifting of the North Atlantic has not proceeded in a symmetrical way from its inception perhaps 200 m.y. ago to the present, or the western side of the proto-Atlantic rift was significantly different from the eastern side due to a process or a history that at present is unclear.

These asymmetries must be accounted for in any attempt to match the geology of the eastern margin of North America with the western margins of Europe and North Africa.

The study of the marine magnetic anomalies has proceeded rapidly, and the areal coverage over the North Atlantic is relatively complete (Pitman and Talwani, 1972). The magnetic roughness characteristic of the central ridge and basin provinces disappears on both sides of the North Atlantic near the foot of the continental shelves, becoming relatively smooth for lateral distances of several hundred kilometers before local structure of the shelf and coastal provinces reintroduce higher amplitude anomalies (Heirtzler and Hayes, 1967; Nafe and Drake, 1969; Vogt *et al.*, 1970). Both sides of the basin exhibit these "quiet zones," but the width in the western Atlantic (400 km) is roughly twice that of the east, corresponding approximately with the difference in widths of the continental shelves. In addition to the difference in width, the western quiet zone may represent a longer time interval than the eastern zone, with older oceanic crust near the western continental margin than near the eastern one, as discussed in Section VII. It is significant that the duration of the hiatus in sedimentation represented by horizon β is greater in the west than the east. Of course, the age of the oceanic basement beneath the western (and eastern) limits of the horizon β is expected to be younger than the basement beneath the corresponding quiet zones. It is tempting to speculate that there was an eastward jump of the ridge spreading axis during the magnetically quiet period in the Late Triassic, producing a wider zone of Late Triassic oceanic basement along the margin of the North American continent. This would have the consequence that sedimentation processes had an earlier beginning along the western continental margin, and have resulted in a thicker total deposit along the western continental margin than in the east. Corresponding to the thicker belt of sediment, the regional gravity anomaly is lower along the western margin. Regional heat flow is lower as well, presumably due to the blanketing effect of a sedimentary layer not thick enough or old enough to produce sufficient internal heat from radioisotope decay to equal that from the underlying mantle. While these results are suggestive of some difference between the conditions operating at the continental margins at the time of the initial rifting of the North Atlantic, since 155 m.y. ago the Atlantic basin appears to have developed up to the present time in a remarkably symmetrical fashion.

X. THE INFERRED HISTORY OF THE NORTH ATLANTIC

The application of sea-floor spreading to the geological history of the North Atlantic has been discussed by Heirtzler *et al.* (1968), Le Pichon (1968), and Dietz and Holden (1970). Pitman and Talwani (1972) have sequentially fitted together magnetic lineations of equivalent ages on either side of the North Atlantic in a manner similar to that used previously to match continental margins (Bullard *et al.*, 1965). Phillips and Forsyth (1972) have applied the method of finite rotations to a reconstruction of the North Atlantic and have in addition considered relevant paleomagnetic data for the surrounding continents. This process gives paleogeographic reconstructions for the North Atlantic continents and oceanic plates valid for the age of a particular pair of lineations. Each magnetic lineation, or isochron of the magnetic reversal time scale, is assumed to have formed at the mid-Atlantic ridge axis. Therefore, since each lineation was at one time a plate boundary, fitting together lineations of equivalent ages from both sides of the ocean ridge reconstructs the relative positions of the continents and discloses the shape of the intervening North Atlantic basin at the particular age of the linear anomalies. This presumes that the growing crustal plates of the North Atlantic have neither been deformed nor subducted on their Atlantic margins or within the growing North Atlantic basin. With the exception of the folding of Jurassic and Cretaceous rocks in western Portugal (Black *et al.*, 1964), there has been little orogeny along the Atlantic margins since early Mesozoic. However, there was considerable deformation in the Paleozoic, so the history of growth of the North Atlantic as depicted by Pitman and Talwani (1972), Phillips and Forsyth (1972), Berggren and Phillips (1971), and Francheteau (1970) applies only for the present episode of oceanic growth during the past 180–200 m.y.

Pitman and Talwani (1972) consider the North Atlantic basin to consist of portions of three plates. Greenland and North America are now part of the same North American plate, including the western Atlantic Ocean. The Eurasian plate consists of the eastern Atlantic Ocean north of the Azores–Gibraltar ridge, Europe, the British Isles, and Russia (at least to the Urals). The African plate includes the eastern Atlantic south of the Azores–Gibraltar ridge and the African continent (at least to the region of the Red Sea and Rift Valley). According to this view, the North Atlantic developed by the gradual separation of the Eurasian and African plates from the North American plate, beginning active lateral drift about 180 m.y. ago.

While there is now a wealth of magnetic anomaly data available for the North Atlantic, there are some difficulties with its interpretation particularly in regard to the timing of the oldest anomalies along the continental margins which are related to the onset of rifting apart of the Permian proto-Atlantic basin. Heirtzler and Hayes (1967), Berggren and Phillips (1971), and Emery *et al.* (1970) supposed that the 400-km-wide magnetic quiet zone along the western basin boundary about lat. 35° N formed during the Kiaman polarity interval in Permian time, 270–220 m.y. ago. This would mean that initial rifting of the North American, African, and European plates occurred during or before Kiaman time. This is not in agreement with the preliminary results from leg XI of the DSDP (1970). These results indicate that the age of the oceanic basement at the seaward margin of the western quiet zone is about 160 m.y., not in excess of 200 m.y. Also, extrapolating the sea-floor spreading rates from other DSDP data in the North Atlantic basin proper suggests that the continental margin itself is 190–200 m.y. old, rather younger than the Kiaman interval (Peterson *et al.*, 1970; Emery *et al.*, 1970; Phillips and Forsyth, 1972). Thus, Phillips and Forsyth (1972) assume that the western quiet zone formed during the Late Triassic interval of constant polarity (DeBoer, 1968; Burek, 1970; McElhinney and Burek, 1971), 190–165 m.y. ago. This is approximately the same period in which Pitman and Talwani (1972) assume the onset of active lateral drift. There may be other later modifications to this interpretation, particularly when the margins of the Atlantic basin are better known.

Assuming that the timing of active rifting is now approximately known, and that the Newark series dikes are an expression of the onset of rifting affecting the eastern continental edge of North American 190–202 m.y. ago (Ericson and Kulp, 1961), there is still some question as to the reason for the difference in widths of the eastern and western continental margins of the North Atlantic basin. (Actually, the definition of the width of the continental margins is something of a problem. Here, the margin is used as the width of the continental shelf as defined in Section II. However, this is somewhat of an accident of present sea level, and the continental margin could be defined as the belt of Cretaceous and younger sediments which blanket both coastal provinces on the eastern and western boundaries of the North Atlantic basin, irrespective of sea level. For the present, the physiographic definition will be followed on the assumption that present sea level is somewhat near the mean for the period represented by the development of the North Atlantic).

There are several alternatives for the present width of the continental margins, in terms of plate tectonic theory for the origin for the basin. Bird and Dewey (1970) assume a Paleozoic closing of an earlier or ancestral Atlantic ocean with an associated zone of subduction seaward of the present Appalachian fold belt. At places beneath the western continental shelf there is evidence of an eroded remanent island arc of Paleozoic volcanics, the offshore "borderland" of classic Appalachian geology, a concept based partly upon the earlier work of Kay (1951). This island arc, which may or may not be capped by younger carbonate reefs, would be related to the conspicuous magnetic anomaly running along the shelf break from Newfoundland south to the

Carolinas. If the island arc were due to subduction of oceanic crust beneath the eastern edge of North America, with a westward dipping Benioff zone, the final episode in the closure of the ancestral Atlantic would have resulted in a miogeosynclinal basin bordering the North American plate, a dead island arc, and a very narrow to absent belt of oceanic crust adjacent to the border of the present European and African plates. Assuming that the Atlas fold belt in North Africa is related to the Appalachian fold belt in mechanism and timing, either another subduction zone must also have operated at the border of the African plate underthrusting it to the east, or actual collision of the continental crust on either plate resulted in Devonian time (Bird and Dewey, 1970). (A similar eastward-dipping subduction zone or collision may have occurred during the formation of the British and Norwegian Caledonides.) This African subduction zone would also leave a fossil island or mountain arc near or at the border of the Paleozoic coast of North Africa, but the author is unaware of any current evidence for a buried island arc off North Africa. Hence, at present the collision hypothesis of Bird and Dewey (1970) seems to fit the observations. If the island arc off the North American coast exists, the present system of crustal spreading in the North Atlantic began seaward of it between Newfoundland and the Bahamas and has persisted for about 200 m.y., leaving a wider continental shelf at the western margin, built upon the Paleozoic architecture of a miogeosynclinal basin absent from the eastern boundary.

A second possibility is that regardless of the mode of origin of the American, African, and European Paleozoic fold belts, the present Atlantic basin began as a rift roughly through the middle of the preexisting Paleozoic fold belt during Late Triassic time, and much of the thick accumulation of sediment was deposited in Triassic graben that formed roughly parallel to the rift axis (Dewey and Bird, 1970). Apparently, graben were better developed along the western half of the rift between lat. 25° and 45° N if this explanation holds for the present margin differences, assuming the spreading of the North Atlantic has been essentially symmetrical throughout the present episode of its development.

A third possibility concerns the rift system itself, in particular the axis of the new spreading system associated with the rifting and subsequent opening of the modern Atlantic basin. About 200 m.y. ago, new oceanic crust would have begun to form new ocean floor in the rift valley between the American and African plates. The formation of new crust may have proceeded for a time and then the axis of the active spreading ridge may have migrated to the east by several hundred kilometers, and continued its formation of new oceanic crust until the present. The time at which this ridge migration occurred would be established by the age of the oldest oceanic crust on the eastern margin of the present basin. It appears that the age of the oldest oceanic crust found so far by the DSDP off the North African coast is about 160 m.y. This is the approximate age of the oceanic crust on both sides of the North Atlantic on the seaward edge of the magnetic quiet zones. According to this model, the 160 m.y. isochron could mark the eastern limit of post-Triassic oceanic crust, but on the western margin, the situation could be more complicated. Assuming spreading had begun in the interplate rift about 200 m.y. ago, and persisted for approximately 40 m.y., the rift axis would be about 160 m.y. old with the then older oceanic crust paralleling the axis in two strips at the border of the Jurassic North Atlantic. A ridge shift 160 m.y. ago to the eastern edge of the Jurassic basin would begin producing new oceanic crust from about 160 m.y. ago to the present. This would leave a simple progression of older and older crust from the present ridge axis to the eastern margin, but on the western side, oceanic crust ages would increase to about 160 m.y., immediately jump to an age of about 200 m.y. (within the western quiet zone) and from there on westward decrease in age to 160 m.y. at the fossil ridge axis, and then increase in age to 200 m.y. again at the western oceanic boundary of the North Atlantic. There are two useful properties to the model. One is that it offers another possibility for the origin of the North American east coast magnetic anomaly: it could be the fossil ridge spreading axis before the westward jump. Second, it provides an asymmetrical North Atlantic basin in that older oceanic crust exists along the western margin, while the present ocean crust is quite symmetrical about the present spreading axis for the past 160 m.y. If this is a possible explanation, current estimates of the age of the magnetic anomalies westward of the western quiet zone are likely to be in error. The significance of the two quiet zones would be different as well. Along the North African margin, the quiet zone would represent the limit of the Jurassic oceanic crust. Along the North American margin, the quiet zone would represent 400 km or more of Late Triassic and Jurassic oceanic crust formed during a 30–40 m.y. interval of constant geomagnetic polarity. The greater width of the western basin would then be the result of this belt of pre-Jurassic ocean crust marking the early rift axis of the North Atlantic. With an overall width of approximately 500–600 km for the quiet zone and older ocean crust lying beneath the western continental shelf, and a supposed time interval of 30–40 m.y. to produce this belt from a spreading center midway between the eastern quiet zone boundary and the Late Triassic shelf foot, a spreading half-rate of the order of 10 km/m.y. (1 cm/yr) is required, similar to the present rate.

XI. CONCLUSIONS

With the geological and geophysical evidence reviewed in this chapter in mind, it would appear difficult to regard the North Atlantic basin as one of the original features of the earth's crust. The data available to us now

TABLE I

Major Consequences of Three Points of View Regarding the Development of the Earth's Crust

A. Static crust	B. Drift, Wegener's mechanism	C. Mobile crust, plate tectonics
The Continents		
1. No intercontinental structural relationships required or expected across major ocean basins. Land bridges postulated for some evolutionary problems, requiring oceanization of these features at later dates.	1. Intercontinental relationships expected across certain rifted boundaries assuming that some major features (fold belts) predated continental breakup. Some important paleontological implications here, particularly with regard to the rapidity of diversification of animals after the breakup of Pangaea.	1. Intercontinental relationships expected, and continental–oceanic age relationships also expected in certain areas. Evolution of continental margins linked to the evolution of ocean basins.
2. Ancient borderlands postulated (e.g., Appalachia) in certain areas to provide sediment supply, now oceanized.	2. Continental crust affected by rifts, rotations, lateral faulting. Accretion by collision.	2. Some oceanic crust incorporated into continental margins, and some continental crust reabsorbed into the mantle, in subduction zones.
3. Paleozoic, Mesozoic, Cenozoic fold belts explained mainly by vertical displacements resulting from loading and subsidence, followed by orogeny and uplift. Lateral displacements originating from vertically oriented processes. Truncation of major fold belts at some continental margins accepted.	3. Paleozoic, Mesozoic, Cenozoic fold belts correlated in some cases. Origin of fold belts generally as in (A). Truncation of some foldbelts explained by continental fragmentation.	3. All mobile belts associated with past or present subduction and/or plate–plate collisions. Truncation of major fold belts explained by rifting apart of a plate by new spreading ridge system. Several more styles of fold belts recognized. Parallel systems of fold belts of differing ages expected.
4. Continental shelves persistent features, expected to preserve a near continuous record of continental erosion since Precambrian. Wide shelves with thick sediments fit this model well. The narrow ones without sediments do not. Difference between Atlantic and Pacific type coastlines used as classification.	4. Two types of shelves expected, the wide and narrow, wide on trailing edge, narrow on sides parallel to motion and bounded by faults, narrow on leading edge. Shelves on corresponding trailing edges expected to have similar shapes at shelf break, with somewhat similar coastlines.	4. Wide shelves on continental borders adjacent to spreading ridges, wide on opposite borders until subduction occurs at leading edge of continent. Following subduction, orogency and fold belt result. Also, lateral fault motions of great magnitude may intersect coastal margin at various angles, truncating or offsetting some older features. Shelves resulting from rift by new spreading ridge expected to show very similar sinuosites. Evaporite

1. Considered original features. Expect to find very old oceanic crust, accumulation of very thick sediments (estimates for oceanic sediment column vary from 3 to 5 km average, prior to recent results from DSDP). Sediment ages should span geologic time, from earliest erosion of cratons. Distribution of sediments expected to be roughly concentric with continents. No great lateral fault displacements postulated for ocean floors.

2. Island arc distribution random, since no expected relation to lateral crustal movement. No requirement for petrologic variation across trench–arc system in a systematic manner.

3. Ocean volcanoes and seamounts explained by subsidence in situ, and changes in sea level. No intervolcano age relationships or trends expected over great distances.

1. Considered as original features, subsequently modified by continental motions through ocean crust. New crust would form on trailing edge, old crust overrode on leading edge. New sediments also collect on new ocean crust, old sediments rode over on leading edge. Corridors of old crust between divergent drift paths. Metamorphism of some ocean crust.

2. Island arcs expected to be destroyed by overriding continent. Left behind on trailing edge. So arcs should be associated with older ocean crust, crust becoming younger from arc to continental margin. Source of magma not related to other processes.

3. Volcanoes destroyed on leading edge of continent. New ones grow at random on trailing edge.

1. Considered as temporary features, at least since current mantle turnover processes began. New crust formed at ridge spreading system. A spreading center rifting through a continent forms new ocean crust between older continental crust. Continents regarded as oldest surviving features of earth's crust. As spreading proceeds, a new crust spreads laterally away from ridge axis. Oldest ocean crust must be at continental margin, farthest from spreading system. On leading edge of plate, which may or may not coincide with continental margin, oldest ocean crust reabsorbed into mantle, sediments included. Much metamorphism at this subduction zone. Sediment distribution in oceans very strictly controlled by spreading history. No very old ocean crust expected. No disruption of sediments on trailing edge, except by transform faults meeting margins. Thin oceanic sediment thickness expected, such as 1 to $\frac{1}{2}$ km actually observed, consistent with post-Jurassic sedimentation rates.

2. Island arc distribution fit subduction zone model at trenches. Mountain arcs fit subduction at continental margins. Sediment in trenches a function of spreading rate and proximity to source. Petrologic trends across arcs should show trend from basaltic to andesitic lavas. Source of magma related to remelting of ocean lithosphere along subduction zone.

3. Volcanoes grow at ridge axis and spread laterally. Oldest furthest from ridge. Subsidence automatic with spreading.

TABLE I (*continued*)

A. Static crust	B. Drift, Wegener's mechanism	C. Mobile crust, plate tectonics
4. Trenches, with associated great negative gravity anomalies explained by depression of oceanic crust from loading. Most deep trenches nearly empty, however.	4. Trenches at leading edge of continent fit overriding idea. Lack of new crust on trailing edge a puzzle.	4. Trenches required to conserve area of total crust. Benioff zone seismicity was original clue for subduction.
5. Oceanic and continental structure modified by geosynclinal development. No requirement for large-scale oceanic fracture systems. Oceanic seismic activity random.	5. Large-scale oceanic faults possibly associated with continental drift. Oceanic seismic activity greatest nearest continents.	5. Large-scale fault displacements required at plate boundaries which are neither ridges or trenches. Fault motion related to age of oceanic crust. Oceanic seismic activity controlled by distribution of plate boundaries, and type of dynamics taking place at the boundary.
6. Evolution of ocean basins primarily that of sedimentation over original crust.	6. Evolution of ocean basins partly controlled by motion of the continents.	6. Evolution of the ocean basins related to evolution of the entire crust by lateral spreading and eventual subduction.

Summary

Development of various continental blocks treated separately, according to same set of principles. Ocean basins play role of boundaries for continents. Histories of each related loosely.	Development of various continents treated together up to time of fragmentation, then separately. Ocean basins in between drifting continents show new histories. Some geological observations related to others.	Development of all crust interrelated. Boundary conditions placed on all crust. Most geological processes related in this approach, and ground rules require unification of interpretations and data. Freedom of choice for many geological interpretations removed. Stabilist view in (A) is an acceptable special case for portions of plates which have not been consumed or fragmented, and preserve a long history of relative coherence. These stable platforms are important for they preserve the longest record of geologic history, including evidence of past collisions of early plates.

are both more comprehensive and more equally spaced over the region of the North Atlantic than other oceanic areas of the world, but detailed geological knowledge is still rather thin for the floors of the basin. While it is probable that we will never know the floors of the world's oceans as well as the continents, marine geology and geophysics are developing at a rapid rate. Experience on the continents reminds us that the more detail one gets, the more difficult it usually becomes to make gross generalizations. The ultimate test of the inferences about the development of the North Atlantic basin will be made as this even more detailed data are obtained, for both the ocean basin and its surrounding margins. The results obtained so far and the inferences based upon them seriously challenge the concept and consequences of a static crust.

Table I summarizes the major consequences of three points of view regarding the development of the earth's crust in general, and the North Atlantic in particular. Column A refers to the static crust view, Column B to continental drift according to Wegener, and Column C to a mobile crust as required by plate tectonics. While the summary is not intended to be complete, it concerns consequences of sufficient generality to warrant comparison in terms of data now available. It appears that while the static crust view may well apply to the development of stable continental shield areas, the ocean basins and mobile belts are more in accord with large-scale lateral movements of the lithosphere, at least during the past 200 m.y.

REFERENCES

Anderson, D. L., 1965, Recent evidence concerning the structure and composition of the earth's mantle: *Phys. Chem. Earth*, v. 6, p. 1–131.

Anderson, D. N., Vogt, P. R., and Bracey, D. R., 1969, Magnetic anomaly trends between Bermuda and the Bahama—Antilles Arc: *Trans. Amer. Geophys. Union*, v. 50, p. 189.

Barazangi, M. and Dorman, J., 1969, World seismicity map of ESSA Coast and Geodetic Survey epicenter data for 1961–67: *Bull. Seismol. Soc. Amer.*, v. 59, p. 369–380.

Baron, J. G., Heirtzler, J. R., and Lorentzen, G. R., 1965, An airborn geomagnetic survey of the Reykjanes Ridge, 1963: unpublished informal report No. H-3-65, U.S. Naval Oceanographic Office, Washington, D.C.

Berggren, W. A. and Phillips, J. D., 1971, Influence of continental drift on the distribution of Tertiary benthonic foraminifera in the Caribbean and Mediterranean regions, in: *Symposium on Geology of Libya*, 1969, C. Grey, ed.: Catholic Press, Beirut.

Birch, F., 1966, Earth heat flow measurements in the last decade, in: *Advances in Earth Science*, P. M. Hurley, ed.: Massachusetts Institute of Technology Press, Cambridge, p. 403–430.

Birch, F., 1967, Low values of oceanic heat flow: *J. Geophys. Res.*, v. 72, p. 2261–2262.

Bird, J. M. and Dewey, J. F., 1970, Lithosphere plate: Continental margin tectonics and the evolution of the Appalachian orogen: *Geol. Soc. Amer. Bull.*, v. 81, p. 1031–1060.

Black, M., Hill, M. N., Laughton, A. S., and Matthews, D. H., 1964, Three nonmagnetic seamounts off the Iberian coast: *Geol. Soc. London Quart. J.*, v. 120, p. 477–517.

Bott, M. H. P., 1967, Terrestrial heat flow and the mantle convection hypothesis: *Geophys. J. Roy. Astr. Soc.*, v. 14, p. 413–428.

Brune, J. N., 1969, Surface Waves and crustal structure, in: *The Earth's Crust and Upper Mantle*: Geophysical Monograph 13, P. J. Hart, ed.: American Geophysical Union, Washington, D. C., p. 230–241.

Bullard, E. C., 1954, The flow of heat through the floor of the Atlantic Ocean: *Proc. Roy. Soc. London*, v. A222, p. 408–429.

Bullard, E. C., 1963, The flow of heat through the floor of the ocean, in: *The Sea*, Vol. III, M. N. Hill, ed.: Interscience, New York, p. 218–232.

Bullard, E. C. and Day, A., 1961, The flow of heat through the floor of the Atlantic Ocean: *Geophys. J. Roy. Astr. Soc.*, v. 14, p. 282–292.

Bullard, E. C., Everett, J. E., and Smith, A. G., 1965, The fit of the continents around the Atlantic: *Phil. Trans. Roy. Soc. London*, v. A258, p. 41–51.

Bullard, E. C. and Mason R. G., 1963, The magnetic field over the oceans, in: *The Sea*, Vol. III, M. N. Hill, ed., Interscience, New York, p. 175–217.

Bunce, E. T., Phillips, J. D., Chase, R. L., and Bowin, C. O., 1970, The Lesser Antilles arc and the eastern margin of the Caribbean Sea, in: *The Sea*, Vol. IV (Part 2), A. E. Maxwell, ed.: Interscience, New York, p. 359–386.

Burek, P. J., 1970, Magnetic reversals: their application to stratigraphic problems: *Amer. Assoc. Petrol. Geol. Bull.*, v. 54, p. 1120–1139.

Cain, J. C., Hendricks, S., Daniels, W. E., and Jensen, D. C., 1964, Computation of the main geomagnetic field from spherical harmonic expansions: Report X- 611-64-316, Goddard Space Flight Center, Green belt, Md.

Cain, J. C., Hendricks, S. J., Langel, R. A., and Hudson, W. V., 1967, A proposed model for the international geomagnetic reference field—1965: *J. Geomag. and Geoelect.*, v. 19, p. 335–355.

Clark, S. P. Jr., 1966, Thermal conductivity, in: *Handbook of Physical Constants*, S. P. Clark, Jr., ed.: Geol. Soc. Amer., New York, p. 459–482.

Clark, S. P. Jr. and Ringwood, A. E., 1964, Density distribution and constitution of the mantle: *Rev. Geophys.*, v. 2, p. 35–88.

Cohen, L. H., Ito, K. and Kennedy, G. C., 1967, Melting and phase relations in an anhydrous basalt to 40 kilobars: *Am. J. Sci.*, v. 265, p. 475–518.

Corry, C., Dubois, C., and Vacquier, V., 1968, Instruments for measuring the terrestrial heat flow through the ocean-floor: *Jour. Marine Res.*, v. 26, p. 165–177.

Cox, A., 1969, Geomagnetic reversals: *Science*, v. 163, p. 237–245.

Cox, A., Doell, R. R., and Dalrymple, G. B., 1963, Geomagnetic Polarity epochs—Sierra Nevada II: *Science*, v. 142, p. 382–385.

Cox, A., Doell, R. R., and Dalrymple, G. B., 1964, Reversals of the earth's magnetic field: *Science*, v. 144, p. 1537–1543.

De Boer, J., 1968, Paleomagnetic differentiation and correlation of the Late Triassic volcanic rocks in the central Appalachians (with special reference to the Connecticut Valley): *Geol. Soc. Amer. Bull.*, v. 79, p. 609–626.

Dewey, J. R. and Bird, J. M., 1970, Mountain belts and the new global tectonics: *J. Geophys. Res.*, v. 75, p. 2625–2647.

Dickson, G. O., Pitman, W. C., III, and Heirtzler, J. R., 1968, Magnetic anomalies in the South Atlantic and ocean floor spreading: *J. Geophys. Res.*, v. 73, p. 2087–2100.

Dietz, R. S., 1961, Continent and ocean basin evolution by spreading of the sea floor: *Nature*, v. 190, p. 854–857.

Dietz, R. S., and Holden, J. C., 1970, Reconstruction of Pangea: Breakup and dispersion of Continents, Permian to present: *J. Geophys. Res.*, v. 75, p. 4939–4956.

Doell, R. R. and Dalrymple, G. B., 1966, Geomagnetic polarity epochs: A new polarity event and the age of the Brunhes–Matuyama boundary: *Science*, v. 152, p. 1060.

Drake, C. L., Ewing, J. I., and Stockard, H., 1968, The continental margin of the eastern United States: *Can. J. Earth Sci.*, v. 5 (4, part 2), p. 993–1010.

Drake, C. L., Ewing, M., and Sutton, G. H., 1959, Continental margins and geosynclines: The east Coast of North America north of Cape Hatteras: *Phys. Chem. Earth*, v. 3, p. 110–198.

Drake, C. L., Heirtzler, J., and Hirshman, J., 1963, Magnetic anomalies off eastern North America: *J. Geophys. Res.*, v. 68, p. 5259–5275.

Drake, C. L. and Nafe, J. E., 1968, The transition from ocean to continent from seismic refraction data, in: *The Crust and Upper Mantle of the Pacific Area*, Geophysical Monograph 12, L. Knopoff, C. L. Drake, P. J. Hart, eds.: American Geophysical Union, Washington, D. C., p. 174–186.

Elsasser, W. M., 1967, Convection and stress propagation in the upper mantle: Tech. Report No. 5, June 15, 1967, Princeton University, Princeton, New Jersey.

Elsasser, W. M., 1971, Sea-floor spreading as thermal convection: *J. Geophys. Res.*, v. 76, p. 1101–1112.

Emery, K. O., Uchupi, E., Phillips, J. D., Bowin, C. O., Bunce, E. T., and Knott, S. T., 1970, Continental rise off eastern North America: *Amer. Assoc. Petrol. Geol. Bull.*, v. 54, p. 44–108.

Erickson, G. P. and Kulp, J. L., 1961, Potassium–argon dates on basaltic rocks: *New York Acad. Sci. Annals.*, v. 91, p. 321–323.

Ewing, J. and Ewing, M., 1959, Seismic refraction measurements in the Atlantic Ocean basins, in the Mediterranean Sea, on the mid-Atlantic ridge, and in the Norwegian Sea: *Geol. Soc. Amer. Bull.*, v. 70, p. 291–318.

Ewing, J. and Ewing, M., 1970, Seismic reflection, in: *The Sea*, Vol. IV (part 1), A. E. Maxwell, ed.: Interscience, New York, p. 1–52.

Ewing, J. and Nafe, J. E., 1963, The unconsolidated sediments, in: *The Sea*, Vol. III, M. N. Hill, ed.: Interscience, New York, p. 73–84.

Ewing, J., Talwani, M., Ewing, M., and Edgar, T., 1967, Sediments of the Caribbean, in: Proc. Int. Conf. Tropical Oceanography, Inst. Marine Sci., University of Miami, v. 5, p. 88–102.

Ewing, J. and Tirey, G. B., 1961, Seismic profiler: *J. Geophys. Res.*, v. 66, p. 2917–2927.

Ewing, J., Worzel, J. L., Ewing, M., and Windisch, C. C., 1966, Ages of Horizon A and oldest Atlantic sediments: *Science*, v. 154, p. 1125–1132.

Ewing, M., Ewing, J., and Talwani, M., 1964, Sediment distribution in the oceans: The Mid-Atlantic ridge: *Geol. Soc. Amer. Bull.*, v. 75, p. 17–36.

Ewing, M., Heezen, B. C., and Hirshman, J., 1957, Mid-Atlantic ridge seismic belts magnetic anomalies (abstract): Comm. 10 bis Assoc. Seismol., Gen. Assembly, IUGG, Toronto.

Ewing, M. and Landisman, M., 1961, Shape and Structure of ocean basins, in: *Oceanography*, M. Sears, ed.: Amer. Assoc. Adv. Sci., Washington, D. C., p. 3–38.

Ewing, M., Saito, T., Ewing, J., and Burkle, L., 1966, Lower Cretaceous sediments from the northwest Pacific: *Science*, v. 152, p. 751–755.

Francheteau, J., 1970, Paleomagnetism and plate tectonics: Ph. D. thesis, University of California at La Jolla, Scripps Institution of Oceanography, Ref. 70-30.

Gass, J. G., 1967, The ultrabasic volcanic assemblage of the Troodos Massif, Cyprus, in: *Ultramafic and Related Rocks*, P. J. Wyllie, ed.: J. Wiley and Sons, New York, p. 121–134.

Gass, J. G., 1968, Is the Troodos Massif of Cyprus a fragment of Mesozoic ocean floor?: *Nature*, v. 220, p. 39–42.

Gerard, R., Langseth, M. G., and Ewing, M., 1962, Thermal gradient measurements in the water and bottom sediments of the Western Atlantic: *J. Geophys. Res.*, v. 67, p. 785–803.

Gilluly, J., 1971, Plate tectonics and magmatic evolution: *Geol. Soc. Amer. Bull.*, v. 82, p. 2383–2396.

Hamilton, E. L., 1959, Thickness and consolidation of deep-sea sediments: *Geol. Soc. Amer. Bull.*, v. 70, p. 1399–1424.

Hart, S. R., 1969, Isotope geochemistry of crust–mantle processes, in: *The Earth's Crust and Upper Mantle*, Geophysical Monograph 13, P. J. Hart, ed.: American Geophysical Union, Washington, D. C., p. 58–62.

Heezen, B. C. and Ewing, M., 1963, The mid-oceanic ridge, in: *The Sea*, Vol. III, M. N. Hill, ed.: Interscience, New York, p. 388–410.

Heezen, B. C. and Menard, H. W., 1963, Topography of the deep-sea floor, in: *The Sea*, Vol. III, M. N. Hill, ed.: Interscience, New York, p. 233–280.

Heezen, B. C. and Tharp, M., 1961, Physiographic Diagram of the South Atlantic Ocean: Geol. Soc. Amer., New York.

Heezen, B. C. and Tharp, M., 1965, Tectonic fabric of the Atlantic and Indian Oceans and continental drift: *Phil. Trans. Roy. Soc. London*, v. A258, p. 90–106.

Heirtzler, J. R., 1969, Geomagnetic studies in the Atlantic ocean, in: *The Earth's Crust and Upper Mantle*, Geophysical Monograph 13, P. J. Hart, ed.: American Geophysical Union, Washington, D. C., p. 430–436.

Heirtzler, J. R., 1970, Magnetic anomalies measured at sea, in: *The Sea*, Vol. IV (part 1), A. E. Maxwell, ed.: Interscience, New York, p. 85–128.

Heirtzler, J. R., Dickson, G. O., Herron, E. M., Pitman, W. C., III, and Le Pichon, X., 1968, Marine magnetic anomalies, geomagnetic field reversals, and motions of the ocean floor and continents: *J. Geophys. Res.*, v. 13, p. 2119–2136.

Heirtzler, J. R. and Hayes, D. E., 1967, Magnetic boundaries in the North Atlantic Ocean: *Science*, v. 157, p. 185–187.

Heirtzler, J. R. and Le Pichon, X., 1965, Crustal structure of the mid-ocean ridges, 3, Magnetic anomalies over the Mid-Atlantic ridge: *J. Geophys. Res.*, v. 70, p. 4013–4033.

Heirtzler, J. R., Le Pichon, X., and Baron, J. G., 1966, Magnetic anomalies over the Reykjanes ridge: *Deep-Sea Res.*, v. 13, p. 427–443.

Heiskanen, W. A. and Vening-Meinesz, F. A., 1958, *The Earth and its Gravity Field*: McGraw-Hill, New York, 470 p.

Helsley, C. E. and Steiner, M. B., 1969, Evidence for long intervals of normal polarity during the Cretaceous period: *Earth and Planetary Science Letters*, v. 7, p. 325–332.

Hersey, J. B., 1962, Findings made during the June, 1961, cruise of Chain to the Puerto Rico trench and Caryn seamount: *J. Geophys. Res.*, v. 67, p. 1109–1116.

Hersey, J. B., 1963, Continuous reflection profiling, in: *The Sea*, Vol. III, M. N. Hill, ed.: Interscience, New York, p. 47–71.

Hess, H. H., 1962, History of the ocean basins, in: *Petrologic Studies: A Volume in Honor of A. F. Buddington*, A. E. J. Engel, H. L. James, and B. F. Leonard, eds.: Geological Society of America, New York, p. 599–620.

Hill, M. N., 1957, Recent geophysical exploration of the ocean floor: *Phys. Chem. Earth*, v. 2, p. 129–163.

Horai, K. and Uyeda, S., 1969, Heat flow in volcanic areas, in: *The Earth's Crust and Upper Mantle*, Geophysical Monograph 13, P. J. Hart, ed.: American Geophysical Union, Washington, D. C., p. 95–109.

Isacks, B., Oliver, J., and Sykes, L. R., 1968, Seismology and the new global tectonics: *J. Geophys. Res.*, v. 73, p. 5855–5899.

Kaula, W. M., 1967, Geophysical implications of satellite determinations of the earth's gravitational field: *Space Sci. Rev.*, v. 7, p. 769–794.

Kay, M., 1951, North American Geosynclines: *Geol. Soc. Amer. Memoir* 48, 143 p.

Keen, M. J., 1963, Magnetic anomalies over the Atlantic ridge: *Nature*, v. 197, p. 888–890.

Keen, M. J., 1968, *An Introduction to Marine Geology*: Pergamon, Oxford, 219 p.

Kivioja, L. A., 1963, The effect of topography and its isostatic compensation of free-air anomalies: Inst. Geol. Photogram. Cartog. Rept. 28, Ohio State University, Columbus, 134 p.

Langseth, M. G., Le Pichon, X., and Ewing, M., 1966, Crustal structure of the mid-ocean ridges, 5, Heat flow through the Atlantic Ocean floor and convection currents: *J. Geophys. Res.*, v. 71, p. 5321–5355.

Langseth, M. G. and Von Herzen, R. P., 1970, Heat flow through the floor of the world oceans, in: *The Sea*, Vol. IV (part 1), A. E. Maxwell, ed.: Interscience, New York, p. 299–352.

Le Pichon, X., 1968, Sea-floor spreading and continental drift: *J. Geophys. Res.*, v. 73, p. 3661–3697.

Le Pichon, X., Ewing, J., and Houtz, R. E., 1968, Deep-sea velocity determination made while reflection profiling: *J. Geophys. Res.*, v. 73, p. 2597–2614.

Le Pichon, X. and Langseth, M., 1967, Comments on paper by F. Birch, "Low values of oceanic heat flow": *J. Geophys. Res.*, v. 72, p. 6377–6378.

Lee, W. H. K. and Uyeda, S., 1965, Review of heat flow data, in: *Terrestrial Heat Flow*, Geophysical Monograph 8, W. H. K. Lee, ed.: American Geophysical Union, Washington, D. C., p. 68–190.

Leyden, R., Sheridan, R., and Ewing, M., 1972, A seismic refraction section across the equatorial Atlantic, in: UNESCO-IUGS Symposium on Continental Drift Emphasizing the History of the South Atlantic Area, Montevideo, Uruguay, 1967: *Trans. Amer. Geophys. Union*, v. 53, p. 171–173.

Lubimova, E. A., 1969, Thermal history of the earth, in: *The Earth's Crust and Upper Mantle*, Geophysical Monograph 13, P. J. Hart, ed.: American Geophysical Union, Washington, D. C., p. 63–77.

Luyendyk, B. P., 1969, Origin of short wavelength magnetic lineations observed near the ocean bottom: *J. Geophys. Res.*, v. 74, p. 4869–4881.

Maher, J. C., 1971, Geologic framework and petroleum potential of the Atlantic coastal plain and continental shelf: Geo. Survey Prof. Paper 659, U. S. Gov. Printing Office, Washington, D. C., 98 p.

Marshall, M. and Cox, A., 1971, Magnetism of Pillow basalts and their petrology: *Geol. Soc. Amer. Bull.*, v. 82, p. 537–552.

Mason, R. G., 1958, A magnetic survey off the west coast of the United States between latitudes 32° and 36°N, longitudes 121° and 128°W: *Geophys. J. Roy. Astr. Soc.*, v. 1, p. 320–329.

Mason, R. G. and Raff, A. D., 1961, A magnetic survey off the west coast of North America, 32°N to 42°N: *Geol. Soc. Amer. Bull.*, v. 72, p. 1259–1265.

Matthews, D. H., 1961, Lavas from an abyssal hill on the floor of the North Atlantic Ocean: *Nature*, v. 190, p. 158–159.

Matthews, D. H., 1967, Mid-ocean ridges, in: *International Dictionary of Geophysics*, S. K. Runcorn, ed.: Pergamon Press, London, p. 979–991.

Matthews, D. J., Vine, F. J., and Cann, J. R., 1965, Geology of an area of the Carlsberg Ridge, Indian Ocean: *Geol. Soc. Amer. Bull.*, v. 76, p. 675–682.

McBirney, A. R., 1963, Conductivity variations and terrestrial heat flow distribution: *J. Geophys. Res.*, v. 68, p. 6323–6329.

McElhinny, M. W. and Burek, P. J., 1971, Mesozoic palaeomagnetic stratigraphy: *Nature*, v. 232, p. 98–102.

McKenzie, D. P., 1967, Some remarks on heat flow and gravity anomalies: *J. Geophys. Res.*, v. 72, p. 6261–6273.

McKenzie, D. P. and Sclater, J. G., Heat flow in the Eastern Pacific and sea-floor spreading: *Bull. Volcanology*, v. 33, p. 101–118.

Menard, H. W., 1969, Growth of drifting volcanoes: *J. Geophys. Res.*, v. 74, n. 20, p. 4827–2837.

Molnar, P. and Sykes, L. R., 1969, Tectonics of the Caribbean and Middle America regions from focal mechanisms and seismicity: *Geol. Soc. Amer. Bull.*, v. 80, p. 1639–1684.

Morgan, W. J., 1968, Rises, trenches, great faults, and crustal blocks: *J. Geophys. Res.*, v. 73, p. 1959–1982.

Nafe, J. E. and Drake, C. L., 1959, Floor of the North Atlantic—Summary of geophysical data, in: *North Atlantic—Geology and Continental Drift*, a Symposium, Memoir 12, M. Kay, ed.: Amer. Assoc. Petrol. Geol., Tulsa, p. 59–87.

Officer, C. B., Ewing, J. I., Hennion, J. F., Harkrider, D. G., and Miller, D. E., Geophysical investigations in the eastern Caribbean: Summary of 1955 and 1956 cruises: *Phys. Chem. Earth*, v. 3, p. 17–109.

Oliver, J. and Isacks, B., 1967, Deep earthquake zones, anomalous structures in upper mantle, and the lithosphere: *J. Geophys. Res.*, v. 72, p. 4259–4275.

Opdyke, N. D., 1968, Paleomagnetism of oceanic cores, in: *The History of the Earth's Crust*, R. A. Phinney, ed.: Princeton University Press, Princeton, p. 61–72.

Oxburgh, E. R. and Turcotte, D. L. 1968, Mid-ocean ridges and geotherm distribution during mantle convection: *J. Geophys. Res.*, v. 73, p. 2643–2661.

Oxburgh, E. R. and Turcotte, P. L., 1969, Increased estimate of heat flow at oceanic ridges: *Nature*, v. 223, p. 1354–1360.

Pautot, G., Auzende, and Le Pichon, X., 1970, Continuous deep sea salt layer along North Atlantic margins related to early phase of rifting: *Nature*, v. 227, p. 351–354.

Peterson, M. N. A., Edgar, N. T., Cita, M., Gartner, S., Jr., Goll, R., Migrini, C., and von der Borch, C., 1970, Initial reports of the Deep-Sea Drilling Project, 2: U. S. Government Printing Office, Washington, D. C.

Phillips, J. D., 1967, Magnetic anomalies over the Mid-Atlantic ridge near 27°N: *Science*, v. 157, p. 920–923.

Phillips, J. D. and Forsyth, O., 1972, Plate tectonics, paleomagnetism and the opening of the Atlantic: *Geol. Soc. Amer. Bull.*, v. 83, p. 1579–1600.

Phillips, J. D., Thompson, G., Von Herzen, R. P., Bowen, V. T., 1969, Mid-Atlantic ridge near 43°N latitude: *J. Geophys. Res.*, v. 74, p. 3069–3081.

Pitman, W. C., III, and Heirtzler, J. R., 1966, Magnetic anomalies over the Pacific-Antarctic ridge: *Science*, v. 154, p. 1164–1171.

Pitman, W. C., III, and Talwani, M., 1972, Sea floor spreading in the North Atlantic: *Geol. Soc. Amer. Bull.*, v. 83, p. 619–646.

Press, F., 1972, The earth's interior as inferred from a family of models, in: *The Nature of the Solid Earth*, E. C. Robertson, ed.: McGraw-Hill, New York, p. 147–171.

Raff, A. D. and Mason, R. G., 1961, A magnetic survey off the west coast of North America, 40°N to 52 1/2°N: *Geol. Soc. Amer. Bull.*, v. 72, p. 1259–1265.

Ratcliffe, E. H., 1960, The thermal conductivity of ocean sediments: *J. Geophys. Res.*, v. 65, p. 1535–1541.

Reitzel, J., 1961, Some heat flow measurements in the North Atlantic: *J. Geophys. Res.*, v. 66, p. 2267–2268.

Revelle, R. and Maxwell, A. E., 1952, Heat flow through the floor of the Eastern North Pacific Ocean: *Nature*, v. 170, p. 199–200.

Ringwood, A. E., 1969, Composition and evolution of the upper mantle, in: *The Earth's Crust and Upper Mantle*, Geophysical Monograph 13, P. J. Hart, ed.: American Geophysical Union, Washington, D. C., p. 1–17.

Ringwood, A. E. and Green, D. H., 1966, An experimental investigation of the gabbro-eclogite transformation and some geophysical consequences: *Tectonophysics*, v. 3, p. 383–427.

Ringwood, A. E. and Green, D. H., 1969, Phase transitions, in: *The Earth's Crust and Upper Mantle*, Geophysical Monograph 13, P. J. Hart, ed.: American Geophysical Union, Washington, D. C., p. 637–649.

Robertson, E. C., ed., 1972, *The Nature of the Solid Earth*: McGraw-Hill, New York, 677 p.

Rona, P. A., Brakl, J., and Heirtzler, J. R., 1970, Magnetic anomalies in the northeast Atlantic between the Canary and Cape Verde Islands: *J. Geophys. Res.*, v. 75, p. 7412–7420.

Schneider, E. D., 1970, Deep sea diapiric structures: *Trans. Amer. Geophys. Union*, v. 51, p. 316.

Schubert, G. and Turcotte, D. L., 1972, One-dimensional model of shallow-mantle convection: *J. Geophys. Res.*, v. 77, p. 945–964.

Sheridan, R. E., Houtz, R. E., Drake, C. L., and Ewing, M., 1969, Structure of continental margin off Sierra Leone, West Africa: *J. Geophys. Res.*, v. 74, p. 2512–2530.

Spiess, F. N. and Mudie, J. D., 1970, Small-scale topographic and magnetic features, in: *The Sea*, Vol. IV (part 1), A. E. Maxwell, ed.: Interscience, New York, p. 205–250.

Stauder, W. S. J., 1962, The focal mechanism of earthquakes: *Advan. Geophys.*, v. 9, p. 1–76.

Sykes, L. R., 1965, The seismicity of the Arctic: *Bull. Seismol. Soc. Amer.*, v. 55, p. 501–518.

Sykes, L. R., 1967, Mechanism of earthquakes and nature of faulting on the mid-oceanic ridges: *J. Geophys. Res.*, v. 72, p. 2131–2153.

Sykes, L. R., 1969, Seismicity of the mid-oceanic ridge system, in: *The Earth's Crust and Upper Mantle*, Geophys. Monograph 13, P. J. Hart, ed.: American Geophysical Union, Washington, D. C., p. 148–152.

Sykes, L. R. and Ewing, M., 1965, The seismicity of the Caribbean region: *J. Geophys. Res.*, v. 70, p. 5065–5074.

Talwani, M., 1970, Gravity, in: *The Sea*, Vol. IV (part 1), A. E. Maxwell, ed.: Interscience, New York, p. 251–297.

Talwani, M. and Le Pichon, X., 1969, Gravity field over the Atlantic Ocean, in: *The Earth's Crust and Upper Mantle*, Geophysical Monograph 13, P. J. Hart, ed.: American Geophysical Union, Washington, D. C., p. 341–351.

Talwani, M., Le Pichon, X., and Ewing, M., 1965, Crustal structure of the mid-oceanic ridges, 2, Computed model from gravity and seismic refraction data: *J. Geophys. Res.*, v. 70, p. 341–352.

Talwani, M., Sutton, G. H., and Worzel, J. L., 1959, Crustal section across the Puerto Rico trench: *J. Geophys. Res.*, v. 66, p. 1265–1278.

Talwani, M., Windisch, C., Langseth, M., and Heirtzler, J. R., 1968, Recent geophysical studies on the Reykjanes ridge: *Trans. Amer. Geophys. Union*, v. 49, p. 201.

Tilley, C. E., 1947, The dunite–mylonites of St. Paul's Rocks (Atlantic): *Amer. J. Sci.*, v. 245, p. 483–391.

Torrance, K. E. and Turcotte, D. L., 1971, Structure of convection cells in the mantle: *J. Geophys. Res.*, v. 76, p. 1154–1161.

Tryggvason, E., 1962, Crustal structure of the Iceland region from dispersion of surface waves: *Bull. Seismol. Soc. Amer.*, v. 52, p. 359–388.

Van Andel, T. and Bowin, C. O., 1968, Mid-Atlantic ridge between 22° and 23°N latitude and the tectonics of mid-ocean rises: *J. Geophys. Res.*, v. 73, p. 1279–1298.

Vening Meinesz, F. A., 1948, Gravity expeditions at sea, 1923–1938, 4: Netherlands Geodetic Commission, Delft.

Vine, F. J., 1966, Spreading of the ocean floor: new evidence: *Science*, v. 154, p. 1405–1415.

Vine, F. J. and Hess, H. H., 1970, Sea-floor spreading, in: *The Sea*, Vol. IV (part 2), A. E. Maxwell, ed.: Interscience, New York, p. 587–622.

Vine, F. J. and Matthews, D. H., 1963, Magnetic Anomalies over oceanic ridges: *Nature*, v. 199, p. 947–949.

Vine, F. J. and Moores, E. M., 1969, Paleomagnetic results for the Troodos igneous massif, Cyprus: *Trans. Amer. Geophys. Union*, v. 50, p. 31.

Vogt, P. R., Anderson, C. N., Bracey, D. R., and Schneider, E. D., 1970, North Atlantic magnetic smooth zones: *J. Geophys. Res.*, v. 75, p. 3955–3967.

Vogt, P. R., Avery, D. E., Schneider, E. D., Anderson, C. N., and Bracey, D. R., 1969, Discontinuities in sea-floor spreading: *Tectonophysics*, v. 8, p. 285–317.

Vogt, P. R., Schneider, E. D., and Johnson, G. L., 1969, The crust and upper mantle beneath the sea, in: *The Earth's Crust and Upper Mantle*, Geophysical Monograph 13, P. J. Hart, ed.: American Geophysical Union, Washington, D. C., p. 556–617.

Von Herzen, R. P. and Langseth, M. G., 1966, Present status of oceanic heat flow measurements: *Phys. Chem. Earth*, v. 6, p. 365–407.

Von Herzen, R. P. and Lee, W. K., 1969, Heat flow in oceanic regions, in: *The Earth's Crust and Upper Mantle*, Geophysical Monograph 13, P. J. Hart, ed.: American Geophysical Union, Washington, D. C., p. 88–95.

Von Herzen, R. P. and Maxwell, A. E., 1959, The measurements of thermal conductivity of deep-sea sediments by a needle probe method: *J. Geophys. Res.*, v. 64, p. 1557–1563.

Von Herzen, R. P. and Uyeda, S., 1963, Heat flow through the eastern Pacific Ocean floor: *J. Geophys. Res.*, v. 68, p. 4219–4250.

Wilson, J. T., 1965, A new class of faults and their bearing on continental drift: *Nature*, v. 207, p. 343–347.

Wilson, J. T., 1970, Continental drift, transcurrent, and transform faulting, in: *The Sea*, Vol. IV (part 2), A. E. Maxwell, ed.: Interscience, New York, p. 623–644.

Wiseman, J. D. H. and Ovey, C. D., 1955, Proposed names of features on the deep sea floor, 2, General principles governing the allocation of names: *Deep-Sea Research*, v. 2, p. 261–263.

Worzel, J. L. and Harrison, J. C., 1963, Gravity at sea, in: *The Sea*, Vol. III, M. N. Hill, ed.: Interscience, New York, p. 134–174.

Worzel, J. L. and Shurbet, G. L., 1955, Gravity anomalies at continental margins: *Proc. National Acad. Sci.*, v. 41, p. 458–469.

Worzel, J. L., 1965, *Pendulum Gravity Measurements at Sea, 1936–1959*: Interscience, New York, 442 p.

Yoder, H. S. and Tilley, C. E., 1962, Origin of basalt magma, an experimental study of natural and synthetic rock systems: *J. Petrol.*, v. 3, p. 362–532.

INDEX